1st International Conference on Food Security and Sustainable Agriculture in the Tropics (IC-FSSAT 2017)

IOP Conference Series: Earth and Environmental Science
Volume 157

Sulawesi Selatan, Indonesia
24 – 25 October 2017

ISBN: 978-1-5108-6397-2
ISSN: 1755-1307

Printed from e-media with permission by:

Curran Associates, Inc.
57 Morehouse Lane
Red Hook, NY 12571

Some format issues inherent in the e-media version may also appear in this print version.

Printed by Curran Associates, Inc. (2018)

For permission requests, please contact the Institute of Physics
at the address below.

Institute of Physics
Dirac House, Temple Back
Bristol BS1 6BE UK

Phone: 44 1 17 929 7481
Fax: 44 1 17 920 0979

techtracking@iop.org

Additional copies of this publication are available from:

Curran Associates, Inc.
57 Morehouse Lane
Red Hook, NY 12571 USA
Phone: 845-758-0400
Fax: 845-758-2633
Email: curran@proceedings.com
Web: www.proceedings.com

Table of contents

Volume 157

1st International Conference on Food Security and Sustainable Agriculture in The Tropics (IC-FSSAT)

24–25 October 2017, Sulawesi Selatan, Indonesia

Accepted papers received: 15 May 2018
Published online: 1 June 2018

Preface

Papers

Crop Production and Environments

Geospatial Agriculture

GMO Food, Food Safety and Product Development

Integrated Pest Management

Marine Biodiversity

Plant Biotechnology and Genetic Engineering

Plant Breeding

Rural and Institutional Development

PREFACE

International Conference on Food Security and Sustainable Agriculture in the Tropic (IC-FSSAT 2017) was held in Makassar, Indonesia on October 24-25 2017. IC-FSSAT 2017 is dedicated to address issues related to food security and sustainable agriculture in the Tropics, hosted by Faculty of Agriculture and Publication Management Centre (PMC), Hasanuddin University.

Food security and sustainable agriculture still becomes a global issue in the tropic. The major goal and feature of the conference was to promote knowledge, science, and technology as well as to address issues in food security and sustainable agriculture in the tropics, by bringing together researchers, scientists, practitioners and scholars in the respected fields. Professors from Japan and France are invited to deliver keynote speeches, together with invited speakers from Australia, United Kingdoms, Malaysia, Finland and Indonesia. Their presence indicates the concern and role of the researchers, scientists and practitioners which are significant in keeping food security and sustainable agriculture.

This proceedings present selected papers submitted to the conference by academics and researches from universities and research institutes. All papers were subjected to rigorous peer-reviews by conference committee members and international reviewers to ensure their compliance to meet the required standard for qualified scientific publication. This volume presents recent researches in the field of Food Security and Sustainable Agriculture in the Tropic covering various related areas: Crop Production and Environment, Geospatial Agriculture, GMO Food, Food Safety and Product Development, Integrated Pest Management, Marine Biodiversity, Plant Biotechnology and Genetic Engineering, Plant Breeding and Rural and Institutional Development.

We would like to appreciate all authors who have contributed to this proceedings, the conference committee, speakers, attendees, organizing committee and sponsors who have made the 1st IC-FSSAT a succces. We wish the conference will have siginificant contribution in field of food security and sustainable agriculture. We are also expecting that this conference proceedings contributes in looking at a new paradigm for food security and sustainable agriculture, especially in the tropic.

Imam Mujahidin Fahmid
Chair of Organizing Commitee ICFSSAT and Director of Publication Management Center
Hasanuddin University.

IOP Conf. Series: Earth and Environmental Science **157** (2018) 011002 doi:10.1088/1755-1315/157/1/011002

Conference Committee

Advisory Committee
Prof. Dwia Aries Tina Pulubuhu, Hasanuddin University, Indonesia
Prof. Junaedi Muhidong, Hasanuddin University, Indonesia
Prof. Budu, Hasanuddin University, Indonesia
Prof. Sumbangan Baja, Hasanuddin University, Indonesia

Conference Chair
Dr. Imam Fahmid Mujahidin, Hasanuddin University, Indonesia

Program Chair
Muhammad Arsyad, Ph.D., Hasanuddin University, Indonesia
Prof. Meta Mahendradatta, Hasanuddin University, Indonesia
Rinaldi Sjahril, Ph.D. Hasanuddin University, Indonesia

Technical Committee
Yoshio Kawamura, Kyoto Prefectural College of Agriculture, Japan
Musrizal Muin, Hasanuddin University, Indonesia
Ingeborg Menzler-Hokkanen, University of Helsinki, Finland
Salengke, Hasanuddin University, Indonesia
Hari Iswoyo, Hasanuddin University, Indonesia
Razak Alimon, Universiti Putra Malaysia
Asmuddin, Hasanuddin University, Indonesia
Christianto Lopulisa, Hasanuddin University, Indonesia
Didi Rukmana, Hasanuddin University, Indonesia
Keith Walters, Harper Adams University, UK
Andi Nasruddin, Hasanuddin University, Indonesia
Ifayanti Ridwan, Hasanuddin University, Indonesia
Kinya Toriyama, Tohoku University, Japan
Muhammad Riadi, Hasanuddin University, Indonesia
Yunus Musa, Hasanuddin University, Indonesia
Hiroshi Ehara, School of Bio-agricultural Sciences, Nagoya University, Japan
Rusnadi Padjung, Hasanuddin University, Indonesia
Rismaneswati, Hasanuddin University, Indonesia
Zaenal, Hasanuddin University, Indonesia
Rahmansyah Dermawan, Hasanuddin University, Indonesia
Heikki M. T. Hokkanen, University of Helsinki, Finland

IOP Conf. Series: Earth and Environmental Science **157** (2018) 011003 doi:10.1088/1755-1315/157/1/011003

Peer review statement

All papers published in this volume of *IOP Conference Series: Earth and Environmental Science* have been peer reviewed through processes administered by the proceedings Editors. Reviews were conducted by expert referees to the professional and scientific standards expected of a proceedings journal published by IOP Publishing.

IOP Conf. Series: Earth and Environmental Science **157** (2018) 012001 doi:10.1088/1755-1315/157/1/012001

Collaboration of liquid bio-ameliorant and compost effect to crop yield and decreasing of inorganic fertilizer utilization for sustainable agriculture

B Rasyid[1,2]

[1]Dept. of Soil Science, Faculty of Agriculture, Hasanuddin University
[2]Tropical Agriculture Research Front Division (TARF) - Japan International Research Center for Agricultural Sciences (JIRCAS)
Jl. Perintis Kemerdekaan, Kampus Unhas Tamalanrea – Makassar, 90245, Indonesia

E-mail: *burrasyid@unhas.ac.id*

Abstract. Soil quality and plant productivity are main issue in agriculture production. The purpose of this research was to obtain sustainable crop management in effort to improve soil quality and increase maize production through collaboration of liquid bio-ameliorant and compost. Field experiment was carried out in two planting season with factorial experimental design replicated three times in 2m x 2m plots. Duncan multiple range test was used to analysis the effect of treatment on all parameters evaluated. The first planting season, treatments were arranged in three factors as: (1) planting space with two spaces, (2) three concentration of liquid bio-ameliorant, and (3) three level of urea fertilizer. The second planting season, treatments were arranged in two factors as: (1) liquid bio-ameliorant (LBA) with four concentrations and (2) compost with four levels. In the first season, result showed in soil quality parameters such as microbial density and soil chemical properties increased approximately 28%. The highest yield of 9.00 ton ha^{-1} was found in application 300 ml l^{-1} LBA + urea 240 kg ha^{-1}. In the second season, collaboration treatment of 250 ml l^{-1} LBA + 10 ton ha^{-1} compost had the highest yield by 10.47 ton ha^{-1}. This study confirmed that collaboration of liquid bio-ameliorant and compost could be used as fertilizer complement and reducing inorganic fertilizer utilization to sustain crop production and soil quality.

1. Introduction

Environmental plant management and implementation of appropriate production technology is the main frame in improvement and development of sustainable plant production. The adoption of better management practices (BMPs) can be practice to improve soil organic carbon (SOC) content, enhance soil quality, restore degraded ecosystems, increase biomass production, improve crop yield, and encourage investment in soil resources for soil restoration [1]. Intensive agricultural practices in conventional farming systems negatively impact the soil biological communities of agroecosystems due to the disturbance induced by chemical fertilizer, pest control measures or soil tillage [2]. Application of green manure and/or crop residues incorporation have been proposed as alternative cropping system in order to reduce SOC when farmyard manure is unavailable [2, 3]. Sugar mill wastes have been used as ameliorant for example, ash applications raise soil pH [4] whilst filter cake adds organic matter as well as raising pH [5]. Recently, combining of organic bio-ameliorant with chemical fertilizer could be alternate practice in optimizing crop production and maintain soil productivity. Benítez-Noyola [6] demonstrated that maize plants fertilized with 90 and 180 kg ha^{-1}N and inoculated with *Paenibacillus polymyxa* extracted from 20 to 28% more nitrogen and produced more grain than plants that were only chemically fertilized. The possibility of maintaining grain maize yields by replacing 87 kg ha^{-1} Urea (50%) of the complete N fertilizer dose (175 kg ha^{-1}) with 300 kg/ha of an organic compost elaborated with fruit and vegetable wastes and enriched with 147g N/kg N compost [7]

Agriculture by product as organic material is often found in agricultural production area which was in many cases it become waste. The other organic material source comes from liquid (sludged) waste of biogas product. These materials can be used to produce liquid bio-ameliorant through anaerobic fermentation process [8]. The basic process is activation of biological community to mineralized organic material which it mentioned as biological processes. It is crucial for the maintenance of soil fertility due to their role in the mineralization of organic matter, enhance plant nutrient uptake, and sustainability of crop production [9]. This is consideration on how it is necessary to study the effect of collaboration of liquid bio-ameliorant and compost to reduce nitrogen utilization and increasing maize production. Better knowledge and considerable technology of maize cultivation management practices through the combination application of liquid bio-ameliorant and compost for achieving sustainable agriculture is the main challenge and purpose of this study.

2. Materials and methods

Field experiment was carried out in Teaching Farm, Faculty of Agriculture, Hasanuddin University. Two planting season of maize was set up with factorial experimental design replicated three times in 2m x 2m field plots. The first planting season, treatments were arranged in three factors as: (1) planting space with two type spaces (P_1 = 75cm x 25cm and P_2 = 50cm x 20cm); (2) liquid bio-ameliorant (LBA) with three concentrations (B_0=0; B_1=100; and B_2=300 ml l^{-1}); and (3) urea with three levels (N_0=0; N_1=240 and N_2=300 kg ha^{-1}). Treatment in the second season were arranged in two factors as: (1) LBA with four concentrations: (B_0=0; B_1=150; B_2=250 and B_2=350 ml l^{-1}); and (2) compost with four levels: (K_0=0; K_1=2; and K_2=5; and K3 =10 ton ha^{-1}). In this season, planting space of 75 x 25 cm was practiced.

Liquid bio-ameliorant was starting applied two weeks after planting and it continued within every two weeks interval during vegetative growth period. Nitrogen fertilizer treatment was applied in split three times: 40% in 7 Days After Planting (DAP), 30% in 30 DAP, and 30% 50 DAP. Furthermore, all plots were supplemented additionally with 150 kg ha^{-1} super phosphate (SP36) and 100 kg ha^{-1} potassium chloride (KCl). Preparation of liquid bio-ameliorant was done by mixing all materials organic containing sludge of biogas, lake organic material sediment, plant by-product. Liquid bio-ameliorant was ready for use after approximately four (4) weeks incubation. Soil sample for soil chemical properties analysis was collected as composite sample.

Liquid bio-ameliorant, soil and plant analysis was conducted at the Laboratory of Soil Chemistry and Fertility, Department of Soil Science, Faculty of Agriculture, Hasanuddin University, Makassar. Statistically analysis of significant difference the experiment using ANOVA and the effect of different treatment was evaluated using Duncan's multiple range test. All statistical analysis was performed using IBM-SPSS ver. 20 computer package.

3. Result and discussion

3.1. Soil, liquid bio-ameliorant and compost properties

Soil type in the field experiment was Typic Hapludalfs [10] which means the soil has moderate nutrient content with high base saturation (>35%), high sum of basic cations and has an argilic horizon (clay accumulation subsurface horizon). Selected of soil and liquid bio-ameliorant properties was summarized in Table 1. Maize cultivation was highly depend on the soil fertility and environmental conditions. In case of soil with low nutrient content, especially nitrogen can decrease maize production significantly. However, application of inorganic fertilizer for long term tend to decreased soil quality and it caused direct effect to the environment and plant production. In this case a combined method to maintain sustainability production could be achieved by supplementing crop production environment using organic material such as liquid bio-ameliorant.

Result of soil and organic matter analysis showed enrichment nutrient more than 50% obtained from LBA and compost. Soil fertility is also increased from additional microbe density of 11x104 cfu as biological component. Increasing of chemical and biological properties have possibility to soil physical properties and its effect will increase quality of plant environmental production.

Table 1. Selected properties of soil, liquid bio-ameliorant, and compost

Properties	Soil	LBA
Chemical properties		
CEC (me/100g)	18.34	-
pH (H_20)	6.12	6.23
P_20_5 (ppm P)	12.76	10.01
N-total (%)	0.06	0.49
C-organic (%)	1.94	0.98
K (cmol(+)kg^{-1})	0.22	0.82
Ca (cmol(+)kg^{-1})	6.36	31.71
Mg (cmol(+)kg^{-1})	2.67	2.78
Na (cmol(+)kg^{-1})	0.33	7.99
Soil fractions		
Sand (%)	20	-
Silt (%)	28	-
Clay (%)	52	-
Microbial density (cfu)		11 x 10^4

3.2. Soil Quality Improvement

Maintaining sustainability or improving soil quality could be achieved by application combination method between chemical fertilizer and organic matter or combining organic material in different type such as liquid bio-ameliorant and compost. Table 2 showed clear result that liquid bio-ameliorant could be applied to improve soil quality.

Table 2. Soil quality improvement on collaboration of liquid bio-ameliorant and nitrogen treatment

Treatments	Soil Chemical Properties								
	C	N	P2O5	K	Ca	Mg	Na	CEC	BS
	%	%	ppm	-----------------------------	cmol(+) / kg			------------	%
N0B0	1.94	0.06	10.99	0.22	6.36	2.67	0.33	18.34	52
N0B1	1.75	0.34	36.87	0.25	8.38	4.87	0.41	22.33	62
N0B2	2.49	0.36	37.82	0.36	7.90	3.50	0.38	24.33	50
N1B0	2.14	0.31	18.48	0.17	6.83	4.63	0.33	28.51	42
N1B1	2.59	0.08	34.99	0.25	7.78	3.03	0.42	22.13	52
N1B2	2.59	0.28	35.74	0.22	7.48	2.79	0.61	24.13	46
N2B0	2.71	0.22	10.00	0.22	6.83	2.85	0.35	22.53	46
N2B1	2.24	0.24	34.99	0.24	8.49	0.42	0.52	21.34	45
N2B2	2.70	0.21	34.04	0.33	8.08	1.60	0.41	21.14	49

Comparison on single treatment of LBA and nitrogen showed LBA give better result than nitrogen. Evaluation in all nutrient content except in carbon content was found higher value in LBA than nitrogen with increasing of LBA value in average of 10.82%.

With regards to combination concentration effect was analyzed treatment of 300 ml l^{-1} LBA and 240 kg ha^{-1} nitrogen have highest value for macro nutrient (N, P and K). Some combination treatment such as 300 ml l^{-1} LBA and 300 kg ha^{-1} nitrogen have high value in carbon organic, potassium and calcium. These two facts were put an important that high concentration of liquid bio-ameliorant and low level nitrogen had highest increase of nutrient content compare to the other treatment. Improvement soil quality could be related to the increasing of mineralization and transformation of organic material with activation of microbial activity.

3.3 Effect of Collaboration Treatment on Plant Performance

Plant density and fertilizer management are two components in cultivation system have significant effect to increase yield. Plant height and biomass were plant performances with clear response affected by planting space, nitrogen, liquid bio-ameliorant and compost application (table 3 and 4). In the first season, plant height showed significant response in the treatment of nitrogen fertilizer, but liquid bio-ameliorant, and planting space have various difference effect on this combination treatment (table 3).

Table 3. Effect of Collaboration Treatment on Plant Height (cm)

First season	Average	Second season	Average
$P_1B_0N_0$	134.37^{a}	**B0K0**	146.71^{a}
$P_1B_1N_0$	137.77^{a}	**B0K1**	179.21^{b}
$P_1B_2N_0$	143.50^{a}	**B1K0**	183.88^{b}
$P_2B_2N_0$	150.55^{a}	**B0K2**	194.63^{b}
$P_2B_0N_0$	153.19^{a}	**B1K1**	207.25^{b}
$P_2B_1N_0$	157.65^{ab}	**B3K0**	216.50^{b}
$P_1B_2N_1$	167.90^{b}	**B3K1**	221.21^{bc}
$P_1B_0N_2$	171.40^{b}	**B2K0**	229.83^{c}
$P_1B_0N_1$	176.03^{b}	**B2K1**	230.46^{c}
$P_1B_1N_2$	177.33^{b}	**B1K2**	231.04^{c}
$P_2B_0N_2$	177.90^{b}	**B3K2**	233.25^{c}
$P_2B_2N_1$	178.86^{bc}	**B0K3**	233.63^{c}
$P_2B_1N_2$	183.00^{c}	**B2K3**	238.25^{cd}
$P_2B_2N_2$	183.55^{c}	**B2K2**	244.71^{d}
$P_1B_2N_2$	183.57^{c}	**B3K3**	263.00^{d}
$P_2B_0N_1$	184.64^{c}	**B1K3**	268.84^{d}
$P_2B_1N_1$	185.60^{c}		
$P_1B_1N_1$	198.07^{c}		

Data followed by different letters indicate significant differences at $P < 0.05$

Plant height showed almost combination treatments between liquid bio-ameliorant and nitrogen has highest value ranged from 177.33cm to 198.07cm. Treatment without nitrogen showed the lowest value ranged from 134.37cm to 157.65cm. The overall evaluation on this parameter was found that combine application of planting space of 75cm x 25cm, 100 ml l^{-1} LBA, and 240 kg ha^{-1} urea had the highest value, but no significant effect to planting space of 50cmx20cm, 300 ml l^{-1} LBA, and 240 kg ha^{-1} urea or planting space of 50cmx20cm, 100 ml l^{-1} LBA, and 300 kg ha^{-1} urea. Result for plant biomass parameter was follow tendency result of plant height. The highest result was shown by combination treatment of planting space of 75cmx25cm, 100 ml l^{-1} LBA, and 240 kg ha^{-1} urea, but it is not significant different result to planting space of 75cmx25cm, 100 ml l^{-1} LBA, and 300 kg ha^{-1} urea. In the second season, liquid bio-ameliorant and compost treatment was observed higher value in combination than single treatment (table 3 and 4). All treatment was significant effect compare to control on plant height. Generally, combination treatment between liquid bio-ameliorant and compost showed highest value in all treatment. Plant height value in this combination ranged from 207.25cm to 268.84cm. The interesting result was found within concentration of liquid bio-ameliorant and compost. The highest value of plant height was 268.84cm for application 150 ml l^{-1} LBA with 10 ton ha^{-1} compost, but it was not significant to the treatment of 250 or 350 ml l^{-1} LBA with 5 or 10 ton ha^{-1} compost. In plant biomass result was found also that all treatment significant effect to plant biomass. Combination treatment between liquid bio-ameliorant of different concentration with 10 ton ha^{-1} compost showed high value compare to the other combination or single treatment.

Table 4. Effect of Collaboration Treatment on Plant Biomass Weight (g)

First season	Average	Second season	Average
$P_2B_0N_2$	51.50^{a}	**B0K0**	68.33^{a}
$P_1B_0N_0$	53.57^{ab}	**B1K0**	90.00^{b}
$P_1B_1N_0$	54.00^{b}	**B0K1**	100.67^{b}
$P_2B_0N_0$	56.83^{b}	**B2K0**	124.00^{b}
$P_2B_2N_0$	57.57^{b}	**B1K1**	139.00^{b}
$P_2B_1N_0$	61.53^{b}	**B3K1**	140.00^{b}
$P_2B_1N_1$	67.27^{b}	**B3K0**	145.67^{bc}
$P_2B_1N_2$	75.47^{b}	**B2K1**	145.67^{bc}
$P_1B_2N_2$	80.00^{b}	**B0K2**	150.67^{c}
$P_1B_2N_0$	82.37^{b}	**B1K2**	152.33^{c}
$P_2B_0N_1$	85.60^{bc}	**B2K2**	156.67^{c}
$P_1B_0N_1$	96.37^{c}	**B0K3**	181.67^{d}
$P_2B_2N_1$	96.97^{c}	**B3K2**	182.33^{d}
$P_1B_2N_1$	98.77^{c}	**B1K3**	184.00^{d}
$P_2B_2N_2$	109.43^{c}	**B2K3**	206.67^{d}
$P_1B_1N_2$	110.83^{c}	**B3K3**	206.67^{d}
$P_1B_0N_2$	119.27^{c}		
$P_1B_1N_1$	124.00^{c}		

Data followed by different letters indicate significant differences at $P < 0.05$

3.4. Plant Production

Plant production is illustrates of the working of interactions of various factors which are determine all production process including soil, fertilizers, and the environmental production. Total yield in this experiment was shown in table 5.

Table 5. Maize production for two planting seasons

First season	Yield (ton ha^{-1})	Second season	Yield (ton ha^{-1})
$P_1B_0N_0$	3.00	**BK0**	4.31
$P_2B_0N_1$	3.12	**BK1**	6.40
$P_1B_1N_2$	5.32	**BK2**	8.02
$P_2B_2N_1$	9.00	**BK3**	10.47

The collaboration treatment effect between liquid bio-ameliorant and compost was shown clear effect on production. Combination treatment of 250 ml l^{-1} LBA and 10 ton ha^{-1} compost showed the highest value of 10.47 ton ha^{-1} compare to the other combination. The result was found also that LBA improving environmental production as reducing nitrogen fertilizer, improving soil quality and affect plan production, directly. The result could be used to confirmed the usefulness of liquid bio–ameliorant as material in improving maize production.

4. Conclusion

Soil quality improvement and increasing maize production was affected by environmental management and soil properties. Plant performance had high results under combination treatments between low liquid bioameliorant concentration and high nitrogen fertilizer application. The combination treatment effect between liquid bio-ameliorant with nitrogen or liquid bio-ameliorant with compost can increase production until more than 10 ton ha^{-1} compare to control with total production around 3 ton ha^{-1}. This result confirmed that liquid bio-ameliorant combining with compost could be used as fertilizer substitution or complement to reduce chemical fertilizer utilization. For the future, liquid bioameliorant has potential as bio material in improving maize production sustainability.

Acknowledgements
Greatly acknowledge Directorate General of Strengthening Research and Development, Ministry of Research, Technology, and Higher Education, Indonesia for financial support. We also acknowledge to JIRCAS, for accepting on visiting research fellowship program. To all students supported in this research, acknowledgment for their assistance in the field experiment and Dept. of Soil Science, Fac. of Agriculture, Hasanuddin University for laboratory facilities in soil and plant samples analysis.

References

[1] Lal R, Kumble J M, Follett R F and Cole C V 1998 *The Potential of U.S. Cropland to Sequester Carbon and Mitigate the Greenhouse Effect* (Chelsea: Ann Arbor Press)
[2] Zhao S, Qiu S, Cao C, Zheng C, Zhou W and He P 2014 Responses of soil properties, microbial community and crop yields to various rates of nitrogen fertilization in a wheat–maize cropping system in north-central China *Agric. Ecosyst. Environ.* **194** 29–37 doi:http://dx.doi.org/10.1016/j.agee.2014.05.006.
[3] Drinkwater L E, Wagoner P and Sarrantonio M 1998 Legume-based cropping systems have reduced carbon and nitrogen losses *Nature* **396** 262–65
[4] Vance E D 1996 Land application of wood-fired and combination boiler ashes: an overview *J. Environ Qual* **25** 937–44
[5] Mokolobate MS and Haynes R J 2002 Comparative Liming Effect of Four Organic Residues Applied to an Acid Soil *Biol Fertil Soils* **35** 79–85
[6] Benítez-Noyola M 2013 *Inoculación con Paenibacilluspolymyxa fertilización nitrogenada en maíz bajocondiciones de temporal* M Sc. Thesis (Mexico: Colegio de Postgraduados Montecillo, Texcoco Edo de México)
[7] Naveed M, Khalid M, Jones D L, Ahmad R and Zahir Z A 2008 Relative efficacy of Pseudomonas spp., containing ACC-deaminase for improving growth and yield of maize (Zea mays l.) in the presence of organic fertilizer *Pakistan J. of Bot.* **40** 1243-51
[8] Setiawan 2010 *Pengolahan Limbah Organik dengan EM sebagai Bahan Baku Pembuatan Kompos* (Semarang: Fakultas Kesehatan Masyarakat UNDIP)
[9] Johansson J, Paul L and Finlay R 2004 Microbial interaction in the mycorrhizosphere and their significance for sustainable agriculture *FEMS Microbiol. Ecol.* **48** 1-13 doi:http://dx.doi.org/10.1016/j.femsec.2003.11.012.
[10] NRCS-USDA 2003 *Keys to Soil Taxonomy, 9th ed.* (Washington: United State Department of Agriculture)

Improving quality and digestibility of cocoa pod with white rot fungi

J Mustabi, Wedawati and A K Armayanti

Faculty of Animal Science, Hasanuddin University, Jl. Perintis Kemerdekaan KM.10 Makassar, 90245, South Sulawesi, Indonesia.

E-mail: *jamila.mustabi@unhas.ac.id*

Abstract. White rot fungi is a type of fungus that is able to degrade lignin in the feed material from waste, so it can be used to increase the added value of cocoa pod as alternative feed ingredients to meet the nutritional needs of cattle. The purpose of this study is to investigate the use of white rot fungi in improving the quality and digestibility cocoa pod as feed. The study consisted of two phases, namely fermentation using three isolates of white rot fungi (*Coprinus comatus*, *Corilopsis polyzona* and *Lentinus torulosus*) on pod husks and quality testing *in vitro* digestibility of fermented. Results of analysis of variance show that the treatment was highly significant on the content of lignin, cellulose and hemicellulose pod husks. Fermented cocoa husks with white rot fungi can degrade lignin content of 1.42% - 12.28% and highly significant improved on *in vitro* digestibility of dry matter and organic matter. The conclusion, isolates of white rot fungi most active in degrading lignin was *Lentinus torulosus* isolates and less ability to degrade cellulose and hemicellulose.

1. Introduction

Exploration and resource types of feed derived from agro-industrial residual, is important in terms of its use as a substitute for the main feed, especially feed costs are the largest expense for farmers. Various studies have been conducted to utilize waste into valuable feed materials of high biological value and provide added value. Mustabi *et al.* [1] has conducted research on the use of rice straw fermented white rot fungi as feed substitution grass in goats.

Residue plant cocoa pod utilization as alternative feed ingredients to meet the nutritional needs of cattle. Agro-industrial wastes as feed material have always been associated with low prices and poor quality. The amount of utilization depends on the potential both in quantity and quality that can be utilized. Quantity aspects related to the amount of waste generated from a percentage of the production process and ration formulate. More emphasis on the quality aspects of nutritional value that can be utilized by livestock to increase production and productivity.

Cocoa Pod utilization as feed is generally limited by several factors such as the low nutritional quality due to high fiber content and their networks on a feed from waste has undergone a process lignification (hardening). Lignin and cellulose are often formed lignosellulose compounds in plant cell walls, which is a strong bond [2]. The older the plant, the higher lignin content consequently digestibility decreases with the increasing lignification. In addition to cellulose and hemicellulose lignin binding also bind to the cell wall proteins.

Some groups of microorganisms were reported capable of degrading lignin compounds, being able to use cellulose as a carbon source for their growth substrate, whereas the lignocellulosic components that can be used by cattle are cellulose and hemicellulose. Problems often arise in the processing of lignocellulosic materials using microorganisms is the loss of organic material substrates used by microorganisms as a source of nutrients in the bioconversion process. Murni *et al.* [3] suggest that microorganisms ideal in bioconversion of lignocellulose into livestock feed is a microorganism that has the great ability to decompose lignin but low power degradation of the cellulose and hemicellulose. Most fungi lignolitic don't have the ability to use lignin as a single source for energy and carbon as well as a lot depends on the polysaccharide easily digested in the substrate. Based on the consideration that white rot fungi degrade lignin is the most active, so as to improve the quality and digestibility cocoa pod as feed for sustainable livestock development.

2. Materials and Method

2.1. Location and Description of Study

The experiment consisted of two stage, namely (1) fermentation using 3 isolates of white rot fungi on the substrate cocoa pod was carried out in laboratory animals valorisasi waste Hasanuddin University Faculty of Animal Husbandry; (2) Analysis of digestibility of dry matter and organic matter digestibility of fermented cocoa pod conducted at the Laboratory of Feed Chemistry Hasanuddin University Faculty of Animal Husbandry.

2.2. Sample Preparation and Treatment

The material used in this research is cocoa pod, isolated white rot fungus (*Coprinus comatus, Cantharellus freisii* and *Lentinus Torulosus*), cocoa pod, water, Lime ($CaCO_3$), plastic polypropylene (PP), aluminium foil and the materials used in the analysis digestibility. While the tools used are glass bottles (height 9 cm and 5 cm in diameter), pipe, measuring cup, oven, autoclave, analytical balance and laboratory tools in the analysis of digestibility.

The first phase of the research was the fermented cocoa pod. Prior to fermentation, first performed seed media making a mushroom growth of substrate coca pod 92%, rice bran 6% and lime 2% [4]. The water is added 70% of the mixture. And then inserted into a glass bottle (9 cm high, 5 cm diameter) 50 g or ± 3/4 of the height of the container, sealed and sterilized into the autoclave at a temperature of 121^0 C with a pressure of 1.1 atmospheres for 20-30 minutes 2 times. This process is done so that all spores and pathogen microbes are really dead. After that, the fungal isolate was inoculated into the planting medium at the level of 10% of the weight of the substrate (W/W), and incubated at 15 and 30 days. Each treatment was repeated 3 times, so there are 54 treatment units. At this stage the data is processed using analysis of variance completely randomized design [5], and then test the contrast [6] to determine the isolates, and the incubation period of the best influence on variables. After incubation, the samples were dried fermented with 55° C oven for 3-4 days, then ground with a hole diameter of 1 mm filter, and then used for the determination of [7] to determine the levels of lignin, cellulose and hemicellulose, the analysis of dry matter and in vitro organic matter [8].

2.3. Statistical Analysis

The data is processed using analysis of variance completely randomized design according to [5]. Contrast tested [6] to determine the isolates, the level and period of incubation of the best effects on variables.

3. Results and Discussion

Effect of fermented cacao pod with 3 white rot fungal isolates with an incubation period of 15 and 30 days on the content of lignin, cellulose and hemicellulose cocoa pod can be shown figure 1.

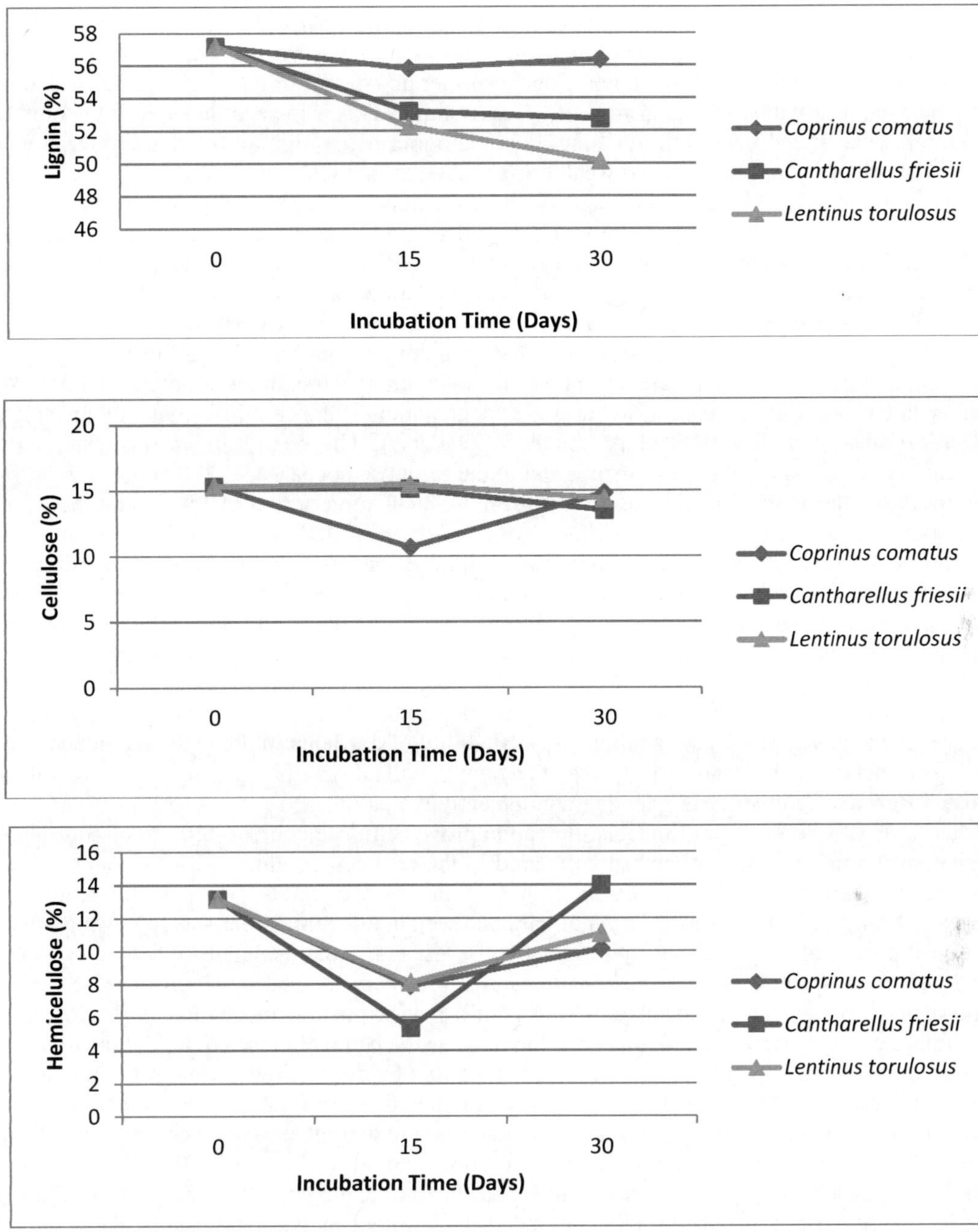

Figure 1. Effect of the fermented cocoa pod with some white rot fungi with an incubation period of 15 and 30 days on the content of lignin, cellulose, and hemicellulose.

Based on the analysis of variance results, treatment showed a highly significant effect on the lignin content of the cocoa pod. The cocoa pod that was not inoculated white rot fungi containing higher lignin compared with the pods that treated with white rot fungi inoculation. Further trials showed isolate C (*Lentinus torulosus*) had lower levels than isolates A (*Coprinus comatus*) and B (*Cantharellus friesii*), while at the 15-day incubation period there was no significant different with an incubation period of 30 days in isolates C (*Lentinus torulosus*).

IOP Conf. Series: Earth and Environmental Science **157** (2018) 012002 doi:10.1088/1755-1315/157/1/012002

Figure 1 shows that at the incubation period of 15 days, the lowest lignin content was shown bu inoculated cocoa pod with isolates C (*Lentinus torulosus*), up to 30-day incubation period. The longer the fermentation, the more lignin degraded. Changes in lignin content of the substrate occur due to the utilization of lignin structure for the growth of white rot fungi. This is in accordance with the opinion of Risdiyanto [9] that stated that white rot fungi degrade lignin has a high ability and a microorganism capable of degrading lignin in the wood weathering processes into CO_2 and H_2O.

In this study, the skin fermented cocoa with white rot fungi can degrade lignin content of 1.42% - 12.28%. This value is higher than the research of Nelson and Suparjo [10] that found a decrease of 9.52% of the lignin content in fermented cocoa pod with *Phanerochaete chrysosporium*, but lower compared to Laconi [11] and Alemawor *et al.* [12] that found a decrease of 18.36% and 17.06%, respectively, in lignin content of cocoa pod treated with *Phanerochaete ostreatus*.

The results of calculation of variance showed that the treatment was highly significant to the cellulose content cocoa pod. Best isolates were less power degradation of the cellulose content. In contrast test showed isolate C (*Lentinus torulosus*) lower power degradation of the cellulose content than isolates B (*Cantharellus friesii*) and isolates of A (*Coprinus comatus*). The results of the analysis were not significant between 15 and 30 days of incubation of the cellulose cocoa pod.

The results of the analysis of the cocoa pod substrate incubation period of 15 days on the cellulose content of the lowest decline in isolates C (*Lentinus torulosus*), while fermented cocoa pod on a 30-day incubation isolate A (*Coprinus comatus*) increased cellulose content. The longer the incubation period, the content of cellulose substrate decreases except in isolates C (*Lentinus torulosus*) increased cellulose content in the incubation period of 30 days. This is caused by fungi outlines cellulose into simpler substances to be used as a nutrient in growth, so the production of enzymes by the fungus is increasing including cellulose digesting enzyme. According to Zeng *et al.* [13] results overhaul lignocellulosic components will be used by the fungus to meaningful growth will suppress the degradation of lignin and degradation activity will occur again if the availability of nutrients in the media is reduced. While according to Nelson and Supardjo [10], that the fungus will utilize the existing nutrients, including cellulose for growth and reduced lignin degradation activity.

In this study of white rot fungi inoculation can improve cellulose content of 0.8% of controls. This suggests that the increase is lower when compared to the research conducted by Nelson and Suparjo [10] using *Phanerochaete chrysosporium* seen that the rind contains cellulose fermented cocoa increased 2.91%. The results of calculation of variance seen that the treatment was highly significant to the hemicellulose content of the cocoa pod. In contrast test shows that isolate B (*Chantharellus friesii*) higher levels than isolate C (*Lentinus torulosus*) and A (*Coprinus comatus*). Hemicellulose content of the cocoa pod declined in the incubation period of 15 days but the incubation period of 30 days hemicellulose content increased. Although not the same as the hemicellulose content of the treatment is not white rot fungi were inoculated (control). This is caused by the ability of white rot fungi isolates to degrade hemicellulose, lignin makes increases with the length of incubation. The amount of lignin is degraded in the fermentation process causes hemicellulose content increased. According to Nelson and Suparjo [10], that the degradation of lignin will open up access to reshuffle the cellulose and hemicellulose. Results overhaul of cellulose to produce cellulase enzymes remodel simple sugars limit the production of most of the enzymes that degrade hemicellulose by white rot fungi [14].

In research with an incubation period of 15 and 30-day fermentation of the cocoa pod with white rot fungi can improve the hemicellulose content of only 6.07% by isolates B (*Cantharellus friesii*). The increase in value is higher compared to studies conducted by hemicellulose content of cocoa pod fermented by using *Phanerochaete chrysosporium* decreased by 4.41% by [10]. Results of analysis of variance showed that highly significant on the in vitro digestibility of dry matter and organic matter. Contrast test showed that cocoa pod fermented by white rot fungi is higher than unfermented cocoa pods (Control Vs A, B, C). This suggests that each inoculation of white rot fungi can improve in vitro digestibility of dry and organic matter cocoa pod.

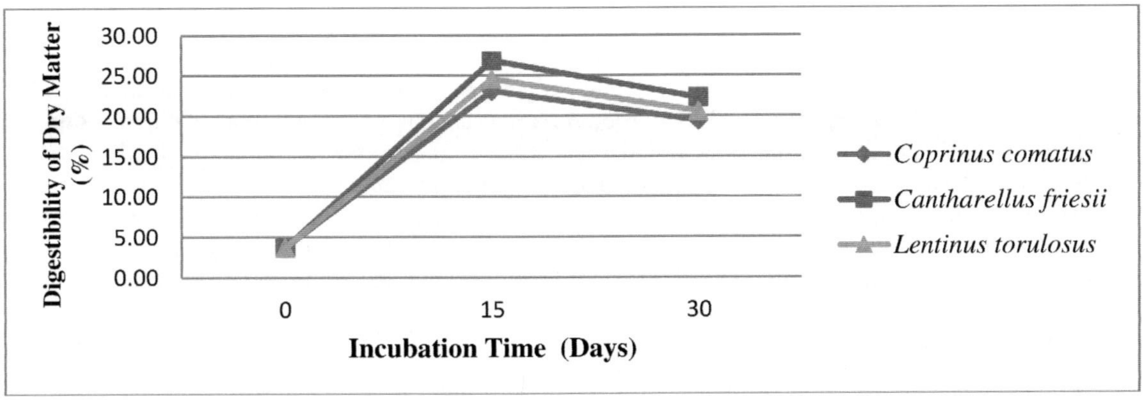

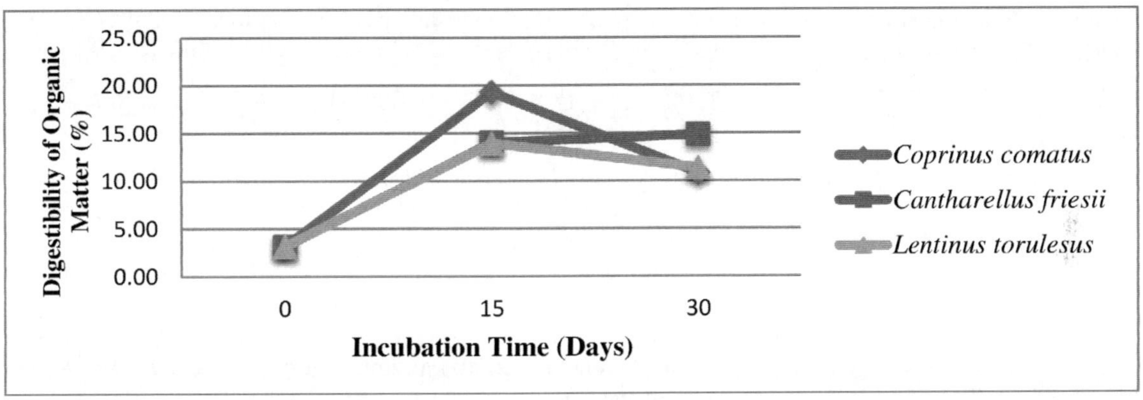

Figure 2. Effect of the fermented cocoa pod with some white rot fungi with an incubation period of 15 and 30 days on the dry and organic matter in vitro.

In figure 2 shows that the incubation period of 15 and 30 days, the highest dry matter of cocoa pod digestibility was in B isolates (*C. friesii*), isolates C (*L. torulosus*), and then isolate A (*C. comatus*). Unfermented cocoa pod (control) has very low dry matter digestibility of 3.71% in contrast to research of Afrijon [15], the dry matter content cocoa pod are used 40.02%. In this study, dry matter digestibility values increased 80 to 86%. It is higher when compared with the results of research on in vitro dry matter digestibility cocoa pod that amoiasi with urea, an increase in dry matter ranges from 22 to 26% [15]. Similarly, studies in vitro dry matter digestibility rind fermented cocoa EM-4 conducted by Abidah [16], increasing the dry matter of 18 to 19%.

The results of this study showed that white rot fungi are fermented with the cacao pod able to increase the value of in vitro dry matter digestibility of 86% allegedly in the process of fermentation produce an enzyme that is able to reduce levels of fiber in the cocoa pod. This is in accordance with the opinion of Winarno *et al.* [17], the fermentation process feed material by microorganisms causes beneficial changes such as improving the quality of feed ingredients of both aspects of nutrition and digestibility and increases the shelf. Fermentation products usually have a higher nutritional value than the original material for their enzyme produced from the microbe itself.Contrast test conducted in this study shows no significant differences were found between the three isolates (A, B and C). While the incubation period finds a difference between 15 vs. 30 days on organic matter digestibility cocoa pod were inoculated isolate A (*C. comatus*), different from the inoculation isolate B and C.

In figure 3 shows that the digestibility of organic matter in the highest 15-day incubation A isolates (*C. comatus*), then isolate B (*C. friesii*) and isolate C (*L. torulosus*). By contrast, the incubation period of 30 days which is the highest organic matter digestibility was inoculated isolate B (*C. friesii*), but the

A isolates (*C. comatus*) and isolate C (*L. torulosus*) experienced a decline in value of 15 to 30 days. Organic matter digestibility cocoa pod used is very low at 3.14% in contrast to research [15], a digestible organic matter that is 46.37%.In this study, organic matter digestibility ranged between 70 to 83%. This value is higher than the in vitro digestibility of organic matter cocoa pod that amoniasi with urea, organic matter digestibility ranged from 18 to 27% [15].

The results showed that the fermentation of the cocoa pod with white rot fungi is able to increase the in vitro digestibility of organic matter by 83%, presumably because the white rot fungi produce enzymes that can degrade crude fiber fermentation process. This is in accordance with the opinion of [18], factors affecting the digestibility of organic material is crude fiber and mineral content of feedstuffs.

4. Conclusion

Lentinus torulosus is the most active isolate in degrading lignin and lacking its ability to degrade cellulose and hemicellulose. Fermented cocoa pod using three white rot fungi isolates namely (*Coprinus comatus, Cantharellus friesii, and Lentinus torulosus*), is able to improve the digestibility of dry matter and organic matter in vitro.

References

[1] Mustabi J, Ngitung R, Islamiyati R, Akhirany N, Natsir A, Jousuf, K, Ismartoyo, Kuswinanti, T, and Rinduwati 2013 Rice straw fermented with white rod fungi as an alternatif elephant grass in goat feeds ISSN 1992-6197 *Global Veterinary* **10** (6): 697-701

[2] Sutardi, T 1980 *Landasan Ilmu Nutrisi Departemen Ilmu Makanan Ternak Fakultas Peternakan* (Bogor: Institut Pertanian Bogor)

[3] Murni, R, Suparjo, Akmal, and Ginting B L 2008 *Buku Ajar Teknologi Pemanfaatan Limbah untuk Pakan* (Jambi: Laboratorium Makanan Ternak Fakultas Peternakan Universitas Jambi)

[4] Achmad, Mugiono, T.Arlianti and C.Azmi 2011 *Panduan Lengkap Jamur* (Jakarta: Penebar Swadaya)

[5] Steel R G D and Torrie J H 1993 *Prinsip dan Prosedur statistika suatu Pendekatan Biometrik* (Jakarta: Terjemahan PT Gramedia)

[6] Gomez K A and Gomez A A 1995 *Prosedur Statistik untuk Penelitian Pertanian* (Jakarta: Terjemahan Universitas Indonesia Press)

[7] Van Soest P J 1976 *New Chemical Methods for Analysis of Forages for The Purpose of Predicting Nutritive Value* Conf. Pref IX International Grassland

[8] Goto, I and Minson 1977 Prediction of the dry matter digestibility of tropical grasses using a pepsin-cellulase assay *Animal Feed Science. Technol.* **2** 247-253

[9] Risdiyanto, H 2007 *Pemilihan spesies jamur dan media imobilisasi untuk Produksi enzim ligninolitik* (Pros. Seminar Nasional Rekayasa Kimia dan Proses)

[10] Nelson and Suparjo 2011 Penentuan lama fermentasi kulit buah kakao dengan *phanerochaete chrysosporium*: evaluasi kualitas nutrisi secara kimiawi *AGRINAK* **01** 1–10

[11] Laconi E B 1998 *Peningkatan Kualitas Kakao melalui Amoniasi dengan Urea dan Biofermentasi dengan Phanerochaete chrysosporium* serta *Penjabarannya dalam Formulasi Ransum Ruminansia* Disertasi (Bogor: Program Pascasarjana Institut Pertanian Bogor)

[12] Alemawor F., V.P. Dzogbefia, E.O.K. Oddoye and J.H. Oidham 2009 Effect of *Pleurotus ostreatus* fermentation on cocoa pod husk composition: Influence of fermentation period and Mn^{2+} supplemention on the fermentation process *Afr. J. Biotechnol.* **8** 1950-1958

[13] Zeng G., M Yu, Y. Chen, D. Huang, J. Zhang, H. Huang, R. Jiang, and Z. Yu 2010 Effects of inoculation with *Phanerochaete chrysosporium* at various time points on enzyme activities during agricultural waste composting *Bioresour. Tech.* 101 222–227

[14] Kirk, T.K., and Cowling, E.B. 1984. *Biological Decomposition of Solidd Wood* (*The Chemistry of Solid Wood*) Ed. Rowell R M (Washington DC: American Chemical Society)

[15] Afrijon 2011 Pengaruh Pemakaian Urea Dalam Amoniasi Kulit Buah Coklat Terhadap Kecernaan Bahan Kering dan Bahan Organik Secara in – vitro. Akademi Pertanian Sumatera Barat *Jur Embrio* **(4)** (1) 1- 5

[16] Abidah 2002 *Daya Cerna In Vitro Bahan Kering dan Bahan Organik kulit buah kakaoyang difermentasi Dengan Effective Microorganisme-4 (EM-4) pada Berbagai Lama Penyimpanan.* (Makassar: Fakultas Peternakan, Universitas Hasanuddin)

[17] Winarno, F G, Fardiaz S and Fardiaz D 1980 *Pengantar Teknologi Pangan* (Jakarta: Gramedia Pustaka Utama)

[18] Ismail, R 2011 *Analisis Bahan Pakan* http:// risman ismail2. wordpress. Com /2011 /11 /19/ analisis -bahan pakan/ Accessed March 7 2017

IOP Conf. Series: Earth and Environmental Science **157** (2018) 012003 doi:10.1088/1755-1315/157/1/012003

The rain water management model on an appropriate hilly area to fulfil the needs of cocoa farm during dry season

M Hasbi[1,2], R Darma[3], M Yamin[2], M Nurdin[2] and M Rizal[2]

[1] Department of Computer Science, Mathematics, Hasanuddin University, Jalan Perintis Kemerdekaan KM 10, Makassar, 90245, Indonesia.
[2] Doctoral Program, Department of Agricultural, Hasanuddin University, Jalan Perintis Kemerdekaan KM 10, Makassar, 90245, Indonesia.
[3] Department of Agicultural Socio-economics, Hasanuddin University, Jalan Perintis Kemerdekaan KM 10, Makassar, 90245, Indonesia.

E-mail: *hasbi.unhas@gmail.com*

Abstract. Cocoa is an important commodity because 90% farmers involved, easily marketed, and potentially harvested along the year. However, cocoa productivity tended to decrease by an average of only 300 kg hectare^{-1} year^{-1} or away from the potential productivity of two tons. Water management was an alternative method to increase its productivity by harvesting rainwater on the hilly cocoa farm area and distributing the water based on the gravity law. The research objective was to describes how to manage rainwater at the hilly cocoa farm area, so that the water needs of cocoa farm were met during the dry season. The important implication of the management was the water availability that supports the cocoa cultivation during the year. This research used qualitative method with descriptive approach to explain the appropriate technical specification of infrastructure to support the rainwater management. This research generated several mathematical formulas to support rainwater management infrastructure. The implementation of an appropriate rainwater utilization management for cocoa farm will ensuring the availability of water during dry season, so the cocoa farm allowed to produce cacao fruit during the year.

1. Introduction

Water was one of the main needs of cocoa plant to produce good bean quality. Therefore, lack of water that exceeds the threshold of the cocoa plant needs would cause the cocoa plant to suffer. Rainwater was one of the water resources in an area that faced water shortages [1]. Studies in several cities in Australia suggest that rainwater use could save up to 29.9% (Perth) and 32.3% (Sydney) water supply [2]. In Jordan, the use of rainwater by residents as an alternative source of clean water reduces water consumption by up to 19.7%.

Cocoa (*Theobroma cacao L.*) is a cultivable plant in the tropics, such as Indonesia. The population of cocoa per hectare ideally is around 1,000 trees with the size of planting of 3 m x 3 m. The specialty of cocoa plants was that they could bear fruit during the year if their water, nutritional and environmental needs were met. Cocoa plants could produce an average of 3 kg of dry beans per tree per year or about 3000 kg year^{-1}.

The main problem of cocoa crops according to the cocoa farmers, from the results of our survey since 2012 to 2014 in Bulukumba and Wajo District, was insufficiency of water to fulfill the needs of cocoa plants durig the dry season. According to them, most of cocoa diseases were caused by the lack of water.

The many aspects associated with rainwater management, this study was limited only to rainwater management using small artificial reservoirs at the hilly area or artificial-mini-hill-basin (AMHB) that applied geomembrane technology. The problem formulation of this research is how to manage rain water in the hilly areas of cocoa farm so that the water needs of cocoa plants are fulfilled during the dry season. The research objective is to describes how to manage rainwater in the hilly areas of cocoa farm so that the water needs of cocoa farm were met during the dry season. With proper water management and utilization would ensure the availability of water for cocoa farm during dry season, so that cacao could produce fruit throughout the year.

2. Literature review and Logical Framework

2.1. Literature Review

Rainwater harvesting system (RHS) was one of the rainwater utilization management (RUM) component. Rainwater harvesting (RH) was a method implemented to collect rainwater that fell on the roofs of buildings, the ground, the street, or the rocky hills that would be used as a source of clean water supplies [3].

RHS components depend on RH objectives. RHS for the purpose of clean water or drinking water consists of the following components: rainwater catchment area (RCA), rainwater channel (RCh) that drains rainwater from rainwater catchment to conveyance, rainwater filter (RF), rainwater storage tank (RST), rainwater sewer (RS), and a rainwater pump (RP) [4, 5].

The rainwater catchment area (RCA) component was where the rainwater were captured and the materials were used in the construction. The materials affected the efficiency and quality of rainwater collection. The materials used for the rainwater catchment surface should not contain toxins and materials that could degrade the quality of rainwater. Generally, the materials used were anti-rust material such as aluminum, galvanized iron, concrete, fiberglass shingles, etc. [3].

RHS implemented after the tsunami disaster in Banda Aceh was a simple, easy and inexpensive RHS constructed. This system was very helpful for the people affected by the disaster and the difficulty of clean water after the tsunami [6]. The rainwater drainage system usually consists of collecting channels or pipes that drain rainwater down on the roof to the storage tank. The collector or pipe channel had the size, slope and mounting in such a way that the maximum volume of rainwater could be accommodated [4]. The size of the container channel depends on the area of the catchment area, usually the diameter of the catching channel measuring 2.0 to 5.0 inc. [4].

According to UNEP [3], there were several types of RHS that could be applied: (1) RHS with roof system, this system uses the roof of the house as a rainwater catchment area. If the system was applied to individual homes, then the possibility of collected water was not significant, but if the system was massively applied, then the water collected was very abundant; (2) land surface catchment areas. this system uses the ground surface with a very simple method to collect rainwater. Compared with the roof system, RWH with the ground surface system allows more collecting rainwater because of its wider catchment area. Rainwater collected with ground-level systems was more suitable for agriculture because of low water quality. The rainwater collected with this system could be accommodated in a farm basin. However, some of the water contained in the farm basin was likely to sink into the soil [3].

The volume of rainwater that could be utilized by cocoa farm depended on the efficiency coefficient of the rainwater storage. The rainwater storage efficiency coefficient was the percentage of rainwater that could be utilized effectively. This coefficient depends on the RHS and its utilization to meet water needs. For indoor utilization needs (such as: shower, washing, toilet, leakage), the efficiency coefficient was (75-90) %, while for outdoor needs (eg irrigation, reservoir (liter / day) was about 50% [3].

2.2. Logical framework

Management of rainwater utilization (MRU) for cocoa farm was the management of rainwater based on the need of cocoa farm. The management model was defined in equation (5). The aim of MRU was to optimize the productivity of cocoa farm throughout the year from an average of 500 kg ha^{-1} $year^{-1}$ to more than 3000 kg ha^{-1} $year^{-1}$.

Farmers' cocoa farm were generally located in hilly areas. The contours of the hilly cocoa farm led to high runoff coefficients. As a result, cocoa farm in such location would quickly dehydrated. On the other hand, one of the advantages of a hilly cocoa farm was that the rainwater could be managed on the highest area. Thus, the water distribution system throughout the cocoa farm was applying the law of gravity.

3. Research methods

The research was a preliminary study of the application of AMHB that were carried out for one week, at the beginning of November 2016. The research areas were conducted in AMHB at "*Sentra Pemberdayaan Tani*" (SPT), Genting Village, Semarang and AMHB Nglanggeran, Nglanggeran Village, Patuk, Gunung kidul, Yogyakarta. The researchers took this location only as an example representing the AMHB built with the SPT concept developed by "*Obor Tani*" foundation.

In this study, researchers used in-depth interview techniques with two technical supports of the two AMHB as the informants, the first was Mr. Pratomo as a senior technical enginer support, and the second was Mr. Cahyo as an assistent of technical enginer. Researcher also use technical documentation of the AMHB to get written data which was considered relevant with this research. This research used qualitative models and some mathematical formulas used to analyze and describe the information obtained logically.

4. Results and Discussion

Based on in-depth interviews conducted by researchers on both informants on how the management of rainwater utilization using AMHB could be applied. Both informants, complement each other, explained systematically and comprehensively the advantages, disadvantages, technical requirements, organization, empowerment, and community assistance associated with the application of the AMHB. From the very comprehensive explanation, in this paper the researcher would only focus on discussing the aspects of advantages and disadvantages in general, the application of main technology aspects, and the required technical specifications of AMHB in more detail.

4.1. Advantages and Disadvantages Aspects

According to the two informants, the implementation of rainwater management using a AMHB relatively required more expensive for initial investment compared to the mini-valley-basin (MVB). The advantage of deploying AMHB was that there is no need for additional investments in water pumps and its operational costs, such as fuel, oil, maintenance costs, and skilled human resources to handle the agricultural machinery system. According to the experience based calculation of the two informants that the additional machinery investment and its operational costs of MVB for three years was equivalent to the investment of AMHB. Thus, the AMHB investment would be cheaper after more than three years.

4.2. Technology and Its Supporting Infrastructure

Management of rainwater utilization using AMHB requires additional technology, which should not be used in MVB, in order for rainwater stored in reservoirs to be controlled and avoid potential landslides. There are two major types of investments needed to construct the AMHB, which was its own development investment and its infrastructure. The investments in the construction of AMHB such as cut and fill, soil compaction using heavy equipment with certain specifications, and other supporting facilities. The main infrastructure investment in additional technology required to complete the AMHB is a geomembrane, some devices for rainwater management system technology.

4.3. Required Technical Specifications

According to the two informants, there are three main technical factors that must be met to build and manage the appropriate AMHB, which are the suitability of the volume of rainwater to be managed, the annual rainfall at the cocoa farm area, and the adequacy of the hill area to meet the needs cocoa farm. The need for rainfall volume to be managed in AMHB is calculated based on the accumulation of maximum volume of water of each productive cacao trees during the dry season, the average evaporation factor throughout the year, and other water requirements. The rainfall factor is needed to calculate the rainwater catchment area to meet the capacity of the AMHB. While the location determination is based on the height of the location and the wide area of available rainwater catchment area.

4.4. Rainwater Utilization Management

The management of rainwater utilization intended in this paper is the management of rainwater based on the needs of the cocoa farm during the dry season. Therefore, the underlying factors of management are the volume of rainwater that falls in the cocoa farm location throughout the year, the percentage of rainwater required by the cocoa farm during the dry season, the percentage of rainfall evaporation from the AMHB for a year. Thus, the volume of rainwater that should be managed in AMHB was the accumulation of cocoa farming needs and evaporation throughout the year.

4.4.1. The water Requirement of Cocoa Farm

The volume of rainwater required to manage a cocoa farm during the dry season depended on the population of the cocoa farm and the maximum water requirements of the most productive cocoa plant at the same environmental conditions. Mathematical formula for determining the volume were as follows:

$$Vc = N * n * V_{max} \quad (1)$$

Where Vc : The volume of water needed by the cocoa farm during the dry season (m^3)
N : The Length of dry season at the area of cocoa farm (day)
V_{max} : The maximum water requirement of the most productive cacao plant at the same environmental conditions during the dry season (liters)
n : The number of cocoa plants.

For example: A cocoa farm had an area of 2 ha, population of 2135 trees, with an average age of 5 years. The cocoa farm was located in village X which had a dry season for 7 months. Suppose the most productive cocoa tree needs an average of 5 liters of water $tree^{-1}$ day^{-1}. For this case, then each cocoa plant for 7 months of dry season required 7 x 30 days x 5 liters $tree^{-1}$ day^{-1} was equal to 1050 liters $tree^{-1}$ or 1.05 m^3 per tree. So the need for rainwater that have to be stored in the AMHB during the year to meet the needs of cocoa plant during the dry season was equal to 2135 trees x 1.05 m^3 per tree or equal to 2241.75 m^3.

4.4.2. The Efficiency Coefficient of AMHB

The efficiency coefficient of AMHB determined the volume of rainwater that must be harvested during the rainy season. In general, there were three main factors that could affect the efficiency coefficient of AMHB, the first was the infiltration factor or permeation of the AMHB, the second was the evaporation factor of the AMHB, and the third was the damage factor of AMHB. The implementation of AMHB system could minimize the water infiltration or permeation factor of the AMHB and the damage factor could be well controlled. In other words, the evaporation factor of AMHB was the only major factor affecting the efficiency coefficient. According to Pratomo [7], the evaporation factor of AMHB was estimated at about 20%. Thus, the efficiency coefficient of the AMHB was about 80%.

4.5. Management of Artificial-Mini-Hill-Basin (AHMB)

There were three types of rainwater catchment area (RCA) that could support the AMHB in cocoa farm, the first was the AMHB itself, the second was the building roof (BR) of the cocoa farm which was higher than the AMHB, and the third was the cocoa farm at the higher level of cocoa farm area (HCFA) than the AMHB.

IOP Conf. Series: Earth and Environmental Science **157** (2018) 012003 doi:10.1088/1755-1315/157/1/012003

4.5.1. Volume of AMHB. For example, the volume of water need of a cocoa farm during the dry season was **Vt** (m^3), then the mathematical model for determining the volume of AMHB needed to meet the needs of the cocoa farm was as follows:

$$Vk = (1 + e)Vt \tag{2}$$

Where ***Vk*** : The minimum volume of AMHB for a year (m^3)
Vt : The volume of rainwater that cocoa farm need during the dry season (m^3)
e :The efficiency coefficient of AMHB (%)

By knowing ***Vk***, the AMHB could be determined its area and depth. According to (Pratomo, 2009), the water depth within the AMHB was about 3m. Factors needed to consider for determining the depth of the AMHB were the physical security of the AMHB itself and its evaporation (Pratomo, 2009).

4.5.2. Rainwater Catchment Area (RCA) Management. Types of RCA that could be utilized to fulfill **Vk** were the AMHB itself, BR, or HCPA: The first priority to fulfill the AMHB was the AMHB itself, the second priority was the building roof (BR) of cocoa farm which was higher than AMHB if any, and the third priority was the HCFA.

The mathematical model for determining the rainwater volume of RCA cached from the AMHB was as follows:

$$Vtk = P * L * Ch \tag{3}$$

Where ***Vtk*** : The rain water volume from AMHB as RCA for a year (m^3)
P : The Length of AMHB (m)
L : The width of AMHB (m)
Ch : The average annual rainfall (mm)

The mathematical model for determining the rainwater volume of RCA cached from the BR was as follows:

$$Vta = Pa * La * Ch \tag{4}$$

Where ***Vta*** : The rain water volume fromBR as RCA for a year (m^3)
Pa : The length of BR (m)
La : The width BR (m)
Ch : The average annual rainfall (mm)

The mathematical model for determining the rainwater volume of RCA cached from the HCFA was as follows:

$$Vts = Ps * Ls * Ch * R \tag{5}$$

Where ***Vts*** : The rain water volume from HCFA as RCA for a year (m^3)
Ps : Length of HCFA (m)
Ls : The width of HCFA (m)
Ch : The average annual rainfall (mm)

Thus, the total area of rainwater catchment required is the accumulation of AMHB, BR, and HCFA areas.

5. Conclusions

- With the right management of AMHB, rainwater can be optimally utilized as a source of cocoa farm water to overcome the lack of water during the dry season.
- Rainwater management, by applying AMHB, had to apply geomembrane and water-controling-system technologies.

- Rainwater management, by applying AMHB, must meet the requirements of the wide areas of each available RCAs in the cocoa farms.
- The mechanism to identify the volume of rainwater management in cocoa farm, by applying AMHB, was started with identifying the total of cocoa plants, the maximum water requirement of the most productive cocoa in the same environmental conditions, the average annual rainfall, the duration of the dry season and the AMHB evaporation coefficient at the cocoa farm area, and AMHB volume, and finally, determine carefully according to the mathematical equations (1) through (5) above.
- The volume of rainwater utilization for cocoa farm was different at each of the different cocoa farm environmental conditions, it was primarily dependent on the annual rainfall average, the duration of the dry season, and the maximum water requirement of the most productive cocoa at the same area during the year.

References

[1] Anie Y 2011 Air hujan sebagai alternatif pengelolaan sumber daya air di perkotaan *Teknologi dan Kejuruan* **34** 105-114

[2] Zhang Y, Chen L, Chen D and Ashbolt S 2009 Potential for rainwater use in high-rise buildings in Australia cities *J. of Env. Manag.* **91** 222-226

[3] UNEP 2001 *International Technology Centre: Rainwater Harvesting* (Australia: Murdoch University of Western Australia)

[4] Abdulla F and Al-Shareef A 2009 Roof rainwater harvesting systems for household water supply *Jordan Desalination* **243** 195-207

[5] Song J, Han M, Kim T and Song J E 2009 Rainwater harvesting as a sustainable water supply option in Banda Aceh *Desalination* **251** 233-240

[6] Amin M and Han M 2009 Water environmental and sanitation status in disaster relief of Pakistan's 2005 earthquake *Desalination* **248** 436-445

[7] Pratomo 2009 *Waduk Mini Penyelamat Ladang Petani Yayasan Obor Tani* http://obortani.com/read/2009/03/27/waduk-mini-desa-genting-selesai-corporate-social-responsibility-csr.html Accessed August 26, 2017

IOP Conf. Series: Earth and Environmental Science **157** (2018) 012004 doi:10.1088/1755-1315/157/1/012004

Species variation in home garden agroforestry system in South Sulawesi, Indonesia and its contribution to farmers' income

S A Paembonan[1], S Millang[1], M Dassir[2] and M Ridwan[3]

[1] Lab. of Silviculture, Fac.of Forestry, Hasanuddin University, Jalan Perintis Kemerdekaan KM. 10, Makassar, 90245, Indonesia.
[2] Lab. of Forest Policy, Fac.of Forestry, Hasanuddin University, Jalan Perintis Kemerdekaan KM. 10, Makassar, 90245, Indonesia.
[3] Lab. of Forest Economic, Fac.of Forestry, Hasanuddin University, Jalan Perintis Kemerdekaan KM. 10, Makassar, 90245, Indonesia.

E-mail: *samuelap@unhas.ac.id*

Abstract. Home-garden is one of the types of agroforestry which is commonly practiced by rural communities in South Sulawesi, Indonesia. The study aimed to determine the diversity levels of the species constituting the home-gardens and their contribution to the farmers' incomes. The variables used in the study were the widths variation of the land owned as the home-gardens and the socioeconomic backgrounds of the community. The study results indicated that in small land, the community cultivated annual crop plants interspersed with agricultural commodities, and the trees as the boundary, while in the wider land they integrated various species plants within the area. The diversity index of the home-gardens was categorized as moderate with a value of 1.25 to 2.18, while species uniformity index was ranging from moderate to high with values of 0.49 to 0.77. The total incomes from home gardens varied greatly from one community to another, and it was largely determined by the composition and density of the constituent species. The contribution of the home-gardens to the income of the farmers amounted to 43.27%–49.06%. The sustainable management of the home-garden agroforestry can give a significant contribution to the farmers' incomes and the preservation of biodiversity and environment.

1. Introduction

Agroforestry is a land-use system that combines trees and crops or animals simultaneously or sequentially to increase overall crop and livestock production on a common land under technological conditions and management measures that are appropriate to the culture of the local community [1, 2].

The agroforestry system consists of several combinations of forestry products and plantation crops with various advantages of each community which in turn is expected to improve the welfare of the community [3]. One of the agroforestry systems that can increase the income of farmers that is widely practiced by the community is home-garden, which is planted with forestry crops and commodity crops simultaneously in a field around the house of village community [4, 5].

The Economic aspect is a critical consideration in applying agroforestry system that is how significant role of agroforestry can give an advantage by utilizing resources and technology owned. In addition, it should also be noted how far agroforestry provides sustainability of the results needed to meet the needs of life.

IOP Conf. Series: Earth and Environmental Science **157** (2018) 012004 doi:10.1088/1755-1315/157/1/012004

The home garden has multi species in one site with different stratum [6, 7, 8] and has the ability as a source of biomass energy for its owner [9, 10]. The presence of home garden agroforestry can also be significant in the context of climate mitigation [11, 12, 13]. In addition, the proper management of home garden can ensure environmental sustainability [7, 14].

Many studies have been reported about the species variability in the home garden and its contribution to farmer income [2, 4, 5, 6, 8, 9, 14]. However, none of the studies have discussed and compared with different cultural background. In this study, we observed the variation of agroforestry species in different locations and different culture in South Sulawesi province, Indonesia. There are two ethnicities in South Sulawesi namely Ethnic Toraja and Ethnic Bugis-Makassar. These two ethnicities with different cultures also have different habits in selecting species developed in the home garden. Based on the above reasons, this study aims to determine the variety of species is grown by the community in South Sulawesi in the home garden agroforestry system and how much its contribution to farmer's income.

2. Research Methods

2.1. The Study Site

The study was conducted in South Sulawesi Province, Indonesia. Determination of observation plot based on two ethnics in South Sulawesi that is in Tana Toraja regency for Toraja ethnic and in Regency of Bone and Bulukumba for ethnic Bugis-Makassar (figure 1).

A total of 45 plots were observed to determine species composition and describe species diversity at each site. Each regency consists of three Districts and is represented by 15 plots chosen purposively. Observation and data collection was done in the plot with a size of 20 m x 50 m.

Interviews with the farmers were conducted to obtain the history of home-garden management and its contribution to total farmer's income.

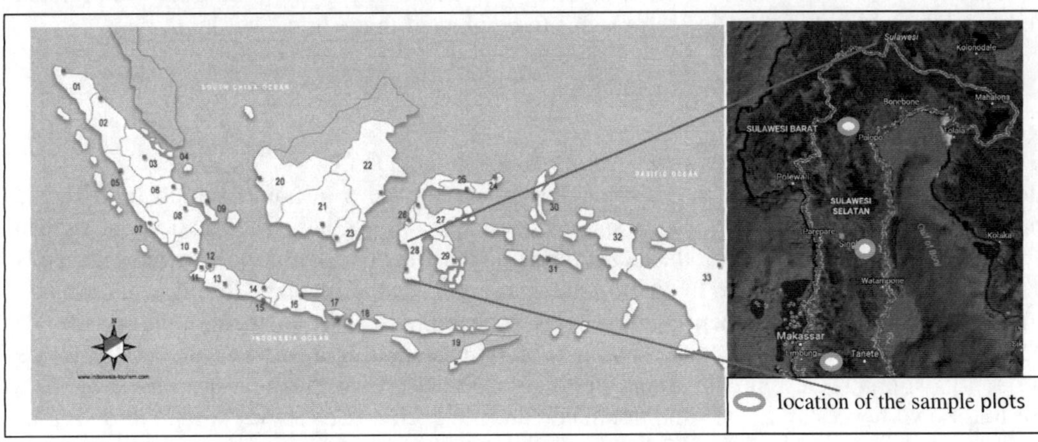

Figure 1. Map of Indonesia and South Sulawesi Province (enlarge).

2.2. Methods of Data Collection

The primary data collection techniques were carried out as follows:

1. All plant species in the plot of 20 m x 50 m were recorded by their species name and numbers.
2. Conducting interviews to farmers to obtain data on the history of home-garden management and income earned by the home-garden if assessed with money.
3. Calculate the contribution of home-garden income to total farmer's income.

Supporting data, in the form of socioeconomic data and other information were collected from the local government and previous related study results.

IOP Conf. Series: Earth and Environmental Science **157** (2018) 012004 doi:10.1088/1755-1315/157/1/012004

2.3. Data Analysis

*2.3.1. Important Value Index (IVI).*Important Value Index was analyzed to determine the dominance level of a species the home- garden, by using the formula of Mueller-Dombois and Ellenberg [15]:

$$IVI = RD + RF + RDo \quad (1)$$

where: RD is Relative Density (RD); RF Relative Frequency; and RDo is Relative Dominance

2.3.2. The diversity index. Calculated by the Shannon-Wiener formula [16]:

$$H' = -\sum Pi \ln Pi \quad (2)$$

where H '=Index of diversity; $Pi = \frac{ni}{N}$; ni = the number of individual species of i

N = total number of individuals

2.3.3. The uniformity index. Calculated by the following formula [16]:

$$J' = \frac{H'}{H'max} \quad (3)$$

where, J '= uniformity index; H '= index of diversity; H'max = ln S; S = number of species.

*2.3.4. Income Analysis.*The analysis is used to know the income obtained from the home garden agroforestry system using the formula:

$$I = TR - TC \quad (4)$$

where: I = Income (IDR/ha/year); TR=Total Revenues (IDR/ha/year);

TC = Total Cost (IDR/ha/year).

2.3.5. The contribution to the total incomes of the Farmer. The contribution of home garden income to the total income of farmers was obtained through comparison as follows:

$$CI = \frac{Hg}{TI} \times 100\% \quad (5)$$

where CI = Contribution of home garden Income (%); Hg = Income in home garden;

TI = Total Farmers Income

The analysis for each of the different cultures is done descriptively to identify the selected plant species and the reason for choosing the species to be planted in the home garden.

3. Results and Discussion

3.1. Composition of Species and Important Value Index (IVI)

Field observations showed that the composition of plant species in home-garden in different places varies from one site to another. The number of species and species specifications at each location differed according to the culture of the local community and the consideration of economic needs. There were about 47 species have been found in home garden in South Sulawesi Province (table 1).

The composition of constituent plant species in each district has its specific characteristics. This difference in composition is influenced by the economic background, customs, and culture of the local community. Toraja society predominantly chooses certain tree species that are linked to cultures to build traditional houses and other traditional ceremonies such as *Casuarina junghuhniana*, *Elmerillia pubescens*, *Palaquium obovatum* and *bamboos*. While the Bugis-Makassar community is more likely to choose plant species associated with traditional wedding ceremonies, building common houses and for daily consumption habits.

Various kinds of home garden production found on the site of this study is in line with the opinion of Widianto *et al.* [5] that the products from the agroforestry system are also diverse, such as fruits, wood for building, firewood, sap, feed, vegetables, tubers, and grains. Variations of species in the home-garden occurred according to the site conditions and wishes of the farmer. Nair [14] explain that the selection of plant species in agroforestry is influenced by economic needs, family preferences, the availability of complementary food or substitutes for staple foods. Sometimes the selection of species is also determined by the market trend without considering the specific land suitability factors or because of the support programs from the local government [4].

The predominant species of agricultural commodities grown by the community are cocoa, coffee, and cloves because they provide a high income and can bear fruit several times a year. This product can significantly increase the income of rural communities.

Table 1. The highest Important Value Index (IVI) of six species in each regency and districts.

Species	Bulukumba			Bone			Tana Toraja		
	1	2	3	1	2	3	1	2	3
Casuarina junghuhniana							76.53	82.65	57.09
Pinus merkusii							-	64.24	85.72
Arenga pinnata			18.25		12.75		15.42		
Gmelina arborea				31.72		22.87			
Elmerillia pubescens							72.37	28.17	21.62
Palaquium obovatum							11.91		
Tectona grandis	56.26	73.69							
Swietenia macrophylla		15.09							
Mangifera indica		13.49							
Toona sureni	14.39								
Gliricidia sepium		36.51	21.13	32.54	35.93	35.72			
Albizia sp.			23.87						
Aleurites moluccana				47.63		22.54			
Artocarpus heterophyllum	45.67								
Arthocarpus altilis				15.23					
Musa paradisiaca					14.55				
Nephelium lappaceum	17.98								
Durio zibethinus	13.93	23.92			26.24				
Syzygium aromaticum			20.11		21.53	18.75		20.40	42.48
Coffea robusta	104.74		100.52	85.74	56.41	65.26			20.03
Coffea Arabica							22.11	15.42	
Theobroma cacao		79.51	38.25	45.34	-	57.27	44.86	21.99	38.52

3.2. The Species Diversity (H') and Uniformity Index (J')
The diversity index of the home garden ranging from low to moderate level, with a value of 1.25–2.18. Thus, the species selected by the community are not very diverse, where only species that needed by the community to be planted. On the other hand, the Uniformity Index (J') in all districts was ranging from moderate to high with values of 0.49–0.77. In this category, the species composition was quite similar across districts and regency (table 2).

Table 2. Species Diversity Index (*H'*) and Species Uniformity Index (*J'*).

District	*H'*			*J'*		
	Bulukumba	Bone	Tana Toraja	Bulukumba	Bone	Tana Toraja
Distric 1	1.25	1.53	1.79	0.49	0.64	0.69
District 2	1.89	2.05	1.87	0.72	0.71	0.70
District 3	1.51	1.72	2.18	0.54	0.57	0.77

The selection of species was heavily influenced by human activities and determines the value of the diversity of these communities.

3.3. Distribution of crops based on the width of land ownership
The results show that the composition of plants within the area was influenced by width area of land ownership. The total area of home-garden ownership varies widely in rural communities, ranging from 0.2 ha to ≥ 1.5 ha. In small land, the community cultivated annual crop plants interspersed with agricultural commodities, and the trees as the boundary, while in the wider land they integrated plants within the area, such as the annual crops, agricultural commodity crops, fruit trees, and timber trees.

The more land owned by the farmer, the greater the proportion of tree crops to annual crops. The planted trees in home gardens do not need further maintenance, while seasonal crops and agricultural commodity crops require intensive maintenance. In general, farmers who have 1 - 1.5 ha of land earn the highest income contribution from home-garden because of the large proportion of agricultural commodity crops. The area of land ownership in this study is not a limiting factor in determining the types of crops grown by farmers, but more closely related to the habits and livelihood needs of local farmers. The result of the study shows that vertical structure of mixed home-garden consists of four strata. The upper stratum is occupied by trees with intolerant species while in the middle stratum of other tree species is partial light demanders, the third stratum by commodity plant, and the rest occupied by shrubs and herbs.

3.4. Contribution to Farmer's Income

The contribution of home garden income is the percentage of the proportion of farmer's income from home garden to total farmer's income. In general, farmers in this study are rural farmers who are not civil servants or private.

The results showed that in addition to the home-garden plantation business, some farmers also have paddy fields and livestock. Annual crops and other commodity crops for livestock were cultivated within the home-garden area. The contribution of the home garden to total income varies widely depending on the composition of plant species being cultivated. Huge contributions were obtained when most types of agricultural commodities are developed such as coffee, cocoa, and cloves.

The income contribution by the home garden to the total income of farmers in average was 49.06%, 43.27%, and 48.5% in Bulukumba, Bone, and Tana Toraja regency, respectively. The contribution has no significant difference of one to another. The contribution of income from home-garden aside determined by the composition of the mixture of plant species was also identified by the size of land ownership (figure 2).

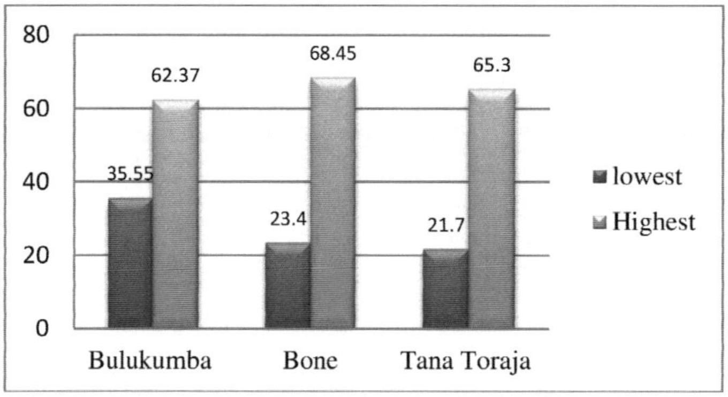

Figure 2. The contribution income of home-garden to the total income of farmers

According to Hartoyo *et al.* [6], the development of agroforestry is one of the models of community land management that can support the survival and welfare of the community, even in some places serves as saving that will be the basis to sustain the life. In the field of economic, agroforestry can increase the income of farmers through the diversification of commodities, and in the field of ecology, people can improve environmental quality.

4. Conclusion

- The home-gardens in South Sulawesi Province have a moderate species diversity index, with high-level species uniformity index.
- The contribution of income from home-garden, besides was determined by the composition of the mixture of plant species, was also influenced by the size area of land ownership.
- The average economic contribution of home-garden agroforestry system to total farmers' income is in the range of 43.27% to 49.06%.

- The selection of tree species to be planted in the home garden is influenced by the culture of the community, that is: in Toraja society to be used as building material of custom house and other traditional customs ceremony, whereas in Bugis-Makassar society to be used in building ordinary house, traditional ceremony of marriage and for daily consumption habits.
- The sustainable management of the home-garden agroforestry can give a significant contribution to the farmers' incomes and the preservation of biodiversity and environment.

References

[1] Hairiah K and Rahayu S 2007 *Pengukuran 'karbon tersimpan' di berbagai macam penggunaan lahan Bogor* [in Indonesian] (Indonesia: World Agroforestry Centre - ICRAF, SEA Regional Office, University of Brawijaya, Unibraw, Indonesia) p 77

[2] Paembonan S A 2013 *Model Agroforestri Berbasis Tongkonan yang Berwawasan Konservasi Lingkungan di Kabupaten Tana Toraja* in Kuswantoro et al. (eds) *Agroforestry for Food and Better Environment-National Seminar of Agroforestry* [in Indonesian with English summary] *Proc. of National Seminar on Agroforestry* 21 May 2013 Malang pp 494–498

[3] Mbile P, Stolle F, Boundzanga G, Mane L, Loumeto J, Homb M, Ifo S, et al 2013 Readiness for reducing emissions from deforestation and forest degradation under uncertain, national circumstances *Int. J. of Eco. and Dev.* **24** (1) 27- 48

[4] Kumar B M and Nair P K R 2004 The enigma of tropical home gardens *Agr. Syst.* **61** 135–152

[5] Widianto, Hairiah K, Suharjito D and Sardjono MA 2003 *The Function and Role of Agroforestry* [in Indonesian] (Bogor: World Agroforestry Centre (ICRAF))

[6] Hartoyo A P P, Siregar I Z, Supriyanto, Prasetyo L B and Thelaide I 2016 Biodiversity, Carbon Stocks and Community Monitoring in Traditional Agroforestry Practices: Preliminary Results from Two Investigated Villages in Berau, East Borneo. *Procedia Environ. Sci.* **33** 376–385

[7] Mattsson E, Ostwald M, Nissanka S P and Pushpakumara D 2015 Quantification of carbon stock and tree diversity of home gardens in a dry zone area of Moneragala district, Sri Lanka *Agrofor. Syst.* **89** 435–445

[8] Utomo B, Prawoto A A, Bonnet S, Bangviwat A and Gheewala S H 2016 Environmental performance of cocoa production from monoculture and agroforestry systems in Indonesia *J. Clean. Prod.* **134** 583–591

[9] Ali A and Mattsson E 2017 Individual tree size inequality enhances aboveground biomass in home garden agroforestry systems in the dry zone of Sri Lanka *Sci. Total Environ.* **575** 6–11

[10] Stas S M, Rutishauser, Chave J, Anten N P R and Laumonier Y 2017 Estimating the aboveground biomass in an old secondary forest on limestone in the Moluccas, Indonesia: Comparing locally developed versus existing allometric models *For. Ecol. Manage.* **389** 27–34

[11] Dhillon R S and von Wuehlisch G 2013 Mitigation of global warming through renewable biomass *Biomass and Bioenergy* **48** 75–89

[12] Mandal S A, Dutta I C, Jha P K and Haque S M 2013 Potential carbon offset in public plantation: offering an opportunity for REDD+ and soil fertility *Int. J. of Eco. and Dev.* **25** (2) 21 – 31

[13] Yamamoto Y, Kubota N, Ogo T and Priyono 1991 Changes in the structure of home gardens under different climatic conditions in Java Island *Japan. J. Trop. Agric.* **35** 104–117

[14] Nair P K R 2012 Carbon sequestration studies in agroforestry systems: A reality-check. *Agrofor. Syst.* **86** 243–253

[15] Mueller-Dombois D and Ellenberg H 1974 *Aims and Methods of Vegetation Ecology* (Canada: John Wiley and Sons Inc.) p 547

[16] Magurran AE 1987 *Ecological Diversity and Its Measurement*. (London: Chapman and Hall) p 179

IOP Conf. Series: Earth and Environmental Science **157** (2018) 012005 doi:10.1088/1755-1315/157/1/012005

Soil physicochemical properties to evaluate soil degradation under different land use types in a high rainfall tropical region: A case study from South Sulawesi, Indonesia

A Ahmad[1], C Lopulisa[1], A M Imran[2] and S Baja[1]

[1]Department of Soil Science, Faculty of Agriculture, Hasanuddin University, Jalan Perintis Kemerdekaan KM. 10, 90245, Makassar, Indonesia.
[2]Department of Geology, Faculty of Engineering, Hasanuddin University, Jalan Perintis Kemerdekaan KM. 10, 90245, Makassar, Indonesia.

E-mail: *asmita.ahmad@agri.unhas.ac.id; asmitaahmad@yahoo.com*

Abstract. Intensive cropping in the tropical region always becomes one of important driving forces of soil degradation. The primary aim of this study is to analyze the states and the dynamics of soil physicochemical properties to evaluate soil degradation in the tropical region a high rainfall on agricultural areas in South Sulawesi. A number of soil characteristics were analyzed for physical and chemical properties, and clay minerals with X-ray diffractometer. The degree of soil degradation is determined using Wischmeier and Smith equation. This study reveals that mean annual precipitation in 1979-2016 ranged from 1853.15 to 2981.30 mm/year. For land used for paddy field, palm oil, cacao and coffee plantation, the texture dominated with silt loam-clay loam, cation exchange capacity was 18.63-26.32 $cmol^+ kg^{-1}$, 0.98-2.91% of C-organic, 32-55% of base saturation, 0.1-3.5 cm h^{-1} of permeability, soil clay minerals were montmorillonite-kaolinite-halloysite, and the index erodibility was 0.3-0.5. Land used for mixed plants and shrubs, the texture dominated with silt loam-sandy clay loam, cation exchange capacity was 18.63-27.12 $cmol^+ kg^{-1}$, 1.09-2.89% of C-organic, 32-55% of base saturation, 0.2-4.9 cm/h of permeability, soil clay minerals were kaolinite-halloysite, and index erodibility was 0.1-0.3. Land use for cultivated in the high intensity of rainfall has changed the physicochemical properties of soils, but cultivated in monoculture has at some degree increased soil erodibility.

1. Introduction

Land with sloping topography in the tropic area had many used for an intensively cropping system for needing the community feed. Tropical region always connected with a high rainfall activity [1], where the period of rainy season from the equator to 30°N in May to October while from the equator to 30°S in November to April [2]. Cooperate rainfall and cultivation increased many processes in soil, such as weathering, alteration, eluviation, and illuviation [3], and as an important agent to decrease soil degradation.

Land use with intensively used giving significance for decreasing soil physicochemical properties [4, 5]. Soil degradation usually disturbed soil organic-carbon content [6], texture [7, 8], and fertility of soil [9]. Intensive cropping in the tropical region always becomes one of important driving forces of soil degradation, especially erosion and landslides.

The primary aim of this study is to analyze the states and the dynamics of soil physicochemical properties to evaluate soil degradation in the tropical region a high rainfall in agricultural areas

2. Materials and Method

2.1. Study site

Soil was taken from a high degree of landslides-prone area and no-landslides area with different land use types in three districts of South Sulawesi, Indonesia 9 (figure 1) based on ESDM data [10]. The coordinates location were; Enrekang District for degraded soil; E1: 119°50'43.69"E and 3°16'52.32"S;

and E3: 119°47'48.34"E and 3°14'37.65"S; while for non-degraded soil were; E2 119°50'44.44"E and 3°16'53.89"S; and E4: 119°47'48.60"E and 3°16'14.36.86"S, North Toraja District for degraded soil; T1: 119°53'13.11"E and 2°52'26.96"S; and T3: 119°53'26.87"E and 2°52'1.96"S, and for non-degraded soil were; T2: 119°53'13.32"E and 2°52'29.51"S; and T4: 119°53'25.09"E and 2°52'5.45"S, and East Luwu District for degraded soil were; L1: 120°48'0.72"E and 2°22'57.06"S; and L3: 120°47'40.19"E and 2°22'23.46"S; and for non-degraded soil were; L2: 120°48'4.33"E and 2°22'59.49"S; and L4: 120°47'36.64"E and 2°22'25.26"S. The area had an elevation of 320-1542 m above sea level (asl) and slope topography range of 40-60%.

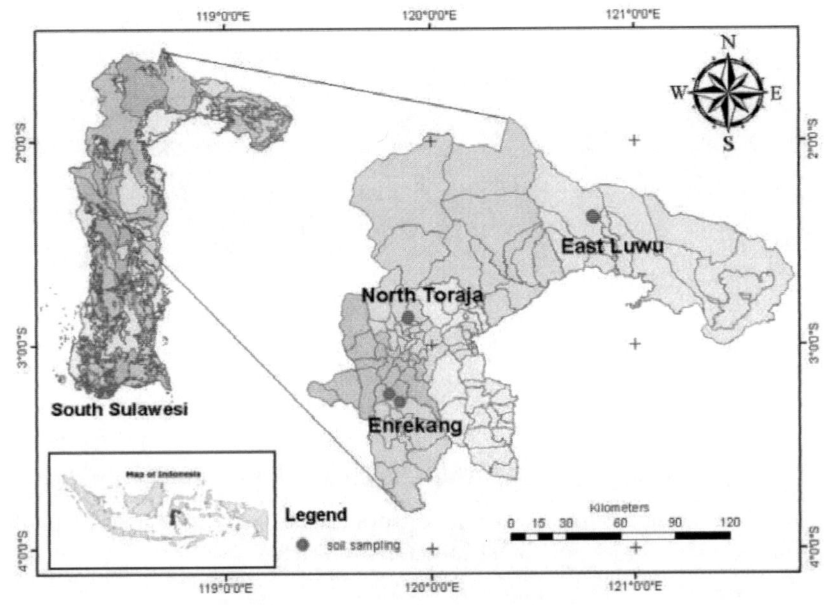

Figure 1. Location map of study area

2.2. *Soil analysis*

The soils analysis were pH with pH meter, texture with hydrometer method, Cation Exchange Capacity (CEC) with 1 M NH_4OAc, cations of Ca, Mg, K and Na done by extracting with ammonium acetate in pH 7 then measured with an Atomic absorption spectrophotometer, base saturation (BS), permeability with permeameter, and C-Organic content with Walkley and Black method. Soil erodibility from [11] modified by [12]was used to evaluate soil degradation degree.
The equation from Wischmeier and Smith (1978) is:

$$100K = 1.292\ [2.1M^{1.14}(10^{-4})(12\text{-}a) + 3{,}25(b\text{-}2) + 2{,}5(c\text{-}3)] \quad (1)$$

where:

K = Soil erodibility
M = percent fines sand + percent silt (0.1-0.02mm) times the quantity 100-percent clay
a = percent organic matter,
b = the soil-structure code used in soil classification, and
c = the profile-permeability class.

The soil temperature in 25 and 50 cm depth were measured with soil temperature tools and soil minerals were analyzed with X-Ray Diffractometer (XRD) Shimadzu XRD-7000.

IOP Conf. Series: Earth and Environmental Science **157** (2018) 012005 doi:10.1088/1755-1315/157/1/012005

3. Results

3.1. Rainfall Data

Mean annual precipitation in 1979-2016 ranges from 1853.15 to 2981.30 mm year^{-1} [13, 14]. The data showed that the distribution of rainfall becoming extreme start from 1991, where Enrekang District has dry season extreme, but in 2000 where North Toraja District and East Luwu had wet season extreme and reached 4500-5200 mm year^{-1} (figure 2). The extreme climate still continued until now. This was a connection to global warming phenomena [15] and impact of green revolution which caused intensively cropping in all region in Indonesia [16].

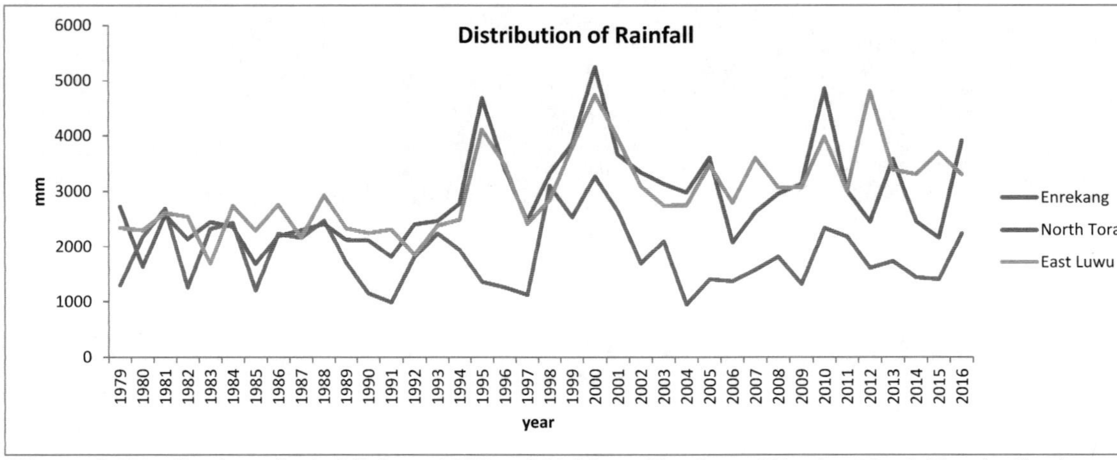

Figure 2. Distribution of rainfall from 1979 to 2016 from Enrekang District, North Toraja district and East Luwu district

3.2. Physicochemical Properties of Soil

Degraded soils at 25 cm depth have a temperature ranging from 24 to 34 °C, and 23-29 °C at 50 cm depth. The area was used for paddy field in North Toraja, palm oil and cacao in East Luwu, crop and coffee plantation in Enrekang District. The range of soil pH was 4.5-6.4, and it indicated increasing of leaching process. Texture dominated with silt loam-clay loam with difference in the fraction presentation, and C-organic ranged of 0.98-2.91% (table 1). Soils with non-degraded had a temperature at 25 cm depth ranging of 26-31°C and 23-29 °C at 50 cm depth. Land used dominating with mixed plants and shrubs. The range of soil pH was 5.0-6.5, texture was very varied and dominated with silt loam, clay loam, and sandy clay loam, and C-organic ranged of 1.09-2.89%. BS and CEC in degraded soil and non-degraded soil had same variation from first layer and second layer and gave not much contribution to soil degradation.

3.3. Soil Erodibility

The permeability of soil degraded ranged of 0.1-3.5 cm/hour with criterion of very slow-moderate and the index erodibility was 0.3-0.5 with soil clay minerals dominated with montmorillonite, vermiculite, illite, kaolinite, and halloysite. Permeability of soil non-degraded ranged of 0.2-4.9 cm hour^{-1} with criterion of very slow-moderate and the index erodibility was 0.1-0.3 with soil clay minerals dominated with illite, kaolinite, and halloysite (table 2).

IOP Conf. Series: Earth and Environmental Science **157** (2018) 012005 doi:10.1088/1755-1315/157/1/012005

Table 1. Physicochemical properties of soil

Soil layer	Soil depth	Sand	Fine sand	Silt	Clay	Texture	pH H_2O	C-Organic	Exchangeable Cations Ca	Mg	K	Na	CEC	BS
	cm	--------- % --------					1:2.5	%	cmol$^+$/kg					%
E1.1	0-10	5	12	41	42	Clay loam	4.5	2.91	6.86	2.14	0.22	0.36	24.62	39
E1.2	10-30	5	12	38	45	Clay	5.6	1.86	6.14	1.86	0.16	0.25	26.25	32
E2.1	0-10	4	17	42	37	Clay loam	6.0	2.41	6.92	2.24	0.28	0.36	24.62	40
E2.1	10-20	4	15	31	50	clay	6.0	1.86	6.78	2.11	0.23	0.25	26.20	36
E3.1	0-10	4	19	40	37	Clay Loam	5.7	2.67	6.85	3.25	0.21	0.21	25.32	42
E3.2	10-30	2	10	39	49	Clay	5.8	2.14	7.42	3.22	0.19	0.25	26.32	42
E4.1	0-10	4	19	38	39	Clay loam	6.1	2.89	6.81	3.27	0.24	0.20	25.21	42
E4.2	10-30	4	13	35	48	Clay	6.1	2.52	7.45	3.23	0.20	0.25	27.12	41
T1.1	0-10	5	11	57	27	Silt loam	6.1	2.83	6.74	2.65	0.25	0.19	20.89	47
T1.2	10-40	4	8	60	28	Silty clay loam	6.2	2.53	7.25	3.41	0.36	0.28	21.85	52
T2.1	0-20	2	5	52	41	Silty clay	6.2	2.55	7.34	2.23	0.26	0.17	20.78	48
T2.2	20-50	4	7	41	48	Silty clay	6.5	2.41	7.41	3.21	0.39	0.18	21.93	51
T3.1	0-20	3	6	71	20	Silt loam	6.4	2.88	5.86	3.42	0.26	0.26	19.63	50
T3.2	20-50	3	8	50	39	Silty clay loam	6.3	2.33	5.41	4.15	0.42	0.35	18.63	55
T4.1	0-10	6	15	41	37	Clay loam	5.9	2.79	5.37	3.53	0.30	0.24	19.56	48
T4.2	10-40	5	9	40	46	Silty clay	6.1	2.14	5.65	3.79	0.39	0.37	18.63	55
L1.1	0-10	18	43	16	23	Sandy clay loam	5.7	1.85	3.75	2.65	0.36	0.25	20.85	34
L1.2	10-40	6	13	49	33	Silty clay loam	5.7	0.98	3.98	2.38	0.25	0.24	19.63	35
L2.1	0-10	15	35	21	29	Sandy clay loam	5.6	1.96	3.17	2.63	0.28	0.19	19.33	32
L2.2	10-20	17	40	18	24	Sandy clay loam	5.0	1.58	3.76	2.88	0.25	0.25	19.79	36
L3.1	0-10	15	16	53	16	Silt loam	5.3	1.76	5.70	2.78	0.34	0.32	19.63	47
L3.2	10-20	12	18	52	18	Silt loam	5.3	1.65	5.65	3.25	0.26	0.26	20.74	45
L4.1	0-10	7	16	42	35	Clay loam	5.9	1.09	5.17	2.92	0.41	0.38	19.61	45
L4.2	10-20	19	43	20	18	Sandy loam	5.8	1.41	5.76	3.31	0.35	0.22	19.99	48

Table 2. Soil erodobility index from different land use types

Soil layer	M	Organic Matter a	Permeability cm/hour	Code of Permeability c	Soil structure	Code of soil stucture b	Erodibility K	Soil Mineral	Land Use types
E1.1	2390.0	5.02	0.5	5	granuler medium-coarse	3	0.2	soil degraded; montmorillonite (28.6%), vermiculite (5.9%)	crop
E1.2	2102.0	3.21	0.1	6	blocky	4	0.3	illite (12.5%), kaolinite (6%), halloysite (47),	crop
E2.1	2663.0	4.15	0.5	5	granuler fine-medium	2	0.2		shrub
E2.1	1565.0	3.21	0.4	6	granuler medium-coarse	3	0.2	soil non-degraded; montmorillonite (0.8%), illite (5.8%),	shrub
E3.1	2539.0	4.60	0.4	6	granuler medium-coarse	3	0.3	chlorite (32.7%), halloysite (52.5), kaolinite (8.2%)	coffe
E3.2	1999.0	3.69	0.2	6	blocky	3	0.3		coffe
E4.1	2337.0	4.98	0.5	5	granuler medium-coarse	3	0.2		mixed palnts
E4.2	1833.0	4.34	0.3	6	blocky	3	0.2		mixed palnts
T1.1	4172.0	4.88	1.3	5	granuler fine-medium	2	0.3	soil degraded; montmorillonite (12.1%), vermiculite (7.4%)	paddy fields
T1.2	4328.0	4.36	0.4	6	blocky	4	0.5	illite (12.1%), kaolinite (23.1%), halloysite (39.4%),	paddy fields
T2.1	3079.2	4.40	0.5	5	granuler fine-medium	2	0.3		shrub
T2.2	2139.0	4.15	0.6	5	granuler medium-coarse	3	0.2	soil non-degraded; montmorillonite (16.6%), illite (12.8%),	shrub
T3.1	5705.7	4.97	1.3	5	granuler medium-coarse	3	0.5	chlorite (10.3%), halloysite (38.3%), kaolinite (22%)	paddy fields
T3.2	3057.8	4.02	0.4	6	blocky	4	0.4		paddy fields
T4.1	2598.1	4.80	0.5	5	granuler fine-medium	2	0.2		mixed plants
T4.2	2169.0	3.69	0.6	5	granuler medium-coarse	3	0.2		mixed plants
L1.1	1274.5	3.19	3.5	4	granuler medium-coarse	3	0.2	Soil degraded; vermiculite (2.6%), illite (0.9%),	cacao
L1.2	3311.1	1.69	0.3	6	blocky	4	0.5	kaolinite (94.7%), halloysite (1.8%),	cacao
L2.1	1535.9	3.38	4.9	4	granuler fine-medium	2	0.1		mixed palnts
L2.2	1398.9	2.72	1.5	5	granuler medium-coarse	3	0.2	Soil non-degraded; vermiculite (0.7%), illite (20.5%),	mixed palnts
L3.1	4468.0	3.03	1.6	5	granuler fine-medium	2	0.4	kaolinite (74.8%), halloysite (4%),	palm oil
L3.2	4282.0	2.84	1.2	5	granuler medium-coarse	3	0.4		palm oil
L4.1	2746.1	1.88	0.4	6	granuler fine-medium	2	0.3		shrub
L4.2	1683.2	2.43	1.9	5	granuler medium-coarse	3	0.2		shrub

4. Discussion

Physicochemical properties of soil on degraded soil and non-degraded soil with different land use types showed the difference in soil texture conditions from the percentage of fine-coarse soil fractions. The difference in the size of the soil fraction affected the cohesiveness of the soil grains and influenced soil erodibility index value [4, 17, 18]. Degraded soils in cultivated soils had larger erodibility index values with greater difference of organic material content between first layer and second layer of soil at same profile, compared with non-degraded soils, the percentage difference between organic matter content in first layer to second layer was very small (table 1). Organic matter had functioned as an aggregate binding of soil fraction, so the soil is difficult to degrade [6].

Differences in the value of the erodibility indexes in the first and second layers of soil profile lead to decrease in soil stability so that the soils were easily eroded (table 2). Differences in the value of the erodibility index generally occur on land used for monoculture plantations, whereas land used for mixed plantations and shrub, the stability of the soil in second layer is higher or equal than layer in first layer so as to prevent soil degradation. According to Dibal *et al.* [17], mixed plants is one of management to retain soil from degradation, especially in the tropical region with high activity of rainfall, where study area had high mean annuals of rainfall categories [14]. Rainfall contributed to increase weathering in soil and produced secondary clay minerals. These minerals can absorb water in their surface crystal and in interlayer space of crystal [19, 20] and caused the soil permeability slower. Variety of clay minerals in soil have contributed to decreasing soil stability [21, 22, 23] and triggering erosion-landslides.

Soils structure in form of blocky shape in the second layer of profile had a relationship to trigger the erosion/landslides. Blocky structure indicated to present of clay minerals fraction in the soil and inhibits water sub-surface movement downward [24]. This support with permeability data, where the value of permeability became very slow downward (table 2).

5. Conclusions

Land use for cultivated in the high intensity of rainfall has changed the physicochemical properties of soils, but cultivated in monoculture has at some degree increased soil erodibility.

Acknowledgments

Thanks to the Ministry of Research, Technology and Higher Education Indonesia for the bppdn scholarship and dissertation grant.

Thanks to the Post-Graduate Program and LP2M of Hasanuddin University for the opportunity to do this research.

References

[1] Rodriguez-Lloveras X, Buytaert W and Benito G 2016 Land use can offset climate change induced increases in erosion in Mediterranean watersheds *Catena* **143** 244–55

[2] Ratan R and Venugopal V 2013 Wet and dry spell characteristics of global tropical rainfall *Water Resour. Res.* **49** 3830–41

[3] Virto I, Imaz M, Fernández-Ugalde O, Gartzia-Bengoetxea N, Enrique A and Bescansa P 2014 Soil degradation and soil quality in Western Europe: current situation and future perspectives *Sustainability* **7** 313–65

[4] Ezeabasili A C C, Okoro B U and Emengini E J 2014 Relative erodibilities of some soils from Anambra basin sky *J. soil Sci. Environ.* **3** 83–90

[5] Zare M, Panagopoulos T and Loures L 2017 Simulating the impacts of future land use change on soil erosion in the Kasilian watershed, Iran *Land use policy* **67** 558–72

[6] Manyiwa T and Dikinya O 2013 Using universal soil loss equation and soil erodibility factor to assess soil erosion in Tshesebe village , north east Botswana *African J. Agric. Res.* **8** 4170–8

[7] Salako F K 2003 Susceptibility of Coarse-textured Soils to Soil Erosion by Water in the Tropics 3–21

[8] Rosenbloom N A, Doney S C and Schimel D S 2001 Geomorphic evolution of soil texture and

organic matter in eroding landscapes *Global Biogeochem. Cycles* **15** 365–81
[9] Sharma U C, Datta M and Sharma V 2014 Soil fertility, erosion, runoff and crop productivity affected by different farming systems *Ecopersia* **2** 629–50
[10] ESDM 2009 *Peta Zona Kerentanan Gerakan Tanah Provinsi Sulawesi Selatan (Map of Soil Susceptibility of South Sulawesi Province)*
[11] Wischmeier W H and Smith D D 1978 *Predicting rainfall erosion losses* (United States Department of Agriculture)
[12] Arsyad S 2010 *Konservasi Tanah & Air (Soil and Water Conservation)* (Bogor: IPB Press)
[13] Global Weather 2017 *Climate data in 1979 -2014 for Enrekang, North Toraja and East Luwu District*
[14] BMKG 2017 *Data curah hujan tahun 2008-2016 untuk Kabupaten Enrekang, Toraja Utara dan Luwu Timur (Rainfall Data in 2008-2016 for Enrekang, North Toraja, and East Luwu District)*
[15] Houghton J, Filho M, Bruce J, Lee H, Callander B, Haites E, Harris N and Maskell K 1995 Climate Change 1994 vol 15
[16] Sopandie D, Poerwanto R and Sobir 2012 *Sustainable Agriculture System Revolutionizing the Green Revolution* (Bogor, Indonesia: IPB Press) p 2012
[17] Dibal J M, Bashir A U, Umara B G and Baraya B 2014 Variability of soil erodibility factor with some soil management practices in a semi-arid agroecological condition, Nigeria ARPN *J. Eng. Appl. Sci.* **9** 2206–11
[18] Centeri C, Szalai Z, Jakab G, Barta K, Farsang A, Szabó S and Bíró Z 2015 Soil erodibility calculations based on di ff erent particle size distribution measurements *Hungarian Geogr. Bull.* **64** 17–23
[19] Schoonheydt R A and Johnston C T 2011 The surface properties of clay minerals *EMU notes Mineral.* **11** 335–70
[20] Hazen R M, Sverjensky D A, Azzolini D, Bish D L, Elmore S C, Hinnov L and Milliken R E 2013 Clay mineral evolution *Am. Mineral.* **98** 2007–29
[21] Regmi A D, Yoshida K, Dhital M R and Devkota K 2013 Effect of rock weathering, clay mineralogy, and geological structures in the formation of large landslide, a case study from Dumre Besei landslide, Lesser Himalaya Nepal *Landslides* **10** 1–13
[22] Yalcin A 2007 The effects of clay on landslides: A case study Appl. *Clay Sci.* **38** 77–85
[23] Kitutu M G, Muwanga A, Poesen J and Deckers J A 2009 Influence of soil properties on landslide occurrences in Bududa district , Eastern Uganda *African J. Agric. Res.* **4** 611–20
[24] Schoonover J E and Crim J F 2015 An Introduction to soil concepts and the role of soils in watershed management *J. Contemp. Water Res. Educ.* **154** 21–47

IOP Conf. Series: Earth and Environmental Science **157** (2018) 012006 doi:10.1088/1755-1315/157/1/012006

The application of parallel wells to support the use of groundwater for sustainable irrigation

Suhardi

Dept. of Agricultural Engineering, Faculty of Agriculture, Hasanuddin University, Jalan Perintis Kemerdekaan KM 10. Makassar, 90245, Indonesia.

E-mail: *suhardi@unhas.ac.id*

Abstract. The use of groundwater as a source of irrigation is one alternative in meeting water needs of plants. Using groundwater for irrigation requires a high cost because of the discharge that can be taken is limited. In addition, the use of large groundwater can cause environmental damage and social conflict. To minimize costs, maintain quality of the environment and to prevent social conflicts, it is necessary to innovate in the groundwater taking system. The study was conducted with an innovation of using parallel wells. Performance is measured by comparing parallel wells with a single well. The results showed that the use of parallel wells to meet the water needs of rice plants and increase the pump discharge up to 100%. In addition, parallel wells can reduce the influence radius of taking of groundwater compared to single well so as to prevent social conflict. Thus, the use of parallel wells can support the achievement of the use of groundwater for sustainable irrigation.

1. Introduction

Groundwater becomes one of the alternatives in the fulfillment of the water needs of rice plants, especially in the dry season. This is supported by the availability of about 25% of the total freshwater available [1]. In the use of ground water for irrigation, it must be done with caution because the amount of water taken is very large. Impacts that can be generated by long-term groundwater harvesting are environmental degradation such as land subsidence [2], costly losses due to decreased pumping capacity due to groundwater impairment [3, 4], and can caused to social conflict between groundwater users due to competition of resource use [5]. Therefore, this study aims to meet the needs of crop water by using sustainable groundwater irrigation.

2. Materials and methods

The study was conducted by testing the method of groundwater extraction using the parallel well method (diameter of pipe 2 and 3 in) and comparing it with conventional method (single well with 2 in diameter pipe). The required data was a debit data generated by each pumping method. The tool used are the pump unit, stopwatch and container of water reservoir. Some software was used in data processing such as excel add-ins such as VBA and solver.

The method of groundwater table fluctuations is one of the methods to estimate the groundwater recharge rate [6], used to estimate groundwater table decrease due to groundwater extraction. The decrease of groundwater table as one of the causes of land subsidence was determined by using 2 dimensional groundwater flow equation [7] shown in equation 1.

$$K.h\left(\frac{\partial^2 h}{\partial x^2}+\frac{\partial^2 h}{\partial y^2}\right)=S_y\frac{\partial h}{\partial t}\pm R(x,y,t) \quad (1)$$

IOP Conf. Series: Earth and Environmental Science **157** (2018) 012006 doi:10.1088/1755-1315/157/1/012006

where:

K = hydraulic conductivity (m/s)
h = head (m)
x,y = coordinate position (m)
t = time (s)
Sy = Safe yield (dimensionless)
R = Recharge or discharge (m^3/s)

The above equations were solved numerically by finite different methods using an implicit approach (*forward finite difference*) in which each space is evaluated for the next time (*t*+1) with the Crank-Nicolson approach [8]. This method has been applied in the investigation of groundwater level [9] with a coefficient of determination of 0.85.

The use of parallel wells was expected to minimize pump operational cost through increasing pump discharge so that the area of irrigation was increased. The area of irrigation depends on the discharge generated by the pump unit, calculated using the equation:

$$A = 0.1\frac{Q}{W_R} \quad (2)$$

where:

A = area of irrigation (ha)
Q = discharge (m^3/d)
W_R = water requirement (mm/d)
0.1 = conversion m^3 to mm*ha.

3. Results and Discussion

The results obtained were the change of groundwater level caused by groundwater extraction. In this case, the use of a single well to irrigate a certain rice field area requires many pump units. The use of several pumping units resulted in a decrease in the groundwater level at depth until the suction capacity of the pump suction depth became widespread. This was shown in the following figure:

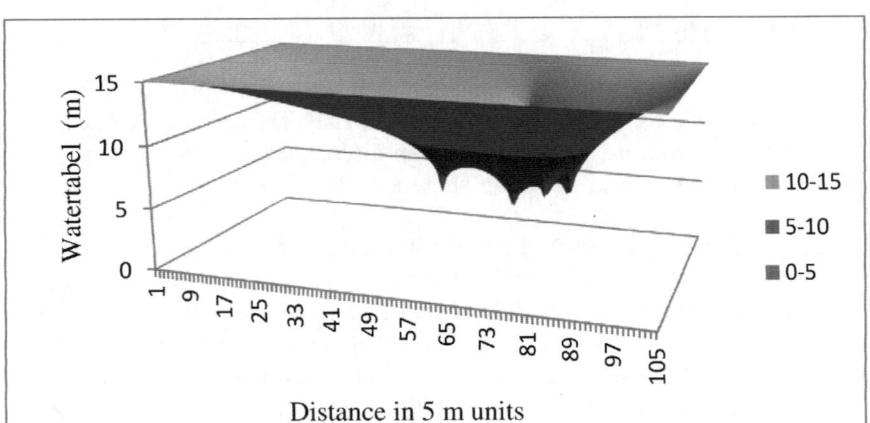

Figure 1. Groundwater profile using single well.

Figure 1 shows that groundwater harvesting with a single well leads to widespread groundwater degradation in the shooting area. If this condition occurs over a long period of time it can cause a land subsidence [2]. In contrast to the use of parallel wells, this method can increase the discharge rate for each pump unit so that the groundwater retrieval point location is reduced. Thus, the decrease of groundwater level does not occur widely as shown in figure 2 below:

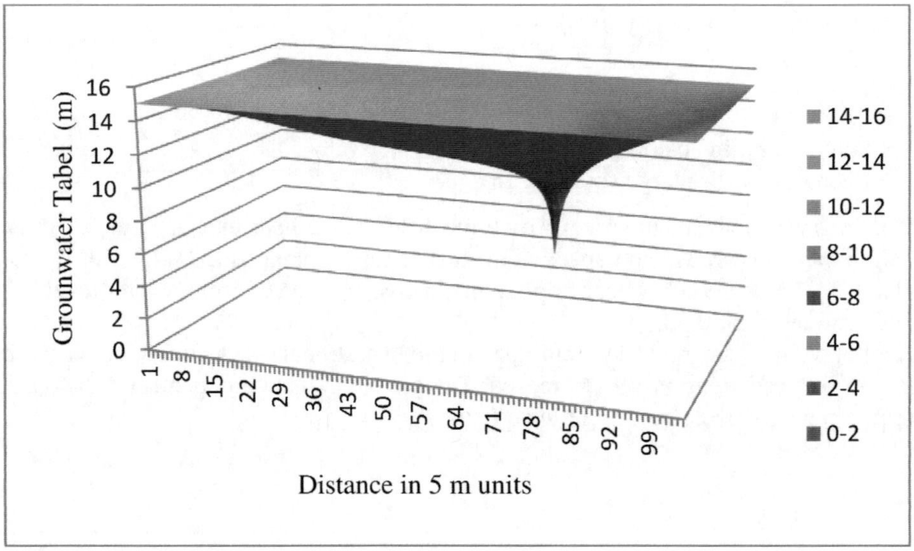

Figure 2. Groundwater profile applying parallel wells.

Because groundwater level reduction was not widespread, the pumping capacity relatively stable because the pump unit was not affected. Another advantage of using parallel wells was that the discharge obtained for one pump unit increases to 100%. The following is the increase of pump discharge data as a result of the use of parallel wells compared to conventional methods.

Table 1. Increased discharge of pumps using parallel wells

Treatment	Discharge			
	average (l/s)	Stdev	Average (m^3/d)	Stdev
Single well, pump 2"	5.27	0.18	455.33	14.22
Parallel wells, pump 2"	6.71	0.15	579.74	12.86
Parallel wells, pump 3"	10.74	0.46	928.11	39.33

Increased discharge for one pump unit can reduce the pumping cost because the cost depends on the price of each lifting of water for a certain volume (m-ha) [10]. These costs include fuel costs, pump unit maintenance costs and operator fees. A large pump discharge reduces costs due to the length of pump operation to get a certain volume of water shorter so that fuel, maintenance costs and operator costs become less.

In addition, the increased discharge of the pump will increase the area of irrigation covered by one pump unit and the decrease of groundwater level for a certain distance becomes small. Thus, the chances of social conflict occur from competition between groundwater users can be eliminated. The area of irrigation for one pump unit using a parallel well increases up to 100% when compared to conventional methods. The amount of irrigation area was calculated with the assumption that the water requirement of rice plant at the research location is 4.60 mm / day [11].

Table 2. Increasing the area of irrigation by using parallel wells

Treatment	Irrigation cover area	Increased area to single wells
	(ha)	(%)
Single well, pump 2"	9.90	-
Parallel wells, pump 2"	12.60	27.27
Parallel wells, pump 3"	20.18	103.84

With increasing irrigation area, irrigation cost can be reduced due to the length of pump operation reduced so as to reduce fuel cost, pump unit maintenance cost and operator cost. Thus, the use of parallel wells can reduce the impact of groundwater use for irrigation on environmental damage, reduce pump operating costs and reduce the impact on social conflicts so that sustainable groundwater irrigation.

4. Conclusion

The application of parallel wells supports the sustainable use of groundwater for irrigation through increased pump discharge thus reducing pump operational costs, reducing environmental impacts through reducing impact areas, and preventing social conflicts through reducing conflict between resource users.

References

[1] Black P E 1996 *Watershed Hydrology*. 2nd edition (New York: Ann Arbor Press, Inc) p 460
[2] Li C, Tang X and Ma T 2006 Land subsidence caused by groundwater exploitation in the Hangzhou-Jiaxing-Huzhou Plain, China *Hydrogeology J.* **14** (8) 1652–1665
[3] Medellín-Azuara J, MacEwan D, Howitt R E, Koruakos G, Dogrul E C, Brush C F, Kadir T N, Harter T, Melton F and Lund J R 2015 Hydro-economic analysis of groundwater pumping for irrigated agriculture in California's Central Valley, USA *Hydrogeology J.* **23** (6) 1205–1216
[4] Bunu M Z 1999 Groundwater management perspectives for Borno and Yobe states, Nigeria *J. of Env. Hydrology* **7** (19) 1-10
[5] Shrestha S 2017 The contested common pool resource: Ground water use in urban Kathmandu, Nepal *The Geographic. J. of Nepal* **10** 153-166
[6] de Silva R P 2004 Spatial Variability of Groundwater Recharge - I. Is it really variable? *J. of Spatial Hydro.* **4** (1) 1-18
[7] Trescott P C and Larson S P 1977 Solution of three-dimensional groundwater flow equations using the strongly implicit procedure *J. of Hydro.* **35** 49 - 60
[8] Carnahan B, Luther H A and Wilkes J O 1990 *Applied Numerical Methods* (Malabar, Florida: Rober E. Krieger Publ. Co) p 604
[9] Suhardi, Setiawan B I, Pawitan H and Waspodo R S B 2011 Optimization of groundwater utilization for irrigation in wajo district, south sulawesi province of indonesia (in Bahasa) *Jurnal Irigasi* **6** (2) 80-89
[10] Kovacs K and West G 2016 The influence of groundwater depletion from irrigated agriculture on the tradeoffs between ecosystem services and economic returns *PLoS ONE* **11** (12) 1-22
[11] Suhardi 2008 *Groundwater Management Model for Irrigation with Single and Multiple Pump Operation. Case: Data Sub-Watershed, Wajo District, South Sulawesi* (in Bahasa). *Dissertation* (Bogor: Bogor Agricultural University)

Compost applicators for horticulture

Iqbal, M Achmad and M T Sapsal

Agricultural Engineering Departement, Hasanuddin University, Jalan Perintis Kemerdekaan KM 10, Makassar, 90245, Indonesia.

E-mail: *iqbaliqma@yahoo.com*

Abstract. Horticulture is the art of planting fruit trees, vegetables, and ornamental or agricultural science that deals with the cultivation of gardens, including planting vegetable plants, fruit, flowers, and shrubs and ornamental trees. Fertilization is one of the important things to increase production, even until now regarded as a dominant factor in agricultural production. The use of compost can provide benefits for soil and plants. Problems that occur at the time of application of compost needed manpower and considerable expense, so it needs an efficient technology in the form of mechanical equipment that is simple and easy to operate. This study aims to modify applicator for sugar cane dry land so that it can be used on horticultural crops (vegetables) land and seeks to increase the efficiency of the applicator compost through modifications the system coupling that can be drawn using the tractor two wheels. The results show that the prototype model of applicator conveyor belt type had been made was functioning properly. Modification is done by replacing the joint connection between the applicator and the tractor. The volume of applicator compost is one meter cubic.

1. Introduction

Horticulture is a science that studies about the cultivation of vegetables, fruits, and ornamental plants. Currently, horticulture becomes a commercial farming enterprise [1]. Horticulture subsector is an important component in agricultural development which continues to grow and develop over time [2].

Factors causing low production of horticultural commodities at the farm level include the low quality of seeds, cultivation techniques are still limited, cost production are relatively large, small-scale enterprises and seasonality.

Fertilization is an important to increase production, even until now regarded as a dominant factor in agricultural production. Until the end of the 20th century fertilization is an important factor to increase production because there is no other alternative to replace [3].

Recently, organic fertilizer (compost) has been considered the heart of organic farming systems [4]. The function of compost is as soil conditioner. In terms of increasing the soil bearing capacity, compost is clearly superior and environmentally friendly than synthetic chemical fertilizers because it can increase the content of organic matter in soil. The content of organic matter in the soil has a very important and the amount of organic material that is often used for directly measuring soil fertility index. The compost can provide benefits for the soil and plants, among others: provide nutrients for the plants, improve soil structure, increase the population and activity of soil organisms, improve the ability to water holding and soil aggregates, increases infiltration, preventing erosion, supporting the deployment and penetration of plant roots, and to strengthen plant resistance to pests and diseases.

Organic fertilizer is useful in improving the physical, chemical and biological properties of the soil [5]. Beside that, soils that are given organic matter will be more easily processed [6]. Organic fertilizers can be made from various types of materials, such as crop residues (straw, corncob, and coconut husk), sawdust, animal waste, mushroom media waste, market waste, household and factory, and fertilizer green [7].

Application compost as a fertilizer on agricultural land must be in large amounts (high dose), this is caused by the nutrient content of the compost is lower compared with inorganic fertilizer so that to get an equivalent dose of the nutrient needs of plants. Problems that occur at the time of application of compost is needed manpower and considerable expense, so it needs an appropriate technology like mechanical equipment that is simple and easy to operate.

Applicator compost is one alternative problems solver in the application of compost on the farm. This study was conducted to modify the dry land sugarcane plant compost applicator pulled by four-wheel tractor to be used in the horticultural crops (vegetables) field pulled by hand tractor. Compost applicator is designed to help the process of fertilization use compost on farm. Compost applicator design previously been performed by Iqbal [8] were used for the process of fertilization on sugarcane plantations. The results of the previous research showed the rationing metering device mechanism using a type of conveyor belt has been able to function properly [8]. The spending rate of compost can be adjusted to the height of door openings regulator or the percentage of door openings regulator. The width of the door opening and the forward speed of the applicator can be adjusted to the rate of expenditure or dose of compost desired when the application of compost in the cane fields.

The application compost equipment or applicator compost in the field is a specifically designed to apply organic fertilizer or compost into the soil or in between the plants with specific dose of compost. Compost applicator is a trailer pulled by a tractor. The components of applicator are auger, pit where the discharge of compost, regulator doors, conveyor belt as a metering device, and fertilizers bin as the main framework. Applicator will be operated by hand tractor.

The general objective of this study was to modify the sugarcane applicator compost on dry land so that it can be used on land horticultural (vegetables).

2. Material And Method

2.1. Tools and Materials

The tools used in design process are welding equipment, tools kit, drill, grinder, hacksaw and gauges. In the process of designing applicators compost materials used are steel, iron plate, iron, welding wire, bolts, wheels, rubber conveyor belt, sprocket, auger, iron shaft, chain, wooden planks, and bearing.

2.2. Applicator Design

The expected of compost applicator is it capable putting compost into soil or horticultural gardens. Compost applicator designed using drawbar pull and able to put the compost on the soil surface. The important component of the applicator is compost metering device.

2.3. Design Method

The manufacturing step of applicator is follows the flowchart of the design. Manufacture begins with calculating the dimensions, the design and selection of technical drawings as well as the purchase of raw materials for the applicator.

Framework is the part that made first. In this framework, there is a section that will connect the coupling rod applicator and the tractor. Prototype manufacture followed by made compost bin of from iron (L type) and trapezoidal shaped. The process of grafting is done by using electric welding.

The next step is to assemble a conveyor belt on the shaft of metering device and coat the inner frame wall with an iron plate. Each shaft of metering device mounted use two bearings placed on each end of the shaft.

Open-close mechanism of the door regulator made of iron plate with open-close system manually. The open-close system works vertically located just above the front conveyor belt end (front axle).

Furthermore, the manufacture of auger shaft parts. As the shaft of metering device the auger shaft installed using two bearings placed on each end of the shaft. Then mounted auger transmission system of compost by taking the power source of motion of the wheel axle applicators are connected by pulleys auger of compost. While the rotary power to drive metering device of conveyor belt, taking the source of compost auger pulley attached to a conveyor belt pulleys. This transmission splicing system uses a chain and sprocket.

3. Result And Discuss

3.1. Analysis of Mechanical Design Applicator

Compost applicator design process is requiring several parameters that constitute the basic data for designing the applicator. Equipment list and the field condition where compost applicator to operate is important that the design can be done well. The data is about the physical characteristics of compost, horticultural cultivation techniques (chili), the land condition, pull the power source, and a dose of fertilization.

3.1.1. Physical Properties and Compost Characteristics. The row material compost is from the residue of the sugarcane (filter cake). It is cut by chopper into small pieces with a length 1-5 cm. The Density of compost is 340 kg/m^3.

3.1.2. Land Condition of Horticultural (chili). Horticultural cultivation system especially chilies are used by farmer in several areas in the southern of South Sulawesi province, there are 2 system. The first system is to use a seedbed (rice fields and plains) with a width of 1 meter beds and the second without beds (the highland and tilt). Spacing average is used is 60-70 cm, and is mediated furrow that can serve as irrigation and drainage channels to a depth of 25 cm and a width of 30 cm. The height of bed on land is 20-25 cm. The second system is planting chili on the garden on a hilly terrain (altitude with terracing and use of plastic mulch to prevent weed growth and erosion. This is planting pattern using a distance of 30 cm between plants and 1 meter between beds.

3.1.3. Hopper. It serves as a container or a compost before it is applied in the field. This box is a tub that can be made with basic materials such as stainless steel, iron plates, wood (board) and plastics. It is can prismatic hollow elongated or truncated cone that tapers toward the drain holes. Outlet can be circular or square usually located at the bottom or at the bottom of the tub wall.

3.1.4. The Metering Device Hole. The function of metering device hole is as a regulator of the output dose of compost before it is applied to land. It used in compost applicator design model is a type of conveyor belt assisted by auger. Type of conveyor belt is a unit that serves issued a fertilizer with a relatively high speed. Fertilizer pushed onto the running belt and the dose is set by adjusting the width of the door opening at the top of the running belt. Auger is the type of metering device that is fully taken manure to drain holes between the rotating screw thread [9]. Dose is set by adjusting the forward speed of the tractor which is connected to the shaft of belt conveyor and auger shaft.

3.1.5. The Transmission. It is a component serves to distribute or move a motion from sources such as motors, PTO or drive wheel. Some types of transmission to be used in the design of these applicators include: sprocket and chains, gears and chains.

3.2. Applicator Prototype

Compost applicator consists of several parts of the framework, like hopper, door openings regulator (dose), metering device (conveyor belt), the shaft conveyor belt, transmission systems, auger and outlet holes compost. Applicators compost has a height of 160 cm (from the ground), a length of 300 cm and a width of 136 cm.

This applicator consists of several parts which are detailed as follows:

3.2.1. Frame. Frame is the first part that made in the design of compost applicator prototype. The size of the frame without wheels applicator is 100 cm high, the length is 200 cm and width is 110 cm.

3.2.2. Compost Bin. The volume of compost applicator bin can be calculated using the equation:

$$V_b = 0.5 * (ab + cd) * ae * ds \tag{1}$$

Where : V_b = Applicator bin volum (m^3)
ab=pq = width bottom side of applicator bin (m)
cd=rs = width of the upper side of the applicator bin (m)
ae=bf=pt=qu = high of applicator bin (m)
ds=cr=bq=ap = length of applicator bin (m)

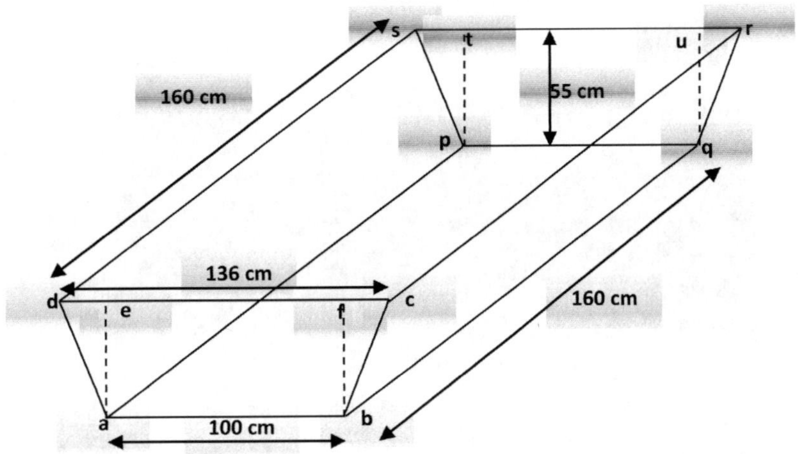

Figure 1. Dimensions of applicator compost bin

The volume applicator based of dimension of the figure above is:
V_b = 0.5 * (136 + 100) cm * 55 cm * 160 cm = 1 038 400 cm^3 = 1 m^3

The compost bin of applicator is trapezium shaped prism and has a volume is 1 m^3, equivalent to 353 kg of compost if the density of compost is 340 kg/m^3 [8]. The function of compost bin is to accommodate the compost ready to be applied to plants. Additionally tub holder compost should be strong for the pivot shaft of metering device because they work in a tub of compost. Because there is a hole of metering device and door openings, it takes place for the belt distributor, auger, and the regulator door of compost. The compost bin is made from steel frame with walls of sheet iron plate.

Doses of compost that will be applied is 10 tons/ha. If the density of compost is 340 kg/m^3, the total volume of compost is 10000 kg divided by 340 kg/m^3 = 29.4 m^3. This is a very large volume and weight (figure 2), so it will affect less well on land as it can cause soil compaction and requires a very large tractor. Therefore, in designing the volume of compost tub, applicators only operate with the compost along the 100 m length of the groove, and then fill up again and so on.

3.2.3. Compost Metering Device. The suitable of metering device of the applicator is conveyor belt type [10]. It could application the compost with big dose. The compost will be distributed by using a conveyor belt leading to the outlet holes, where spending or compost dose adjustment can be set with the door (regulator dose) above the belt. This applicator will use power from the wheel axle applicator that is transmitted through the chain. Distribution of fertilizer can be divided into multiple streams in accordance with the need to use an auger. The materials used for the conveyor belt is a hard rubber sheet is capable of supporting heavy loads such as compost. Auger as a distributor of compost is made from iron.

3.2.4. Regulatory Door Openings. The door is located directly above the end of the conveyor belt, can be increased or decreased. The function of this component is to adjust the amount or volume of compost which will be allocated on crop lands or as a regulator of doses of compost. Door openings regulator may include an iron plate or board that height of the openings can be adjusted as desired dosage of compost.

3.2.5. Transmission system. The power source to applicate the compost is from axle applicator. So that is needed a transmision system can adjust the spin speed required and the direction of rotation. There are several things that must be caution in making transmission systems such as the treatment is simple, inexpensive, easily available and cheap. The transmission system consists of chain, gear, belts and pulleys.

Figure 2. Applicator compost for horticulture

4. Conclusions

The results showed that the prototype of applicator conveyor belt type had been made and modified according the hand tractor specification. It is functioning properly. The shape of applicator compost bin is prism trapezium and the volume is about 1 m^3 or equivalent 352 kg of compost.

References

[1] Zulkarnaen 2009 *Basics of Horticulture* (Jakarta: Bumi Aksara)
[2] Djatnika I, Syah M J A, Widiastoety D, Yufdy M P, Prabawati S, Pratikno S and Luthfiyah O 2015 *Horticulture Innovation People Increase Income Improvement* (Jakarta: Agricultural Research And Development Agency, IAARD Press)
[3] Rosmarkam, Afandie and Yuwono N W 2002 *Soil Fertility Science* (Yogyakarta: Kanisius)
[4] Hoitink H A and Keener H M 1993 *Science and Engineering of Composting: Design, Environmental, Microbiological, and Utilization Aspect* (Wooster: Ohio Agric. Res. Dev. Or., The Ohio State Univ. OH) 728 pp.
[5] Simanungkalit R D M, Suriadikarta D A, Saraswati R, Setyorini D and Hartatik W 2006 *Organic Fertilizer And Biofertilizer* (Jakarta: Indonesian R & D Center for Agricultural Land Resources Research and Development Agency)
[6] Mastur, Mandang T, Haridjaja O and Karama A S 1993 Manipulation of physical properties and mechanization of soils by providing organic materials and soil treatment in relation to energy and instrument feedback *Jurnal Keteknikan Pertanian* **3** (1) 9-20
[7] Yang S S 2001 *Recent advances in composting in Proc. of Issues in the Management of Agricultural Resources* (Taiwan: Food & Fertilizer Technology Center) pp 6-12
[8] Iqbal 2014 The design of the applicator compost for sugarcane crop dry land *Jurnal Keteknikan Pertanian* **28** (1)
[9] Bainer R, Kepner R A and Barger E L 1978 *Principles of Farm Machinery* (New York: John

Wiley and Sons Inc.)
[10] Srivastava A K, Carroll E G, Roger P R and Dennis R B 2006 *Engineering Prinsiple of Agricultural Machine* 2nd ed. ASAE Textbook Number 6 (New York: American Society of Agricultural Engineers)

Soil water retention and plant growth response on the soil affected by continuous organic matter and plastic mulch application

B Rasyid[1,2], M Oda[1] and H Omae[1]

[1]Tropical Agriculture Research Front (TARF) Japan International Research Center for Agricultural Sciences (JIRCAS) 1091-1 Maezato-Kawarabaru Ishigaki, Okinawa, 907-0002, JAPAN
[2]Dept. of Soil Science, Fac. of Agriculture, Hasanuddin University-Indonesia
Jl. Perintis Kemerdekaan, Kampus Unhas Tamalanrea – Makassar, 90245, Indonesia

E-mail: *burrasyid@affrc.go.jp*

Abstract. Soil-water and plant growth interaction is a primary key to develop environmental plant production system. The objective of this research is to evaluate change in soil water retention characteristics and plant response as the effect of continuous organic matter and plastic mulch application. The experiment was conducted in the plastic house field with plot size of 5 m (length) x 1 m (width). The field had treatments of plastic mulch type (mesh and poly) and no mulch, nitrogen (0, 10 and 40 kg N ha^{-1}), and 2 ton ha^{-1} organic matter (incorporated into all plots). Water retention measurement using sand box method for low suction and pressure plate apparatus was applied for high suction. Completely randomized block experimental design and Duncan-MRT were used to analysis the effect of treatment on the parameters. Soil organic carbon and nitrogen increased slightly in both mulch types, but C:N ratio decreased in poly mulch which had the lowest value during two planting season. Various change in soil water retention was shown in different mulch type with mesh mulch had the highest result on lower suction, and control was the lowest water retention on the high suction. Soil water availability was highest in mesh mulch type followed by control and poly mulch type. This study could conclude that continuous incorporation of organic matter and mesh-plastic mulch was useful in achieving environments to improve soil C:N ratio and soil water retention.

1. Introduction

Soil-water and plant growth interaction has become a crucial issue in soil-water and nutrient management to maintain sustainable plant productivity. Nutrient dissolution and movement process from rhizosphere to plant uptake could be well done supported by soil water. Therefore, soil water availability had to be criticized as the limitation factor to optimize crop production. Soil texture and organic matter in soil have been considered the key components that determine soil water retention [1]. Soil organic matter content and composition affect both soil structure and adsorption properties; so, water retention may be affected by changes in soil organic matter [2]. However, the effect of continuous crop residues incorporation is still under discussion, especially if change of soil water retention characteristics were taken into account.

Incorporation of organic matter had been put into practice as soil amendment in protect the soil from solar energy to reduce evaporation and aid in moisture retention to decrease soil water loss [3]. Several studies conclude that soil moisture, rather than temperature, has the most significant influence on soil organic matter [4]. Moreover, soil organic matter had been applied as an important factor to predict soil water retention characteristics [5, 6]. Combining method between organic matter incorporation and mulching is still an interesting challenge to be explored in order to improve of soil water loss prevention and characterized soil water retention. Mulch, itself could be act as barriers to movement of moisture out of the soil [3], ensure maintenance of a lower water content in soils from zone with high rainfall, reduce nutrient leaching, and increased yield [7]. In many cases, type and kind of mulch materials were most consider to be applied in crop production. Filipov [8] explained that the wide variety of plastic mulch has different properties: reflect, absorb or transmit sunlight, determine a specific microclimate around a crop plant and inside of high tunnels or greenhouses, is effective at reducing weed growth (except in the planting holes).

The most popular type of plastic mulch was polyethylene (PE) which has excellent tensile strength as the requirement for mechanical application of the plastic mulch and to resist tearing when exposed to strong win and foot traffic. Beside it, mulch color could be influence on the microclimate. In this study plastic mulch become consideration to be evaluate because there are wide variety of plastics have been used in the field, recently. Specially in Japan agriculture, growing plant in greenhouse or in "high plastic tunnels/house" field with plastic covering and mulching soil surface have been popular and usual practices. Nowadays, two types of black polyethylene *i.e.* high- density and low-density of polyethylene were most used mainly in the plastic house field. Low-density polyethylene type is called "mesh plastic" because this type has micro pore which has capacity to drain and evaporate water. The objective of this study is to evaluate change in soil water retention characteristics and plant growth response as affected by continuous incorporation of organic matter and plastic mulch. Better knowledge and considerable information from this study result could be applied as strategic technology within soil-water and nutrient management.

2. Materials and methods

The experiment was conducted in two plastic houses field with plot size of 5 m (length) x 1 m (width) of Tropical Agriculture Research Front (TARF) – Japan International Research Center for Agricultural Sciences (JIRCAS), Ishigaki, Okinawa Prefecture. Soil type is considered as Ultisols [9]. The experiments was set up using a completely randomized block design with treatment two types of plastic mulch (mesh and poly) and control (no mulch), nitrogen (0, 10, and 40 kg N ha^{-1}). Rose grass and Okra (*Abelmoschus esculentus*) was growth to evaluate plant growth response on incorporated soil by 2 ton ha^{-1} of organic matter. Disturbed soil sample was collected for chemical properties analysis and undisturbed sample using ring cores sample (2.5 cm in height and 50 cm^3 in volume) for water retention measurement.

The hanging water principle is applied on sand box method for low suction of pF 0.5 to 1.5. Measurement of high suction of pF: 2.0, 2.5, 3.0, 3.5, and 4.2 was using pressure plate apparatus. Measurement preparation was done by saturated all samples in the tray with tap water for 48 h. During this process, capillary raised was perform in all samples by adding water gradually into tray until around 5 mm below the core's upper edge to exclude air trapped in soil. Fully saturated sample is then transferred to sand box chamber for low suction and pressure plate chamber for high suction measurement.

The water content at each suction step was calculated when equilibrium was attained at each suction. In sand box chamber, equilibrium was attained when water was stop to drain until the position of height suction reached. The equilibrium in pressure chamber was reached until no water move out from chamber drain pipe. After equilibrium in each suction, soil sample and then weighed as defined as wet mass. Dried mass sample was obtained after measurement of the pF 4.2, the cores were oven dried in temperature of 105 °C for 48 h and then weighed as defined as dried weighed. Water content calculation was done by gravimetric method and the results were converted to volumetric basis, since the amounts of water remaining in the soil at any equilibrium state is a function of the size and volume of water filled pores.

Total soil organic carbon and nitrogen content was analyzed using an Automatic High Sensitive NC Analyzer, Sumigraph NC-22F. Plant response was determined based on dry weight of plant biomass. The mean values of the results were analyzed using SPSS ver. 20. General linear regression further analyzed the amount of variance explained by the dependence of one variable on another where treatments relationship was implied. Significant differences tested with RMS-Duncan post-hoc test.

3. Results and discussion

3.1 Change in Soil Organic Matter, Nitrogen Content and C:N Ratio

Incorporation of organic matter into soil could be contribute to increase soil organic matter and nitrogen. Total soil organic matter, nitrogen content and C:N ratio in each treatment in the soil with continuous organic matter incorporation is shown in table 1. Slightly difference was found among treatment on soil organic matter, nitrogen content and C:N ratio. It showed for nitrogen increased in average only 0.001% and soil organic matter increased in average 0.07%. Nitrogen changed almost same in all treatment except in addition 40 kg N with increasing of 0.004% N, however in case of soil organic matter the highest value of 0.09% C was found in 10 kg N treatment. Change in C and N value had directly implication to C:N ratio. In respect to the effect of nitrogen and mulch on C:N ratio showed decreasing of C:N ratio during two planting season.

Table 1. Average of soil organic matter, nitrogen and C:N ratio on two planting season

Treatment	Soil N		Soil OM		C:N ratio	
	1st (%)	2nd (%)	1st (%)	2nd (%)	1st season	2nd season
Control	0.125 (0.034)	0.126 (0.028)	1.102 (0.20)	1.174 (0.287)	9.486 (0.382)	8.795 (0.588)
Mesh mulch	0.120 (0.027)	0.121 (0.022)	1.060 (0.17)	1.118 (0.214)	9.404 (0.382)	8.786 (0.420)
Poly mulch	0.125 (0.030)	0.126 (0.022)	1.080 (0.17)	1.156 (0.251)	9.315 (0.359)	8.629 (0.280)
0 kgN	0.118 (0.027)	0.119 (0.020)	1.054 (0.15)	1.106 (0.225)	9.458 (0.400)	8.893 (0.476)
10 kgN	0.132 (0.030)	0.130 (0.026)	1.131 (0.22)	1.220 (0.264)	9.279 (0.232)	8.696 (0.214)
40 kgN	0.120 (0.032)	0.124 (0.025)	1.056 (0.15)	1.122 (0.252)	9.468 (0.445)	8.621 (0.549)

(…) standard deviation

Nitrogen treatment in relation to C:N ratio for the first season of 40 kg N was the highest value, but it became the lowest in the second season. For mulch treatment was maintained the lowest value in poly mulch type during both the season. This result informed an important function of poly mulch type combine with high dose nitrogen in mineralization of organic matter when it applied into soil continuously. The organic matter could be higher value in poly mulch type as the maintaining of temperature conditions during all growth season. For mulch treatment was maintained the lowest value in poly mulch type during both the season.

3.2. Evaluation of Plant Response to the Applied Treatment

Biomass of pasture (rose grass) and okra was used to evaluate plant response in treatment of plastic mulch and nitrogen after two cropping seasons. Statistical analysis showed there is no significant effect among all treatments in biomass of rose grass (table 2). The highest result was performed in combination treatment of mesh plastic-mulch type and 40 kg ha^{-1} N with biomass weight of 15.23 kg. However, on single treatment of plastic mulch was shown poly mulch had the highest value of 12.83 kg in total average of biomass weight. In case of nitrogen was found biomass of 12.84 kg as the highest result came from treatment of 10 kg ha^{-1} N. Even biomass yield on this first season was shown no significant among the treatment, but the possibility of nutrient support to the second planting season could be adapted.

Table 2. Biomass of rose grass and okra after two cropping season

Plant	Nitrogen		
	0	10	40
Rose grass	11.00±2.1	12.84±1.6	11.92±2.6
	10.56^{ns}	12.37^{ns}	12.83^{ns}
	Mulch		
	Control	Mesh	Poly
	10.56±2.3	12.50±1.9	13.16±1.6
	11.0^{ns}	11.93^{ns}	12.84^{ns}
Okra	Nitrogen		
	0	10	40
	4.73±2.2	3.75±1.9	4.71±2.1
	3.75^{ns}	4.71^{ns}	4.73^{ns}
	Mulch		
	Control	Mesh	Poly
	2.55±0.5	3.82±0.9	6.81±1.2
	2.55*	3.82*	6.81*

Value reported as mean, standard deviation and subset of Duncan-MR test

With regards to decreasing of C:N ratio, there is possibility additional nutrient from decomposition of organic matter. This expectation can be proven and explained based on the second cropping result. In the second cropping season with Okra as experiment plant was shown significant different in mulch treatment. The result in total average showed increasing of Okra production more than twice with application of poly mulch compare to control and mesh treatment. In the combination treatment, the highest value of 8.27 kg was found by poly mulch without nitrogen (0 kg N) application. In case of fertilizer, there is tendency of nitrogen treatment have no effect to increase biomass yield. General biomass yield evaluation between two house field has found that almost biomass yield in both house field has difference value range 10% - 26.5% (unpublished raw data). Comparing result tendency between single, combination treatment and total average could be generated explanation as the differences of environmental characteristic of two house field. However, increasing of yield is still need to be evaluated regarding on biomass yield using continuous incorporation of organic matter.

3.3. Soil Water Retention in Plastic Mulch and Nitrogen Treatments

Change of soil water retention for type of plastic mulch and nitrogen treatment with increasing of pressure suction is presented in figure 1. Mulch application has clear effect to water retention compare to none mulch or fertilizer treatment in both cropping season. Water retention of mesh mulch (46.8% and 13.3%) and 10 kg N ha^{-1} (43.7% and 11.8%) in each low to high suction had high moisture content compare to other result for single treatment. Figure 2a and b show the different tendency of moisture content result in mulch and nitrogen treatment for the both cropping season. The highest water retention of 32.48 % could be retained by mesh type while none mulch (control) only 26.16 % vol. basis on the suction range of pF 0.5 to 4.2. Nitrogen treatment has no different moisture content in all evaluated suction with average value range of 28.23% to 30.26 % vol. basis. Water retention in combination treatment between mesh mulch and 10 kg N ha^{-1} was shown the optimum moisture content with 47.4% in the low suction and 14.6% in the high suction.

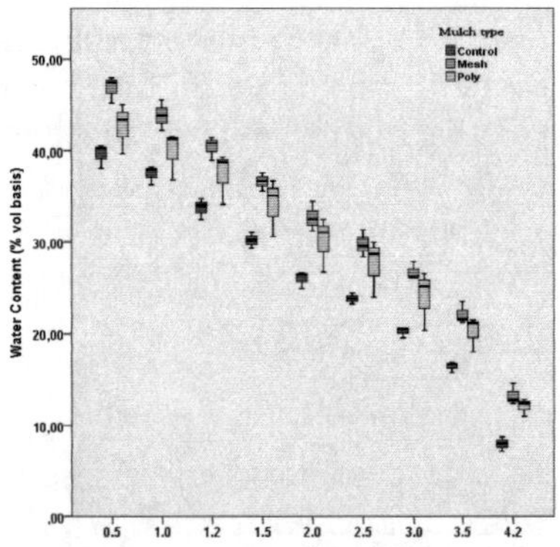

(a) Mulch first cropping

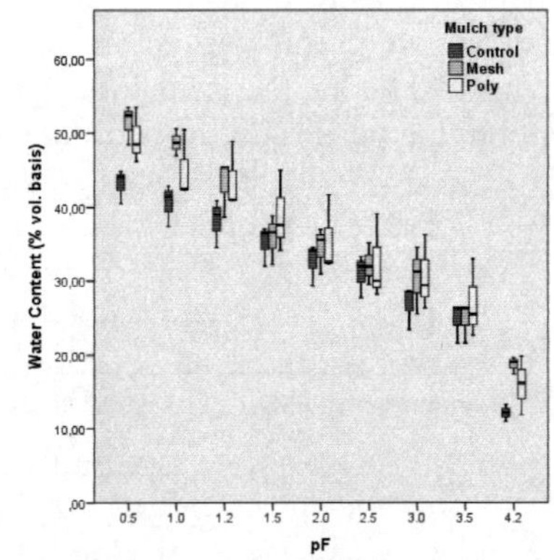

(b) Mulch second cropping

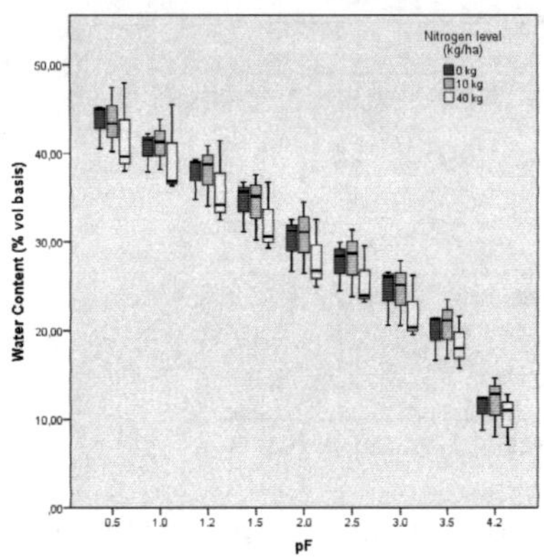

(a) N- first cropping

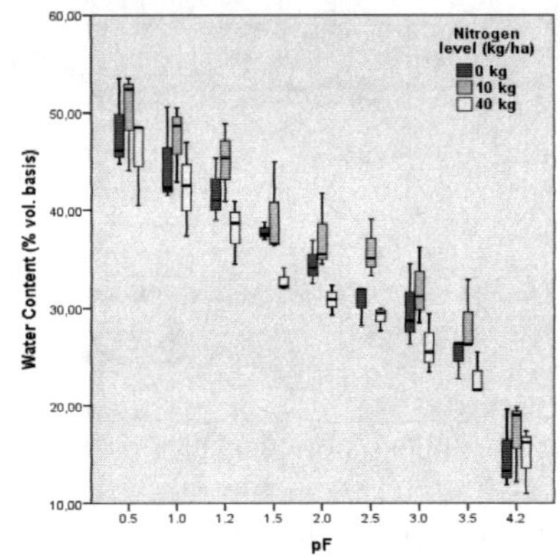

(b) N- second cropping

Figure 1. Soil water retention for mulch and nitrogen with increasing of pressure suction

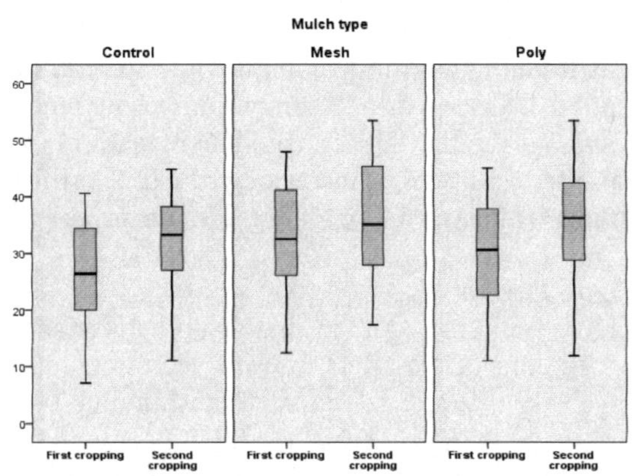

(a)

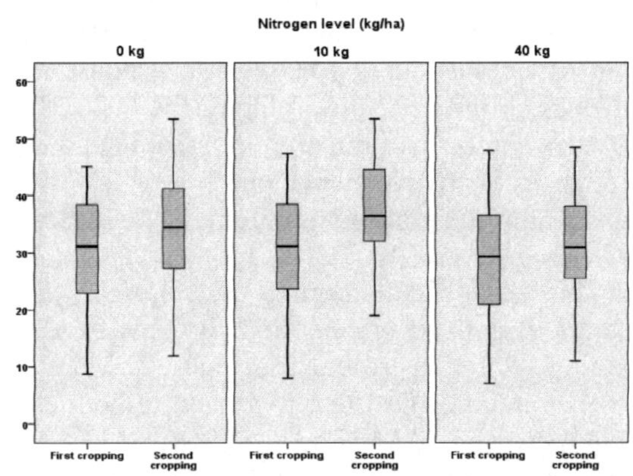

(b)

Figure 2. Water retention after two cropping season in treatment: (a) mulch type, and (b) nitrogen

4. Conclusion

The result of this study was confirmed the function organic matter incorporation into soil continuously in improving soil water retention and plant growth environment. The effect of plastic mulch and nitrogen was not significant in soil organic matter, nitrogen, and C:N ratio. But there was tendency improving of these properties using poly mulch type combine with 40 kg N ha^{-1}. Water retention generally increased after twice addition of organic matter into soil. Moreover, application mesh mulch and 10 kg N ha^{-1} had higher water retention as compared to non-nitrogen, 40 kg N ha^{-1} or non-mulch and poly mulch. Future study are being need to investigate effect of different amount and type of organic matter incorporation to soil bio-physical characteristics.

Acknowledgement
This study was conducted during visiting research fellowship program in Tropical Agriculture Research Front (TARF) – Japan International Research Center for Agricultural Sciences (JIRCAS). We would like to thank for all facilities, support, and cooperation to conduct this experiment and the acceptance for the fellowship program.

References

[1] Bescansa P, Imaz M J, Virto I, Enrique A and Hoog-moed W B 2006 Soil water retention as affected by tillage and residue management in semiarid Spain *Soil Till. Res.* **87** 1 19-27.

[2] Rawls W J, Pachepsky Y A, Ritchie J C, Sobecki T M and Bloodworth H 2003 Effect of soil organic carbon on soil water retention *Geoderma* **116** 1 61–76 Elsevier Science B.V. doi:10.1016/S0016

[3] Jamieson T, Gordon R, Cochrane L, and Patterson G 2011 *Soil Moisture Conservation* (Truro: Nova Scotia Department of Agriculture and Fisheries)

[4] Yang Y H, Fang J Y, Tang Y H, Ji C J, Zheng C Y, He J S and Zhu B 2008 Storage, patterns and controls of soil organic carbon in the Tibetan grasslands *Glob. Change Biol.* **14** 7 1592–99 http://dx.doi.org/10.1111/j.1365-2486.2008.01591.x

[5] Wösten J H M, Pachepsky Y A and Rawls W J 2001 Pedotransfer functions: bridging the gap between available basic soil data and missing soil hydraulic characteristics *J. Hydrol.* **251** (3) 123–50 http://dx.doi.org/10.1016/S0022-1694(01)00464-4

[6] Saxton K E and Rawls W J 2006 Soil water characteristic estimates by texture and organic matter for hydrologic solutions *Soil Sci. Soc. Am. J.* **70** 5 1569–78 http://dx.doi.org/10.2136/

[7] Henry G T and Barbara C S 2000 *Effect of Red Plastic Mulch on Early Tomato Production* (Ames: Department of Horticulture) http://www.public.iastate.edu/~taber/

[8] Filipov F 2013 The effect of the drip irrigation and plastic mulch on the compaction of greenhouses soil *Recent Adv.in Energy, Env. and Geology* ISBN 978-960-474-338-4

[9] Soil Survey Staff 2014 *Keys to Soil Taxonomy, 12th ed.* (Washington: USDA-Natural Resources Conservation Service)

IOP Conf. Series: Earth and Environmental Science **157** (2018) 012009 doi:10.1088/1755-1315/157/1/012009

Cocoa based agroforestry: An economic perspective in resource scarcity conflict era

S Jumiyati[1], M Arsyad[2], Rajindra[1], D A T Pulubuhu[3] and A Hadid[4]

[1] Muhammadiyah University of Palu, Jl. Jabal Nur No.1, Talise, Mantikulore, Kota Palu, 94118, Central Sulawesi, Indonesia.
[2] Department of Agriculture Socio Economics, Faculty of Agriculture, Hasanuddin University, Jl. Perintis Kemerdekaan KM 10, Makassar, 90245, South Sulawesi Indonesia.
[3] Department of Sociology, Faculty of Social and Political Sciences, Hasanuddin University, Jl. Perintis Kemerdekaan KM 10, Makassar, 90245, South Sulawesi Indonesia.
[4] University of Tadulako, Jl. Soekarno Hatta KM.9, Tondo, Mantikulore, Kota Palu, 94148, Central Sulawesi, Indonesia.

E-mail: *srijumiyati@unismuhpalu.ac.id*

Abstract. Agricultural development towards food self-sufficiency based on increasing production alone has caused the occurrence of environmental disasters that are the impact of the exploitation of natural resources resulting in the scarcity of resources. This paper describes the optimization of land area, revenue, cost (production inputs), income and use of production input based on economic and ecological aspects. In order to sustainability farming by integrating environmental and economic consideration can be made through farmers' decision making with the goal of optimizing revenue based on cost optimization through cocoa based agroforestry model in order to cope with a resource conflict resolution.

1. Introduction

Agricultural development was proclaimed by the government no longer focus on the problem of production alone. However, the agenda of self-sufficiency sustainable is the main agenda apart from others such as diversification of food, increase of added value, competitiveness, export and improvement of farmer's prosperity. Program of income increasing and ability of farming and agribusiness management towards Good Agriculture Practices (GAP) and Good Farming Practices (GFP), science and technology application, and concern for conservation of resources both physical and genetic and environment become the main priority agenda. The agricultural and plantation sector not be separated from the utilization of resources has value to meet an increasingly diverse of human needs. Resources commonly used in agriculture are land resources, human resources and capital [1]. These issues should direct to resource conflict resolution based.

The problems of agricultural economic development related to the scarcity of the main resources are decline of quality and quantity of agricultural land resources, the weakness of human resources, that is farmers in adopting technology and innovation affecting the productivity, efficiency, the lack of access to business services, especially that affect farmers in financing their farms business so that the productivity achieved is still below of the potential productivity [2]. The various problems faced to need the development of farming activities with the absorption of low production inputs as a form of sustainable agriculture [3].

The scarcity-resources era sweeping the globe also haunts Indonesia's famous country with abundant resources [4]. Agroforestry as a technique of land management that applying combination of forest trees with agricultural crops is an innovation model of land use that is more efficient in land use and production inputs which aim to optimizing the production and income of per unit area that refers to the principle of sustainable yields and addressed at which multipurpose, optimal and sustainable production under the positive influence of improved edaphic and microclimate conditions with imitating forest condition, and with the technique of management matching with cultural attitude of local communities [5].

Thereby, it is really necessary to conduct a study to know the excellence of agroforestry system application in bridging farmer importance from an economic perspective in resources-scarcity era as a global reality, which is also challenge for sustainable agriculture development. It is hoped that the application of cocoa-based agroforestry can maintain ecosystem sustainability, reduce the risk of harvest failure and price fluctuation caused by diversity commodities. On the other hand, high productivity can be maintained [6].

2. Method

The research was conducted in Balinggi Sub-district, Parigi Moutong District, Central Sulawesi Province, Indonesia. Sampling was done purposively by taking sample of 16 farmers who have a land area of 7 ha then simulated with the aim of optimizing cocoa-based agroforestry. The optimization of cocoa-based agroforestry farming in Parigi Moutong Regency was conducted quantitatively with the formulation of *Linear Programming* model formulated to as follows:
Maximize objective function (Revenue) Max:

$$Z_{Max} = \sum_{oe=1}^{n} C_o X_e \quad (1)$$

X_e = Decision Variable (optimum revenue) value of cocoa-based agroforestry production.
C_o = Parameter of objective function (cocoa based agroforestry farming income).
with constraint function

$$\sum_{j=1}^{bi} a_o X_e = b_o \quad (2)$$

$a_o X_e$ = Parameter of constraint function o for decision variable e
b_i = constraint capacity o [7].

3. Results and Discussion

3.1. Optimization of Land Area

Indonesia has a land potential of 186.20 million ha which has been largely utilized for various purposes in both agricultural and non-agricultural sectors [8]. Parigi Moutong Regency in the last 6 (six) years has increased the area of 208.68 ha of seasonal crops, 49.43 ha of plantation and 6.27 ha of fish pond. Based on the Spatial and Regional Planning (RTRW), the plantation area is less than 1,766.01 ha from the expected target [9]. The increase of plantation area for farmers to adopt profitably and sustainably is by agroforestry. The recommended optimization of cocoa-based agroforestry farming area with the objective of maximizing the revenue on 7 ha of land is by applying a monoculture system of *nyatoh* (*Palaquium* spp) 2 ha, *palapi* 1 ha, cocoa 1 ha, agroforestry system of cocoa with *nyatoh* 2 ha and agroforestry cocoa with *palapi* 1 ha (figure 1).

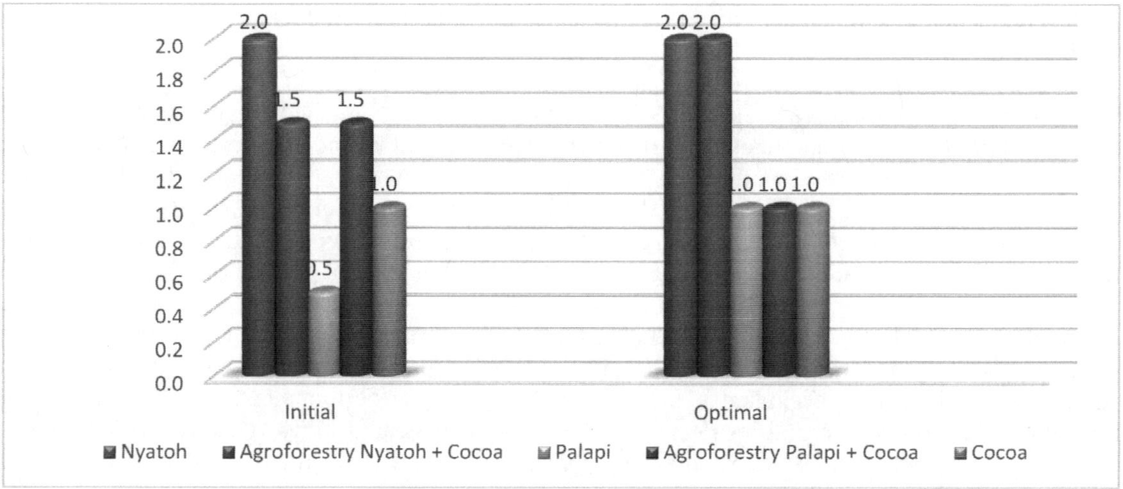

Figure 1. Optimizing of land area

3.2. *Optimization of Revenue*

Optimization of farm revenue on 7 ha of land can be achieved by increasing the revenue of agroforestry cocoa with *nyatoh* farming. This is because agroforestry cocoa with *nyatoh* farming can provide additional revenue because the production comes from two types of plants also reduces production costs. It can efficient allocation of production input. Thus, the optimum recommendation for farming revenue is to optimize the revenue from agroforestry cocoa with *nyatoh* farming with the increase of revenue from IDR 490.4 million to IDR 653.8 million with an increase rate of IDR 163.5 million (25%) (figure 2).

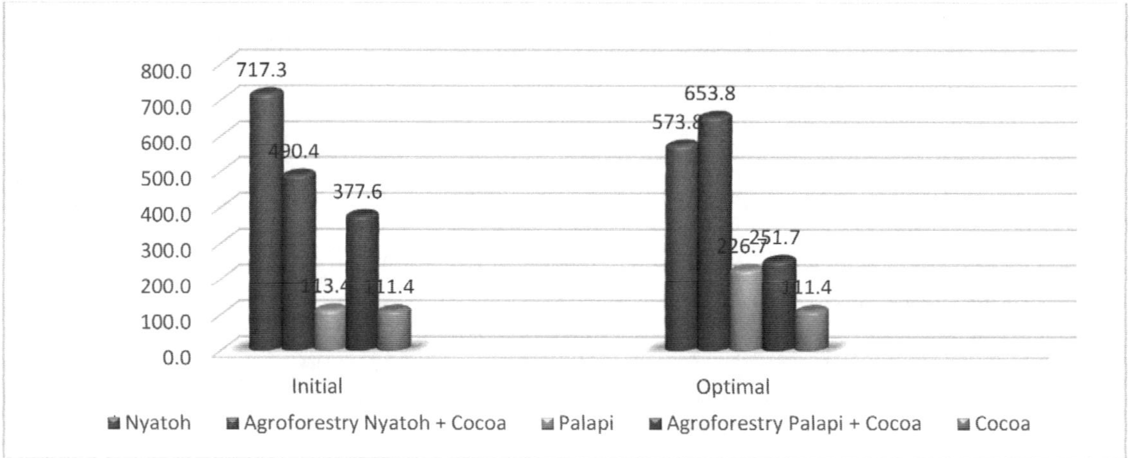

Figure 2. Optimization of revenue

3.3. *Optimization of Cost*

Costs required are the production cost in annual crops and annual crop maintenance. Various agroforestry farming costs with different cropping patterns depend on the type of cultivated plant [10]. The cost optimization recommendation is to increase the production cost at the agroforestry of cocoa with *nyatoh* farming with the initial cost of IDR 137.5 million to IDR 183.4 million up by IDR 163.4 million (25%) (figure 3).

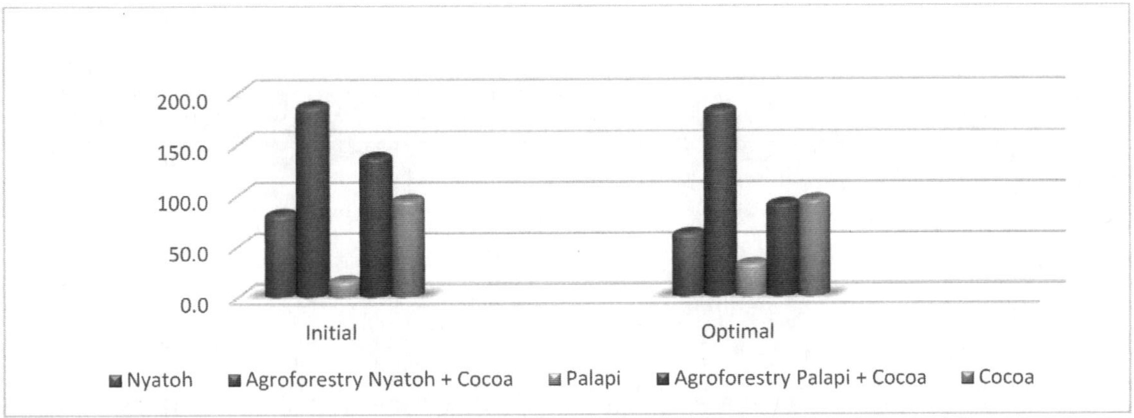

Figure 3. Optimization of cost

3.4. Optimization of Income

Economic factor is the priority of farmers in the selection of crops in cultivating of agroforestry land. For farmers, economic factors directly affect farmers' incomes then ecological factors become a priority after economic factors [11]. Total farm income increased from the initial income of IDR 1.341.4 million to IDR 1.348.9 million with an increase of IDR 7,500 million. The contribution of agroforestry income was IDR 630.4 million (46.7%) of the total optimum revenue. This shows that agroforestry farming has a prospect for increasing of farmer's income compared to monoculture farming (figure 4).

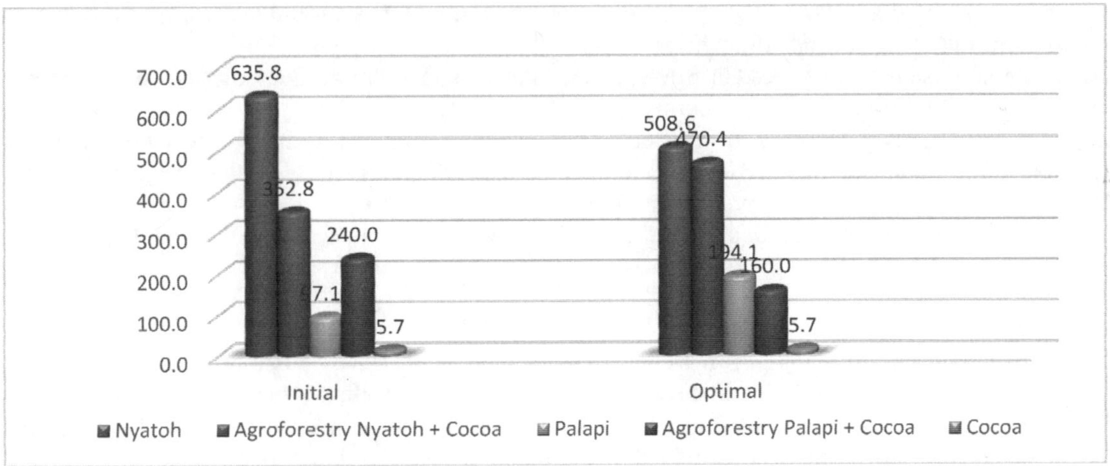

Figure 4. Optimization of Income

3.5. Residual Value of Production Input and Income Increase at Optimization

At the time of farming optimization, the input production used is all used up. This shows that the application of agroforestry techniques by farmers leads to efficiency and effectiveness of input production use. Thus. It is not necessary to add input because it will not cause an increase in income but result in residual value which can increase production cost. On the other hand, if the farmer adds 1 unit of tree seedlings there will be an increase of IDR 52,000 (table 1).

Table 1. Residual value of production input and increase revenue at optimization of cocoa based agroforestry

No	Production inputs	Residual value	Increase revenue (IDR)
1	Land area	0	0
2	Seed of tree	0	52,000
3	Seed of cocoa	0	0
4	Hole of planting	0	0
5	Planting	0	0
6	Manure	0	0
7	Insecticide	0	0
8	Maintenance 3x a year	0	0
9	NPK	0	0
10	Urea	0	0
11	KCL	0	0
12	TSP	0	0
13	Pruning	0	0
14	Harvesting of cacao	0	0
15	Harvesting og cocoa	0	0

4. Conclusion and Suggestion

It can be concluded that (1) optimization of optimum land use is by applying monoculture of cocoa 1 ha, *nyatoh* 2 ha , *palapi* 1 ha, agroforestry of cocoa with *nyatoh* 2 ha and cocoa with *palapi* 1 ha; (2) The optimum revenue, cost and income came from the agroforestry of cocoa with *nyatoh* farming with the increase of revenue (25.0%), cost (25.0%) and income (46.7%); (3) The application of cropping pattern of cocoa-based agroforestry in the management of farming can minimize the use of production inputs and optimize income and competitiveness.

Some possible suggestions are: (1) Site-specific land management as one solution to handle gap between government policy and community based land model with local wisdom value owned: (2) Changes in land use patterns as a form of economic adaptation must also be in line with ecological adaptation in the framework of efficiency, optimization and competitiveness so that sustainability can be guaranteed; (3) The need for government appreciation that not only supports increased production but also creates market opportunities for farmers who apply environmentally friendly and sustainable cultivation techniques.

Acknowledgments

The authors would like to thanks Prof. M. Abu Bakar Lahjie, Prof. Made Antara and Prof. Muchlis Rahmat for their valuable discussion in developing idea and earlier drafts of this paper. Special thank is going to the farmers who participated in interview process. Further thanks to Sofya A. Rasyid, S.P., M.P. for her helps a lot in the process of collecting data and Dr. Hendra Pribadi, S.P., M.P who has assisted in the processing of data and various parties that can't be mentioned one by one.

References

[1] Mahendra F 2009 *Sistem Agroforestri dan Aplikasinya* (Yogyakarta : Graha Ilmu)
[2] Usman A 2010 *Agroforestri, Solusi Sosial dan Ekonomi Pengelolaan Sumberdaya Hutan* , (Bandung: Alfabeta)
[3] Simon H 1994 *Social Forestry and Sustainable Forest Management (Proc. of the Seminar on the Development of Social Forest Management)* Yogyakarta August 29-2 September, 1994
[4] Lahjie A M 2004 *Teknik Agroforestri* (Samarinda: Fakultas Kehutanan Universitas Mulawarman)
[5] Debertin D L 1986 *Agricultura Production Economics* (New York: MacMilan Publishing Company)

[6] Guntara 2013 *Agroforestri sebagai Alternatif Pemanfaatan Lahan Bawah Tegakan untuk Peningkatan Pendapatan Petani di Kabupaten Lumajang*
[7] Budiasa I W 2011 *Materi Usulan Penelitian Strategis Nasional. "Pengembangan dan Optimasi Teknologi Produksi Pangan Sistem Usahatani Terintegrasi (SIMANTRI) untuk Pertanian Berkelanjutan: Pendekatan Linear Programming* (Denpasar: UNUD)
[8] Mulyani A, Ritung S, Las I 2011 Potensi dan ketersediaan sumber daya lahan untuk mendukung ketahanan pangan *J. Litbang Pertanian* **30** (2) 73-80
[9] Andresi B 2014 Pemetaan perubahan penggunaan lahan Kecamatan Sausu Kabupaten Parigi Moutong Tahun 2007 dan 2013 *E-Journal Geo Tadulako UNTAD* 1-17
[10] Yuwariah Y 2015 *Potensi Agroforestri untuk Meningkatkan Pendapatan, Kemandirian Bangsa dan Perbaikan Lingkungan* Seminar Agroforestry, Fakultas Pertanian, Universitas Padjajaran, Bandung
[11] Nicholson W 1991 *Micro Economics Theory: Basic Principle and Extension* 4th Ed. (Hindsdale, Illinois-USA: The Dryden Press)

IOP Conf. Series: Earth and Environmental Science **157** (2018) 012010 doi:10.1088/1755-1315/157/1/012010

Evapotranspiration and water balance in a hot pepper (*Capsicum frutescens* L.) field during a dry season in the tropics

S Laban[1], H Oue[2] and D A Rampisela[1]

[1]Departement of Soil Science, Faculty of Agriculture, Hasanuddin University, Jl. Perintis Kemerdekaan KM 10, Makassar, 90245, Indonesia.
[2]Graduate School of Agriculture Sciences, Ehime University, 3-5-7 Tarumi, Matsuyama, Ehime, 790-8566, Japan.

E-mail: *ika-l@unhas.ac.id*

Abstract. Evapotranspiration and water balance in a hot pepper (*Capsicum frutescens* L.) field during the 2nd dry season were analyzed in this study. Actual evapotranspiration (ET) was estimated by Bowen Ratio Energy Budget (BREB) method, potential evaporation (EP) was calculated by Penman method, and irrigation volume of water was measured manually. Meteorological instruments were installed in the experimental field during hot pepper cultivation. Leaf area index increased during the growing stages where the highest LAI of 1.65 in the generative stage. The daily average of ET was 1.94 and EP was 6.71 mm resulting in low Kc. The Kc values were significantly different between stage to stage under T-test analysis (α = 0.05). Moreover, Kc in every stage could be related to soil water content (SWC) in logarithmic function. Totally, ET during hot pepper cultivation was 179.19 mm, while rainfall was 180.0 mm and irrigation water was 27.42 mm. However, there was a water shortages during vegetative and generative stages. This study suggested that consumptive water of hot pepper was complimented by soil and groundwater under the condition of water shortages in the vegetative and generative stages during the 2nd dry season.

1. Introduction

Chili is one of the strategic commodities in Indonesia besides rice, corn, soybean, beef, and sugar cane. There are two types of chili commonly cultivated in Indonesia, namely chili pepper (*Capsicum annum)* and hot pepper (*Capsicum frutescens* L.). Hot pepper is the most important low elevation vegetable in terms of production area and value but has the major production constraints such as seed material quality, high production costs, fluctuating market prices, and farmers' lack of knowledge [1].

Nationally, the production of hot pepper in 2015 was 869,954 tons produced from various provinces in Indonesia. However, the hot pepper productivity has decreased in some province in 2016, such as 10.89% in West Java; 6.94% in Aceh; 10.36% in North Sumatera; 4.77% in South Sulawesi; 23.38% in Central Sulawesi; and 12.42% in Lampung [2]. The decreasing of chili production may due to the limiation in amount of water supply [3] where traditionally, farmers in Indonesia cultivated hot pepper during the dry season as complementally crop to rice especially in the 2nd dry season on September to the middle of December. In this season, rainfall was limmited and surface irrigation water was stopped.

IOP Conf. Series: Earth and Environmental Science **157** (2018) 012010 doi:10.1088/1755-1315/157/1/012010

Irrigation should be available [4] to overcome the decreasing of production during dry season caused by drought conditions. Therefore, information about the amount of hot pepper water consume is required. Understanding the consumptive water use and water balance is critical to evaluate water stress. Therefore, the aims of this study were to estimate the water consume of hot pepper and to analyze the water balance on hot pepper cultivation. The output of this study is necessary due to the planning of irrigation, improving irrigation practices, and assisting in irrigation scheduling [5, 6, 7].

2. Method

2.1. Study site

This study was conducted in the experimental field, which is administratively located in Tanabangka village, Bajeng Barat district, Gowa Regency, South Sulawesi, Indonesia. There are two seasons in this area; one is the rainy season from December to April, and the other is the dry season from May to November. The total annual rainfall varies from 2,467 mm to 3,294 mm [8]. The study field was under Renggang Water User Association (P3A Renggang) coordination. The total agriculture area of Renggang was 74.40 ha. Land use during the 2nd dry season 2013 were used for 38.8 ha (51.54%) of green beans, 4.91 ha (6.60%) of rice, 0.96 ha (1.29%) of maize, 3.55 ha (4.77%) of vegetable, 0.20 ha (0.27%) of soy bean, 0.11 ha (0.15%) of hot pepper, and 26.39 ha (35.47%) fallow. Hot pepper (*Capsicum frutescens* L.) var. Baskhara F1 was cultivated in the field during the 2nd dry season from September 11th to December 11th, 2013. The crops was transplanted in the field after 37 days seedling and the growing stage was divided in three; vegetative stage (September 11 – October 2), generative stage (October 3 to November 22), and harvesting stage (November 23 – December 11). The area of hot pepper field was 40.8 m x 27.37 m. Each plot of 20 m x 0.7 m was separated by 0.40 m space and the planting density was 50 cm x 50 cm. The study field was surrounding by green beans field, fallow, and corn.

2.2. Data

Meteorological instruments were installed in the field on September 7 before hot pepper transplanted. Rainfall data were collected by automatic rain gauge (ECRN-100, Decagon). The station for the microclimate observation consisted of CNR-4 (Kipp and Zonen, Netherland) for net radiation at 2.0 m height; psychrometers HMP-45A (Vaisala Inc. Helsinki, Finland) for air temperature and relative humidity observation at 0.5 m and 2.0 m for both, respectively; anemometers 014A (Met One, USA) for wind speed observation; heat plate PHF-03 (PREDE) for soil heat observation; and thermocouple for soil temperature observation at soil surface (at 0 cm). All of the instruments were connected to the CR23X logger (Campbell). Soil moisture at 5 cm soil depth was observed by 5TE (Decagon) sensor. Automatic rain gauge and 5TE sensors were connected to the EM5b logger. All of the microclimate data were recorded every 10 minutes and were averaged in every one hour. The plant height, number of leaf and leaf area index (LAI) of hot pepper were measured weekly. There was no surface irrigation water in the 2nd dry season. Therefore, the irrigated water was pumped up from a well. The farmer pumped ground water, collected it in a big tank and irrigated to every hot pepper plant manually like a manual drip irrigation method. In this study, we measured manually the volume of pumping water by volumetric method in each event of irrigation and daily groundwater level was measured by scale [9].

2.3. Data analysis

2.3.1. Evapotranspiration. The actual evapotranspiration as the consumptive water use was estimated by using Bowen Ratio Energy Budget method [10, 11, 12]:

$$ET = [(Rn - G)/(1 + Bo)]/(2499 - 2512\ T)$$

where ET is actual evapotranspiration (mm/d), Rn is net radiation (W m^{-2}), G is soil heat flux (W m^{-2}), Bo is bowen ratio, T is air temperature (°C). Bowen ratio was calculated by equation:

$$Bo = \gamma\ (\Delta T/\Delta e)$$

where γ is pscychrometer constant, ΔT (°C) and Δe (kPa) are differences in air temperature and vapor

pressure between two height above the crop canopy.
EP was estimated by Penman method [13].

2.3.2. Crop coefficient (Kc). Crop coefficient (Kc) was calculated as the ratio of actual evapotranspiration to potential evaporation (EP). The differences of Kc in every stage of hot pepper growth was analysed by parametric statistical analysis. On the other hand, the relationship of soil water content (SWC) and Kc was analysed by logarithmic function.

2.3.3. Water balance. The water balance was analysed by using the simple equation:

$$R + I = ET + P + \Delta S$$

where R is rainfall, I is irrigation practiced, ET is actual evapotranspiration, P is percolation, and ΔS is changes of store water within the soil layer.

3. Result and Discussion

3.1. Microclimate Condition

The event of irrigation, daily rainfall and the daytime of microclimate conditions in the hot pepper field are shown in figure 1. During the 2nd dry season, irrigation event was done eight times with the total of irrigation was 27.42 mm while the total rainfall was 180.0 mm.

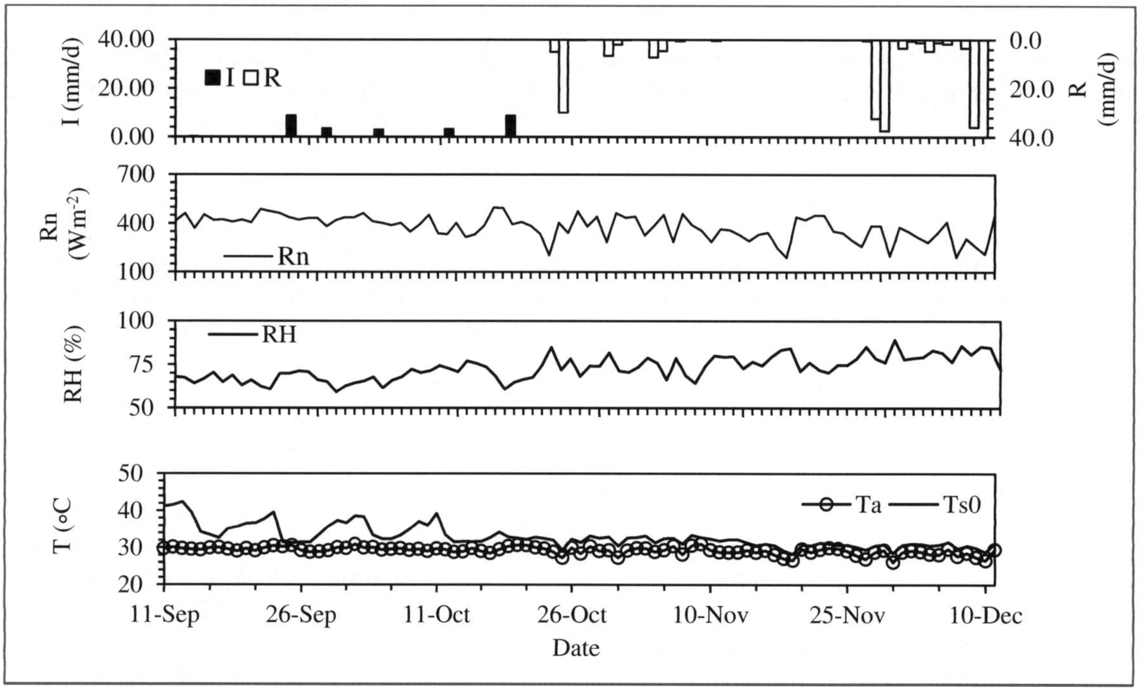

Figure 1. Variation of volume of irrigation (I), daily rainfall (R), and day time average of microclimate condition (net radiation; Rn, relative humidity; RH, air temperature; Ta, and, soil temperature at 0 cm depth; Ts0) in hot pepper field, 2013.

The daytime average of net radiation was 379.21 Wm^{-2} and the average relative humidity was 72.49%. Relative humidity increased from 65.79% (vegetative stage) to 72.23% (harvesting stage). Soil temperature at 0 cm depth was higher than air temperature during the hot pepper cultivation.

3.2. Groundwater Level

Groundwater level fluctuation during the dry season are shown in figure 2. The groundwater level was 69.0 cm from the soil surface in July and gradually increased until it reached to the peak; 170.0 cm from the surface on October 22.

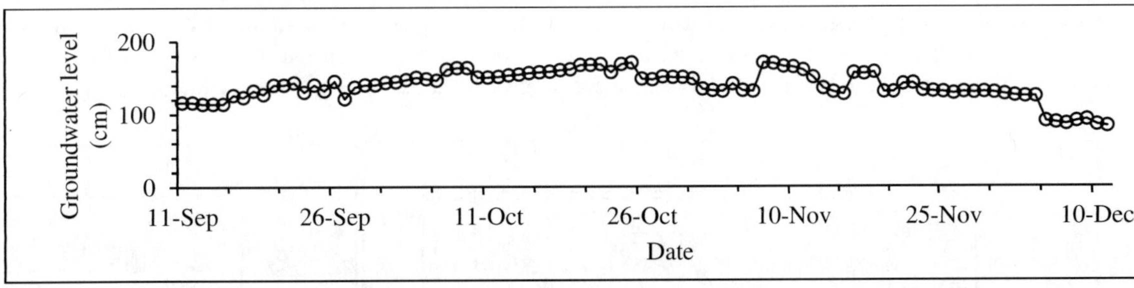

Figure 2. Groundwater level fluctuation during the dry season, 2013.

3.3. Plant Height, Number of Leaf and Leaf Area Index (LAI)

The variation of plant height, number of leaf and LAI during the generative stage in the experimental field are shown in figure 3. These values increased due to the plant growth. The highest plant height was 44.9 cm, the number of leaf was 135.5 and LAI was 1.65.

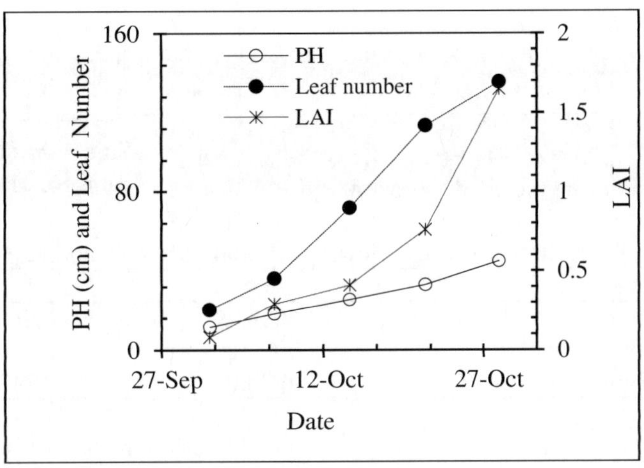

Figure 3. Variation of plant height (PH), number of leaf, and LAI of hot pepper during the generative stage.

3.4. Crop Coefficient (Kc)

The daily consumptive water use and crop coefficient (Kc) of hot pepper are shown in figure 4. Based on estimation, ET was lower than EP where the daily average of ET and EP were 1.94 mm and 6.71 mm, respectively. The result of ET in this study was simmilar to the study that conducted by Qiu [14]. We assumed the lower ET caused by the low plant density and resulting in the low Kc [15]. The crop coefficient of hot pepper increased from the vegetative stage to the harvesting stage where the average of Kc was 0.25 and 0.33, respectively. Statistically, Kc in each stage was significantly different between stage to stage as shown in figure 5. This study also founded Kc could be related to soil water content at 5cm depth (SWC5) by a logarithmic function where the R^2 for vegetative, generative, and harvesting stages were 0.86, 0.81, and 0.85 respectively (figure 4). This relationship on every stage of hot pepper growth was expressed by Kc = a *ln*(SWC5) + b as shown in table 1.

3.5. Water Balance of Hot Pepper Field

Water balance in the hot pepper cultivation in the 2nd dry season is shown in table 2. Totally, the amount of water supply by rainfall and irrigation was higher than evapotranspiration. However, the water supply during vegetative and generative stages was less than evapotranspiration. This water balance suggest that the consumptive water of hot pepper could be supplemented by soil and ground water, which accounted for 70.78% in vegetative stage and 28.02% in generative stage.

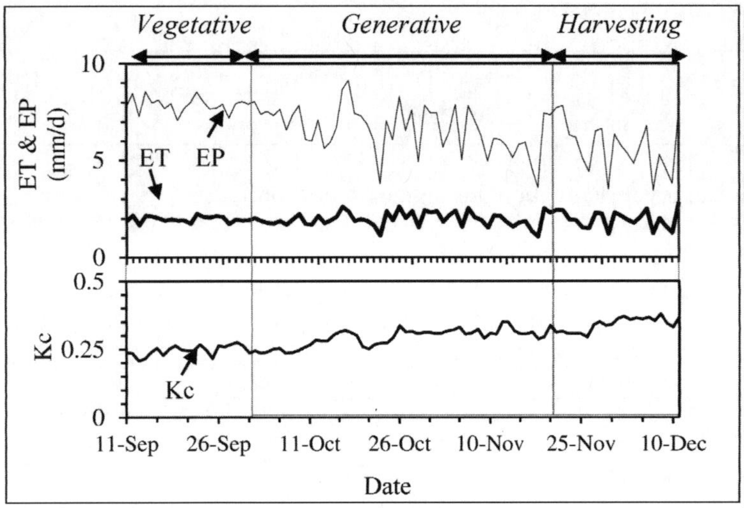

Figure 4. Daily variations of ET, EP and Kc of hot pepper crops during the 2nd dry season.

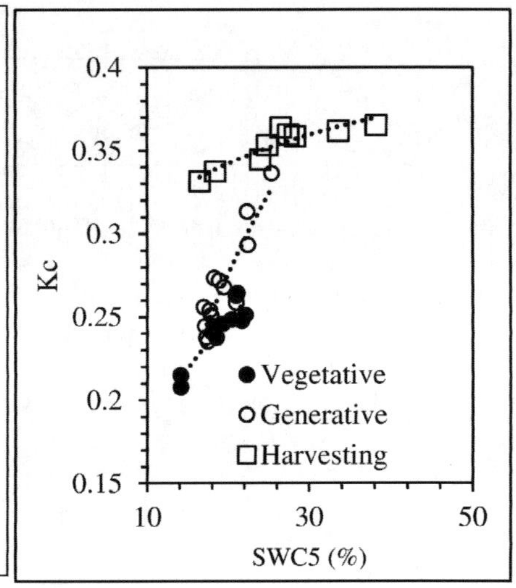

Figure 5. The relationship of SWC5 (%) and Kc of Hot Pepper in every stages.

Table 1. Parameters of logarithmic function for the relation of SWC5 and Kc in the hot pepper field.

Crop stages	a	b
Vegetative	0.1038	- 0.0626
Generative	0.216	- 0.3714
Harvesting	0.0429	0.2137

Table 2. Water balance in the hot pepper field during the 2nd dry season, 2013.

Plant stages	Days (Date, 2013)	Water supply (mm)		Evapotranspiration		P+ΔS
		Rainfall	Irrigation	(mm)	(mm/d)	(mm)
Vegetative	22 (11Sept-2 Oct)	0.00	12.39	42.41	1.93	-30.02
Generative	51 (3 Oct - 22 Nov)	57.20	15.03	100.35	1.97	-28.12
Harvesting	19 (23 Nov - 11 Dec)	122.80	0.00	36.43	1.92	86.37
Total	92 (11 Sept - 11 Dec)	180.00	27.42	179.19		28.23

4. Conclusion

Evapotranspiration and water balance in the hot pepper field were analyze in this study. Daily evapotranspiration was highest in generative stage and was lower than potensial evaporation throughout the growing season, which resulted to the low crop coefficient (Kc). This study also showed that Kc was different between the growing stage to stage, changing from 0.25 to 0.33. We also found that the Kc in each growing stage could be individually related to soil water content (SWC) with a logarithmic function. Water balance analysis showed that the hot pepper could grow even under the water shortage during vegetative and generative stages. It was suggested that the consumptive water of hot pepper could be supplemented by soil and ground water, which accounted for 70.78% in vegetative stage and 28.02% in generative stage.

Acknowledgment
This study was supported partly by the project "Designing Local Frameworks for Integrated Water Resources Management" under "Water Management Project". The authors acknowledge the Research Institute for Humidity and Nature (RIHN), Kyoto, Japan. The authors also would like to thank NGO PELANGI's staff for their assistance in the field experimental data collection.

References

[1] Vos J G M and Duriat A S 1995 Hot pepper (Capsicum spp.) production on Java, Indonesia: toward integrated crop management *Crop Prot.* **14** 205–13
[2] Central Bureau Statistics of Indonesia 2014 *Indonesian Statistic 2016* 113–33
[3] Darko R O 2013 Water requirement, deficit irrigation and crop coefficient of hot pepper (Capsicum frutescens var legon 8) using irrigation interval of two (2) days ARPN *J. Agric. Biol. Sci.* **8** 139–46
[4] Durigon A and de Jong van Lier Q 2013 Canopy temperature versus soil water pressure head for the prediction of crop water stress *Agric. Water Manag.* **127** 1–6
[5] Erie, L.J., O.F. French, D.A. Bucks and K H 1981 *Consumptive use of water by major crops in the South western United States*. (United States Departement of Agriculture, Conservation Research Report No. 29, 42p.)
[6] Gercek S, Comlekcioglu N and Dikilitas M Effectiveness of water on yield and water use efficiency on hot pepper (Capsicum annuum L .) Scientia Horticulturae **120** 325–9
[7] Lozano D, Ruiz N and Gavilán P 2016 Consumptive water use and irrigation performance of strawberries *Agric. Water Manag.* **169** 44–51
[8] BPS S of G R 2016 *Gowa Regency in Figures 2016* (BPS-Statistics of Gowa Regency)
[9] Laban S, Oue H and Rampisela A 2015 Irrigation practice and its effects on water storage and groundwater fluctuation in the first dry season in the rice cultivation region, South Sulawesi, Indonesia *Procedia Environ. Sci.* **28** 271–9
[10] Esmaiel Malek 1993 Comparison of the Bowen Ratio Energy Balance and the water balance methods for the measurement of evapotranspiration *J. Hydrol.* **146** 209–20
[11] Abtew W and Assefa M 2013 *Evaporation and Evapotranspiration: Measurements and Estimations* (London, New York: Springer)
[12] Peacock C E and Hess T M 2004 Estimating evapotranspiration from a reed bed using the Bowen ratio energy balance method *Hydrol. Process.* **18** 247–60
[13] Muniandy J M, Yusop Z and Askari M 2016 Evaluation of reference evapotranspiration models and determination of crop coefficient for *Momordica charantia* and *Capsicum annuum Agric. Water Manag.* **169** 77–89
[14] Qiu R, Kang S, Li F, Du T, Tong L, Wang F, Chen R, Liu J and Li S 2011 Energy partitioning and evapotranspiration of hot pepper grown in greenhouse with furrow and drip irrigation methods *Sci. Hortic.* (Amsterdam). **129** 790–7
[15] Alemayehu Y A, Steyn J M, Annandale J G, Alemayehu Y A, Steyn J M and Annandale J G 2009 FAO-type crop factor determination for irrigation scheduling of hot pepper (*Capsicum annuum* L.) South African *J. Plant Soil* **26** 186–94

IOP Conf. Series: Earth and Environmental Science **157** (2018) 012011 doi:10.1088/1755-1315/157/1/012011

Tolerance limits of Indonesian rice varieties to drought and salinity in germination phase using PEG and NaCl as selection agents

M Farid and I Ridwan

Department of Agronomy, Faculty of Agriculture, Hasanuddin University, JalanPerintisKemerdekaan KM 10, Makassar 90245, South Sulawesi, Indonesia.

E-mail*: farid_deni@yahoo.co.id*

Abstract. This research aims to study tolerance limit of various Indonesian rice varieties to drought and salinity in germination phase using PEG and NaCl as selection agents and to obtain a resistance benchmarks of the rice varieties on drought and or salinity. Two sets of experiments were conducted, both set using Split Plot design, to analyse the plant tolerance to drought and salinity stress, respectively. Four levels of salinity i.e. Control, EC 4, 6 and 8 mmhos/cm, respectively were used in the first experiment and four levels of drought condition in the second experiment i.e. control, -0.33, -0.67 and -1.0 MPa, respectively. 25 varieties used in both experiments were germinated using wet papers previously applied with PEG and NaCl. High genetic diversity was shown by the rice varieties with the highest heritability index (H>0.7) on the PEG and NaCl with concentration of -0.67 MPa and 6 mmhos/cm, respectively. The tolerance limit of the rice varieties to salinity was on the NaCl concentration of 6 $g.L^{-1}$, while to drought was on the PEG concentration of 100 $g.L^{-1}$ (-0.67 MPa). Parameters that can be used as benchmarks for tolerance to drought and salinity in the germination phase were vigor index, radicle and plumula length.

1. Introduction

Rice is a strategic commodity that plays an important role in the economy and national food security so that it becomes the main base in the revitalization of agricultural development in the future. Efforts in developing the agricultural sector are always faced with a dynamic domestic and international environment. As a country located in the tropics, Indonesian agriculture has a comparative advantage. Such comparative advantage is the basis for the building of national and international competitive advantage.

Indonesian Agricultural Research Agency has been working to produce innovations to increase rice production. Through the implementation of intensive research and innovation of Integrated Crop Management and Resources (ICMR), adoption of suitable technology for local conditions that can improve production, yield quality, and preserve the environment can be conducted by farmers to increase farmers' income. However, the approach taken is still partial for each region [1].

Based on data from Indonesian Agricultural Research and Development Agency [2] the national rice productivity for the last 5 years only ranges from 4.98 to 5.34 t ha^{-1}, while the yield potential of the released varieties ranges from 8 to12 t ha^{-1} (table 1). This shows that the yield potential of the developed varieties was far from the expected target. Thus, the use of locally specific varieties is necessary to be implemented, especially on marginal land with appropriate package of cultivation technology.

Table 1. Land area, production and productivity of Indonesian national rice in 2011-2015.

Indicator	Unit	Year 2011	2012	2013	2014	2015
Harvest Area	Ha	13,203,643	13,445,524	13,835,252	13,793,640	14,117,000
Production	T	65,756,904	69,056,126	71,279,709	70,846,465	75,398,000
Productivity	t ha^{-1}	4.98	5.14	5.15	5.14	5.34

In order to meet Indonesian needs for food, rice cultivation can not only be planted on fertile land and available water, but also needs to be taken further in marginal lands. To increase productivity, utilization of marginal lands that suffer from drought stress, salinity, acidity, and its combination is necessary to be considered. Data of 2012, show that Indonesia has dry land of about 148 million ha (78%) and wetland area of 40.20 million ha (22%) of 188.20 million ha of total land area [3]. Swamp land in Indonesia reaches 33,393,570 million ha which spread in some areas of Sumatra, Kalimantan, Sulawesi, and Papua. In Indonesia, dry land is quite large with estimated about 60.7 million ha or 88.6% of the land area, while the rice field area is only 7.8 million ha or 11.4%. However, the expansion of cultivation in the area of new openings often faces ecological limiting factors such as drought and salinity or both.

Drought and salinity stress reduce rice yield. Reduction of the rice yield caused by drought can be ranged from 20 to 100% in rainfed areas while yield reduction from salinity stress can range from 20 to 50% [4]. Salinity interferes with rice growth and development [5] include in the earlier stages such as germination phase [6]. However, the response of plant to salinity were varied between varieties and level of salinity [7].

2. Methodology

The research was carried out in the Laboratory of Agronomy Department, Faculty of Agriculture Hasanuddin University. Two sets of trial were set both in a Split Plot design with main plot for each trial was salinity and drought levels, respectively and rice varieties as sub plot. Drought conditions were established using PEG 5, 10 and 15% to represent media water potential of -0.33, -0.67 and -1.0 MPa. NaCl concentration of 3, 6 and 9 g L^{-1} was used to establish three salinity conditions which consisted of EC 4, 6 and 8 $mmhos.cm^{-1}$, respectively. Varieties used were 25 Indonesian rice varieties ie. Cimalaya Muncul, Inpari 4, Situ Bagendit, Ciliwung, IR 29, Pokali, IR 58443-6B-10-3, Limboto, Inpago 6, Inpago 10, Cigeulis, Inpari 35, Inpari 32, Inpari 33, Inpari 39, Inpari 43, Ciherang, Mekongga, Inpari 19, Inpari 10, MSP 8, MSP 9, MSP 13, MSP 14 and MOD I F 1.

A test on paper method was employed in the trial to germinate the 25 varieties of rice on various concentration of NaCl and PEG. Parameters measured were percentage of germination, germination rate, radicle and plumule length, plant fresh weight and vigor. Heritability was calculated for each of parameter. Concentration of NaCl and PEG that result in highest heritability will be used as concentration to test the varieties at growth and production stages in a further test hydroponically in screen house.

3. Results and Discussion

Table 2 shows that the germination percentage, germination rate, plumule length and vigor on the test on paper decreased with increasing Polyethylene glycol (PEG) concentration. The highest percentage of reduction in growth compared to normal condition (P0) was on seedling vigor with reduction reached 25% when water potential dropped to -0.67 MPa. Further reduction in vigor to more than 50% was shown in PEG concentration that resulted in water potential of -1.0 MPa. This justifies the method used in establishing the drought condition. The more PEG absorbed by the seed the less amount of water imbibed into the seed, because the PEG compound has polar properties, which is a characteristic that can bind water and priming seeds to control imbibition. This is in line with Hardegree and Emmerich [8] that Polyethylene glycol (PEG) is a compound used in priming where PEG has the properties of controlling the imbibition and hydration of the seed. Water is an indispensable factor in germination. The presence of water is essential for enzyme activity and decomposition of food reserves, nutrient translocation, metabolism/ biosynthesis, cell division, growth and other physiological processes [9]. Decrease in water potential increased the root length of the rice seedlings. This might be caused by the response of the plant as one of the mechanisms of resistance to drought stress. Increase in volume and root length is one of the mechanisms of plants to cope with drought stress [10]. Drought will cause the plant to respond widely ranging from gene expression to metabolism in its growth [11].

Table 2. Growth of various Indonesian rice varieties on different concentration of Polyethylene glycol (PEG).

Parameter	P_0	PEG concentration					
		P_1 (-0.33 MPa)		P_2 (-0.67 MPa)		P_3(-1.0 MPa)	
			A		B		C
Percentage of Germination (%)	91.47	78.93	13.71	73.51	19.63	66.22	27.60
Germination Rate	39.08	35.16	10.02	30.02	23.17	23.60	39.61
Root Length (cm)	5.13	5.90	13.05	6.67	30.02	7.08	38.01
Plumula Length (cm)	6.33	5.72	9.64	5.19	18.01	3.46	45.34
Plant Fresh Weight (g)	0.06	0.07	12.86	0.08	26.23	0.07	16.39
Vigor	11.50	11.00	4.35	8.65	24.78	5.51	52.09
Average	25.59		63.62		23.64		36.51

A, B, C: The percentage decrease in the growth of each parameter observed from the control (P0).

Table 3 shows that the percentage of germination, germination rate, plumula length and vigor parameters of the rice varieties decreased with increased in concentration of Sodium Chloride (NaCl) in the test on paper. The higher the concentration of NaCl as selection agent for salinity tolerance from EC of 4 to 8 mmhos cm^{-1} resulted in reduction of seedlings growth that range from 30 to 70%.

Table 3. Growth of various Indonesian rice varieties on different concentration of Sodium Chloride (NaCl).

Parameter	P_0	NaCl concentration					
		P_1(EC 4 mmhos cm^{-1})		P_2(EC 6 mmhos cm^{-1})		P_3(EC 8 mmhos cm^{-1})	
			A		B		C
Percentage of Germination (%)	91.47	77.96	14.77	72.98	20.21	66.22	27.60
Germination Rate	39.17	35.16	10.24	30.37	22.47	23.77	39.32
Root length (cm)	2.50	4.98	99.20	7.43	197.20	6.15	146.00
Plumula Length (cm)	6.80	5.44	20.00	3.90	42.65	1.03	84.85
Plant Fresh Weight (g)	0.04	0.05	31.71	0.06	65.85	0.06	51.22
Vigor	12.51	11.13	11.03	8.57	31.49	6.68	46.60
Average	25.42		31.16		63.31		65.93

A, B, C: The percentage decrease in the growth of each parameter observed from the control (P0).

Salinity or salt stress can cause poisoning in plants so that the germination phase and growth of the seedlings become inhibited. In the germination phase, the growth of sprouts will be hampered due to the amount of Cl^- content dissolved with water has been seeded by sprouts [12]. Some anions such as Cl^- can cause severe cell membrane damage in excessive amounts and cause leakage of cell membranes. High salt levels inhibit seed germination, plant height, yield quality, production and damage to plant tissue [13]. As for the parameters of root length and wet weight of the plant, it increases with increasing concentration of NaCl. Root growth exposed to the saline environment is usually less affected [14]. This is thought to be due to improved balance by maintaining water-absorbing ability.

Heritability analysis is a quantitative benchmark to determine whether the phenotypic differences of a character are caused by genetic or environmental factors, so it can give an idea whether the observed character is more influenced by genetic or environmental factors. The highest heritability value as plant response to the concentration of PEG shown by the plant vigor character (95.28), while response to NaCl concentration was shown by germination rate character (90.24) as shown in table 4. Characters with high heritability will increase the effectiveness of selection in tolerance test because the observed character is a reflection of the influence of genetic factors compared with environmental influences [15]. Quantitative characters that have high heritability will produce a selection progress for desirable traits, whereas if low heritability is less effective to be used as selection material.

Table 4. Heritability values (%) of parameters as response of various Indonesian rice varieties to different concentration of Sodium chloride (NaCl) and Polyethylene glycol (PEG).

Parameter	Heritability (%)			
	NaCl	Criteria	PEG	Criteria
Percentage of Germination	79.31	High	69.45	High
Germination rate	90.24	High	91.01	High
Root Length	59.42	High	54.13	High
Plumula Length	61.94	High	56.41	High
Plant Fresh Weight	67.94	High	72.64	High
Vigor	77.64	High	95.28	High

Notes: $0 < h^2 \leq 20$ (low), $21 < h^2 \leq 50$ (medium), $50 < h^2 \leq 100$ (high)

4. Conclusion

The tolerance limit of rice resistance to drought was at PEG concentration of 100 $g.L^{-1}$ (-0.67 MPa) and to salinity was at NaCl concentration of 6 $g.L^{-1}$ (EC 6 $mmhos.cm^{-1}$). There is a high genetic diversity of Indonesian rice varieties to drought and salinity with high heritability (H> 0.7) at water potential of -0.67 MPa and salinity of EC 6 $mmhos.cm^{-1}$.

References

[1] Syakir M 2007 *Pemantapan inovasi dan diseminasi teknologi dalam memberdayakan petani.* (Badan Penelitian dan Pengembangan Pertanian Jakarta Selatan)
[2] Indonesian Agricultural Research and Development Agency 2017 *Luas lahan, produksi dan produktivitas padi nasional tahun 2011-2015* (Deptan).
[3] Indonesian National Spatial Planning Coordination Agency 2012 *Menata kawasan hutan dan mempertahankan lahan pertanian* (Sekretariat Tim Pelaksana BKPRN Jakarta Indonesia)
[4] Wang Y, Zhang L, Nafisah A, Zhu L, Xu J and Li Z 2013 *The Crop J.* **1** 134-142
[5] Reddy I N B L, Kim B K, Yoon I S, Kim K H and Kwon T R 2017 Salt tolerance in rice: Focus on mechanisms and approaches *Rice Sci.* **24** (3) 123-144
[6] Zeng L H and Shannon M C 2000 Salinity effects on seedling growth and yield components of rice *Crop Sci.* **40** (4) 996–1003
[7] Hariadi Y C, Nurhayati A Y, Soeparjono S and Arif I 2015 Screening six varieties of rice (*Oryza sativa*) for salinity tolerance. *Proc. Env. Sci.* **28** 78 – 87 (The 5th Sustainable Future for Human Security SustaiN 2014)
[8] Hardegree S P and Emmerich W E 1992 Seed germination response of four southwestern range grasses to equibration at subgermination matrix potential *Agron. J.* **84** 994-998
[9] Abidin Z 1994 *Dasar-dasar pengetahuan tentang zat pengatur tumbuh* (Penerbit Angkasa, Bandung)
[10] Jones H G and Corlett J E 1992 *J. of Agric. Sci.* **119** (3) 291-296
[11] Darmawan J and Baharsjah S J 1998 *Dasar-dasar fisiologi tanaman* (SITC Jakarta)

[12] Staples R C and Toennissen G H 1984 *Salinity tolerance in plants strategies for crop improvement* (John Wiley and Sons Canada)
[13] Levitt J 1980 *Response of plants to environmental stresses vol. 2 Water, radiation, salt and other stresses* (Academic Press New York) 606 p
[14] Sipayung 2003. *Stres garam dan mekanisme toleransi tanaman* Online http://www.library.usu.oc.id/modulus.php.com accessed on 08 May 2017
[15] Basir M 2001 Pemanfaatan nilai heritabilitas dan koefisien korelasi untuk menentukan indikator seleksi *J. Agrivigor* **1** (1) 1-6

IOP Conf. Series: Earth and Environmental Science **157** (2018) 012012 doi:10.1088/1755-1315/157/1/012012

The effect of water regime and soil management on methane (CH_4) emission of rice field

O Naharia[1], P Setyanto[2], M Arsyad[3], H Burhan[4] and M Aswad[5]

[1]Department of Biology, Manado State University, Jalan Tondano, Koya, Tondano Selatan, 95618, North Sulawesi, Indonesia.
[2]Research Station for Agricultural Environment Preservation, Jakenan 05 Pati, 59182, Central Java, Indonesia.
[3]Department of Agriculutural Social-economics, Faculty of Agriculture, Hasanuddin University, Jalan Perintis Kemerdekaan KM 10, Makassar, 90245, Indonesia.
[4]Syngenta Research and Development, Station Box 08, Cikampek, 41303, West Java, Indonesia.
[5]Universitas Sulawesi Barat, Jalan Prof. Dr. Baharuddin Lopa, S.H., Baurung, Banggae Timur, Baurung, Banggae Tim., Kabupaten Majene, 91412, Sulawesi Barat, Indonesia.

E-mail: *arsyad@unhas.ac.id*

Abstract. Mitigation of CH_4 emission of rice field is becoming a serious issue. The Agricultural Environment Preservation Research Station in Central Java conducted a field study to investigate the effect of water regime and soil tillage on CH_4 emission from paddy fields. Treatments consisted of two factors. The first factor was water regime, *e.g.*, 1) continuously flooded 5 cm, 2) intermittent irrigation and 3) saturated water condition at 0-1 cm water level. The second factor was soil management, *e.g.,* 1) normal tillage, 2) zero tillage + 3 sulfosate ha^{-1} and 3) zero tillage + 3 L paraquat ha^{-1}. Most of treatments gave a significant reduction of total CH_4 emission between 34 – 85% during the wet season crop as compared to normal rice cropping practice, while in the dry season the CH_4 reduction ranged between 16 – 92%. No-tillage with non-selective herbicides combined with intermittent/saturated irrigation system significantly reduced methane emission without significantly affecting rice productivity as compared to normal tillage with continuous flooding (farmers practice)

1. Introduction

Irrigated lowland rice system account for about 80% of the world harvested rice area and 92% of total rice production. Methane (CH_4) is one of the gases released from an anaerobic decomposition of soil organic matter. Flooded rice soil contributes as much as 25% or ~ 100 Tg CH_4 on an annual basis [1]. The projected increase of rice production during the coming decades [2] is expected to result in further increase in CH_4 fluxes to the atmosphere if prevalent cultivation practices continue [3, 4, 5] Current recommendations to minimize CH_4 emission in rice are based mostly on adapted rice varieties, intermittent irrigation and management of crop residues under full cultivation. No research has been carried out on the effect of no-tillage on CH_4 emission. Further, Indonesian data on the effect of water regime on CH_4 emission still need to be further validated under different soil and climatic condition.

This study reports the results of two experiments on the effect of no-tillage on CH_4 emission of a paddy field under different water regimes.

2. Materials and Method

2.1. Research Site

Two field experiments were conducted during the wet season (November– March), and dry season (April-July), at Jakenan, Central Java, Indonesia. The soil properties were relatively high acidity, low CEC and low organic matter content. The soil was classified as Incept sol with a silt loam texture.

2.2. Experimental layout and Soil Management

The experiments covered two cropping cycles (wet and dry season) with a short fallow period in between. Rice was grown under irrigated lowland conditions during the two consecutive seasons. Treatments consisted of two factors: water regime (A) and soil management (T). The factor A treatments consisted of three different water regime *e.g.*, 1) continuously flooded 5 cm (A1), 2) intermittent irrigation (A2) and 3) saturated water condition at 0-1 cm water level (A3). An illustration of the water treatments is shown in figure 1. Factor B treatments were: 1) normal tillage (T1), 2) zero tillage + 3 L ha^{-1} (TOUCHDOWN SL480) (T2) and 3) zero tillage + 3 L ha^{-1} paraquat (GRAMOXONE SL200) (T3). The treatments were arranged in 3 x 3 factorials and the experimental design was randomized complete block with three replicates. The experimental plot size was 6 m x 4 m. IR 64 rice cultivar was used in this study and transplanted at 25 days after nursery sowing.

Soil cultivation for T1 treatment was carried out one day before rice transplanting. Herbicides sulfosate and paraquat were sprayed at 9 and 4 days before rice transplanting, respectively. Inorganic fertilizer in the form of urea, SP-36 and KCl was applied at the rate of 120 kg N ha^{-1}, 60 kg P ha^{-1} and 90 kg K ha^{-1}, respectively. Water level in the plots was controlled daily.

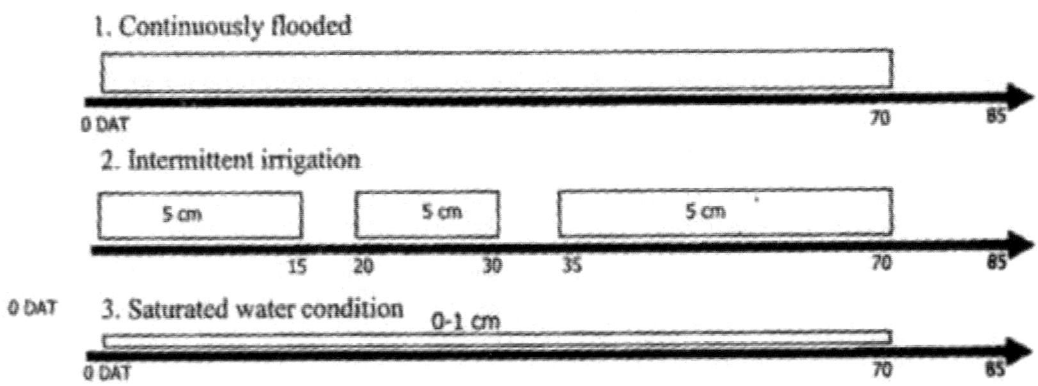

Figure 1. Illustration of water treatments *i.e.* 1) continuosly flooded 5 cm, 2) intermittent irrigation and 3) saturated water condition at 0-1 cm water level

2.3. Methane Flux Measurement

Methane fluxes were recorded every four days using the closed chamber method originally described [6]. Gas samples from each of the plots were collected using a 5 mL plastic syringe at four different intervals i.e. 3, 6, 9 and 12 minutes. Methane gas concentration inside the syringe was analysed using gas chromatograph Simadzu GC 8A equipped with flame ionization detector and a 3 m length and 1 mm diameter of porapak N column. The GC performance required for such analyses are; 1) column temperature: 75^0C and 2) injector/detector temperature: 90^0C. During gas sampling, the temperature increase and the headspace of the chamber were also recorded. This parameter is important for CH_4 flux calculation. Methane flux calculation was derived from the equation described [7].

2.4. Data Analysis

Data of CH_4 emission and yield parameters from field experiment i.e. rice production, plant height, tiller number and biomass, were analysed using analysis of variance (ANOVA). The treatment means were compared using Duncan Multiple Range Test (DMRT) and Least Significant Difference (LSD).

3. Results and Discussion

3.1. Season CH_4 Flux

The patterns of CH_4 flux from rice field as affected by water regime in two consecutive seasons are shown in figure 2 and figure 3.

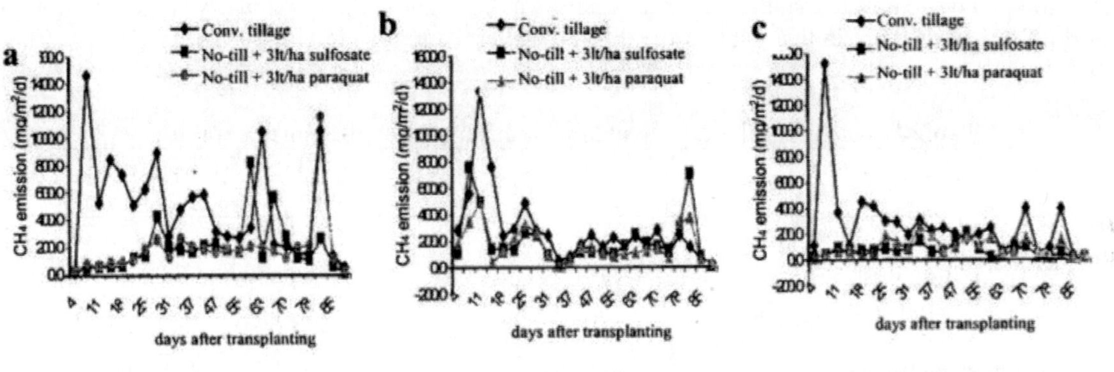

Figure 2. Seasonal CH_4 flux pattern of three different water management practices: a) continuously flooded, b) intermittent irrigation, and c) saturated water condition during wet season.

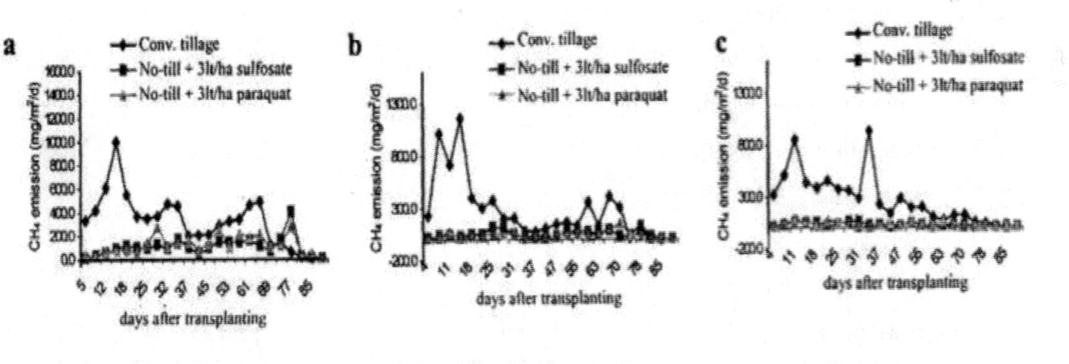

Figure 3. Seasonal CH_4 flux pattern of three different water management practices: a) continuously flooded, b) intermittent irrigation, and c) saturated water condition during dry season.

Methane flux for continuously flooded treatment (A1) started to increase within the first two weeks after flooding with an average 29.7 mg CH_4 m^{-2} h^{-1} in the wet season and 22.8 mg CH_4 m^{-2} h^{-1} in the dry season. After two weeks, the fluxes slightly decreased with an average 16.2 and 13.6 mg CH_4 m^{-2} h^{-1} for the respective season. Other irrigation treatments (A2 and A3) showed the same pattern but with different intensities. For intermittent (A2) the average flux in the first two weeks was the same in both seasons i.e. 26.5 mg CH_4 m^{-2} h^{-1}, while in the latter stage it emitted as much as 7 mg CH_4 m^{-2} h^{-1} and 6.5 mg CH_4 m^{-2} h^{-1} respectively. Saturated irrigation (A3) was showing the lowest flux with the average of

IOP Conf. Series: Earth and Environmental Science **157** (2018) 012012 doi:10.1088/1755-1315/157/1/012012

9.4 and 11.2 mg CH_4 m^{-2} h^{-1} in the two respective seasons. The average CH_4 emission reduction by implementing A2 and A3 treatment as compared to A1 was 54.7% (23.95%) and 80.1% (11.25%) respectively during the wet season crop, while in the dry season it reduced as much as 55.7% (12.15%) and 79.4% (14.07%). A study conducted [8] on different water management also showed similar results. Two or three peaks have been usually observed in most field studies on CH_4 emission [9]. The first peak is associated with the decomposition of soil organic matter or plant materials from the previous season. The second and third peaks are associated with the rice plants since they are not observed in unplanted fields [10].

3.2. *Total Seasonal CH_4 Emission, Grain Yield and Plant Growth Parameters.*

Table 1 shows that the water regime significantly influenced CH_4 emission. Continuously flooded water regime (A1) showed the highest seasonal CH_4 emission compared with intermittent irrigation (A2) and saturated 0-1 cm water depth (A3). The pattern was constant in the wet and dry season period. The result of this study were in line with observations of several authors [11, 12, 13, 14, 15, 16, 17, 18, 19, 20].

Table 1. Total CH_4 emission treated with different water regime as soil tillage

Treatment	(kg CH_4 h^{-1} $season^{-1}$)	
	Wet Season	Dry Season
A1 = Continuous flooding	303.08 a	255.24 a
A2 = Intermittent (5 cm)	132.31 b	55.09 b
A3 = Saturated (0 – 1) cm	137.56 b	53.11 b
T1 = Normal tillage	253.95 a	160.81 a
T2 = No-tillage + 3 L h^{-1} sulfosate	189.07 b	105.98 b
T3 = No-tillage + 3 L h^{-1} paraquat	129.92 b	96.65 b
A1T1 = Continuous flooding + Normal tillage	422.66 a	285.27 a
A1T2 = Continuous flooding + No-tillage 3 L h^{-1} sulfosate	158.33 c	241.05 ab
A1T3 = Continuous flooding + No-tillage 3 L h^{-1} paraquat	180.85 c	239.41 b
A2T1 = Intermittent + Normal tillage	246.47 b	91.90 c
A2T2 = Intermittent + No-tillage 3 L h^{1} sulfosate	177.05 c	43.74 cd
A2T3 = Intermittent + No-tillage 3 L h^{-1} paraquat	143.71 cd	23.69 d
A3T1 = Saturated + Normal tillage	240.10 bc	105.26 c
A3T2 = Saturated + No-tillage 3 L h^{-1} sulfosate	61.54 d	33.18 d
A3T3 = Saturated + No-tillage 3 L h^{-1} paraquat	88.12 d	26.83 d

Number in the same column followed by common latter for treatment A and T are not significantly different ($P<0.05$) by LSD, and interaction by DMRT.

Treatment A1T1 showed the highest seasonal CH_4 emission compared with the other treatments. In the same A1 treatment with different T treatments, *i.e.*, T2 and T3, total CH_4 emission was suppressed. The same pattern was also recorded with A2 an A3 treatments. Those treatments increased methane emission if combined with T1, and the decreased it when combined with T2 and treatments. These situations were recorded in both wet and dry season experiments. Very treatments were showing significant differences on rice grain yield indicating that the treatments did not significantly affect rice for productivity (table 2).

Table 2. Yield (kg ha^{-1}; mc14%) at different water regime and soil tillage

Treatment	Soil Tillage						Average	
	T1		T2		T3			
Water Regime	WS 2002/03	DS 2003	WS 2002/03	DS 2003	WS 2002/03	DS 2003	WS 2002/03	DS 2003
A1	5167.3	5144.0	5123.6	4656.7	4723.0	5274.3	5004.7 a	5025.0 a
A2	4814.3	4543.0	4725.7	4435.7	4535.3	4323.0	4701.7 b	4535.9 b
A3	4597.7	4321.7	4665.3	4243.7	4571.0	4372.0	4611.3 b	4312.b
Average	4859.8	4670.6	4838.2	4445.3	4619.8	4754.4		

Number in the same column followed by common latter for treatment A and T are not significantly different ($P<0.05$) by LSD

4. Conclusion

The result of the two field studies shows that no-tillage following the application of non-selective herbicides such as paraquat (GRAMOXONE SL200) or sulfosate (TOUCHDOWN SL480) can significantly reduce methane emission from Indonesia rice paddy fields. The reduction is higher during the dry than the wet season. Methane emission are further reduced when no-tillage is combined with intermittent or saturated irrigation. In the two experiments no tillage did not affect rice productivity significantly as compared to normal tillage with continuous flooding (current farmer's practice). Reduction of methane emission is one further potential benefit of no-tillage rice in the tropics. This combined with short term economic benefits for farmers such as lower water consumption, lower cultivation cost and increased planting index could lead to reconsider opportunities for no-till rice in tropical Asia.

References

[1] Houghton J T, Calandar BA and Varney S K 1992 *Climate Change* (*The supplementary report to the IPCC scientific assessment*) (Cambridge: Cambridge University Press)

[2] IRRI 1994 *World Rice Statistics* (Manila, Philippines) p 42

[3] Nungkat P, Kusuma Z, and Handayanto E 2015 Effects of organic matter application on methane emission from paddy fields adopting organic farming system *J. of Deg. and Mining Lands Manag.* **2** (2) 303-312

[4] Anastasi C, Dowding M and Simpson V J 1992 Future CH_4 emission from rice production *J. Geophys Res.* **97** 7521-7525

[5] Wihardjaka A 2015 Mitigasi emisi gas metana melalui pengolahan lahan sawah *J. Litbang Pertanian* **34** (3) 95-104

[6] Schutz H A, Holzapfet-Pschorn, Conrad R, Rennenberg H and Seiler W 1989 A 3 – year continuous record on the influence of daytime, seaso, and fertilizer treatment on methane emission rates from Italian rice paddy *J. Geophys. Res.* **16** 405-416

[7] Lantin R S, Aduna J B and Javellana A M J 1995 Methane measurements in rice fields. Instruction manual and methodologies, maintenance and troubleshooting guide (A joint undertaking by: IRRI, US-EPA and UNDP)

[8] Wang Z Y, Xu Y C, Li Z, Wassman R, Neue H U, Lantin R S, Buendia L V, Ding Y P and Wang Z Z 2000 A four-year record of methane emission from irrigated rice fields in Beijing region of China *Nutr. Cycl. Agroecosyst* **58** 55-65

[9] Yagi K and Minami K 1990 Effect of organic matter application on methane emission from some Japanese paddy fields *Soil Sci. and Plant Nutr.* **36** 599-610

[10] Yang X Y, Cai Z C, Ohara T and Akimoto H 2003 Methane emission from rice Fields in mainland China: amount and seasonal and spatial distribution J. of Geophy. Res.: Atmospheres **108** p.4505

[11] Husin Y A 1994 *Methane Flux from Indonesia Wetland Rice: The Effects of Water Management and Rice Variety* (PhD Thesis) Bogor Agricultural University

[12] Setyanto P and Kartikawati R 2008 Sistem pengelolaan tanaman padi rendah emisi gas metan *J.*

Pen. Tan. Pangan **27** 154-163
[13] Wihardjaka A 2015 Mitigasi emisi gas metana melalui pengolahan lahan sawah *J. Litbang Pertanian* **34** 95-104
[14] Hutabarat T R, Anas I, Idris K and Sugyanta 2011 *Polasi Mikroba Tanah, Emisi Metan dan Produksi Padi pada Enam Kombinasi Aplikasi Pemupukan dengan Budidaya Padi SRI* (Bogor: IPB)
[15] Suprihati 2007 *Populasi Mikroba dan Fluks Metana (CH_4) serta Nitrous Oksida (N_2O) pada Tanah Sawah: Pengaruh Pengelolaan Air, Bahan Organic dan Pupuk Nitrogen* (Bogor: Disertasi Sekolah Pascasarjana, Institut Pertanian Bogor)
[16] Husny Z 2010 Methane (CH_4) flux in rice cultivation of system of rice intensification and convensional in fresh water swamp, tidal swamp, and irrigated rice fields *Agroekotek J.* **2** 38-45
[17] Mulyadi and Wihardjaka A 2014 Emisi gas rumah kaca dan hasil gabah dari tiga varietas padi pada lahan sawah tadah hujan bersurjan *J. Penelitian Tan. Pangan* **33** 116-121
[18] Naharia O 2004 *Teknologi Pengairan dan Pengolahan Tanah pada Budidaya Padi Sawah untuk Mitigasi Gas Metana (CH_4)* (Bogor: Sekolah Pascasarjana IPB)
[19] Najamuddin M 2014 Strategi mitigasi emisi gas metan pada budidaya padi sawah *J. Agribisnis* **8** 171-188
[20] Muttaqin A, Miftahudin and Rusman I 2016 Bacteria as greenhouse gases reducing agents from paddy plantation *J. Sumberdaya Hayati* **2** 45-51

IOP Conf. Series: Earth and Environmental Science **157** (2018) 012013 doi:10.1088/1755-1315/157/1/012013

Initial assessment on the use of cocoa pulp in complete feed formulation: *in vitro* dry matter and organic matter digestibility

A Natsir[1], A Mujnisa[1], M Z Mide[2], N Purnomo[2] and M F Saade[3]

[1]Animal Nutrition Department, Faculty of Animal Science, Hasanuddin University, Jl. Perintis Kemerdekaan KM 10, 90245, Makassar, South Sulawesi, Indonesia.
[2]Animal Feed Industry Laboratory, Faculty of Animal Science, Hasanuddin University, Jl. Perintis Kemerdekaan KM 10, 90245, Makassar, South Sulawesi, Indonesia.
[3]Post graduate student, Faculty of Animal Science, Hasanudin University, Jl. Perintis Kemerdekaan KM 10, 90245, Makassar, Indonesia.

E-mail: *asmudin_natsir@unhas.ac.id*

Abstract. Cocoa pulp is a by-product from cocoa industry which is produced in large quantity, but very limited study has been carried out in utilizing it as energy source in animal feed. The purpose of this study was to assess the *in vitro* dry matter (IVDMD) and *in vitro* organic matter digestibility (IVOMD) of complete feed containing different levels of cocoa pulp. The experiment was carried out according to completely randomised design consisting of four treatments and three replications. The treatments were P0 = Complete feed containing 0% cocoa pulp, P1 = Complete feed containing 5% cocoa pulp, P2 = Complete feed containing 10% cocoa pulp, and P3 = Complete feed containing 15% cocoa pulp on dry matter basis. The results of the study indicated that the average IVDMD was 567, 538, 566, and 526 g kg^{-1} DM, while the average IVOMD was 522, 491, 502, and 461 g/kg DM, respectively for treatment P0, P1, P2, and P3. Statistical analysis indicated that increasing levels of coca pulp in the feed significantly affected ($P<0.05$) the IVDMD and IVOMD of the feed. In conclusion, cocoa pulp is potential to be used up to 10% in complete feed with corn cobs as the fibre source.

1. Introduction

Indonesia is one of the top six countries of cocoa production in the world. The cocoa production in Indonesia reached 661,243 tons in 2015 [1]. Among the provinces in Indonesia, South Sulawesi is one of the largest cocoa production region with an area of 246,223 ha producing around 143,237 tons of cocoa beans [2]. In addition to the main products which are cocoa butter and cocoa powder, the processing of cocoa fruit also resulted in byproducts such as cocoa pods, cocoa pods husk, and pulp which are usually disposed off as waste. In the cocoa processing industry, only 10% cocoa butter or cocoa powder can be derived from of the fresh cocoa and roughly 68.5% of fresh cocoa mass will become waste. The amount of cocoa pulp, one of the byproducts form cocoa processing, could reach up to 40% the fresh weight of cocoa beans [3]. Another study reported that processing of 1 kg of cocoa beans could yield 100-190 ml of cocoa pulp [4]. Actually, cocoa pulp has been used for various purposes, such as for acetic acid production [5], organic herbicides for the plants [6, 7], jam for human consumption, juice and fermented beverages [8, 9, 10]. However, a large proportion is not utilized and considered as a waste.

One alternative on the utilization of cocoa pulp is by utilizing it as animal feed especially for ruminants such as cattle and goat. Based on its physical form, cocoa pulp is potentially useful as feedstuff if combined or mixed with other ingredients, especially those that could absorb water, such as corn cobs meal and rice bran. Chemically, cocoa pulp contains high percentage of sugar [11], and can be used to substitute molasses in complete feed or in the urea molasses block (UMB). Use of complete feed has some advantages, for example it has been shown to increase feed consumption, daily body weight gain of Bali cattle [12] and improved the quality and yield of goat milk [13]. Moreover, the use of complete feed formulated from materials such as corn cobs resulted in a good performance of local goats in South Sulawesi [14].

In order to optimally use the cocoa pulp as a feed ingredient in a complete feed it is important to have initial information related to its biological value when used as a part of the complete feed. Many methods have been applied to evaluate the nutritive value of a feedstuff. One of those is *in vitro* technique [15]. This study was carried out to determine the effects the increased levels of cocoa pulp in a complete feed using corn cobs as the main fibre source, on *in vitro* dry matter digestibility and *in vitro* organic matter digestibility.

2. Materials and Methods

2.1. Formulation of Complete Feed

Formulation of the complete feeds for this study was carried out according to the following protocols: All feed ingredients having large or rough particle size, such as corn cobs, coconut cake meal was groundto smaller particles using a hammer mills. Each feed ingredient was weighed according to the formulations in each dietary treatment (four different treatments). All feedstuff in each treatment were mixed thoroughly using a hand mixer. The mixture then was made into cube form using simple press equipment. The feed was then dried in oven at 65^{0}C for 48 hours. One hundred g samples of each treatment diet was then ground to pass 1-mm sieve for both chemical analysis (proximate and fibre analysis) and for the *in vitro* study. The composition of each dietary treatment is presented in table 1.

Table 1. Feedstuff composition of the complete feed according to the treatment

Feedstuff (%)	Treatments			
(DM basis)	P0	P1	P2	P3
Corn cobs	50.0	50.0	50.0	50.0
Rice bran	18.0	18.0	18.0	18.0
Coconut cake meal	7.5	7.5	7.5	7.5
Shrimp waste meal	5.0	5.0	5.0	5.0
Cocoa pulp	0.0	5.0	10.0	15.0
Molasses	15.0	10.0	5.0	0.0
Urea	1.5	1.5	1.5	1.5
Salt	1.0	1.0	1.0	1.0
Mineral Mix	2.0	2.0	2.0	2.0
Total	100.0	100.0	100.0	100.0

P0 = complete feed containing 0% cocoa pulp, P1 = complete feed containing 5% cocoa pulp, P2 = complete feed containing 10% cocoa pulp, and P3 = complete feed containing 15% cocoa pulp.

IOP Conf. Series: Earth and Environmental Science **157** (2018) 012013 doi:10.1088/1755-1315/157/1/012013

2.2. *In vitro Study*

The *in vitro* study was carried out according to completely randomized design consisting of four treatments, P0, P1, P2, and P3 and three replications for each treatment giving a total of 12 experimental units [16]. The treatments were P0 = complete feed containing 0% cocoa pulp, P1 = complete feed containing 5% cocoa pulp, P2 = complete feed containing 10% cocoa pulp, and P3 = complete feed containing 15% cocoa pulp. In the formulations the molasses were replaced by cocoa pulp. The pepsin-cellulase *in vitro* digestibility technique [15, 17] was used in order to determine the *in vitro* dry matter disappearance (IVDMD) and *in vitro* organic matter disappearance (IVOMD) of the samples. This technique is a two-stage procedure requiring the addition of cellulase in an acid buffer and subsequent addition of pepsin in an acetate buffer into the samples incubated at 50 °C for a total of 120 hours.

2.3. *Laboratory Analysis and Calculation*

Sample DM content was determined by drying at 105°C in the oven for 24 h (Memmert, Germany). The percentage of ash was determined by combustion of samples in the furnace (Nabertherm, Germany) for 6 h at 550°C. Organic matter (OM) was calculated as 100-%ash (DM basis). Total N content was determined by the Kjeldahl procedure [18] and percentage of crude protein (CP) was calculated as total N*6.25. Fibre composition (NDF and ADF) was analysed according to the procedure of Goering and Van Soest [19].

IVDMD and IVOMD were calculated as follows:
IVDMD = (DM incubated (g) - DM residue (g)) x 100% / DM incubated (g)
IVOMD = (OM incubated (g) - OM residue (g)) x 100% / OM incubated (g)

2.4. *Data Analysis*

All the experimental data were analysed using analysis of variances according to the completely randomised design with experimental model: $Y_{ij} = \mu + T_j + \epsilon_{ij}$, where Y_{ij} = observation value, μ = general mean, T_i = effects of treatment (i = 1, 2, 3, 4), and ϵ_{ij} = experimental error (i = 1, 2, 3, 4 and j = 1, 2, 3).

3. Results and Discussion

The Proximate analysis indicated that the crude protein content, ether extract, and ash of the treatments increased as the cocoa pulp levels in the diets increased, while fibre content and nitrogen free extract components decreased as the levels of cocoa pulp increased (table 2). Based on Van Soest analysis, Neutral detergent fibre (NDF) and acid detergent fibre (ADF) of the complete feed decreased as the levels of cocoa pulp in the diets increased (table 2). Statistical analysis was not carried out for chemical components of the diets. The change of the chemical components of the diets as the levels of cocoa pulp increased is understandable as the chemical component of molasses, in which the cocoa pulp intended to substitute was not exactly the same. For instance, both ingredients are different in their protein and fibre contents. Therefore, the fibre content of the complete feed for instance increased when cocoa pulp increased due to crude fibre content of cocoa pulp was somewhat higher than that of molasses The change of chemical components of the complete feed was merely a consequence of substituting molasses with cocoa pulp in the diets, in which both feedstuffs do not share exactly similar chemical components. In general, however, the chemical component of the complete feed (all treatments) was quite satisfactory for ruminants with the crude protein content of the diets varied from 12.7% (P0) to 16.8% (P3) (table 2). Ruminant animal requires at least 7.5% crude protein in their diet [20].

Table 2. Chemical components of the experimental diets[a]

Components (g/kg DM)	Treatments			
	P0	P1	P2	P3
Proximate components				
Crude protein	127	131	166	168
Ether extract	32	45	59	57
Crude fibre	178	177	174	161
Nitrogen free extract	582	548	487	484
Minerals	82	99	114	129
Fibre components				
NDF	528	469	395	448
ADF	251	238	233	225

[a]Animal Feed Chemical Laboratory, Hasanudin University, 2017. P0 = complete feed containing 0% cocoa pulp, P1 = complete feed containing 5% cocoa pulp, P2 = complete feed containing 10% cocoa pulp, and P3 = complete feed containing 15% cocoa pulp.

The *in vitro* technique is commonly used to evaluate the nutritive value of feeds for ruminant animals. The differences or variation of the chemical components of the diets affected the parameters observed. Statistical analysis indicated that increasing levels of cocoa pulp in the complete feed affected (P<0.05) the IVDMD and IVOMD of the whole feed. The average IVDMD value ranged from 526 g kg^{-1} DM (P0) to 567 g kg^{-1} DM (P3), and from 461 g kg^{-1} DM (P3) to 522 g/kg DM (P0). Further statistical analysis indicated that the average percentage of IVDMD was the lowest for the treatment P3, while the IVDMD for the treatment P2 was similar (P>0.05) to P0 and much higher (P<0.05) than that for P1. Based on the IVOMD, the average value for treatment P0 was much higher (P<0.05) than that for P1, P2, and P3. Moreover, IVDMD for P2 was much higher (P<0.05) than that of P1 and P3, while IVOMD for P1 was higher than P3.

Table 3. The means of IVDMD and IVOMD of the experimental diets

Parameters	Treatments				Significant (P<)
	P0	P1	P2	P3	
IVDMD (g/kg DM)	567^{a}	538^{b}	566^{a}	526^{b}	P<0.05
IVOMD (g/kg DM)	522^{a}	491^{b}	502^{c}	461^{d}	P<0.05

^{a,b,c,d}means at the same row sharing different superscript were different at 5% level of significant. IVDMD= *In vitro* dry matter disappearance, IVOMD = *In vitro* organic matter disappearance, P0 = Complete feed containing 0% cocoa pulp, P1 = Complete feed containing 5% cocoa pulp, P2 = Complete feed containing 10% cocoa pulp, P3 = Complete feed containing 15% cocoa pulp.

The data (table 3) indicates that the change of IVDMD and IVOMD values of the diet did not follow the pattern of increasing levels of cocoa pulp in the complete feed, in which the change in the value of IVDMD and IVOMD of the four treatments was not consistent. From this study it seems that the optimal level for inclusion of cocoa pulp in the complete feed was up to 10%. There is no exact explanation why the value of IVDMD and IVOMD was fluctuated and did not follow the pattern of increased levels of cocoa pulp. The most possible reason for this phenomenon, however, might be related to associative effects. This effect is known when the apparent digestibility of a mixture of feeds is not equal to that of the sum of the individual component feeds. Associative effects are broadly categorised into negative and positive associative effects [21, 23]. Even though both ingredients, molasses and cocoa pulp, have similar characteristics in terms of provision of degradable carbohydrate, differences in their fat content might be also contribute to this differences. The fat content of the feed increases as the level of pulp cocoa in the diet increase. It seems that fat content of the diet containing 15% cocoa pulp has passed the tolerable level of fat for this particular diet. As a result, the extent of IVDMD and IVOMD for the feed containing 15% pulp was the lowest compared to other treatments [22]. Other factor that might have an effect on digestibility of the diet is the fibre source of the diet [24, 25]. Akhirany *et al* [26] reported that different forages has its own rumen degradation characteristics and may have an impact on affecting the digestibility when combining with other feedstuff as a complete feed. Therefore it is interesting to further explore the digestibility of the diet when cocoa pulp is combined with other fibre sources such as rice straw, corn stalk, or peanut straw, or soy bean straw.

4. Conclusion

Cocoa pulp has the potential to be used as an alternative feedstuff to substitute the use of molasses in complete feed using corn cobs as the main fibre source. This was based on the fact that the inclusion 10% (dry matter basis) in the complete feed has similar IVDMD and IVOMD with the complete feed containing 15% molasses. However, further study is required in order to determine whether the complete feed containing cocoa pulp is acceptable by the ruminant animal and could result in a comparable effects on the animal performance fed on conventional diet.

Acknowledgments
The authors would like to thank Universitas Hasanuddin that fully funded the study through the *Lembaga Penelitian dan Pengabdian Masyrakat* (LP2M) Universitas Hasanuddin.

References

[1] Ditjenbun 2015 *Statistics of Indonesian 2014-2016* (Jakarta: Directorate General of Plantation, Indonesia)
[2] Statistical Center Bureau 2015. *South Sulawesi in Number* (South Sulawesi: Statistical Center Bureau)
[3] Gildemberg A L Jr, Luiz H G, Priscilla E, de Almeida F C, Tavares and Figueira A 2008 *FEMS Yeast Res.* **8** 788–98
[4] Tri-Panji, Suharyanto, Rakayan E and Hasim 1995 *Menara Perkebunan* **63** 114-122
[5] Pratiwi M B N 2013 *Widyariset* **16** 425–34
[6] Pujisiswanto H 2012 *Jurnal Penelitian Pertanian Terapan* **12** 13-19
[7] Pratama A F, Susanto H and Sembodo D R J 2013 *J. Agrotek Tropika* **1** 80 – 85
[8] Duarte W F, Dias D R, Oliveira J M, Teixeira J A, Silva J B D and Schwan R F 2010 *Food Sci. and Technol.* **43** 1564-72
[9] Takrama J F, Kumi W O, Otoo G, Addo K and Camu N 2015 *J. of Agric. Sci. and Food Technol.* **1** 22-33
[10] Dias D R, Schwan R F, Freire E S and Santos Serôdio R 2007 *Int. J. of Food Sci. and Technol.* **42** 319-29
[11] Haryadi M and Supriyanto 2001 *Processing of Cocoa for Food* (Yogyakarta: Gajah Mada University) pp. 56-70
[12] Bahar S and A. Natsir 2011 *JITP* **1** 180-188
[14] Natsir A, Harfiah and Mide M Z 2014 *Proc. National Seminar 1 Optimalization of Local Resource on Technology Based Typical Farming System* 9 Oktober 2014 (Makassar: Hasanuddin University)

[13] Natsir A, Dagong M I A, Mide M Z, Islamiyati R, Irwan and Basri H 2015 *Proc. of Int. Seminar on Improving Tropical Animal Production for Food Security* 3-5 November 2015 (Kendari: Universitas Halu Oleo)
[15] McLeod M N and Minson D J 1978 *Anim. Feed Sci. Technol.* **3** 277-87
[16] Steel, R G D and Torrie J H Torrie 1981 *Principles and Procedures of Statistics. A Biometrical Approach* 2nd Ed. (New York: McGraw-Hill International Book Company)
[17] McLeod M N and Minson D J 1980 *Anim. Feed Sci. Technol.* **5** 247-50
[18] AOAC 1990 *Official Methods of Analysis* 15th ed (Washington, DC: Association of Official Analytical Chemists)
[19] Goering, H K and Van Soest P J 1970 *Forage Fibre Analysis (Apparatus, reagents, procedures, and some application)* Agric. handbook 379 ARS (Washington, DC: USDA)
[20] Parakkasi A 1999 *Ruminant Nutrition* (Jakarta: Indonesia University Press)
[21] Dixon R M and Stockdale C R 1999 *Aust. J. Agric. Res.* **50** 757-73
[22] McDonald P, Edwards R A, Greenhalgh J F D, Morgan C A, Sinclair L A and Wilkinson R G 2010 *Animal Nutrition* 7th ed (New York: Prentice Hall)
[23] Natsir A 2012 *Fibre Utilization by Ruminants* (Makassar: Masagena Press)
[24] Natsir A, Egan A R, Leury B J, Brandon M and Cunningham C 2001 *The Australian Journal of Dairy Technology* **56** 160
[25] Syahrir S, Wiryawan K G, Parakkassi A, Winugroho M and Natsir A 2012 *Media Peternakan* **35** 123-27
[26] Akhirany N, Ismartoyo, Jusoff K, Natsir A and Ako A 2013 *World Applied Science Journal* **26** 60-66

IOP Conf. Series: Earth and Environmental Science **157** (2018) 012014 doi:10.1088/1755-1315/157/1/012014

Effectivity of *Azotobacter chroococcum* and arbuscular mycorrhiza fungi on physiological characteristics and growth of cocoa seedlings

Nasaruddin and I Ridwan

Department of Agronomy, Faculty of Agriculture, Hasanuddin University, Jalan Perintis Kemerdekaan KM 10 Makassar 90245 Indonesia.

E-mail: *nnasaruddin@gmail.com*

Abstract. This study aims to study the effectiveness of *Azotobacter chroococcum* bacteria and Arbuscula mycorrhiza on some physiological characteristics and growth of cocoa seedlings. The study was conducted from March to October 2015, designed in the form of a two factors experiment based on the Randomized Block Design in a screen house. Inoculation of *A chroococcum* as the first factor consisted of control, inoculation of 10^4 CFU ml^{-1} water and 10^6 CFU ml^{-1} water per tree given as much as 40 ml. Inoculation of arbuscula mycorrhiza as a second factor consisted of control, inoculation of 3.0 g, 6.0 g and 9.0 g per tree, respectively. The experimental results show that inoculation of *Azotobacter chroococcum* 10^6 CFU ml^{-1} water.tree^{-1} and the arbuscular mycorrhizal fungi 6.0 g tree^{-1} resulted in higher chlorophyll a, b and total leaf chlorophyll content, increased light absorption rate, leaf stomatal conductance and better seedling growth.

1. Introduction

Production of cocoa in South Sulawesi, as one of the main centres of cocoa in Indonesia, has decreased significantly in the past 10 years, due to aging crops, decline in the ecological quality of the land and soil fertility, the simple application of cultivation techniques, the shortage of fertilizer and increasingly the high price of fertilizer [1]. One of the efforts made to improve cocoa production and quality is the improvement of the ecology of the land using soil bio-technology (soil microbial services and natural fertilizer technology).

Effective bacteria colonize the roots commonly called Plant Growth Promoting Rhizobacteria (PGPR). PGPR has the ability to protect plant parts on the ground against viral, fungal and bacterial disease by inducing systemic resistance (ISR) [2], can accelerate germination, stimulate root and shoot growth [3], increase leaf chlorophyll levels, increase crop tolerance to drought and salt and may delay leaf aging [4]. One of the important PGPR bacteria in the soil ecosystem is the *Azotobacter chroococcum* (*A chroococcum*) known as biologic fixing agent N_2 by converting N_2 to ammonium through electron reduction and nitrous protonation [5]. Several studies have shown that inoculation of *Azotobacter* can improve the growth and absorption rate of N in annual crops such as in pepper plants [6] and vanilla plants [7].

A chroococcum may be symbiotic with Arbuscular Mycorhiza and has been widely used in cultivation systems in both single inoculation and double inoculation with *A chroococcum*. Arbuscular mycorrhizas are symbiotic with plant roots and capable in increasing the uptake of N, P and K, increasing the efficiency of groundwater use, increasing the osmotic pressure value of plant cells on soil whose moisture content is low enough so that the plant can maximizes the utilization of CO_2 and solar energy to increase the rate of vegetative growth and crop production [8]. Efforts to improve the production and productivity of long-term and sustainable cocoa crops are needed to provide plant material (seeds), especially in the rejuvenation of old crops and damaged crops. Cocoa plant growth in the field is determined by the growth of plants during the nursery. The roots of plant seedlings infected by arbuscular mycorrhizae and symbiotic with *Azotobacter* in the rhizosphere were able to adapt to extreme cropping environments [9]. Symbiosis of bacteria *A chroococcum* with Arbuscula mycorrhizal fungi is very important for plants and can significantly increase the growth of host plants [10]. Understanding such interactions, particularly in cocoa farming areas with tropical climatic conditions and limited resources of fertilizers and simple cultivation systems are crucial in efforts to improve the production and quality of cocoa

This study aims to determine and study the effectiveness of *A chroococcum* and arbuscular mycorrhizae on some ecophysiological characteristics and growth of cocoa seedlings and to determine the best dosage of *A chroococcum* and mycorrhiza Arbuscula for the growth of cocoa seedlings. The results of the study are expected to provide information on the potential utilization of *A chroococcum* and Arbuscula Mycorrhiza which can streamline the utilization of nutrients especially N and P and for cocoa plants in the effort of rehabilitation and rejuvenation of plants.

2. Materials and Methods

The research was conducted in the form of a three factors factorial experiment based on the Randomized Block Design pattern. The first factor is the seed source consisting of clones of Sulawesi I (k1), Sulawesi 2 (k2) and Mukhtar 1 (k3). The second factor was the inoculation of *A. chroococcum* which consisted of three levels *i.e.* without inoculation as control (a0), 10^3 CFUs (a1) and 10^6 CFUs (a2), each was given 40 mL tree^{-1}. The third factor was the treatment of Arbuscular Mycorrhizae fungi (AM) *Glomus* sp in the form of a Zeolite culture, consisting of 4 levels ie without treatment (m0), 3.0 g tree^{-1} (m1), 6.0 g tree^{-1} (m2) and 9.0 g tree^{-1} (m3). Thus each experiment contained 36 treatment combinations that were repeated three times. Leaf chlorophyll and nutrient content were obtained through leaf tissue analysis. The observation of light energy absorption (%) was performed using Miniature leaf spectrophotometer, while the stomatal conductance (mmol m^{-2} s^{-1}) were observed using Prometer. Data was analysed using a factorial ANOVA test, and to determine the best treatment, data was tested using honest significant difference test (Tukey's).

3. Result

3.1. Content of Chlorophyll a, b and Total leaf Chlorophyll

Statistical analysis showed that seed source and AM and its interaction had no significant effect on leaf chlorophyll content. Inoculation of *A chroococcum* and AM had significant effect on leaf chlorophyll b content but its interaction was not significant. The interaction of seed source treatment with inoculation of *A chroococcum* and interaction of inoculation of *A chroococcum* with AM had significant effect on total leaf chlorophyll, but other interactions were not significant.

HSD test (α 0.05) in table 1 shows that inoculation of *A. chroococcum* 10^6 CFU mL water^{-1} per tree shows the highest leaf chlorophyll a and b content and significantly different from control. Inoculation of AM 6.0 g tree^{-1} showed the highest leaf chlorophyll content and was significantly different from the seedlings in control treatment. Seedlings sourced from Mukhtar 1 clone seeds inoculated with *A. chroococcum* 10^3 CFUs mL water^{-1} per tree showed the highest total leaf chlorophyll and significantly different from the other *A chroococcum* inoculation and seeds source treatment interactions. The interaction of the inoculation treatment of *A. chroococcum* of 10^6 CFU.mL water^{-1} per tree with AM 3 g tree^{-1} showed the highest total leaf chlorophyll and significantly different from the interaction of the other *A. chroococcum* and AM.

Table 1. Mean of leaves chlorophyll content of Cocoa Seedlings Clone Sulawesi 1, Sulawesi 2 and Mukhtar 1 at 4 months after treatment.

Chlorophyll types	Treatment	*Inoculation of Arbuscular Mycorrhizae (AM)* (M)				Mean	HSD (a)
		m0	m1	m2	m3		α 0.05
Chlo a	0 CFU ml^{-1} (a0)	0.112	0.117	0.121	0.123	0.118_y	
	10^3 CFU ml^{-1} (a1)	0.121	0.124	0.126	0.119	0.122_x	0.004
	10^6 CFU ml^{-1} (a2)	0.122	0.120	0.126	0.123	0.123_x	
Chlo b	0 CFU ml^{-1} (a0)	0.291	0.299	0.302	0.298	0.297_y	
	10^3 CFU ml^{-1} (a1)	0.300	0.302	0.305	0.303	0.303_x	0.003
	10^6 CFU ml^{-1} (a2)	0.302	0.306	0.308	0.304	0.305_x	
	Mean	0.297b	0.302a	0.305a	0.302a		
	HSD (m) α 0.05		0.005				
Total Chlorophyll	Seeds Sources	*A. chroococcum (CFU mL^{-1}water)*				Mean	HSD (s)
		0 (a0)	10^3 (a1)		10^6(a1)		α 0.05
	Sulawesi 1 (s_1)	0.292_a^y	0.292_a^z		$0.291a_b^y$	0.292	
	Sulawesi 2 (s_2)	0.300_a^x	0.297_a^y		0.294_b^x	0.297	0.003
	Mukhtar 1 (s_3)	0.295_b^y	0.303_a^x		0.296_b^x	0.298	
	HSD (a) α 0,05			0.003			HSD (m)
	Arbuscular Mycorrhizae						α 0.05
	0 g (m0)	0.289_b^y	0.293_b^y		0.293_a^y	0.292	
	3 g (m1)	0.296_b^x	0.293_b^y		0.303_a^x	0.297	0.005
	6 g (m2)	0.292_a^x	0.301_a^x		0.300_a^x	0.298	
	9 g (m3)	0.289_b^y	0.300_a^x		0.298_a^{xy}	0.296	
	HSD (a) α 0.05			0.003			

The numbers followed by the same letter in the columns (x, y, z) and on the lines (a, b, c, d) are not significant at the 5% confidence level.

3.2. Energy Radiation Absorption and Stomatal Conductance

Statistical analysis showed that seed source, inoculation of *A. chroococcum* and AM treatment had significant effect on light energy absorption at wavelength 746 nm of cocoa seedlings on age of 4 months after treatment (table 2). However, the no interaction effects was found. Clonal type treatment did not affect the leaf stomatal conductance, but inoculation of *A. chroococcum* and AM fungi had significant effect, while treatment interaction was not significant at four months after treatment.

HSD α 0.05 showed that cocoa seedlings with seed sourced from clone Mukhtar 1 had the highest radiation absorption rate and significantly different from the seedlings of clone Sulawesi 1, but not significantly different from the clone Sulawesi 2 seedlings. Inoculation of *A chroococcum* 10^6 CFU ml^{-1} water showed the highest percentage of radiation absorption and significantly different from control seedlings, but not significantly different from the seeds inoculated with *A chroococcum* of 10^3 CFU ml^{-1} water. Inoculation of Arbuscular Mycorrhizae of 6.0 g $tree^{-1}$ showed the highest percentage of radiation absorption and highest stomatal conductance and significantly different from other AM dosages.

3.3. Dry weight of stem and leaves, total roots and lateral roots

Statistical analysis showed that inoculation of *A. chroococcum* had significant effect on dry weight of stem and leaf of cocoa seedlings, but seed source, AM inoculation treatment and interaction between treatments had no significant effect (table 3). Inoculation of *A. chroococcum* and AM significantly affected the total root dry weight of cocoa seedlings, while seed sources and treatment interactions had no significant effect.

Table 2. Mean of Radiation Adsorption at wavelength of 746 nm (%) and stomatal conductance (mmol m^{-2} s^{-1}) of cocoa seedlings at four months after treatment due to effect of inoculation of *A. chroococcum* and Arbuscular Mycorrhizae.

Parameter	Treatment	Arbuscular Mycorrhizae (M)				Mean	HSD (s) α 0.05
		0 g(m0)	3 g(m1)	6 g(m2)	9 g(m3)		
	Seed Source						
Radiation Energy Absorption	Sulawesi 1 (s1)	27.67	30.85	33.98	31.79	31.07x	0.96
	Sulawesi 2 (s2)	30.76	31.74	33.45	33.14	32.27y	
	Mukhtar 1 (s3)	30.03	31.66	34.73	32.76	32.29y	
	A. chroococcum						HSD (a) α 0.05
	0 CFU ml^{-1} (a0)	28.75	29.78	33.95	32.16	31.16x	0.96
	10^3 CFU ml^{-1} (a1)	29.06	31.81	34.86	32.68	32.10xy	
	10^6 CFU ml^{-1} (a2)	30.65	32.65	33.34	32.85	32.37y	
	Rata-rata	29.49 a	31.42b	34.05c	32.56b		
	HSD (m) α 0.05		1.42				
Stomatal Conductance	*A. chroococcum*						HSD (a) α 0.05
	0 CFU.ml^{-1} (a0)	23.76	25.9	28.57	25.52	25.94 y	
	10^3 CFU.ml^{-1} (a1)	25.23	26.17	28.04	26.76	26.55 y	0.83
	10^6 CFU.ml^{-1} (a2)	26.64	29.08	28.75	28.81	28.32 x	
	Rata-rata	25.21 c	27.05 b	28.45a	27.03 b		
	HSD (m) α 0.05		1.22				

The numbers followed by the same letters in the columns (x, y) and on rows (a, b) are not significantly different at the 5% confidence level.

Table 3. Mean dry weight of stems and leaves (g), total roots (g) and lateral root (g) of cocoa seedlings age four months after inoculation treatment of *A chroococcum* and Arbuscular Mycorrhizae

Parameter	*Treatment*	Inoculation of Arbuskular Mycorrhizae (M)				Mean	HSD (a) α 0.05
		0 g(m0)	3g(m1)	6g(m2)	9 g(m3)		
Stem and Leaf	0 CFU.ml^{-1} (a0)	9.53	25.66	32.59	23.26	22.76 b	4.15
	10^3 CFU.ml^{-1} (a1)	18.22	24.14	38.16	23.31	25.96 ab	
	10^6 CFU.ml^{-1} (a2)	19.23	29.13	36.69	28.57	28.41 a	
	Mean	15.66 z	26.31y	35.81x	25.05 y		
	HSD (m) α 0.05		6.11				
Total Root	0 CFU.ml^{-1} (a0)	1.62	4.38	6.13	3.48	3.90^x	0.84
	10^3 CFU.ml^{-1} (a1)	3.06	4.59	6.4	4.67	4.68^{xy}	
	10^6 CFU.ml^{-1} (a2)	2.86	4.23	6.65	5.46	4.80^x	
	Mean	2.51c	4.40 b	6.39 a	4.54 b		
	HSD (m) α0.05		1.24				

The numbers followed by the same letters in the columns (x, y) and on rows (a, b) are not significantly different at the 5% confidence level.

HSD α 0.05 test in table 3 showed that inoculation of *A. chroococcum* 10^6 CFU mL^{-1} water per tree had the highest dry weight of stems and leaves and total of roots weight and differed significantly with control, but not significantly different from inoculation *A. chroococcum* of 10^3 CFU mL^{-1} water per tree. Inoculation of AM 6.0 g $tree^{-1}$ showed the highest average of stems and leaves dry weight and the highest total roots weight and differed significantly from other AM inoculations.

4. Discussion

Growth and development of plants is a function of genetic and environmental factors. The experimental results show that the seed source has significant effect on the observed growth parameters. Seed sources from Sulawesi 1 clone seeds have a lower chlorophyll a, b and total chlorophyll content than the seed sources of clones Mukhtar 1 and Sulawesi 2. This will result in lower absorption of light energy and stomata conductance of plant leaves in cocoa seedlings of Sulawesi 1 clone compared to seedlings originating from clones Mukhtar 1 and Sulawesi 2.

Cocoa plants have a considerable genetic variation in their morphological and physiological properties [11]. In generative propagation of this plant, leaf development occurs in a rhythmic fashion and is relatively unrecognized by environmental factors, so growth is entirely under endogenous control [12], and the assimilate content of the seeds is reflected in seed size [1] but there are intra- and inter-specific variations [13]. The pattern of vegetative plant organ development will greatly determine the rate of further plant growth and plant response to the environment. Thus the size and weight of seeds will determine the next plant growth.

Microbes can act as plant growth promoting agents that produce a variety of growth hormones, as well as the various organic acids needed in the growth of root feathers and are able to improve the biological health of the soil. Plants have different responses to the changing conditions of the Rhizosphere reflected from the plant genotype. The experimental results showed that the seed source of the Mukhtar 1 clone inoculated with *A chroococcum* 10^6 CFU mL^{-1} water showed a better average mean of stem and leaf dry weight, roots and lateral root development compared to Sulawesi 1 and Sulawesi 2 clones.

Azotobacter produces growth hormone and phosphate solvent compounds that have been used as favorable plants bio-inoculant [14]. Inoculation of *A chroococcum* can reflect root development and influence linearly to plant growth as reflected in dry weight loss. Utilization of *Azotobacter* in non-legume plants, effectively promoting growth and has proven useful for woody plants on dry land [15]. *Azotobacter* is able to produce anti-bacterial and anti-fungal compounds, producing vitamins and growth hormones of IAA and GA3 in the rhizosphere so as to improve root growth which will further improve overall growth.

The experimental results showed that inoculation treatment of mycorrhizal fungi had a good effect on growth parameter. Arbuscular mycorrhiza is an association between certain fungi with plant roots by forming a complex interactional fabric. Mycorrhiza is known as soil fungi because its habitat is in the soil and is in the root zone of the plant (rhizosphere). Mycorhrhizae forms a symbiotic relationship with plant roots in the same way as the nodule root bacteria [16]. AM is generally associated with more than 80% mutualism of vascular plants, including cacao plants and plays an important role in sustainable agriculture [17]. AM enhances the ability of plant root systems to absorb mineral nutrients through the expansion of mycelium, plays an important ecological role in plant nutrient uptake [18], increased soil aggregation through the production of hydrophobic glycoproteins (glomalin) exempt from extra-radical hyphae [19]. The results showed that inoculation of CMA 6.0 $g.tree^{-1}$ can increase chlorophyll a and total chlorophyll leaf, light energy absorption and leaf stomatal conductance and total dry weight.

5. Conclusion

- Cocoa seedlings from clones Mukhtar 1 and Sulawesi 2 have better ecophysiological characteristics and growth compared to clone Sulawesi 1.
- Inoculation of *A chroococcum* of 10^6 CFU mL^{-1} water per tree in cocoa seedling seeds can improve the seedling growth.
- Inoculation of Arbuscular Mycorrhizae fungi of 6.0 g $tree^{-1}$ in cocoa seedlings can improve the growth of cocoa seedlings.

References

[1] Nasaruddin 2012 *Efektifitas pemanfaatan A chroococcum dan mikoriza arbuskula terhadap*

pertumbuhan dan ketersediaan hara tanaman kakao (Disertation) (Makassar: Program Pasca Sarajana Universitas Hasanuddin)
[2] Kloepper J W, Leong J, Teintze M and Schroth M N 1980. Enhanced plant-growth by siderophores produced by plant growth-promoting rhizobacteria *Nature* **286** 885–886
[3] Yeole R D and Dube H C 1997 Increased plant growth and yield through seed bacterization *Ind. Phytopathol.* **50** (3) 316-319
[4] Singh T and Dixit S K 2001 *Use of bio-fertilizers for higher crop yield* (New Delhi: Farm InformationUnit, Directorate of Extension MOAGOI Krishi Vistar Bhavan Pusa)
[5] Kizilkaya R 2009 *J. Environ* **30** 73-82
[6] Ruhnayat A 2007 Effect of *Azotobacter*, bat guano and glyricidia compost on the growth of bushy black pepper (*Piper nigrum* L.) Proc. Seminar XIII Persada (Bogor: Fak. Kedokteran Hewan IPB) p 249-252
[7] _________ 1999 *Pemanfaatan Azotobacter dan mikroba pelarut P sebagai sumber hara N dan P pada tanaman lada* (Laporan Teknis Penelitian Balittro Buku II) p 235 – 244
[8] Thangadurai D C and Mohamed H A B 2010 *Mycorrhizal biotechnology* (Florida: CRC Press Taylor & Francis Group)
[9] Kumar V, Solanki A S and Sharma S 2011 *Thai J. of Agr. Sci.* **44** (1) 53-60
[10] Miransari M 2010 Arbuscular Mycorrhiza and Soil Microbes *Biotechnology* eds. Thangadurai D *et al* (New Hampshire: an imprint of Edenbridge Ltd., British Channel Islands Enfield)
[11] Daymond AJ, Hadley P, Machado R C R and Ng E 2002. Canopy characteristics of contrasting clones of cacao (*Theobroma cacao*) *Exp. Agric.* **38** 359-367
[12] Vogel M 1975 Recherche du déterminisme du rythme de croissance du cacoayer *Café, Cacao, Thé* **19** 265-290
[13] Garcia J R and Nicolella G 1985 Correlação entre algumasmedidas dendrométricas, origem genética e produção de frutos em cacaueiros *Rev. Theobroma* **15** 113-124
[14] Zaidi A, Khan MS and Amil M 2003 Interactive effect of rhizotrophic micro organisms on yield and nutrient uptake of chickpea (*Cicer arietinum* L.) *Eur. J. Agron.* **19** (1) 15-21
[15] Bajpai R D 2003 *Non-legumes too can fix their own nitrogen.* Pers. Comm. Department of Agronomy, J.N. Krishi, Vishwa Vidyalaya, Jabalpur, Madhya Pradesh, India.
[16] Siddiqui Z A and Ryota 2011 Mycorrhizal inoculants: progress in inoculant production technology *Microbes and Microbial Technology* eds. I Ahmad *et al* (New york: Springer Science Business MediSpringer)
[17] Siddiqui Z A and Pichtel J 2008. Mycorrhizae: an overview *Mycorrhizae: Sustainable Agriculture and Forestry* eds. Z A Siddiqui *et al* (The Netherlands: Springer Dordrecht) pp 1-35
[18] Cavagnaro T R, Jackson L E, Six J, Ferris H, Goyal S, Asami D and Scow K M 2006 *Plant and Soil* **282** 209-225
[19] Rillig M C and Mummey D L 2006 Mycorrhizas and soil structure *New Phytol.* **171** 41-53

Towards sustainable agricultural production: Growth and production of three varieties of shallot with some various Nitrobacter bio-fertilizer concentrations

Saharuddin, N E Dungga, E Syam'un and A R Amin

Agronomy Department, Faculty of Agriculture, Hasanuddin University, Jl. Perintis Kemerdekaan KM 10, Makassar, 90245, Indonesia.

E-mail: *ndungga@agri.unhas.ac.id*

Abstract. Organic production is important for health and eliminates the risk of chemical residues. Taking part on sustainable agriculture production, this research was conducted in January to March 2017. Located in Palajau Village, in Jeneponto Regency, altitude 120 m above sea level, South Sulawesi. The research was aimed to determine the growth and production of shallot varieties and their interaction with Nitrobacter bio-fertilizer. The study conducted in the form of a two-factor factorial experiment using Randomized Block Design as environmental design. The first factor was varieties (Bima, Bangkok and Tajuk), and the second factor was concentration of Nitrobacter bio-fertilizer; control-without fertilizer, 30 and 60 mL of fertilizer per 3000 mL of water. Results show that the varietal treatment of Tajuk gave a good response to the observed parameters, and the Nitrobacter bio-fertilizer treatment of 60 mL of fertilizer 3000 mL of water gave good results on the number of bulbs per cluster, bulbs and dried bulbs per plot (of tons per ha bulbs). Bangkok variety combined with Nitrobacter bio-fertilizer concentration of 60 mL of fertilizer per 3000 mL of water per plot showed the highest yield of bulb that is 9.13 tons per ha compared with Bima and Tajuk varieties.

1. Introduction

Shallot is one of important food commodities in Indonesia. Production centers have only concentrated on several places, such as Central Java province, East Java, West Java and East Nusa Tenggara. Consumption of shallot for household tends to increase from year over year. National shallot production in 2012 amounted to 964,195 tons, and in 2013 the production increased to 1,010,773 tons. In 2014 the production of shallots continues to increase to 1,233,984 tons. However, in 2015 production decreased to 1,229,184 tons, and in 2016 production increased by 1,427,885 tons [1].

Varieties of shallot that usually grown in the lowlands are Maja Cipanas, Medan, Sumenep, Bima, Bangkok, Headland and Ampenan varieties. Bima variety is very famous for its production, which is high enough to reach 10 ton ha^{-1}. Bima variety has wide and thick leaf shape. It also has Medium round bulb shape is dark red, resistant to rotten leaf disease. Harvest age could range from 55-60 days after planting [2]. Bangkok variety has a good quality and resistant to low temperatures. While Tajuk variety has a round bulb shape with the widest diameter close to the tip of the root [3].

Nitrobacter bio-organic fertilizer is a microorganism that can improve soil structure and restore the damaged soil structure as before. Plants will thrive on fertile soil and will increase crop yields. Biological fertilizers can also reduce fertilizer costs significantly because the application in the field does not require any additional fertilizer. The results showed that Nitrobacter (some kind of liquid organic fertilizer) can improve plant growth better than chemical fertilizers [4].

Nitrobacter bio-organic fertilizer serves as an auto-immune system in plants. Research results from [4], on papaya plants that use biological fertilizer, showed that plants will grow healthier and larger fruit and sweeter taste, less disease resistance, and better and more durable stored fruit. Nitrobacteria as a natural pesticide is able to overcome plant diseases such as *Fusarium, phytophthora, pythium* and all kinds of microorganisms and pathogens in soil and water, causing root rot, wilted leaves, and dead plants [4].

Organic vegetable production is not only important for human health but also eliminates the risk of chemical residues from the environment. It is really important to take part in sustainable agriculture production. For that reason, this paper aims to determine the growth and production of several varieties of onion and the interaction of onion varieties with Nitrobacter biological fertilizers.

2. Material and Methods

This study was conducted in farmers' fields Palajau village, district Arungkeke, Jeneponto regency, with altitude 120 m above sea level, South Sulawesi, started from January to March 2017. The study was conducted in the form of a two-factor factorial experiment using Randomized Block Design arranged in factorial consisting of two factors. The first factor was varieties with 3 types, namely Bima variety (V1), Bangkok variety (V2), Tajuk variety (V3), and the second factor was the concentration of Nitrobacter bio-organic fertilizer (K) with 3 levels, namely control-without fertilizer (K0), 30 mL of fertilizer/3000 mL of water (K1), 60 mL of fertilizer/3000 mL of water (K2). Variable of shallots growth were observed was: plant height, number of tillers per cluster, number of leaves, weight of dried bulbs per plot and tuber production. The data were analyzed by using the analysis of variance (ANOVA) to see whether the application of various Nitrobacter bio-fertilizer concentration, profitable for some shallot variety. A further test was performed using Honestly Significance Difference Test (HSD) if one of the treatments showed significant effect.

3. Results

Factor of of bio-fertilizer giving significant effect on growth of the tuber number per cluster while varieties factor showed the opposite. The effect on the result of the interaction between the two factors shows a significant effect (table 1). In contrast, the total parameters of the number of bulb per cluster, the effect of both factors on the wet bulb weight per plot (kg), the weight of the dry bulb per plot, tuber yield per plot had significant effect on the growth and onion production but the interaction of the two factors on the three parameters observed (table 2, 3 and 4) have no significant effect.

Table 1. The average number of bulbs per cluster of various shallot varieties and Nitrobacter bio-organic fertilizer application

Concentration of Nitrobacter bio-organic fertilizer (mL/3 L water)	Varieties of shallot			Average	HSD
	V1	V2	V3		
K0 (0)	1.42	1.50	1.67	1.53 **y**	0.11
K1 (30)	1.52	1.66	1.71	1.63 **y**	
K2 (60)	1.68	1.83	1.80	1.77 **x**	
Average	1,60 **b**	1,66 **ab**	1,73 **a**		

Note: The numbers followed by the same letter are not significantly different at the 0.05 Honestly Significant Difference (HSD) Test level

Table 2. Average weight of fresh bulb per onion plot (kg) of various varieties of shallot and Nitrobacter bio-organic fertilizer application.

Concentration of Nitrobacter bio-organic fertilizer (mL/3 L water)	Varieties of shallot			Average	HSD
	V1	V2	V3		
K0 (0)	1.42	1.50	1.67	1.53 **y**	0.11
K1 (30)	1.52	1.66	1.71	1.63 **y**	
K2 (60)	1.68	1.83	1.80	1.77 **x**	
Average	1,60 **b**	1,66 **ab**	1,73 **a**		

Note: The numbers followed by the same letter are not significantly different at the 0.05 HSD Test level

Table 3. Average weight of dried bulbs per plot (ton plot^{-1}) on various shallot varieties and Nitrobacter bio-organic fertilizer application.

Concentration of Nitrobacter bio-organic fertilizer (mL/3 L water)	Varieties of shallot			Average	HSD
	V1	V2	V3		
K0 (0)	1.33	1.45	1.58	1.45 **y**	0.13
K1 (30)	1.46	1.54	1.62	1.54 **y**	
K2 (60)	1.62	1.75	1.74	1.70 **x**	
Average	1.47 **b**	1.58 **ab**	1.65 **a**		

Note: The numbers followed by the same letter are not significantly different at the 0.05 HSD Test level

Table 4. Average yield of tuber production (ton ha^{-1}) on various varieties of shallot and Nitrobacter bio-organic fertilizer application.

Concentration of Nitrobacter bio-organic fertilizer (mL/3 L water)	Varieties of shallot			Average	HSD
	V1	V2	V3		
K0 (0)	7.12	7.48	8.37	7.66 **y**	0.53
K1(30)	7.60	8.28	8.55	8.14 **y**	
K2 (60)	8.40	9.13	9.02	8.85 **x**	
Average	7.71 **b**	8.30 **a**	8.64 **a**		

Note: The numbers followed by the same letter are not significantly different at the 0.05 HSD Test level

4. Discussion

4.1. Interaction of Shallot Varieties and Nitrobacter bio-organic fertilizer

The result of statistical analysis showed that the interaction between onion varieties and nitrobacter biological fertilizer had a significant effect on bulbs bulk parameter per cluster. The varieties of Bima onion (V1) combined with nitrobacter bio-fertilizer concentration of 60 mL per 3 L water per plot (K1) produced the highest number of tubers per cluster from treatment (V1K2) by 21.80 tubers per cluster, and the lowest is found in the control treatment that produced 12.70 bulbs number of tubers per cluster. It is suspected that the varietal response is very helpful to the number of tubers and tillers. The results of the research conducted by Kartapradja and Putrasamedja [5] in Sukamandi showed that the number of tillers that can be formed by this interactions are 6-8 tillers, the magnitude of these variations is made possible by the genetic influence of the cultivars being tested. In certain cases, the number of tubers produced by a variety is closely related to the number of leaves because more number of leaves will experience more photosynthesis [6].

4.2. Varieties of Shallot
The experimental results showed that shallot varieties have a real impact on wet tuber weight per plot, weight of dried tuber per plot and tuber yield per ha of shallot, but no significant effect on the number of tubers per cluster. The result of statistical analysis on the growth and production of shallot crops showed that the Tajuk variety gave a significant effect of the increase of wet weight per plot and the weight of the tuber of ton per ha, while Bangkok variety showed the highest number of dry bulb weight per plot. This was presumably because these three varieties have different growth responses. This is in accordance with Gardner et al. [7], that plant growing power and the plant growth rate are strongly influenced by external and internal factors.

Bangkok variety that applied with Nitrobacter bio-fertilizer concentration of 60 mL per 3 L water gave good effect to onion bulb production for 9.13 ton per ha and has reached productivity potency of Bangkok variety in general which is 9.4 ton per ha . Biological fertilizers can improve yields, improve the quality, increase efficiency and reduce the dosage of artificial fertilizer, improve the physical structure, chemistry, and biology of the soil, suppress pest and disease attacks, make the balance of flora and fauna in the soil and created good environment on agricultural cultivation [8].

4.3. Nitrobacter Bio-organic Fertilizer
The result of statistical analysis showed that the application of bio-organic fertilizer had no significant effect on the parameters of plant height, number of leaves, tuber diameter, fresh bulb weight per sample, and weight of dried tuber per sample. Presumably, the Nitrobacter bio-organic fertilizer given was slowly available to the plant, therefore the plant has not given a significant response. As well as the existing bacteria in the nitrobacter bio-organic fertilizer can not symbiosis well on the soil causing the lack of biological fertilizer influence on plant growth. This is in accordance with Damanik *et al.*, [9] which found that the difference of biological fertilizer with chemical fertilizer is slowly response of the plants, the supply of indirect nutrients, the environmental impact is absent so that the biological fertilizer is slowly available for the plant, therefore the plant has not responded have a real impact. As well as existing bacteria in the biological fertilizer, had lack of influence on plant growth because the application of biological fertilizer in the soil can be washed out due to the occurrence of extreme weather.

5. Conclusions

- Varieties of shallot that had good effect on growth and production on observed parameters which were wet bulbs weight per plot, weight of dried bulb per plot and production of ton per ha.
- Nitrobacter bio-organic fertilizer with concentration of 60 mL per 3 L water gave good influence on observed parameters which were bulbs per cluster, the weight of wet bulb per plot, weight of dried bulbs per plot and production of ton per ha.
- Interaction between Bangkok variety and application of Nitrobacter bio-organic fertilizer with concentration of 60 mL per 3 L water per plot produced the highest production by 9.13 tons per ha.

6. Suggestion

Based on the results of the research, it is necessary to develop the planting of Bangkok varieties with concentration of 60 mL 3 litre water per plot.

References

[1] Central Bureau of Statistics and Directorate General of Horticulture 2016 Production of Red Onion by Province
[2] de Putter H and Adiyoga W 2013 *Improving the Shallot and Hot Pepper Cultivation System in the Coastal Plain of Northern Java vegIMPACT Report 1* (The Netherlands: Wageningen UR)
[3] Priyantono E 2016. *Changes in Quality of Red Onions (Allium Ascalonicum L. Bima Brebes*

Varieties, Headlines And Bali Rubber Saved At Low Temperatures Thesis (Bogor: Graduate School Bogor Agricultural University)

[4] Janardi T 2012 *Nitrobacteric spray soil* http: //www.Nitrobakter.net .id / ind / kampung-media-ginte.htm Accessed May 11, 2016

[5] Kartapradja R and Putrasamedja S 1990 Experiment of onion varieties in Sukamandi *Hort. Res. Bull.* **18** (1) 57-60

[6] Carora A F, Wicaksono K P and Heddy Y B S 2014 The effect of application bioactivator on the growth and yield of shallot *(Allium ascolanium* L.*) J. Prod. Tan.* **2** (5) 434-442

[7] Gardner F P, Pearce R B and Mitchell R I 2013 *Physiology of Crop Plants* (New Delhi: Scientific Publisher) p 327

[8] Chalimah S, Muhadiono, Aznam L, Haran S and Nurita T 2007 *Propagation of Gigaspora sp and Acaulospora sp with Pot Culture at Rumah Kaca* (Bogor: Institut Pertanian Bogor)

[9] Damanik M M B D, Hasibuan B E, Fauzi, Sarifuddin and Hamidah H 2011 *Soil Fertility and Fertilization* (Medan: USU Press)

Advanced yield potential test on synthetic genotype of maize tolerant to drought and low nitrogen

Y Musa and M Farid

Department of Agronomy, Faculty of Agriculture, Hasanuddin University, Jalan Perintis Kemerdekaan KM 10, 90245, Makassar, Indonesia.

E-mail: *yunusmusa@yahoo.com*

Abstract. Increased productivity of maize with intensification can be achieved through the release of tolerant varieties to drought and or low nitrogen conditions while supporting extensification to marginal land. During 2013-2015, previous studies had obtained three candidates of maize new superior varieties (NSV) tolerant to drought and low nitrogen (N) (Syn2-4, Syn2-8, and Syn2-16); and six varieties tolerant to drought or low N (Syn2-1, Syn2-2, Syn2-4, Syn2-8, Syn2-15, and Syn2-16). The potential yield of these genotypes under drought or low N conditions was 6.5 t Ha^{-1}. To examine the potential yield of these 6 genotypes, advanced test was conducted using three maize varieties (Lamuru, Sukmaraga and Bisma) as comparison. The results show that the genotypes of synthetic maize which gave the highest and better yield potential compared to the three comparators varieties of Lamuru, Sukmaraga and Bisma were Syn2-4 (8.13 t Ha^{-1}), Syn2-15 (8.21 t Ha^{-1}), and Syn2-16 (9.23 t Ha^{-1}). While the other three genotypes only gave higher yields than the varieties of Bisma and Lamuru with potential yield of Syn2-1, Syn2-2 and Syn2-8 were 6.74, 7.06 and 7.78 t Ha^{-1}, respectively. Increased production was more dominant due to the difference in the number of cob at harvest and weight of 1000 seeds.

1. Introduction

Corn demand constantly increases every year with population and with the progress of food processing technology industries and poultry farms use corn as feeds. During the period of 2011-2015 the average export volume was 23.96 thousand tons, while import volume was much higher at 2.50 million tons. Negative balance between these volumes are indicated by much smaller export compare to import with deficit of averagely 2 million tons more in import volumes from year 2011 to 2015 [1]. This shows that the dependence of imported maize will increase in recent years. Indonesian corn production in 2011-2015 period has been fluctuated. The lowest production was 17.64 million tons in 2011 and increased in 2014 by 19.0 million tons. An increase in production growth rate of 3.18% in 2015 resulted in production reached 19.61 million tons with an average productivity of 5.18 t Ha^{-1} [2]. To fulfill the requirement, use of synthetic corn variety which is tolerant to drought stress and low N fertilization with high productivity and acceptable to farmer is necessary. The synthetic corn variety can support a larger corn crop and involves less-capitalized farmers, thus contributing to increased production toward sustainable self-sufficiency in maize.

Selection activities require information regarding the morphology and physiology of corn crops tolerant to low drought and N stress, so that the selection process can be carried out efficiently and effectively. The reported morphological and physiological characteristics associated with tolerant

IOP Conf. Series: Earth and Environmental Science **157** (2018) 012016 doi:10.1088/1755-1315/157/1/012016

properties of maize plant to drought stress include deep root system [3, 4, 5, 6, 7, 8], water use efficiency, rate of water loss through transpiration, stomatal density [6], and ability to protect chloroplast apparatus and senescence [9, 10, 11], accumulation of proline [12], leaf rolling sensitivity [13], interval timing of male and female flowers (anthesis interval silking or ASI) and male flower size [4]. Based on a series of studies that have been conducted [14, 15] it was obtained three candidates of superior varieties of corn tolerant to drought and low nitrogen (Syn2-4, Syn2-8, and Syn2-16); and six candidates tolerant to drought or low N (Syn2-1, Syn2-2, Syn2-4, Syn2-8, Syn2-15, and Syn2-16). The potential yield of these candidates was 8 t Ha^{-1} under optimum conditions and 6.5 t Ha^{-1} in drought or low N conditions. A further study needs to be undertaken to study the potential yield of the 6 synthetic maize genotypes compared with three varieties (Lamuru, Sukmaraga and Bisma).

2. Methodology

The study was conducted in Maros Regency, South Sulawesi from June to September 2017 as experimental study set in a randomized block design with three replications. The treatment consists of six genotypes of synthetic corn, namely Syn 2-3, Syn 2-4, Syn 2-5, Syn 2-6, Syn 2-7, Syn 2-8, and 3 comparison varieties i.e. Bisma, Lamuru, and Sukmaraga. Maize were planted in 5 m x 3 m plot using plant spacing of 70 cm x 20 cm. Plots were applied with Urea, SP36 and KCl 300 fertilizers with dosages of 300, 200 and 100 kg Ha^{-1}, respectively. Urea and KCl fertilizers were given twice at the age of 7 and 35 days after planting (DAP), while SP36 was applied at 7 DAP. 10% of the total population were taken as samples with parameters observed were leaf angle, chlorophyll content using Chlorophyll Meter (calculated according to [16], weights of cob, yield, number of harvest crop, number of cob harvest, cob harvest weight and productivity at moisture content of 14% using Grain Moisture Tester PM-410. Data was analysed using analysis of variance followed by Least Significance Difference (LSD) test at $p<0.05$. To determine the relation between growth characters and production, regression and correlation analysis were conducted while heritability in the broad sense using the formula of Syukur *et al.* [17]:

$$KKG = \frac{\sqrt{\sigma^2 g}}{\bar{x}} X\ 100\%$$

Heritability values are grouped according to Bahar and Zen [18] *i.e.* low (<0.2), medium (0.2-0.5) and high (>0.5).

3. Results and Discussion

Table 1 shows that the leaf angle of genotype 3 (Syn 2-5), 5 (Syn 2-7) and 6 (Syn 2-8) showed the highest leaf angle and differ significantly ($p<0.05$) with control genotypes of Lamuru and Sukmaraga (g8 and g9). High angle of the leaf shows upright canopies that are parallel to the sunlight so that the sunlight received by the leaves becomes the optimum compared to the low leaf angle that indicates overlapping leaves results in lower light reception. This is in accordance with Kartahadimaja and Eka [20] stating that plants with narrow leaves and leaf shrubs smaller than 60° will receive less light when compared to more than 60° leaf angle. According Sumajow *et al.* [21], an upright leaf angle is a character favoured by corn farmers. The use of sunlight will be more effective and efficient, result in better photosynthesis activities. Narrow leaf corner character with upright leaf growth type can also increase plant population in unit of plantation area, this is because spacing can be used more closely by not disturbing penetration of light reaching to body part of the plant.

Higher radiation received by the leaves will trigger the optimum process of photosynthesis to produce assimilates that can be distributed to the generative organ to support the high production of corn crops. The weight of cob is influenced by the distribution of assimilates from the photosynthesis process and the availability of water and nutrients available in the soil, the narrow spacing apart from the non-optimum sunlight by the leaves can also detain root development due to nutrient competition causing lack of nutrients for corn so the most prominent impact in this case is the leaves due to lack of raw materials to make leaf chlorophyll so that with the lack of chlorophyll content in the leaves will affect the amount of assimilates produced to be distributed for the formation of seeds and seed filling. This is in line with the opinion of Sumajow *et al.* [21] which states that drought and nutrient deficiency greatly affect the growth and development of cobs, and will even decrease the number of seeds in one cob because of the shrinking of the cob, which consequently decreases the yield.

Table 1. Average leaf angle (°), chlorophyll content (mg cm^{-2}), cob weight (g) and rendemen (%) in some synthetic maize genotypes.

Genotype	Leaf Angle (°)	Chlorophyll Content (mg cm^{-2})	Cob Weight (g)	Rendemen (%)
g1 (*Syn 2-3*)	57.87^{abc}	0.3061	200.20^{bc}	77.23^{a}
g2 (*Syn 2-4*)	59.93^{ab}	0.3048	216.60^{ab}	76.20^{abc}
g3 (*Syn 2-5*)	60.13^{ab}	0.3060	217.93^{ab}	73.84^{d}
g4 (*Syn 2-6*)	60.20^{ab}	0.3086	199.00^{bc}	77.06^{ab}
g5 (*Syn 2-7*)	60.00^{ab}	0.3062	199.47^{bc}	76,37^{abc}
g6 (*Syn 2-8*)	62.67^{a}	0.3044	217.27^{ab}	77.01^{abc}
g7 (Bisma)	59.27^{ab}	0.3055	185.80^{c}	75.02^{cd}
g8 (Lamuru)	55.87^{bc}	0.3054	214.60^{ab}	73.95^{d}
g9 (Sukmaraga)	53.67^{c}	0.3048	234.07^{a}	75.11^{bcd}
LSD $_{0.05}$	4.84	ns	19.99	2.02

Values followed by same letter in the same column are not significantly different ($p<0.05$).

Table 2. Average number of harvested crops (plant plot^{-1}), number of cobs at harvest (cob plot^{-1}), cob weight at harvest (kg plot^{-1}) and productivity (ton Ha^{-1}) of some synthetic maize genotypes.

Genotype	Number of harvested crops (plant plot^{-1})	Number of cobs at harvest (cob plot^{-1})	Cob weight at harvest (kg plot^{-1})	Productivity (ton Ha^{-1})
g1 (*Syn 2-3*)	67.67^{cd}	76.33^{bc}	13.86^{c}	6.74^{bc}
g2 (*Syn 2-4*)	69.00^{cd}	78.00^{abc}	15.16^{b}	7.06^{bc}
g3 (*Syn 2-5*)	70.00^{bcd}	88.00^{ab}	18.46^{ab}	8.13^{ab}
g4 (*Syn 2-6*)	87.00^{a}	93.33^{ab}	16.35^{abcd}	7.78^{ab}
g5 (*Syn 2-7*)	85.67^{ab}	93.33^{ab}	17.85^{abc}	8.21^{ab}
g6 (*Syn 2-8*)	80.33^{abc}	96.33^{a}	20.22^{a}	9.23^{a}
g7 (Bisma)	81.33^{abc}	88.33^{ab}	12.36^{d}	5.63^{c}
g8 (Lamuru)	62.00^{d}	62.00^{c}	12.71^{d}	5.79^{c}
g9 (Sukmaraga)	77.00^{abcd}	81.67^{ab}	17.78^{abc}	7.87^{ab}
LSD $_{0.05}$	16.14	19.09	4.05	1.67

Values followed by same letter in the same column are not significantly different ($p< 0.05$).

Table 2 shows the genotypes of Syn 2-5, Syn 2-7 and Syn 2-8 gave the highest potential on the number of harvest cobs, the weight of the cobs and the productivity. This suggests that these genotypes showed good potential to be developed when compared to the comparative genotypes. According to Syukur *et al.* [17], the quantitative character of the plant is affected by a number of genes, each having a small contribution to its phenotypic appearance. There is an interaction between genes and the environment concerning factors in plant cells.

Genotypes Syn 2-5, Syn 2-7 and Syn 2-8 which are the origin of the count of elders who have extensive genetic viability with the Balance Composite method produce a genotype superior to the elder. This is in agreement with Budiarti [22] which states that the development of improved varieties is geared towards obtaining high yield potential, resistant or tolerant to biotic and abiotic stress, high yield quality, according to consumer tastes and market demand. This is in line with the objectives of plant breeding in Indonesia prioritized to increase genetic yield potential, shorten plant life, improve plant resistance to important diseases, such as rust, leaf blight, virus and nematodes, improve resistance to important pests, such as bean flies and pest sucking pests, improving crop tolerance to physical environmental stresses, such as low pH, dryness, shade and improved seed quality especially colour, size and quality of saving [23]. Longer selection is done in the hope of higher potential than previous elders.

Table 3 shows that characters having wide genetic diversity coefficients and high heritability values were harvest cob weight and productivity. This indicates a potential character enhancement that developed in the next generation. High heritability values show a more significant genetic effect when compared with the environment. This is in agreement with Martono [24] which suggests that extensive genetic diversity suggests a more dominant genetic influence than environmental influences. In addition, Hikam [25] states that the greater the genetic diversity present in a crop population the easier it is for breeders to select the best desired genotypes.

Table 3. Genetic diversity coefficient analysis and heritability values.

No	Character	Genetic Diversity Coefficient		Heritability	
1	Leaf angle	3.63	Narrow	63.65	High
2	Chlorophyll content	0.18	Wide	20.11	Medium
3	Cob weight	6.12	Narrow	78.70	High
4	Rendemen	2.25	Wide	48.66	Medium
5	Number of harvested plants	9.11	Narrow	62.06	High
6	Number of cob at harvest	10.82	Narrow	65.67	High
7	Weight of cob at harvest	14.80	Wide	75.65	High
8	Productivity	14.10	Wide	77.78	High

4. Conclussion

Genotypes Syn 2-8, Syn 2-7, and Syn 2-5 are genotypes that have better potency with productivity for each genotypes were 9.23, 8.21 and 7.78 ton Ha^{-1}, respectively. Characters that have wide genetic diversity coefficients and high heritability are harvest cob weight and productivity. Characters show wide genetic diversity coefficient were chlorophyll content, yield, harvest cob weight and productivity.

References

[1] Indonesian Bureau of Statistics (IBS) 2015 *Data Produksi Tanaman Pangan dan Hortikultura*. (Jakarta:Pusat Data Statistik Pertanian)
[2] Indonesian Bureau of Statistics (IBS) 2016 *Data Produksi Tanaman Pangan dan Hortikultura*. (Jakarta:Pusat Data Statistik Pertanian)
[3] Grzesiak S, Hura T, Grzesiak M T and Pienkowski S 1999 The impact of limited soil moisture and waterlogging stress conditions on morphological and anatomical root traits in maize (*Zea mays* L.) hybrids of different drought tolerance *Acta Physiol. Plant.* **21** (3) 305-315
[4] Bänziger M, Edmeades G O, Beck D and Bellon M 2000 *Breeding for Drought and Nitrogen Stress Tolerance in Maize From Theory to Practice* (Mexico: CIMMYT)
[5] Bohn M, Novais J, Fonseca R, Tuberosa R and Grift T E 2006 Genetic evaluation of root complexity in maize *Acta Agro. Hungarica* **54** (3) 1-13
[6] Blum A 2005 Drought resistance, water-use efficiency, and yield potential: are they compatible, dissonant, or mutually exclusive? *Aus. Agri. Res.* **56** 1159–1168
[7] Vadez V, Krishnamurthy L, Kashiwagi J, Kholova J, Devi J M and Sharma K K 2007 Exploiting the functionality of root systems for dry, saline, and nutrient deficient environments in a

changing climate *J. of SAT Agr. Res.* **4** 1-61

[8] Trachsel S 2009 *Genetic analysis of root morphology and growth of tropical maize and their role in tolerance to desiccation, aluminum toxicity and high temperature* (Switzerland: ETH Zurich) p 113

[9] Prochazkova D, Sairam R K, Srivasta G C and Singh D V 2001 Oxidative stress and antioxidant activity as the basis of senescence in maize leaves *Plant Sci.* **161** 765-771

[10] Mittler R 2002 Oxidative stress, antioxidants and stress tolerance *Trends Plant Sci.* **50** 405–410

[11] Messmer R, Fracheboud Y, Bänziger M, Stamp P and Ribaut J M 2011 Drought stress and tropical maize: QTLs for leaf greenness, plant senescence, and root capacitance *Field Crops Res.* **124** 93–103

[12] Moussa H R and Abdel-Aziz S M 2008 Comparative response of drought tolerant and drought sensitive maize genotypes to water stress *Aus. J. Crop Sci.* **1** (1) 31-36

[13] Edmeades G O, Bolan˜ os J, Chapman S C, Lafitte H R and Banziger M 1999 Selection improves drought tolerance in tropical maize populations: I. Gains in biomass, grain yield, and harvest index *Crop Sci.* **39** 1306–1315

[14] Efendi R, Sunarti S, Musa Y, Farid M, Rahim D and Azrai M 2015 Selection of homozygosity and genetic diversity of maize inbred using simple sequence repeats (SSRs) marker *Int. J of Curr. Res. in Biosci. and Plant Bio.* **2** (3) 19-28

[15] Efend R, Musa Y, Farid M, Rahim D, Azrai M and Pabendon M 2015 Seleksi jagung inbrida dengan marka molekuler dan toleransinya terhadap kekeringan dan nitrogen rendah. *J. Pen. Pert. Tan. Pangan* **34** 43-54

[16] Silla F, Gonzalez-Gil A, Gonzalez-Molina M E, Mediavilla S and Escudero A 2010 *Estimation of Chlorophyll Using a Portable Chlorophyll Meter* (*For Sci.*) (Spain: Universidad de Salamanca) p 67-108

[17] Syukur M, Sujiprihati S and Rahmi Y 2012 *Teknik Pemuliaan Tanaman* (Bogor: Penebar Swadaya)

[18] Bahar H and Zen S 1993 Parameter genetik pertumbuhan tanaman, hasil dan komponen hasil jagung *Zuriat* **4** 4-7

[19] Atma W A, Hana D R, Adi O R H, Meddy R and Agung K 2015 Penampilan karakter agronomi 16 genotip kedelai (*Glycine max* L. Merrill) pada pertanaman tumpangsari dengan jagung (*Zea mays* L.) pola 3:1 *J. Agro* **2** (2)

[20] Kartahadimaja J and Eka E S 2013 Penampilan karakter fenotipik 15 galur inbred jagung selfing ke-14 (S-14) rakitan polinela *J. Agrotrop.* **18** (2) 46-51

[21] Sumajow A Y M, Johannes E X R and Selvie T 2016 Pengaruh pemangkasan daun bagian bawah terhadap produksi jagung manis (*Zea mays var. saccharata Sturt*) *J. ASE* **12** (1A)

[22] Budiarti S G 2001 Skrining plasma nutfah jagung terhadap kekeringan *Bul. Agron.* **29** (1) 19-22

[23] Marame F L, Desalegne, Harjit-Singh, Fininsa C and Sigvald R 2008 Genetic components and heritability of yield and yield related traits in hot pepper *Res. J. Agric. & Biol. Sci.* **4** (6) 803-809

[24] Martono B 2009 Keragaman Genetik, Heritabilitas dan korelasi antar karakter kuantitatif nilam (*Pogostemon* sp.) hasil fusi protoplas *J. Littri* **15** (1) 9-15

[25] Hikam S 2010 *Teknik Perancangan dan Analisis Pemuliaan Tanaman Bahan Kuliah TPAPT* (Bandar Lampung: Fakultas Pertanian. Universitas Lampung)

IOP Conf. Series: Earth and Environmental Science **157** (2018) 012017 doi:10.1088/1755-1315/157/1/012017

Shading effect on generative characters of upland red rice of Southeast Sulawesi, Indonesia

Muhidin[1], E Syam'un[2], Kaimuddin[2], Y Musa[2], G R Sadimantara[1], Usman[1], S Leomo[1] and T C Rakian[1]

[1]Department of Agrotechnology, Faculty of Agriculture, Universitas Halu Oleo, Kendari, Southeast Sulawesi, 93212 Indonesia

[2]Department of Agronomy, Faculty of Agriculture, Hasanuddin University, Makassar, South Sulawesi, 90245 Indonesia

E-mail: *muhidinunhalu@gmail.com*

Abstract. Upland red rice (*Oryza sativa*) contains anthocyanin, a phenolic compounds that can act as antioxidants and functional food for human dietary. The content of functional food on upland red rice is influenced by shading condition, but the production is also influenced by environmental condition, especially the availability of light. The study aims is to assess and analyze the effect of shade on the growth and production of upland red rice. The research was conducted using the quantitative method to obtain the optimal shading condition that can increase the rice anthocyanin content and relatively high production. The research was arranged in split plot design, with shade as main plot and the different of cultivar as sub plot with three replications. The shading treatment consist of 4 levels as follows: n1=shade level < 25 %, n2=shade level 25-50%, n3=shade level 50-75% and n4=shade level > 75%. The cultivar tested were (v_1) = *Labandiri*, (v_2) = *Jangkobembe*, (v_3) = *Ranggohitam,* and (v_4) = *Paedara.* The rice planted in between teak wood trees with different age and level of canopy. The research reveals that shades had an effect in decreasing plant production (the higher level of shade, the higher the decrease level of production), but the shades can improve the quality of red rice through the increase of anthocyanin content.

1. Introduction

Rice is very important staple food in Indonesia. Demand for rice as a staple food continuously increase from year to year, due to high population growth and changes in consumption patterns [1]. To achieve stable production conditions, the increase in rice production must be at least in line with the growth of population [2]. Indonesia together with China, India, Bangladesh and Vietnam are major rice producers [3, 4], but Indonesia is also the main rice consumer [5].

Various efforts have been done to increase rice production. The main strategy in maintaining rice sustainability is through the increasing rice production [3, 6] and lowering rice consumption while promoting local food such as sago [7]. Upland rice can be developed on sub-optimal land, either on (1) dry land through the development of upland rice-based farming or (2) as an insertion plant under plantation stands or forestry crops in intercropping or agroforestry patterns [5, 8, 9, 10, 11].

The development of plant under the stand of plantation crop or forestry will be limited by solar radiation that will in turn affect growth and production [12, 13, 14, 15]. Low light intensity may cause decrease in performance of generative character but can affect in the increasing of anthocyanin levels

[16]. Therefore, the research is very important to get right combination between anthocyanin content increase with the least growth and rice production decrease to compensate the increase.

2. Material and Method

This research was conducted in Teak Wood Botanical Garden Halu Oleo University from April to October 2014. The research conducted in split plot design, with shade as main plot and the different of cultivar as sub plot, with three replication. The shading treatment consisted of 4 levels as follows: n1=shade level < 25 %, n2=shade level 25-50 %, n3=shade level 50-75% and n4= shade level >75%. The rice planted in between teak wood trees with different age and level of canopy. The cultivar tested were (v_1) = *Labandiri*, (v_2) = *Jangkobembe*, (v_3) = *Ranggohitam*, and (v_4) = *Paedara*. Several generative character included number of productive tiller, filled grain content, yield and anthocyanin content were examined. Statistical analyses were performed using two ways of analysis of variances (ANOVA) using the Statistical Package of Social Sciences (SPSS) program version 20 for Windows (Chicago, IL, USA). If the test result showed a significant difference, then tests of treatment differences were performed using Duncan's multiple range test (DMRT) at α=0.05.

3. Results and Discussion

3.1. Productive Tiller

The result showed that shade level and cultivar have significantly affected number of productive tillers and there is no interaction between shade and cultivar on productive tillers. Generally the productive tillers decreased with higher shade levels and this occurred in all four cultivars tested (table 1).

Table 1. Effect of shade and cultivar on the productive tillers and percentage relative comparison (PRC) [1]

Shade treatment	Number of productive tiller				Percentage relative comparison (PRC)			
	v_1	v_2	v_3	v_4				
n_1	11.49 a p	10.51 a q	9.88 a r	10.00 a qr	100.00	100.00	100.00	100.00
n_2	7.88 b r	7.74 b r	10.03 a p	8.72 b q	67.59	73.56	97.89	76.88
n_3	5.75 c qr	5.34 c r	7.20 b p	6.19 c q	48.66	55.88	67.03	74.76
n_4	4.45 d q	4.34 d q	4.80 c pq	5.24 d p	37.85	40.21	48.93	52.62

* Number followed by the same index in the same row, are not significantly different at Duncan's multiple Range Test (DMRT).

** n_1 = shade level < 25 %; n_2 = shade level 25-50 %; n_3 = shade level 50-75 % and n_4 = shade level > 75%. v_1=Cultivar *Labandiri;* v_2=Cultivar *Jangkobembe;* v_3=Cultivar *Ranggohitam;* v_4=Cultivar *Paedara*

Based on table 1 it is apparent that the highest number of productive tillers was found in *Labandiri* cultivars (11.49 tillers), followed by *Jangkobembe* cultivar (10.51 tillers), *Paedara* cultivar (10.00 tillers) and *Ranggohitam* cultivar (9.88 tillers) at the shade level under 25%. Based on the percentage relative comparison (PRC), using a light shade treatment (n_1) as a control, it appears that the higher shade level, the smaller the number of productive tiller and it as also indicated by the smaller number PRC.

3.2. Percentage Filled Grain and Yield

The result showed that at the shade level under 25%, the highest percentage of filled grain was found in *Labandiri* cultivar (93.4%), followed by *Ranggohitam* (93.2%), *Jangkobembe* (90.6%) and the lowest in *Paedara* (73.9%) as seen on table 2. The highest production found on *Jangkobembe* cultivar (4.11 t ha^{-1}), followed by *Ranggohitam* (3.74 t ha^{-1}), *Labandiri* (3.63 t ha^{-1}) and *Paedara* (1.91 t ha^{-1}).

Shading reduced the percentage of filled grain where the highest reduction occurred at *Paedara* cultivar with PRC 73.34%, followed by *Labandiri* cultivar (PRC 89.51%), *Ranggohitam* cultivar (PRC 89.5%) and *Jangkobembe* cultivar with PRC value of 90.18%.

Table 2. Effect of shade and cultivar on the percentage filled grain, production and percentage relative comparison (PRC) [1]

Shade treatment	Percentage filled grain				Percentage relative comparison (PRC)			
	v_1	v_2	v_3	v_4				
n_1	93.4	90.6	93.2	73.9	100.0	100.0	100.0	100.0
n_2	83.6	81.7	83.4	54.2	89.51	90.18	89.48	73.34
n_3	66.4	47.0	54.6	48.8	71.09	51.87	58.58	66.04
n_4	42.4	38.5	37.8	28.2	45.40	42.49	40.56	38.16
	Production (ton ha^{-1})				Percentage relative comparison (PRC)			
	v_1	v_2	v_3	v_4				
n_1	3.63	4.11	3.74	1. 91	100.0	100.0	100.0	100.0
n_2	2.34	2.60	2.92	0.94	64.4	63.26	78.07	49.11
n_3	1.28	0.97	0.83	0.66	35.26	23.60	22.19	34.55
n_4	0.49	0.51	0.34	0.30	13.49	12.40	9.09	15.71

* Number followed by the same index in the same row, are not significantly different at Duncan's multiple Range Test (DMRT).

** n_1 = shade level < 25 %; n_2 = shade level 25-50 %; n_3 = shade level 50-75 % and n_4 = shade level > 75%. v_1=Cultivar *Labandiri;* v_2=Cultivar *Jangkobembe;* v_3=Cultivar *Ranggohitam;* v_4=Cultivar *Paedara*

Base on PRC at higher shade levels (n_3 and n_4), there was a decrease in the percentage of filled grain drastically by nearly half, except in the *Labandiri* cultivar with PRC of 71.09%. It appears that the percentage filled grain decreases with the higher shade level. Further results in Table 2 showed that the percentage filled grain decreases due to increases in shade level. On yield parameter, a similar pattern also found. Based on Table 2, in the shade level 25-50% (n_2), it is apparent that the highest crop production founded in *Ranggohitam* cultivar (v_3) with grain production of 2.92 ton ha^{-1}, then followed by *Jangkobembe* cultivar (v_2) of 3.74 ton ha^{-1}, *Labandiri* (v_1) 2.34 ton ha^{-1} and *Paedara* (v_4) of 1.91 ton ha^{-1}. Based on percentage relative comparison (PRC) of filled grain, it appears that in medium shade level (n_2), grain production decreased and the biggest percentage decrease occurred in *Paedara* cultivar with PCR value 49.1%, and then followed by *Jangkobembe* cultivar with PRC value 63.3%, *Labandiri* with PRC relatively 64.4% and the smallest decrease in *Ranggohitam* with PRC value of 78.1%.

3.3. Anthocyanin Content

Shading increase the anthocyanin levels of upland rice and the level of increase depends on the type of cultivars (figure 1). The highest increase of anthocyanin levels due to shade occurred in *Jangkobembe* cultivar. For the other three cultivars, the anthocyanin content also increased as an effect of shading, but the increase was not as large as in the *Jangkobembe* cultivar. The increase anthocyanin levels as a result of shade level also occurred in the *Labandiri* cultivars, followed by *Ranggohitam* and *Paedara* cultivars.

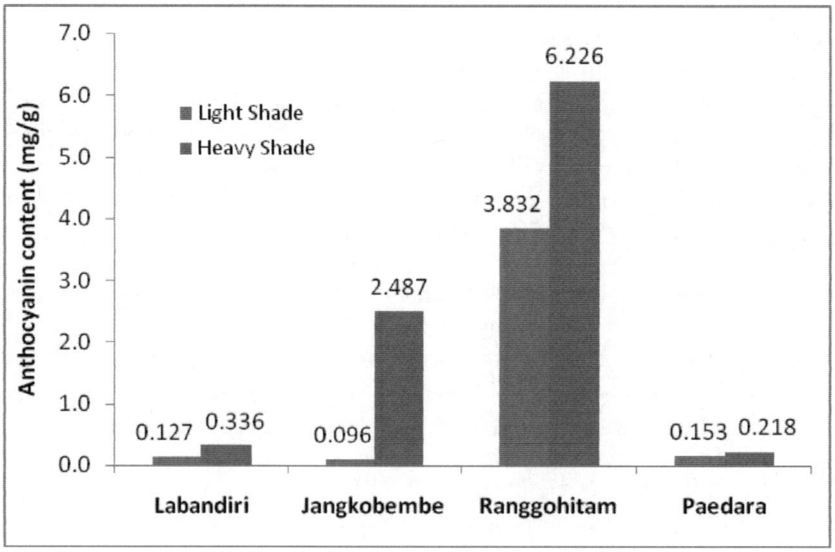

Figure 1. The effect of shade on the anthocyanin content of upland rice

4. Conclusion
The results showed that shade might decrease the growth and production of rice, but at the same time the anthocyanin content increase in line with the increasing of shade level. The shade level decrease the productive tiller and yield on *Labandiri, Jangkobembe* and *Paedara* cultivar. At the same time, the shade increase anthocyanin content on *Jangkobembe* and *Ranggohitam* cultivar. The best cultivar for yield and anthocyanin under shade condition found in *Ranggohitam* cultivar.

References
[1] Muhidin 2015 *Morphological Characterisation and the Quality of Red Rice Irradiated with Gamma Rays in Various Kinds of Shade and the Application of Bacterial Fixation in Dry Land in Southeast Sulawesi* (Makassar: Disertation on Postgraduate Program Hasanuddin University Makassar)

[2] Syaiful S A, Syam'un E, Dachlan A, Jusoff K and Haerani N 2013 The effect of inoculating nitrogen fixing bacteria on production of rice *World App. Sci. J.* **26** (26) 94-99

[3] Sutariati G A K, Khaeruni A, Pasolon Y B, Muhidin and Mudi L 2016 The effect of seed bio-invigoration using indigenous rhizobacteria to improve viability and vigor of upland rice (*Oryza sativa* L.) seeds *Int. J. PharmTech Res.* **9** (12) 565-73

[4] FAOSTAT 2014 *Food and Agricultural Commodities Production/Countries by Commodity* (Rome : Food and Agriculture Organization of the United Nations Statistics Division)

[5] Muhidin, Jusoff K, Syam'un E, Musa Y, Kaimuddin, Meisanti, Sadimantara G R, Baka L R 2013 The development of upland red rice under shade trees *World App. Sci. J.* **24** (1) 23-30

[6] Kadidaa B, Sadimantara G R, Suaib, Safuan L O and Muhidin 2017 Genetic diversity of local upland rice (*Oryza sativa* L) genotypes based on agronomic traits and yield potential in marginal land of North Buton Indonesia *Asian J. of Crop Sci.* **9** (4) 109-117

[7] Muhidin, Leomo S, Alam S and Wijayanto T 2016 Comparative studies on different agroecosystem base on soil physicochemical properties to development of sago palm on dryland *Int. J. of ChemTech Res.* **9** (8) 511-18

[8] Sasmita P, Purwoko B S, Sujiprihati S, Hanarida I, Dewi I S and Chozin M A 2006 Growth and production evaluation of shade tolerant doubled haploid lines of upland rice in an

intercropping system *Bul. Agron* **34** (2) 79 – 86

[9] Kisman, Khumaida N, Trikoesoemaningtyas, Sobir and Sopandie D 2007 Leaf morpho-physiological, markers for adaptation of soybean to low light intensity *Bul. Agron* **35** (2) 96–102

[10] Ginting J, Sengli B, Damanik J, Sitanggang J M and Muluk C 2015 Effect of shade, organic materials and varieties on growth and production of upland rice *Int. J. Scien and Tech. Res.* **4** (1) 68-74

[11] Sopandie D, Chozin M A, Sastrosumarjo S, Juhaeti T and Sahardi 2003 Toleransi padi gogo terhadap naungan *Hayati* **10** (2) 71-75

[12] Las I 1982 *Efisiensi Radiasi Surya dan Pengaruh Naungan Fisik terhadap Padi Gogo* (Bogor : Tesis Fakultas Pasca Sarjana IPB)

[13] Marler TL 1994 Developmental light level effect of growth, morphology and leaf physiology of young carambola trees *J. Amer. Soc. Hort. Sci.* **199** (4) 711-18.

[14] Cruz P 1997 Effect of shade on growth and mineral nutrition of a C4 perenial grass underfield condition *Plant and Soil* **188** 227-37

[15] Khumaida N 2002 *Studies on Adaptability of Soybean and Upland Rice to Shade Stress* (Tokyo : PhD Thesis The University of Tokyo)

[16] Woodall G S and G R Stewart 1998 Do anthocyanin play a role in UV protection of the red juvenile leaves of syzygium? *J. of Exp. Bot.* **49** (325) 1447-1450

IOP Conf. Series: Earth and Environmental Science **157** (2018) 012018 doi:10.1088/1755-1315/157/1/012018

Economic valuation of erosion

Marupah[1], H Zubair[2], D Rukmana[3] and S Baja[2]

[1] Faculty of Engineering, Muhammadiyah Makassar University, Jalan Sultan Alauddin No. 259, Makassar, 90221, Indonesia.
[2] Soil Science Department, Faculty of Agriculture, Hasanuddin University, Jalan Perintis Kemerdekaan KM 10, Makassar, 90245, Indonesia.
[3] Agriculture Social Economics Department, Faculty of Agriculture, Hasanuddin University, Jalan Perintis Kemerdekaan KM 10, Makassar, 90245, Indonesia.

E-mail: *marufahaan47@gmail.com*

Abstract. Various results of erosion research on highland vegetable farming land indicate that the erosion level is classified as dangerous. This condition cannot be tolerated, because it will cause economic problems in the future both society and government. For farmers who are actively processing potatoes, the longer the farming the greater the cost of production so that the rate of profit gained tends to decrease. For the government, environmental degradation will reduce the possible use of the budget for the development of social welfare because the available funds are used to finance the prevention and handling of environmental disasters such as floods and landslides. The purpose of this study is to find out how many profit opportunities are lost due to erosion occurring in potato farming, using the method of analysis of agricultural systems and then calculate the opportunity price of potato farming system. The results of this study indicate that the value needed to reduce erosion by 0.54 $t.ha^{-1}$ in one harvest season was IDR. 5,605,556.-. The opportunity to earn a profit of IDR. 5,600,000.- will be lost if farmers do not apply conservation techniques to potato farming in sub-districts Tinggimoncong.

1. Introduction

Limitations of agricultural land, especially for horticultural commodities, cause the highlands to become the preferred land for agriculture [1, 2]. Utilization of agricultural land in the highlands has higher management complexity and lower productivity compared to lowland farmland [3, 4]. Based on the inherent qualities of highland land such as slope and soil sensitivity to erosion and high rainfall, the chances of erosion, especially on land used for vegetable cultivation are high [4, 5, 6]. Farm management practices currently applied by farmers have caused multifunction (positive externalities) of vegetable farming, especially in erosion and flood control are not optimal. The problem of business land degradation and vegetable production leveling-off is generally related to the fragmentation of business land, ie small family farming areas resulting in inefficient management of farming [7] and land management systems that are not insightful of physical conservation as well as fertility [8, 9, 10].

Results of erosion research on highland vegetable farms with slope slopes above 15% indicate that the erosion rate between 87 - 652 ton ha^{-1} $year^{-1}$ [11, 12, 13, 14]. When referring to the tolerable limits of erosion of about 13.5 ton ha^{-1} $year^{-1}$ [15], the extent of soil erosion is considered harmful [16, 17].

Conditions like that cannot be allowed, because it will cause economic problems in the future. Not only for the general public and government but also for horticulture farmers, especially the potato farmers themselves. For farmers of active potato cultivation, for example, the longer the production cost of farming will be greater, the profit rate will decrease. Long term consequences will reduce the farmer's investment capability in managing land assets, so that at one time the land can no longer be planted with potatoes or no longer able to generate profits. For the government, severe environmental damage will reduce the possibility of using the budget for social welfare development due to limited funds to finance the prevention and handling of environmental disasters such as floods and landslides.

Therefore, holistic considerations are needed in making the policy of restoration of the Jeneberang River Basin, the complexity of the problem is very high with various causes ranging from economic, social and institutional causes of society and government, to the quality of human resources and the availability of science and technology. Even in many cases the causes are related to each other. The economic valuation of land resources for the Jeneberang watershed area can be an input for policy intervention to solve problems related to sustainable farming economy. Therefore, economic valuations of ecological benefits and losses for damage to land and forest resources in the Jeneberang River Basin are necessary.

Various environmental issues combined with farmer behavior in managing potato farming need further attention to find a solution to solve the problem. Therefore, this research tries to do some fundamental analysis to the farming business analysis based on the application of different conservation techniques.

The conservation farming system is an agricultural system that integrates land conservation techniques into existing farming systems with the aim of increasing the income and welfare of farmers while reducing erosion, so that the agricultural system can continue on a continuous basis. Conservation farming system is characterized by: 1) agricultural production and income is high enough, 2) agrotechnology applied acceptable and applied by farmers in accordance with its ability continuously, 3) commodities cultivated in accordance with local biophysical condition, accepted by farmers and sell well in market, and 4) minimal erosion so that the productivity of the land can be maintained on an ongoing basis [8].

The purpose of this research are to calculate the economic value of potato farming as measured by the rate of profit of production and harvesting that is influenced by the applied conservation pattern and to calculate the internalization of farmers to reduce erosion and surface runoff using the Opportunity Cost approach.

2. Research Methods

This research was conducted in Pattapang Village, Tinggimoncong Sub-district, Gowa Regency, which is the center of potato production. Research method with interview to get information about farming system and saprodi used. Method of erosion plot to obtain production data. The farmers interviewed were farmers who did not apply randomly selected conservation techniques.

In conducting potato cultivation, cost and income analysis is the beginning in determining the attitude of conducting potato cultivation. Calculation analysis was conducted to give an idea about production and sales price which will ultimately affect farmers income in potato cultivation business in Kabupaten Tinggimoncong Gowa regency.

2.1. Analysis of Potato Farming

Research on the analysis of potato farming was conducted in potato garden by making 12 observation plots of 88 m^2 with the following characteristics:

1. Plot 1: Plots with potato cultivation conditions are usually carried out by farmers (P1) as controls (Conventional Plots).
2. Plot 2: Plots with potato cultivation conditions treated with soil mechanical conservation consisted of (1) slope cutting (P2); (2) planting cut the slopes of 2 plants with drainage channels (P3); (3) planting cut the slope with rorak every 5 meters. (4) In this plot the planting was set cut the slope, then the plot is called the Conservation Plot (P4).

The study was conducted with three replications, the data collection was started after the plot was completed so that the planting of land for planting could begin. Some of the data taken were the number of seeds used, the amount of fertilizer applied and also the number of work day (HOK) was calculated from land preparation to harvest.

2.2. Opportunity Cost Analysis

This method was conducted to calculate the costs incurred to conserve profits, rather than value the profit itself. For example, to assess the magnitude of the economic benefits to be sacrificed in the event of a change so that the quality of the environment cannot be restored as it was. Stages of implementation were:

1) Identify the opportunity lost due to another activity / change.
2) Assess the magnitude of each type of economic benefit lost.
3) Sums up the magnitude of all the lost economic benefits.

According to Soekartawi [18], there are three variables that need to be considered in the financial analysis of farming, namely (1) farm income, (2) farm costs and (3) net farm income.

1. Farm income, is the multiplication of production obtained with the selling price, the equation as follows:
$$TR = \Sigma\, Yi.Pyi \quad (1)$$
where: TR = total revenue; Yi = production obtained in one i-plant season (kg); Pyi = i commodity price (IDR).
2. Farm Costs, is the value of all inputs used in one growing season during the production process, either directly or indirectly, with the following equation:
$$TC = \Sigma\, Xi.Pxi \quad (2)$$
where: TC = Total cost; Xi = physical number of farm inputs; Pxi = input price i (IDR) and i = kinds of commodities developed in a farm.
3. Net Farm Income, is the difference of total revenue to total expenditure.
$$PU = TR - TC \quad (3)$$
where: PU = farm income (Rp); TR = total revenue (IDR); and TC = total cost of farming (IDR).

The analysis of financial feasibility of farming based on R/C ratio (Return Cost Ratio) [18] calculated by the following formula:
$$RCR = Income\ /\ Production\ Cost \quad (4)$$
Value of RCR = 1 means that the farming can only return capital.

3. Results and Discussion

3.1. Analysis of Potato Farming

The final result of analysis of production cost and harvest of potato crops for both erosion plots is presented in the following table:

Table 1. Cost analysis of potato farming plots

Needs every ha	Treatment			
	P1	P2	P3	P4
Number of Plants (plants)	292	275	251	261
Seeds (IDR)	9,954,864	9,375,300	8,557,092	9,102,564
Manure (IDR)	3,977,400	3,579,660	2,943,275	3,341,016
Chemical fertilizers (IDR)	6,775,000	6,775,000	6,775,000	6,775,000
Plant medicines (IDR)	6,889,000	6,889,000	6,889,000	6,889,000
Labour (IDR)	3,400,000	3,400,000	3,400,000	3,400,000
Additional Workforce (IDR)	3,400.000	3,400,000	3,400,000	3,400,000
Total Cost (IDR)	30,996,264	30,168,960	28,689,368	29,657,580
Production (tonnes)	9.92	10.91	11.29	9.89
Production Value (IDR)	55,681,818	56,477,273	56,401,515	42,848,909
Profit (IDR)	24,685,554	26,308,313	27,712,147	20,191,329
R/C Ratio	1.79	1.87	1.96	1.64

Note: family labour is not taken into account

Calculation of the total selling price of Potato per hectare shows that the total price of Potatoes in a conventional plot (P1) for one hectare can reach IDR. 55,681,818.- and on the conservation plot reached IDR. 56,477,273.- (P2), IDR. 56,401,515.- (P3) and IDR. 42,848,909.- (P4). There is a difference in selling price of IDR. 1,622,000.- for P2 treatment and about IDR. 3,026,000.- for the treatment of P3 at the price of potato crops in conventional plots (P1). In the P4 conservation plot, there is a difference in selling price with conventional plot (P1) of IDR. 11,494,000.- lower, this is because in conservation plot P4 there is excess labour cost and production in plant maintenance when compared to others. Selling price of potatoes varies for each type, where potato selling price for size criteria of AB and DN are IDR. 5000.- and 7000.- per kg, respectively and small potatoes have a selling price of IDR. 2000.- per kg. The results of this study show that about 75% produced were AB size potatoes and about 7 - 10% were in DN size and 15-20% were categorized as small potatoes.

Calculations in table 1 show that the R/C ratio for conventional plots is 1.79 while the R / C ratio for conservation plots P2 is 1.87, P3 is 1.96 and P4 is 1.44. It means that potato farming on conventional plot with capital of IDR. 1000. - will get sale result of IDR. 1790.- while potato farming in conservation plot of treatment of P3 with capital of IDR. 1000.- will get result of sale equal to IDR. 1960.-. This shows that from the feasibility aspect of farming, P3 conservation plot is very feasible to be developed because it can give bigger advantage when compared with conventional plot (P1).

Application of conservation techniques can provide greater benefits to potato farming in Tinggimoncong sub-district, which reduces the erosion rate of 0.54 ton.ha^{-1} per season. However, farmers are still reluctant to apply conservation techniques to potato cultivation due to consideration of longer cultivation time. Reducing the rate of erosion in preserving the environment is an obligation that must be done by every farmer in the sub-district of Tinggimoncong upstream of Jeneberang watershed. By not applying conservation techniques to potato farming, farmers will lose the opportunity to gain profits because the resulting production is less due to the rate of erosion.

3.2. Opportunity Cost analysis

Table 2 shows that the value required to decrease 0.54 tons of erosion during one harvest season was IDR. 5,605,556.-. The value is derived from the difference in harvest yields from conventional potato farming (P1) with conservation potato farming (P3). The resulting difference is then divided by the value of the erosion difference from conventional potato farming with conservation potato farming.

The results of this study indicate that soil conservation requires less seed, increasing the production of potatoes when compared with conventional soil treatments. Conventional soil processing requires more seeds but does not produce large production, because the soil layer is eroded so that the nutrients needed by the plants drift along with the eroded soil. Soil productivity decreases, input production is higher in the next planting season.

Table 2. Opportunity cost value of potato farming on the conventional plot and conservation plot

Component	P1	P3	Difference
Profit (IDR)	24,685,000	27,712,000	3,027,000
Erosion (tonnes)	3.07	2.53	0.54
OP (IDR)	-5,605,556		

P1: conventional plot
P3: conservation plot

From the above calculation, then the value of opportunity cost is obtained if the farmers want to change the technique of soil processing on potato soil from soil processing techniques in the direction of the slope into a technique of soil processing with planting two lane value of cutting the slope of IDR. 5,605,556.- hectare^{-1} for each harvest.

IOP Conf. Series: Earth and Environmental Science **157** (2018) 012018 doi:10.1088/1755-1315/157/1/012018

4. Conclusion

Based on the results of the discussion it can be concluded that the value of economic valuation of erosion on potato crop is IDR-5,605,556.- (a negative value indicates a loss of revenue). This means that farmers will experience a loss of income of IDR. 5,605,556.- per hectare due to the rate of erosion that occurred.

References

[1] Agus F, Irawan, Nurida NL, Dariah A, and Husen E 2006 Agricultural land conversion as a threat to food security and environmental quality. *Proc. of Multifunctional Seminar and Agricultural Revitalization* (Agency for Agricultural Research and Development Bogor) pp. 101-121

[2] Pasandaran E 2006 Alternative control policy on irrigated wetland conversion in indonesia. *J. Agric. Res. and Dev.* **25** (4) 123-129

[3] Prasetyo B H and Suriadikarta D A 2006 Characteristics, potential, and ultisol soil processing technology for dryland agricultural development in indonesia. *J. Agr. Res. and Dev.* **25** (2) 39-46

[4] Dariah A 2007 Agricultural Cultivation on Mountains Land. *Ind. Agr. Res. and Dev. Rep.* **29** (1) 7-10

[5] Hogarth W L, Parlange J Y, Rose C W, Sander G C, Stenhuis T S and Barry A 2004 Soil erosion due to rainfall impact with inflow : an analytical solution with spatial and temporal effect. *J. Hidrology*, **295** (14) 140–148

[6] Gangcai L, Zhang J, Tian G, and Wei C 2005 The effects of land uses on purplish soil erosion in hilly area of sichuan province, china. *J. Mount. Sci.* **2** (1) 68–75

[7] Adiyoga W, Ameriana M, Suherman R, Soetiarso T A, Udhiarto B K and Sulastrini I 2000 Production system of some types of vegetables in Indonesia. *J. Hort.* **9** (2) 258-265

[8] Sinukaban N 2007 Building Agriculture into a Sustainable Industry with Conservation Farming. In: Sinukaban N., *Soil and Water Conservation Key to Sustainable Development. Directorate General of RLPS.* (Ministry of Forestry. Jakarta) pp. 226-241

[9] El Kateb H, Zhang H, Zhang P, and Mosandl R 2013. Soil erosion and surface runoff on different vegetation covers and slope gradients: a field experiment in southern Shaanxi China. *J. Catena* **105** (6) 1-10

[10] Lihawa F, Partuti I M, and Nurfaika 2014 Distribution of spatial aspect type of avalanches in Alo River basin Gorontalo Province. *J. Man and the Env.* **21** (3) 277-285

[11] Arsanti I W and Boehme M 2006 Vegetable Farming System in Indonesia: Appreciation of Multifunction Agriculture, Economy and Environmental Externalities. *Agricultural Multifunction Seminar*. (Agency for Agricultural Research and Development. Agriculture department) Lido, 26 - 27 June 2006

[12] Firmansyah, 2007. Prediction of red yellow podsolic erosion of soil based on usle methods in various farming systems: case study in north Barito district and gunung mas. *J. Assess. and Dev. of Agr. Tech.* **10** 2-29

[13] Zhou P, Luukkanen O, Tokola T, and Nieminen J 2008 effect of vegetation cover on soil erosion in a mountainous watershed. *J. Catena* **75** (4) 319–325

[14] Sutrisna N, Sitorus S R P, Pramudya B and Harianto 2010 Alternative model of vegetable crop conservation farming in upper sub cikapundung river basin. *J. Hort.* **20** (3) 223-240.

[15] Arsyad, S 2010 *Soil and Water Conservation*. Second Edition. (IPB Press. Bogor)

[16] Shukle M K, and Lal R 2005 Erosional effect on soil physical properties in an on-farm study on alfisols in West Central Ohio. *J. Soil Sci. Am* **170** (6) 445–456

[17] Sutono S 2008 Farmers loss due to erosion. *Land Resources Report* **3** (4) 1-5

[18] Soekartawi 2006 *Farming Analysis* (University of Indonesia Press. Jakarta)

IOP Conf. Series: Earth and Environmental Science **157** (2018) 012019 doi:10.1088/1755-1315/157/1/012019

Effectiveness of bio-slurry on the growth and production of soybean (*Glycine max* (L.) Merrill)

Rafiuddin, A Mollah and H Iswoyo

Departement of Agronomi, Faculty of Agriculture, Hasanuddin University, Jalan Perintis Kemerdekaan KM 10, Makassar, 90245, Indonesia.

E-mail: *rafiuddin.syam@yahoo.co.id*

Abstract. This research was aimed to determine the effectiveness of bio-slurry fertilizer on the growth and production of soybean plants which was conducted in the Pucak village, Tompobulu District, Maros Regency, South Sulawesi from July to October 2016. The research was set in randomized block design (RBD) with 8 treatments replicated three times. Treatment used were the application of bio-slurry consisted of 8 level of concentrations, namely: control (0 mL.liter $^{-1}$ of water), 3, 5, 7, 9, 11, 13 and 15 mL.liter $^{-1}$ of water. The variables measured were plant's height, number of pods, weight of 100-seed, and soybean seeds' yield per hectare. The results of research shows that the application of bio-slurry effectively improved growth and yield of soybean (pod's number, 100-seed's weight and seed yield per hectare). Optimal concentration of liquid bio-slurry to obtain maximum results were 9.27 mL.liter $^{-1}$ of water for the highest number of pods (68.49 pods); concentration of 8.75 mL.liter $^{-1}$ of water for heaviest weight of 100 grains (14.22 grams); and the concentration 8,12 mL.liter $^{-1}$ of water for the highest production of seed per hectare (23.20 quintal).

1. Introduction

Soybean (*Glycine max* L.) is one of national main beans' commodities because it is the source of important phyto-protein for food's diversification in contributing the national food's tenacity. In fact, although soybean is Asia's original plant, but ironically Asia's countries become importers of soybeans from the outside Asia's area. Indonesia includes on the prominent soybeans' producer, but still imports seeds, soybeans residues after extracting its oil, and soybean's oil. Nowadays, Indonesia very depends on import soybeans because the less production of soybeans.

The need of soybean in Indonesia reaches around 2.5 million ton per year, while the production of soybeans in country on 2015 only around 0.963 million ton, thus to fulfill the need, it was needed import. Strategy plan for Indonesia's agriculture development is hoped to reach self-sufficient of soybeans on 2020 [1]. The government makes serious efforts to take down the dependence on import, thus soybeans include as one of three food's commodities which is targeted can be self-sufficient beside rice and corn. The main problem is national production of soybeans lower rather than the need of country, thus it always experiences deficit. Therefore, without the efforts and special policy, the national need of soybeans will still depend on import. National productivity of soybeans only reaches 1.56 ton.ha^{-1} with milling around 0.8 – 24 ton.ha^{-1}

IOP Conf. Series: Earth and Environmental Science **157** (2018) 012019 doi:10.1088/1755-1315/157/1/012019

in the farmer's level, while in the research level had achieved 1.7-3.2 ton.ha^{-1}, depends on the field condition and applied technology [2].

The consumption of soybeans leans to increase thus the import process is also increased. According to Subandi [3], there are 5 strategies which have to be conducted to assure the success of increasing soybeans' national production namely amelioration of sale price; utilization of potential area; intensification of plantation process; amelioration of production process; and program consistency and tenacity of agriculture field officers.

Agriculture area which is generally available and used on plantation process of soybeans had been decreased in quality (degradation of land) that caused by utilizing of land resource without counterbalancing the land optimal restitution. The use of chemistry fertilizer continuously in great quantities is one of the degradation land's causes. Supadma [4] stated that since 1984 the use of synthetic fertilizer by Indonesian farmer more and more increased to make maximal the agriculture yield obviously and fast. To solve that problem, society needs to be comprehended about the importance of agricultural system which friendly for environment and the use of organic fertilizer to increase the plant production. Organic fertilizer is the final result of changing or disentangling the remains of plant or animal, consists of almost all unsure such as; N, P, and K even Ca, Mg, and S, although the availability of those unsure usually in a bit quantities. P unsure in compost is almost excessively from dense feces, while N and K are from liquid feces [5].

One of organic fertilizer which can be used to increase the production of soybean plants is bio-slurry fertilizer. Bio-slurry or biogas waste is the product from the result of biogas processing of livestock feces and water through process without oxygen (anaerobic) inside of closed room. Bio-slurry fertilizer have excess such as; can improve the nature of soil physically, chemistry, and biologically and decrease the negative effect of making biogas such as; the smell, the environment contamination which can be the disease source. Bio-slurry also consists of probiotic microbe which is useful to increase the fertility and the health of agriculture land thus, it is hoped will affect on the increasing of yield's quality and quantity [6].

The utilization of bio-slurry on plants should concern on the dosage and concentration. It is due to the possible unfavorable effects on plants when the concentration is too little or too much. Therefore, the selection of appropriate concentration needs investigation through this research [7].

2. Research Method

This research was conducted at Pucak village, Tompobulu subdistrict, Maros regency, South Sulawesi province from July to October 2016. The tools used were hand tractor, sprayer, mattock, gage, syringe, analytic weights and camera, while the materials are the soybean's seeds of Anjasmoro's variety and liquid bio-slurry.

The research was in form of experiment by using randomized block design (RBD), the treatment was the bio-slurry concentration (mL.litre^{-1} of water) consisted of 8 levels *i.e.* 0, 3, 5, 7, 9, 11, 13, and 15. Each treatment was repeated 3 times.

The seeds of soybean were planted in plot 5 m x 4 m by using plant spacing of 25 cm x 25 cm. The treatment of liquid bio-slurry fertilizer was conducted 2 times, namely at 2 days after planting (DAP) and in the phase of 50% flowering.

The obtained data was then analyzed using software of SPSS ver. 23.0 and a further test using Duncan Multiple Range Test (DMRT). The selection of bio-slurry fertilizer's optimal concentration was determined using regression analysis.

3. Results and Discussion

The result of statistical analysis showed that the treatment of liquid bio-slurry fertilizer was effective and could increase the growth and production of soybeans in some observed variables ($p<0.05$). This is because of bio-slurry will increase the availability of macro and micro nutrition unsure which are found on that fertilizer and can be used maximally by soybean plants thus the growth, blooming, and production of plants will be more optimal. Liquid bio-slurry fertilizer is good because it contains various minerals needed by plants, such as; N, P, Mg, Ca, K, Cu, and Zn. Poerwowidodo in [8] stated that organic fertilizer contains N, P, and K thus will accelerate the process of amino acid and protein synthesis, so it will accelerate the growth of plants. Parwata *et al.* [9] stated that the giving of bio-slurry fertilizer can increase the growth of plants rather than without bio-slurry, because the plants can utilize nutrition unsure which are obtained from that bio-slurry to their growth optimally thus can give the maximal yield.

Table 1. the height of plants, the number of pods, the weight of 100-seeds and the production of soybean seeds' yield per hectare on the various liquid bio-slurry concentration.

Liquid Bio-slurry (ml.liter of water^{-1})	Variables			
	Height of Plants (cm)	Number of Pods	Weight of 100-Seeds (gram)	Production of Seeds per Hectare (quintal)
0 (B_0)	63.33 a	44.07 a	10.93 a	11.36 a
3 (B1)	61.83 a	51.40 ab	12.48 abc	17.77 bc
5 (B2)	54.50 a	58.00 b	13.26 bc	20.57 c
7 (B3)	62.33 a	71,53 cd	13.28 bc	23.83 d
9 (B4)	58.67 a	80.80 d	14.71 c	25.58 d
11 (B5)	63.83 a	68.53 c	14.61 c	19.92 c
13 (B6)	61.33 a	56.73 b	13.81 bc	17.37 bc
15 (B7)	57.00 a	47.77 ab	11.74 ab	15.75 b

Note: the number which was followed by similar alphabet means not significantly different on DMRT test level 0.05.

Sutresnawan *et al.* [10] stated that the giving of liquid bio-slurry fertilizer leans to produce the increasing number of leaves, and also the more great number of branch because in forming process of bio-slurry will occur fermentation in biogas waste which changes food substance to be available for plants so, it makes easier the absorption of nutrition unsure that finally can increase the growth and production of plants. N Unsure in liquid bio-slurry more excessive and easy to be absorbed by plants thus can increase the growth of plants.

The result of DRMT test on table 1 shows that the effectiveness of liquid bio-slurry was seen from the highest result of greatest pods' number (80.80 pods) which were found in the giving of liquid bio-slurry on the concentration 11 mL.liter $^{-1}$ of water but on the variable the weight of 100-seeds and production seeds per hectare precisely the highest result (14.71 gram and 25.58 quintal) are on the concentration 9 mL.liter $^{-1}$ of water and still similar on the concentration 11 mL.liter $^{-1}$ of water. This shows that the need of soybean in kicking the growth of pods and the development of seeds needs liquid bio-slurry on the concentration 9-11 mL.liter $^{-1}$ of water. Nitrogen of liquid bio-slurry on that concentration affects the growth and increases the number of pods, the weight of 100-seeds and production of seeds per hectare.

The higher concentration of liquid bio-slurry which is given on soybean plants then the forming and growth of pods and seeds of soybean more increased because soybean plants are able to utilize those nutrition unsure to their generative growth optimally so can give the maximal result, but if the concentration was increased more than 11 mL.liter $^{-1}$ of water, in fact the soybean plants decreased in the growth rapid of

100-seeds' weight and production of seeds per hectare. This is caused by the concentration (11-15 mL.liter $^{-1}$ of water) soybean plants gets excess where actually the excess is not needed again by them.

The content of phosphor unsure on bio-slurry fertilizer influences toward the growth of flowers and pods of soybean plants. This is suitable with the analysis result of bio-slurry unsure which shown that bio-slurry fertilizer contains higher C-organic No.28/Permentan /OT.140/2/2009 that is bigger than 12, and also the content of nitrogen, phosphor, potassium nutrition which are appropriate with quality standard of organic fertilizer that is under 6%. The higher concentration of bio-slurry which is given to soybean plants then the greater of available nutrition unsure supplement and can be absorbed by the plants thus the greater photosynthesis' results which are resulted to be transferred further into pods, the filling and growing of soybean seeds which finally increase the production of soybean seeds [11].

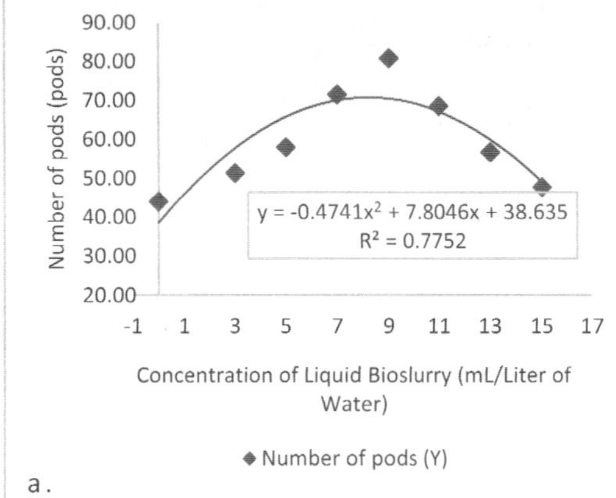

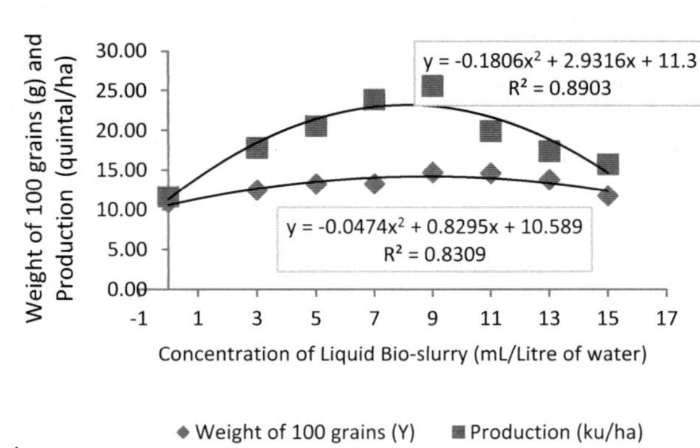

Figure 1. The relation between concentration of liquid bio-slurry with the number of pods (a), the weight of 100-seeds and production of soybean seeds per hectare (b).

The result of regression analysis shows that there are real positive relation between bio-slurry concentration with the number of pods and have quadratic characteristic where the equation is y = -0.3483x2 + 6.459x + 38.548 (R^2 = 0.606). Based on that equation, the optimum concentration of bio-slurry fertilizer is 9.27 mL.liter $^{-1}$ of water which resulted maximal pods around 68.49 pods. The similar condition also occurred between bio-slurry concentration with the weight of 100-seeds and seeds' production that both have quadratic characteristic (y = -0.0474x2 + 0.8295x + 10.589 (R^2 = 0.830) and y = -0.1806x2 + 2.9316x + 11.3 (R^2 = 0.890), so optimum concentration of bio-slurry fertilizer is 8.75 mL.liter $^{-1}$ of water to produce maximum weight of 100-seeds around 14.22 g, while on the variable of production seeds per hectare, the optimum concentration is 8.12 mL.liter $^{-1}$ of water to produce 23.20 quintal. Based on those three variables was known that optimal concentration of bio-slurry for soybeans is 8.12 – 9.27 mL.liter $^{-1}$ of water. The content of the higher N, P, and K unsure on bio-slurry will stimulate the generative growth on soybean plants thus able to accelerate and increase the form of flowers, pods, and seeds. The high of N unsure on bio-slurry fertilizer influences to the increasing of N unsure at the soil thus it will be also absorbed more excessive. Maspray [12] stated that liquid organic fertilizer contains the high N and K unsure and also the growth stimulating substance such as IAA. Furthermore, Nur, Erlina and Nasih (2007) suggested that liquid bio-slurry is applied through leaves thus it fast to be used by soybean plants. This fertilizer can increase the forming of soybeans' leaves chlorophyll and root nodule until the plant's ability more increases to fixate the air of N and to do photosynthesis. Besides that, bio-slurry can improve the nature of soil physically, chemistry and biologically, increase the quantity and quality of production, and also decrease the use inorganic fertilizer [13].

4. Conclusion

- Liquid bio-slurry is effective to increase the growth and production of soybean plants (the number of pods, the weight of 100-seeds and production of seeds per hectare).
- The optimal concentration of liquid bio-slurry to obtain maximal yield were 9.27 mL.liter $^{-1}$ of water to the greatest number of pods (68.49 pods); concentration 8.75 mL.liter $^{-1}$ of water to the heavier weight of 100-seeds (14.22 g); concentration 8.12 mL.liter $^{-1}$ of water to the higher production seeds per hectare (23.20 quintal).

References

[1] Badan Pusat Statistika 2016 *Data of Food's Plants and Horticulture Production* (Jakarta: The Center of Agriculture Statistic Data)

[2] Agung 2016 *The Increasing of National Target of Soybean Production* (Jakarta: Department of Agriculture Research and Development)

[3] Subandi 2007 *Five Strategies in Growing Soybean* (Jakarta: Sinar Tani)

[4] Supadma 2006 *Fertility of Soil and Fertilization of Agriculture Land* (Bandung: Pusaka Buana)

[5] Suryono, Widyatmani and Sumarno 2014 The utilizing animal husbandry waste in the solid agriculture concept *Agron. J.* **29** (2) 95-100

[6] Nur F, Erlina A and Nasih 2007 The effect of dosage and frequency in giving liquid organic fertilizer toward the growth and yield of low land String bean *Agron. and Env. J.* **7** (1) 43 -53

[7] Rahim and Jumiati 2007 The influence of concentration and time of spraying liquid organic fertilizer toward growth and yield of sweet corn *Agritrop. J.* **26** (3)105-109

[8] Parwata I N A, Kusumawati N N C and Suryani N N 2016 The growth and production the green telang flower on various level of application in *bio-slurry* fertilizer *Trop. Animal Husb.* **4** (1) 142 -155

[9] Hamdani 2016 *The Seeding of Rice Plants by Using Various Compost Dosage and Local Microorganism Species through System Mina's Rice Plants* Thesis (Makassar: Graduate Program Hasanuddin University)

[10] Sutresnawan, Kusumawati and Trisnadewi 2015 The growth and production of telang flower *(clitoria ternatea)* which is given various kind and dosage of organic fertilizer *Tropical Animal Husb.* **3** (3) 586-596

[11] Husma M 2010 *The Influence Organic Material and Potassium Fertilizer toward The Growth and Production of Melon Plant (Curcumis melo L.) Thesis* (Kendari: Agronomi Study Program, Haluoleo University)

[12] Maspray 2011 *The Easy Ways to make Fermentation of Cow's Urine for Liquid Organic Fertilizer* http://www.gerbangpertanian.com Accessed 8 March 2017

[13] Parman S 2007 The influence of giving liquid organic fertilizer toward the growth and production of potato plant *(Solanum tuberosum* L.*) Anatomy and Physiology* **25** (2)

IOP Conf. Series: Earth and Environmental Science **157** (2018) 012020 doi:10.1088/1755-1315/157/1/012020

Assessment on the use and availability of rice certified seeds in Bone regency, South Sulawesi province

M Farid, H Iswoyo, I Ridwan, Nasaruddin and R Dermawan

Department of Agronomy, Faculty of Agriculture, Hasanuddin University, Jalan Perintis Kemerdekaan KM 10, Makassar 90245 Indonesia

E-mail: *farid_deni@yahoo.com*

Abstract. Use of certified seeds is necessary in rice farming system to ensure high production and productivity. In South Sulawesi, as one of the main center of Rice in Indonesia, increase in the regional harvest areas and land productivity was not followed by significant increase in rice productivity. Seeds policy implemented by the government not only covers the technology applied in the seeds production but also quality insurance, availability and its distribution system. Despite these efforts, use of certified seeds in the field by farmers are still limited, therefore a survey was conducted to study the use and availability of rice certified seeds in Bone regency. Farmer groups in 9 districts were sampled to obtained data concerned to the use and availability of rice seeds including type, planting season, cropping pattern and availability at planting. Results show that the use of certified seeds in Bone regency were relatively high, however level of seeds supply was low or delayed at time of planting. To overcome the problem of seed supply, most farmers used its own seeds from previous harvest. Some suggestions to resolve this condition are discussed.

1. Introduction

South Sulawesi is one of the largest rice production centers in eastern Indonesia. The availability of new superior rice varieties that have been released has been considerable, but rice seed breeding efforts to support seed availability are still very limited, especially in rice production centers. To overcome these problems, development activities through seed breeding these high yielding seeds is very necessary to support the availability of the superior varieties seeds.

Currently, rice productivity in South Sulawesi only reaches an average of 4.7 tons Ha^{-1}. However, from 2013-2015 the development of harvested area and land productivity for rice increased by 2% and 10.98% respectively [1]. There is a trend of increasing production every year, but the trend is not significant compared to the productivity capabilities of plants. One of the factors causing low productivity of rice in South Sulawesi is the inadequate application of farming technology (cultivation). In addition, there are social factors and land cultivation conditions that also become an obstacle in the effort to increase rice production.

To overcome these challenges, the government provides assistance in the form of production facilities, including subsidized rice seeds through the budget allocated by DIPA Directorate General of Food Crops, Provincial Work Unit, and some Regency / Municipal Work Unit in 2017. In line with the stabilization of food self-sufficiency program, the use of quality seeds is very important to note in increasing the production and productivity of food. This is because seed has a strategic role as a means of bringing new technology, in the form of excellence of varieties with various specifications of

IOP Conf. Series: Earth and Environmental Science **157** (2018) 012020 doi:10.1088/1755-1315/157/1/012020

excellence; high yield, pest resistance, early age to improve the crop index, the superiority of crop yield to suit the consumer's taste.

One of the factors causing the low level of availability of quality seeds is the level of awareness of farmers to use high quality seeds is still very low. In general, farmers only set aside part of the harvest to be used as seed for the next planting season. Hence, the seed used are not in guaranteed quality. This is because farmers cannot afford to buy seeds that are considered expensive and there is a decrease of farmers' trust in certified seed quality, where there is no match between the content of the labels and the reality in the field [2, 3].

Provision of quality seeds through seed revitalization program is expected to support the achievement of four successful agricultural development as has been proclaimed by the government. Germination program not only involves the application of technology in its propagation activities, but also includes quality assurance, availability and distribution system. According to Andri [4], the successful provision of quality seeds cannot be separated from the four seeding subsystems are: 1) research, breeding and release of varieties; 2) production and distribution; 3) quality control and seed certification; and 4) supporting facilities and infrastructure such as institutional and infrastructure.

Some problems in distributing quality seeds are availability, price, quality, and quantity of seed. The availability of quality seeds with sufficient quantities, timely, and easily obtained by farmers plays an important role in improving food productivity. The continued availability of superior seeds is inseparable from the role of seed breeders, resulting in continuous continuity between sustainable seed producers which are essential in the sequence of seed formation and development of a variety.

Therefore it is necessary to prepare a survey to see the fulfillment of seed needs at the farm level and accuracy of its availability.

2. Methodology

Research conducted was a survey study covering 9 districts in Bone Regency, South Sulawesi Province *i.e.* Barebbo, Amali, Lappariaja, Sibulue, Mare, Bengo, Ponre, Taneriattang, China, and Awangpone. Selection of the sub-district is based on the development of rice areas that represent the western, southern, northern and central regions Bone district. Data used in this study were primary and secondary data. Primary data were collected through interviews to farmers, extension workers, seed distributors, traders, farmer groups, seed producers / seed companies, and related Local Government Work Units (Bappeda, Dinas Pertanian and related agencies and stake holders) to the provision of superior seeds. Secondary data owned by districts as the study areas were also collected, such as the extent of rice cultivation, the need for superior rice seeds, the realization of the crops and the availability of seeds for each growing season. Data on use and availability of seeds were collected from farmers during Planting Season (PS) I and II of 2016 and PS I of 2017. Availability of superior seeds is assessed in terms of the types of superior seeds developed, planting time, planting patterns, availability of seeds at planting and seed sources.

Some indicators of this research include information obtained about the use of high yielding varieties of rice and the level of seed availability at planting in Bone District. The collected data is tabulated and analyzed using qualitative and quantitative descriptive analysis.

3. Results and discussion

3.1. Farm management

Astronomically Bone Regency lies in the position of 4°13' - 5°6' South Latitude and between 119°42'-120°30' East Longitude. Its location that close to the equator makes the Regency Bone has tropical climate. Throughout 2014, air humidity ranges from 77 to 86% with air temperature of 24.4 °C - 27.6 °C [5].

The comparison of both farmer group status is dominated by non-breeder farmers (89.18%) (table 1) and only about 11.82% of the respondent farmers are breeder farmers. Table 1 shows that in conducting farming, most farmers in Bone District have a high dependence on external supply of seeds due to the low number of farmers who have the authority to produce and distribute certified quality seeds.

Table 1. Respondent farmers status in Bone Regency

Farmer status	N	(%)
Breeder	33	11.82
Non-breeder	272	89.18
Total	305	100

Source: primary data, 2017

Based on interviews with breeders, the lack of breeder farmers in Bone Regency is based on the marketing constraints of the certified seed produced by seed farmers. Certified seed sales, particularly in Bone regency, compete with subsidized seeds supplied by the government through seed and fertilizer assistance programs. This causes farmers became reluctant to switch to breeder farming and rely only on the cheaper supply of seeds from the government.

From the information collected, it can be arranged several possible models of cropping patterns implemented by farmers in Bone regency. Cropping patterns applied by farmers varied based on the type of land cultivated especially for Rice, Corn and Soybean (figure 1). The number of possible variations of cropping pattern applied by sample farmers in the research area is strongly influenced by the type of land cultivated. The type of land is determined by soil characteristics and the availability of water sources. In general, the type of irrigated rice field has a dam facility that can accommodate rainwater so as to provide a source of water for planting up to the third planting season. However, there is a possibility that the amount of rainfall received in this area is reduced so that even irrigation facilities are available, the water flow is reduced and will affect the farmer's decision to cultivate the cultivation on this land. This may be true when there is prolonged drought.

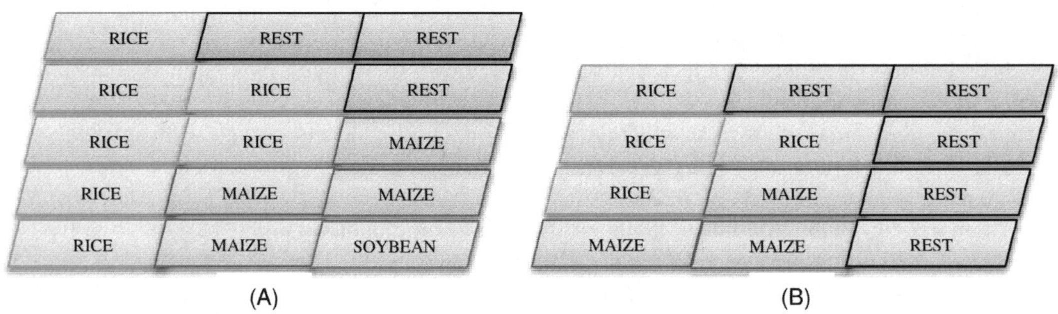

Figure 1. Planting pattern applied by farmer in Bone Regency on irrigated paddy (A) and rainfed (B)

The start of the planting time chosen by the sample farmers to plant the crops during the planting season I, II and III based on the type of land is presented in figure 2. Based on figure 2 it is known that for all types of land, April and May were the time where most of the initial planting for first and second crop were chosen by farmers. In irrigated rice fields farmers tend to plant their land for the first planting season in January, April and May. For the second crop or planting season II, farmers in irrigated rice fields planted in May and some planted their second crop in November. For the third plant was generally carried out between August to December. Implementation of planting time on irrigated land is not constrained by water availability so it is seen that the planting can be done throughout the year by farmers in Bone Regency.

In contrast to irrigated rice fields, rainfed lowland areas only depend on water sources for planting to the available rainfall. This leads to the determination of planting time on this land to refer to the pattern of rainy season and the beginning of the rainy season. Planting for the first planting season on rainfed lowland is only done in April and May. As for the second planting season, most farmers set

November as the beginning of the growing season. The third plant is very rarely done on this land, though it is planted in November.

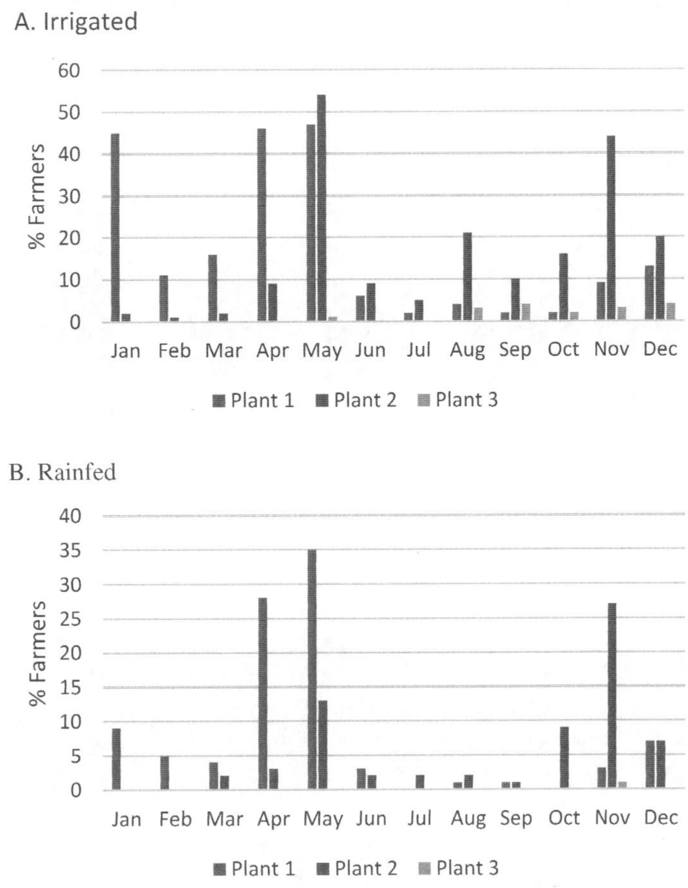

Figure 2. Planting time applied by respondent farmers in Bone Regency per planting season on irrigated paddy (A) and rainfed (B)

The determination of planting time and selection of plant species is applied by sample farmers in Bone district seems to be based on the pattern of climate spread, especially rainfall in this area. Bone district area is divided into two types of rain that is moonson rain type and local rain type. Moonson rain type has the highest rainfall when the Asian monsoon blowing winds are in January and February. This type covers the western part of Bone County. The second type has an inverted rain pattern criteria with monsoon pattern, ie the highest rainfall occurs in May - June. This type covers most of the Bone Regency.

In addition to these two areas, there are Bontocani and Libureng districts, partly following the western region and partly following the eastern region. The amount of monthly rainfall in the Bone Territory varies with an annual average of 201.25 mm. The highest rainfall occurred in June that is 638 mm with the number of rainy days as much as 23 days.

The implications of the information obtained from sample farmers related to planting time is the importance to regulate the supply of seeds for use by farmers. Thus, when the farmers plan planting in the first, second or third cropping season, it is expected that superior planting is available so as not to delay the best planting time where water is sufficient to support plant growth.

3.2. Use and availability of certified seeds

The average rate of seed per hectare per sample farmer per planting season is relatively high. The use of seed per hectare in the 2017 planting season reached 62.25 kg Ha^{-1} in the first planting season of 2017, while in the first and second plant season of 2016 equivalent to 40.62 kg Ha^{-1}, except in the third planting season in 2016 only 20.84 kg Ha^{-1} (figure 3).

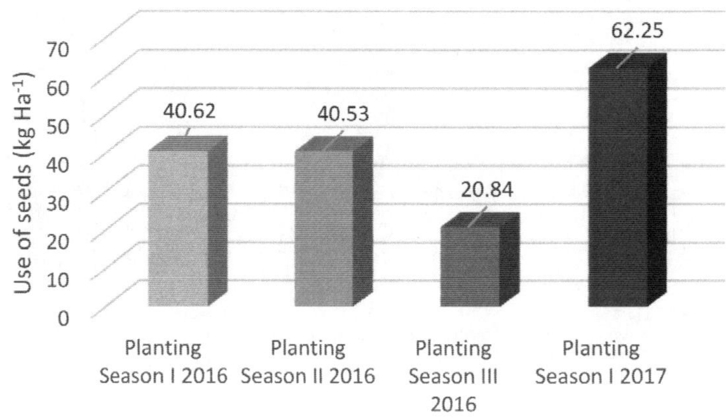

Figure 3. Average amount of rice seed used by respondent farmers in Bone Regency per hectare in planting season 2016 – 2017

The variation of seed use per hectare in the study area varies greatly with the varieties planted. Ciherang and Situ Bagendit varieties showed the highest seed used per hectare reached 57.54 kg Ha^{-1} and 50.50 kg Ha^{-1} respectively, and the lowest was Mekongga variety 37.05 kg Ha^{-1} (figure 4). Despite this, the use of seeds by sample farmers in Bone Regency was still very high compared to the recommended use of seeds for recommended varieties of only 25.00 kg Ha^{-1}.

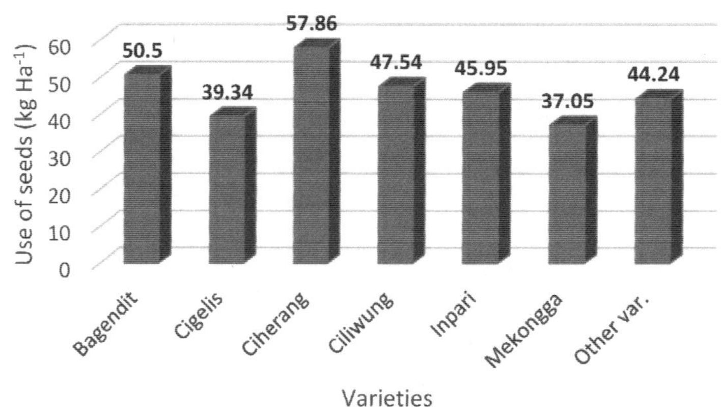

Figure 4. Average of seed use by sample farmers based on variety per hectare in Bone Regency

Rice seeds used by farmers in every growing season were generally open pollinated seeds both certified and non-certified, but dominated by certified seed. Farmers who used hybrid seeds are relatively fewer compared to the use of open pollinated seeds (figure 5).

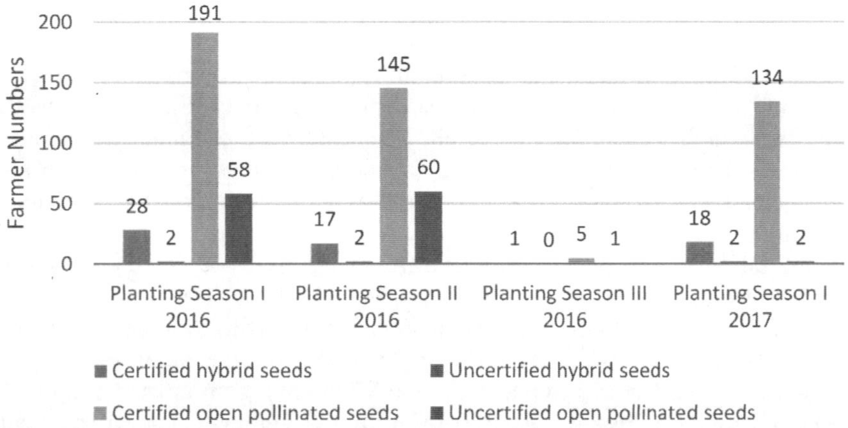

Figure 5. Type of seed used by sample farmers each planting season in 2016 – 2017

The level of superior seed availability at planting in Bone Regency per planting season varies considerably. From the results of the survey conducted on the sample farmers, most of the farmers stated that the superior seeds were available at each growing season (table 2). However, in the first planting season, there were approximately 26.76% of the sample farmers stated that they experienced a delay in obtaining superior seeds at planting and 17.6% of the sample farmers did not get superior seeds in the market. Similarly, during planting season II, although the seed availability was slightly increased, 59.48% of the respondent farmers stated that superior seeds were available, but there were still sample farmers who experienced delays and unavailability of superior seeds ie 26.13% and 14.41% of the sample farmers.

Table 2. Availability of superior varieties seed at planting per planting season in 2016 in Bone Regency.

Availability of superior variety seeds	Planting Season I		Planting Season II		Planting Season III	
	n	%	n	%	n	%
Available	158	55.63	132	59.46	14	73.68
Delayed	76	26.76	58	26.13	3	15.79
Unavailable	50	17.61	32	14.41	2	10.53

Source: primary data, 2017.

Table 3. Solution taken by sample farmers to overcome seed availability in Bone Regency.

How to overcome seed unavailability at planting	Number of farmers n	%
a. Use own seed	237	87.78
b. Delay planting/seeding	17	6.30
c. Look in other regencies	16	5.93

Source: primary data, 2017.

Delays in supply of superior seeds resulted in most sample farmers in the study area (87.78%) switched to use their own seeds saved from previous season harvesting (table 3). The problem that often arises in the use of farmer' own seed is the quality of seeds that do not match the superior seed quality. Quality of seeds below optimal will affect the vigor and seed germination rate at the beginning of plant growth. In addition, seeds that are not qualified often produce plants that are susceptible to pest disease and low production.

Rice seeds used by the community of more than 60 percent come from the informal sector in the form of grain from some previous season's harvest results repeatedly [6]. This means that rice farmers have not responded to the superior rice seeds well. The problems faced in the acceleration of the use of improved varieties is the information system of the existence of the source seed is still weak so that the user knowledge about the superior variety is still limited, in addition the availability of superior varieties is also limited [7].

The use of the farmers' own seeds can reduce the productivity of rice crops. Implication of other production input and means become less effective in the absences of qualified seeds of high yielding variety [8]. The use of high yielding varieties, responsive to fertilization and tolerance to major pest attacks has been shown to increase productivity [9]. Specifically, the use of high quality seeds has an impact on good crop growth and high yields. The qualified seed requirements are: (1) pure and known varieties; (2) high growing power (at least 80%) and good vigour; (3) seeds are healthy, pithy, not wrinkled, harvested when the beans are fully ripened; (4) harvested from healthy plants, uninfected diseases (fungi, bacteria, and viruses); and (5) seeds not mixed with other seeds or grass seeds [10].

The existence of quality seeds is an absolute necessity in the crop, both in terms of quantity, quality, and timeliness. From the results of discussions of farmers and FGDs at the district level, several important issues that need to be discussed include building a local seed breeder system, utilization of locally produced seeds by existing local breeders and provision of trust by local government to increase the breeder's interest, establishing inter-regional and inter-season seeds business from several districts, and more specifically within the Bone regency region itself and providing access and production facilities to local breeders, and ensure the smooth market of seed production.

4. Conclusion

The high utilization of certified rice seeds in Bone Regency is high but availability of this superior rice seed is not sufficient when needed, especially in the second growing season in April-May. The quality of seeds obtained by farmers from government assistance is less than 40% and only 60% of good quality, although overall it is certified by blue label (spread seed).

References

[1] Indonesian Bureau of Statistics 2016 *Produktivitas Padi, Jagung, dan Kedelai Provinsi Sulawesi Selatan Tahun 2015* (Makassar: Badan Pusat Statistik Provinsi Sulawesi Selatan).

[2] Wirawan B and Wahyuni S 2002 *Memproduksi Benih Bersertifikat* (Jakarta: Penebar Swadaya)

[3] Qamara W and Setiawan S A 1995 *Produksi Benih* (Jakarta: Bumi Aksara)

[4] Andri K B 2010 *Pengkajian Sistim Penyediaan (>90%) Kebutuhan Benih Unggul Bermutu (Padi, Jagung, Kedelai) yang Lebih Murah (>20%) secara Berkelanjutan untuk Mendukung Program Strategis Peningkatan Produksi Padi (>10%), Jagung (>20%), dan Kedelai (>20%) di Wilayah Jawa Timur* Research report Indonesian Agricultural Research and Technology Agency (Malang: BPTP)

[5] Indonesian Bureau of Statistics 2015 *Bone dalam Angka 2015* (Bone: Badan Pusat Statistik Kabupaten Bone)

[6] Daradjat A A, Agus S, Makarim A K and Hasanuddin A 2008 *Padi-Inovasi Teknologi Produksi vol 2* (Jakarta: LIPI Press)

[7] Wahyuni S 2011 *Teknik Produksi Benih Sumber Padi Work. Evaluasi Kegiatan Pendampingan SL-PTT 2001 dan Koordinasi UPBS 2012 tanggal 28-29 November 2011* (Balai Besar Penelitian Tanaman Padi)

[8] Asri N 2013 *Pengaruh Penggunaan Benih Bersertifikat Terhadap Produksi dan Pendapatan Usaha Tani Padi di Kabupaten Cianjur* (Bogor: Departemen Agribisnis Fakultas Ekonomi dan Manajemen Institut Pertanian)

[9] Nugraha U S 2004 Legalisasi, kebijakan, dan kelembagaan pembangunan perbenihan Perkembangan Teknologi *TRO.* **16** (1) p 61
[10] Wahyu W 2015 *Penyediaan dan Penyebar luasan Benih Sumber Varietas Unggul (VUB) Melalui Unit Pengelolaan Benih Sumber (UPBS) Di Provinsi Bengkulu BPTP Bengkulu Research Report* (Bengkulu: Badan Penelitian dan Pengembangan Pertanian)

IOP Conf. Series: Earth and Environmental Science **157** (2018) 012021 doi:10.1088/1755-1315/157/1/012021

Harvest index and yield components of aerobic rice (*Oryza sativa*) under effect of water, varieties and seed priming

H A Elkheir[1], Y Musa[2], M Muslimin[3], R Sjahril[2], M Riadi[2] and H Gunadi[4]

[1]Department of Agronomy, Faculty of Agriculture, Omdurman Islamic University, Omdurman, Alfteibah, PO BOX 382, Sudan.
[2]Department of Agronomy, Faculty of Agriculture, Hasanuddin University, Jalan Perintis Kemerdekaan KM 10, Makassar 90245, Indonesia
[3]Department of Soil Sciences, Faculty of Agriculture, Hasanuddin University, Jalan Perintis Kemerdekaan KM 10, Makassar 90245, Indonesia
[4]Department of Agrotechnology, Faculty of Agriculture, Christian University of Indonesia Paulus, Jalan Perintis Kemerdekaan KM 13, Makassar 90245, Indonesia.

E-mail*: ahamedelkheir@yahoo.com*

Abstract. This research was conducted to investigate harvest index and yield components of aerobic rice (*Oryza sativa*) under effect of water, varieties and seed priming in the tropical region. The experiment was set in split-split plot design. The treatments were priming seed with poly ethylene glycol (PEG), treated (P_1) and untreated (P_0) as sub-sub-plot, Three aerobic rice cultivars as sub-plot consisted of Inpago 8, new potential one (V_1), IR64, susceptible aerobic rice cultivar (V_2) and Situbagendit, a normal aerobic rice cultivar (V_3) as sub-plot, three irrigation intervals namely irrigated every 5 days (W_1), irrigated every 10 days (W_2) and irrigated every 15 days (W_3) set as main plot. Yield components and harvest index parameters were calculated. Results indicated significant and insignificant difference, in spite these, there was no significant difference the date of W1 and primed seed revealed the highest data compared with W2, W3 and non-primed seed. High harvest index revealed by $W_1V_1P_0$ was 0.827 and the smallest by $W_1V_3P_1$. Grain yield ranging between 4.49-1.60 ton ha^{-1} and straw about 15.39-5.33 ton ha^{-1} depend upon aerobic rice cultivars and water availability. More research about aerobic rice technology need to face water problem in the future.

1. Introduction

Harvest index is defined as the economical yield (grain yield) of the crop expressed as a decimal fraction of the total biological yield, but the clearly meant total aboveground dry matter production. Otherwise, harvest index define as a ratio between economical yield and biological yield [1]. There are many biotic and abiotic factors play role to increase and decrease different value of harvest index. These factors include type of varieties, plant population density, time and method of planting, fertilization dose especially nitrogen fertilization, soil water contents, water stress, temperature, sun radiation, season of cultivation short dry season and long wet season [2].

Hydro and osmo-priming recently playing an advancing technology practiced by farmer to cultivate crops under drought and salinity conditions. Furthermore, Poly ethylene glycol (PEG) is a substance like growth regulators which have ability to adapt seed and plant through seed priming technology to improve its response to be tolerance against water scarcity, drought and salinity condition. Otherwise, many researchers reported that PEG has famous role to accelerate crops establishment and crop

IOP Conf. Series: Earth and Environmental Science **157** (2018) 012021 doi:10.1088/1755-1315/157/1/012021

flowering and increase the germination behaviors and habituate seed and plant for best growth when suffer from water scarcity, drought and salinity condition. Furthermore, the positive beneficial effects of seed priming have been conducted for many field crops such as aerobic rice [3, 4]. However, the negative results indicated that seed priming decreased germination percentage in barley and corn [5] and reduced rate of seedling emergence and plants per unit area in sunflower [6].

Aerobic rice technology is a new technology of cultivating rice in the world to facing the problem of water scarcity. A scientist have modified aerobic rice through the advantage of "improved upland rice" that cultural practice under aerobic rice technology will lead to save 50% of irrigation water, less methane gas emission, reducing production costs, and finally Aerobic rice practices gives great hope to farmers who do not have access to enough water to grow flooded lowland rice [7, 8].

Aerobic rice technology using alternative irrigation system welting and drying is best practices which helping framers cultivated rice especially under water scarcity condition thus, understanding the ability of soil water holding capacity and duration of soil water contents between field (FC) capacity and wilting point (WP) which is varied according to soil properties (clay, sand, loam and silt) and abiotic condition that relate with water loss mechanisms such as evaporation, transpiration and evapotranspiration will guide farmers to water schedule and irrigation management furthermore, Many researchers recorded that the differentiation in growth, Harvest index and yield of crops especially rice depend upon soil water contents. This research investigated the harvest index and yield components of aerobic rice (*Oryza sativa*) under effect of water, varieties and seed priming.

2. Methodology

2.1. Research location

The research was conducted at Indonesian Center for Cereal Crops, in Bajeng, Limbung district, Gowa regency, South Sulawesi province. It is located -5.14 latitude and 119.42 longitudes and it is situated at elevation and altitude of 8 meters above sea level. The soil is Sand Clay Loam with pH about 5.7. The climate also characterized by long hot and wet summer, Moderate rain, high relative humidity 73-98% and Northwest wind with speed about 30 km hr^{-1}. Rice seed used were Inpago 8, a new potential one (V1), IR64, susceptible aerobic rice cultivars (V2) and Situbagendit, a normal aerobic rice cultivar (V3) obtained from Indonesian Center for Rice Research (ICCR) in Sukamandi, Indonesia.

2.2. Experimental Design

The experiment was set in split-split plot design with three replications and the treatments were priming seed with poly ethylene glycol PEG with concentration 200 g L^{-1}, treated (P1) untreated (P0) as sub-sub-plot. The three cultivars mentioned above was set as sub-plot. The irrigation intervals were: irrigated every 5 days (W1), irrigated every 10 days (W2) and irrigated every 15 days (W3) set as main plot. Based on treatments and replication, a total of numbers of 54 experimental units were used in the experiment (3W X 3V X 2P = 18) × 3 replication).

3. Results

Effects of treatments, water interval, varieties, seed priming and their interaction on harvest index of aerobic rice was shown in table 1. In addition, response of the plant to the treatments indicated by some vegetative growth parameters and yield are also shown in table 1.

3.1. Harvest Index

The largest number harvest index had found that recorded by the treatment $W_1V_1P_0$ that about 0.827 and the smallest one by the treatment $W_1V_3P_1$ was 0.495. The maximum harvest index reported by W_1 was 0.68 and the minimum by W_3 about 0.57. Harvest index of V_2 variety found to be the highest equal 0.62 and V_1 the shortest one equal 0.60. Although seed priming P1 had the smallest was 0.57 and P0 the greatest was 0.64 (table 1).

Table 1. Effects of water interval (W), varieties (V), seed priming (P) and their interaction (W × V × P) on some vegetative growth parameters of aerobic rice

Treatment	Harvest Index	Day of 50% Flowering (DAS)	Number of Reproductive Tillers/hill	Panicle Length (cm)	Number of Grain per Panicle	Weight of 1000 Seed (g)	Grain Yield (ton ha^{-1})	Straw Yield (ton ha^{-1})
Water Intervals								
W_1	0.68^a	89.17^b	22.65^a	23.35	99.37	20.44^a	4.1^a	11.58^a
W_2	0.58^b	97.17^a	18.25^b	22.68	92.84	19.26^b	3.15^b	9.19^b
W_3	0.57^b	93.89^a	15.56^c	22.5	82.55	18.26^b	2.26^c	6.7^c
Varieties								
V_1	0.61	94.61	18.65	25.48^a	110.4	19.56	3.7^a	11.23^a
V_2	0.62	91.78	18.31	21.46^b	78.33	19.18	3.2^b	7.38^b
V_3	0.61	93.83	19.5	21.59^b	85.95	19.23	2.7^c	9.06^c
Seed Priming								
P_0	0.65	94.07	18.32	22.44	89.95	19.59	3	8.28^b
P_1	0.58	92.74	19.32	23.241	93.23	19.05	3.34	10.17^a
SE±	**0.051**	**3.42**	**1.36**	**0.92**	**6.11**	**0.85**	**0.38**	**1.32**
C.V%	**12.02**	**6.33**	**12.5**	**6.96**	**11.55**	**7.64**	**20.54**	**24.72**

DAS = days after sowing;
Values not sharing the same letters in same column differ significantly at $p < 0.05$

3.2. Day of 50% Flowering

The fastest and latest area reached to days of 50% flowering recorded by $W_3V_2P_1$ and $W_2V_3P_0$ after 90.33 and 102.67 days from sowing date respectively (table 1). Days of 50% flowering substantially longest for W_2 was 97.17 compared with the shortest W_1 about 89.17 days. V_1 was equal 94.61 as longest time and V_2 equal 91.78 as shortest time for days of 50% flowering (table 1). Beside that P_0 the latest period and P_1 the earliest one were 94.04 and 92.74, respectively.

3.3. Number of Reproductive Tillers per Plant

Table 1 shows that the $W_1V_2P_1$ treatment indicate that the maximum number of reproductive tillers/plant about 24.67 tillers and $W_3V_2P_0$ treatment was the minimum one which was 13 tillers. On the factor of water interval management Number of reproductive tillers/plant was highest by W_1 was 22.65 and smallest by W_3 was 15.56. The largest Number of reproductive tillers/plant recorded for V_3 was 19.5 compared with V_2 the lowest one was 18.31. For seed priming management highest number of reproductive tillers/hill about 19.32 by P_1 and the lowest about 18.31 by P_0.

3.4. *Panicle Length*

Panicle length substantially longest for treatments $W_1V_1P_1$ was 26.99 compared with shortest one that was 20.06 recorded by the treatment $W_2V_3P_0$ (table 1). The different between water concept by W_1 and W_3 as highest panicle length (cm) was 23.34 and the shortest about 22.49 cm. V_1 and V_2 agreed the longest and shortest which were 25.48 and 21.4 days respectively. P_1 raised panicle length (cm) about 23.24 cm and P_0 the lowest about 22.44 cm.

3.5. Number of Grain per Panicle

Biggest number of seed per panicle obtained by treatments $W_1V_1P_1$ was 129.24 seeds/panicle and smallest by $W_3V_2P_0$ about 70.50 seeds/panicle (table 1). W_1 and W_3 varied on their recorded thus W_1 equal 99.37 and W_3 was 82.55 seed/ panicle, V_1 recorded 110.48 and V_2 was 78.33 seed per panicle/ plant. P_1 cleared the biggest number but P_0 the smallest about 93.23 and 89.95 seeds, respectively.

3.6. Weight of 1000 Seed

$W_1V_3P_1$ reported the highest value of weight of 1000 seed (g) about 21.87g and $W_3V_3P_0$ treatment was the lowest one which was 16.47 g with the range about 5.4 g. The biggest weight of 1000 seed (g) has reported by W_1 that was 20.44 g and smallest by W_3 was 18.26 g (table 3). Among varieties V_1 was 19.56 compared with the light value was 19.18g by V_2. Seed priming management for P_1 found the highest score of weight of 1000 seed (g) was 19.59 and P_0 about 19.05 g the lowest one (table 1).

3.7. Grain and Straw Yield

Grain yield ton ha^{-1} substantially highest for treatments $W_1V_1P_1$ that 4.40 ton ha^{-1} compared with lowest one that was 1.60 ton ha^{-1} recorded by the treatment $W_3V_2P_0$. For the effected of water interval W_1 registered the highest grain yield ton ha^{-1} equal 4.1 and W_3 the lowest one equal 2.26 ton ha^{-1}. V_1 and V_3 had the biggest and smallest performance of grain yield ton ha^{-1} about 3.7 and 2.7 respectively. Seed priming management P_1 the highest weight was 3.34 and P_0 the lowest weight was 3 ton ha^{-1}.

Straw yield ton ha^{-1} was highest value by the treatment $W_1V_1P_1$ was 15.36 ton $ha^{-1}$and smallest value obtained by $W_3V_2P_1$ was 5.33 ton ha^{-1} (table 1). The maximum Straw yield ton ha^{-1} wrote by W_1 about 11.58 and the minimum by W_3 was 6.7 ton ha^{-1}. V_1 had been the largest Straw yield ton ha^{-1} was 11.23 and 7.38 by V_3. Seed priming increased Straw yield ton ha^{-1} in P_1 was 10.17 and decreased in P_0 equal 8.28.

4. Discussion

The result of the effects of water interval (W), varieties (V) and seed priming and their interaction on harvest index and some yield components of aerobic rice recorded significant and insignificant difference, in spite of, there was no significant difference in the date of W_1 (irrigated every 5 days) and primed seed revealed the highest data compared with W_2 (irrigated every 10 days), W_3 (irrigated every 15 days) and non-primed seed and depend upon aerobic rice cultivars.

The significance of water interval on reproductive parameters may be due to significant effect of water management on the number of tillers, this result confirmed with [3] who applied that water regime management significantly affected number of tillers and the interaction between water regime management and varieties was also significant. Among straw and grain yield the significant differences may be due to the small total dry matter which related to the relatively shallow root system and stomata closure and reduced photosynthesis in response to surface soil drying which agreed with Lafitte and Benett [9] and Kato *et al.* [6].

The significant effect of varieties on reproductive parameters explained with genetic constitution beside sometimes interacts with environment that cleared with Amano *et al.* [10] and Cheng and Min [11] aerobic or super-rice cultivars have been released in China, which have shown a markedly high yield potential when planted in the aerobic conditions. However, these aerobic rice cultivars are commonly characterized by unstable yields across environments (locations and years) [12, 5]. Therefore, it is imperative to clarify the responses of their growth and yield formation to environments so it is well known that final yield (phenotype) of a given cultivar is dependent on both genetic factors and growth conditions

The insignificants effect of varieties on harvest index and some yield components may be due to similarity in morphological aspects of vegetative growth between aerobic rice cultivars such as time to heading initiation and duration of the grain heading for example, biomass accumulation in the form of stem biomass reaching at heading and decline during grain filling which finally effect yield output. These results followed same results of Chen *et al* [13] the rice cultivars shown obvious similarity in the change of stem biomass between aerobic rice cultivars and little increase after booting stage, whereas different in duration from booting stage to heading stage

Seed priming during reproductive period of rice clearly reported its effect on harvest index and reproductive stage components attributes may be to pre-germination metabolic activities making the seed ready for radical protrusion which lead to good crops establishment and resistance to different treatments, confirmed with reported that number of tillers of primed seed with PEG increased compared with control [14], but not followed the same result of time of flowering initiation, time of maturity because non-primed faster than primed seed, thus this fact was true so priming techniques help plant to be tolerant against bad conditions especially water-stress, drought and water scarcity and this consist with the work of Desclaux and Roumet [15] who noted that switch from vegetative growth to reproductive growth was earlier under water-stress than under wetter condition.

By applying appropriate irrigation management in rice cultivation, a large volume of water can be saved which would help to bring more land under cultivation using the same available amount of water. This will not only enhance food security but water security as well.

5. Conclusion

Depend upon genotype, water interval management to maintain suitable soil water contents before later irrigation between 44.48 to 18.20% will lead to cultivate aerobic rice under drought (water scarcity) conditions with assistant of seed priming technology using PEG solution 200 g L^{-1} concentration not more.

Acknowledgement

The authors acknowledge the financial support of Hasanuddin University (UNHAS) and Cereal Crop Research Institute (BALITJAS), Indonesia under the indigenous Ph.D. scheme for the conduction of these field experiments.

References

[1] Donald C M and Hamblin J 1976 The biological yield and harvest index of cereals as agronomic and plant breeding criteria. *Adv. Agron.* **28** 361–405

[2] Khush G S 1993 *Breeding Rice For Sustainable Agriculture Systems(International Crop Science I)* (Madison: Crop Science Society of America) pp 189-199

[3] Abdallah E H A, Musa Y, Mustafa M, R Sjahril, Kasim N and Riadi M 2016 Seed germination behaviors of some aerobic rice cultivars (*Oryza sativa* L) after priming with Polyethylene glycol-8000 (PEG-8000) *Intl. J. Sci. Tech. Res.* **5** (2) 227-234

[4] Abdallah E H, Musa Y, Mustafa M, Sjahril R and Riadi M 2016 Comparison between hydro- and osmo-priming to determine period needed for priming indicator and its effect on germination percentage of aerobic rice cultivars (*Oryza sativa* L.) AGRIVITA J. of Agr. Sci. **38** (3) 222-230

[5] Ying J F, Peng S B, He Q R, Yang H, Yang C D, Visperas R M and Cassman K G 1998. Comparison of high yield rice in tropical and subtropical environments I. Determinants of grain and dry matter yields *Field Crops Res.* **57** (1) 71-84

[6] Hussain M, Farooq M, Basra S M A and Ahmad N 2006 Influence of seed priming techniques on the seedling establishment, yield and quality of hybrid sunflower *Int Agric Biol* **8** 14-18

[7] Parthasarathi T, Vanitha K, Lakshamanakumar P and Kalaiyarasi D 2012 Aerobic rice- mitigating water stress for the future climate change *Int. J. of Agronom. and Plant Prod.* **3** (7) 241-254

[8] Lal B A, Nayak K, Priyanka G, Rahul T, Teekam S and Katara J L 2013 Aerobic rice: A water saving approach for rice production *Popular Kheti* **1** (2) ISSN:232 1-0001

[9] Lafitte H R and Bennett J 2002 *Requirements for aerobic rice: Physiological and molecular considerations (Water-wise Rice Production)* ed. B A M Bouman et al (Los Banos: International Rice Research Institute) pp 259-274

[10] Amano T, Shi C, Qin D, Tsuda M and Matsumoto Y 1996 High-yielding performance of paddy rice achieved in Yunnan Province, China: I. High yielding ability of Japonica F1 hybrid rice, Yu-Za 29 *Japan. J. of Crop Sci.* **65** (1)16-21

[11] Cheng S H and Min S K 2001 Super-rice breeding in China *Chine. Rice Res. Newsl.* **9** (2) 13-15

[12] Horie T, Ohnishi M, Angus J F, Lewin L G, Tsukaguchi T, and Matano T 1997 Physiological characteristics of high-yielding rice inferred from cross-location experiments *Field Crops Res.* **52** (1-2) 55-67

[13] Chen S, Zeng F, Pao Z and Zhang G 2008 Characterization of high-yield performance as affected by genotype and environment in rice journal of Zhejiang University Science B **9** (5) 363-370
[14] Arif M, Tariqjan M, Ullahkhan N, Ahmad K, Khan M J and Munir A 2010 Effect of seed priming on growth parameters of soybean *Pak. J. Bot.* **42** (4) 2803-2812
[15] Desclaux D and Roumet 1996 Impact of drought stress on to phenology of two soybeans (*Glycine max* L. Merr.) cultivars *Field Crop Res.* **46** 61-70

IOP Conf. Series: Earth and Environmental Science **157** (2018) 012022 doi:10.1088/1755-1315/157/1/012022

Abundance of arbuscular mychorrizal fungi in rehabilitation area of nickel post-mining land of Sorowako, South Sulawesi

M A Akib[1], K Mustari[2], T Kuswinanti[3] and S A Syaiful[2]

[1] Graduate School, Hasanuddin University, Jalan Perintis Kemerdekaan KM 10, Makassar, South Sulawesi, 90245, Indonesia.
[2] Department of Agronomy, Faculty of Agriculture, Hasanuddin University, Jalan Perintis Kemerdekaan KM 10, Makassar, South Sulawesi, 90245, Indonesia.
[3] Department of Plant Pests and Diseases, Faculty of Agriculture, Hasanuddin University, Jalan Perintis Kemerdekaan KM 10,Makassar, South Sulawesi, 90245, Indonesia.

E-mail: *koeswinanti@yahoo.com*

Abstract. Acceleration management of land rehabilitation in nickel post-mining in Sorowako has been main attention of Vale Indonesia. This acceleration can be done by utilizing of natural resources, especially indigenous endomycorrhiza. Endomycorrhiza also called arbuscular mycorrhizal has got a lot of attention for its ability to form a mutualistic symbiosis with 80% - 96% of plant species. This study aims to determine the dominance of indigenous endomycorrhiza spores and its potential to accelerate the management of land rehabilitation post-mining of nickel, which is carried out in three stages; sampling rhizosphere, trapping spores, isolation and identification of the arbuscular mycorrhizal spores types. The results showed that the dominance of indigenous endomycorrhiza were *Acalauspora* sp (75.1%), *Gigaspora* sp (19.4%) and *Glomus* sp (5.6%). Research on the effectiveness of indigenous endomycorrhiza using *Acalauspora* sp in land rehabilitation of nickel post-mining is still ongoing.

1. Introduction

Vale Indonesia is a nickel-mining company located in Sorowako village, Malili sub-district, Luwu Timur district, South Sulawesi Province, Indonesia. This company has contracted land for mining in Sulawesi with total area of 218,528.99 hectares. Those areas are spread across Sulawesi, such as Sorowako in South Sulawesi (118,387.45 hectares), Pomala in Southeast Sulawesi (635,060.18 hectares) and Bahodopi in Central Sulawesi (36,635.36 hectares) [1]. There are 4,973.15 hectares of area in South Sulawesi that have been explored by the company. Moreover, there are 3,975.91 hectares of the area that are rehabilitated and reclaimed at the end of 2014 [2].

The twelve-year revegetation process has brought positive impacts on both physical and chemical parameters of soil, which are increased significantly [3]. Besides, the growth rate of plants in the revegetation area is around 95-99%. Most of the plants that are planted in this area are pioneer type such as *Paraserianthes falcataria*, *Eucalytus eurograndis*, *Enterolobium macrocarpum*, also local species namely *Melochiaumbellata*, *Sandoricum kacappeae* and *Elmerelia* sp. [4].

In general, the majority of various plants are involved in a symbiotic relationship with another organism. For instance, plants with arbuscular mycorrhizal (AM) are benefitted by each other through mutualistic symbiosis. Specifically, 80% of plant's roots are associated with the fungi [5]. There are 2 types of this association, namely endomycorrhiza and ectomycorrhiza [6]. Furthermore, the major genera found in nickel post-mining land were *Glomus* sp., *Acaulospora* sp., dan *Gigaspora* sp. [7]. However, their abundance level and population growth rate in Sorowako as nickel post-mining area are not revealed yet.

2. Materials and Methods

This qualitative research was carried out in three stages. The first stage was the taking of rhizosphere samples from rehabilitation area in nickel post-mining land owned by Vale Indonesia in Sorowako Village, that have been rehabilitated for less than 2 years in Sumasang 1; 2 to 4 years in Sumasang 2 block A and more than 4 years in Sumasang 2 block B. The sampling method was done by developing the method undertaken by [8, 9]. Mycorrhizal spores were isolated from the host plant rhizosphere using the wet screening method of Gerdemann and Nicolson (1963) which has been compared with other methods by [12]. The second stage was trapping cultures of spora were performed to obtain high spore viability, using trapping culture methods from [10] and [11] in nursery agroplastid in Parepare. Finally or third stage, the population of AM spore from *Glomus* sp., *Acaulospora* sp. and *Gigaspora* sp. were observed in Laboratory of Microbiology of Research and Development Center for Environment and Forestry Makassar. Microscopic characters of spore were matched with identification guide from [13] to identify genus of mycorrhizal fungi. Data are presented in a pie chart and bar chart.

3. Results and Discussion

Root is a suitable environment for a microbe to live, which interaction could benefit for both sides. According to scientists, the contact areas for a microbe in the plant are divided into phyllosphere (above-ground) and rhizosphere (under-ground) [14]. On the other side, the microbes that interact with plants are divided into phyllosphere (on the surface of plant organs) and epiphyte (colonialize the inner tissue of the plant) [15]. Moreover, microbes that interact with plants under the ground are further divided into rhizoplane (attach to root) and endophyte (live inside root cells) [16]. Those microbes help the plant to supply the nutrition or act as an anti-pathogenic agent that can be harm for the host. Next, the host plant provides the habitat and food supplies for microbes to live. In other words, this interaction is mutualistic for both sides.

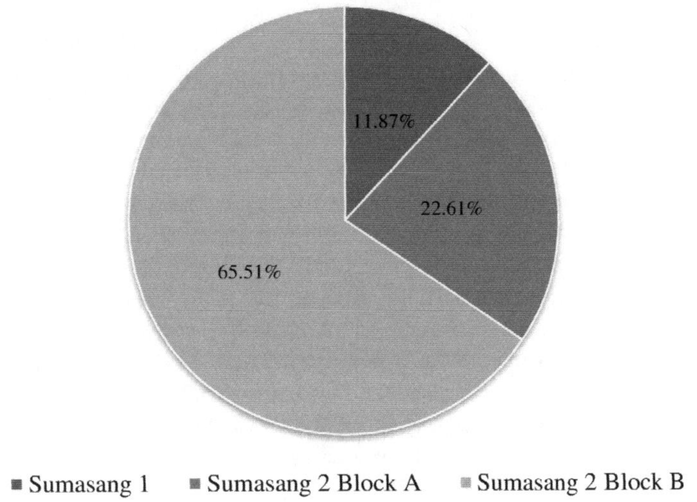

Figure 1. Population abundance of arbusculas mycorrhizal spores at various ages of nickel post-mining rehabilitation land.

Arbuscular mycorrhizal spore was found abundance in the area with more than 4 years of rehabilitation. The abundance level of the three genera was 65.51% (figure 1). As a result of crop adaptation to the environment, the energy supply needed for the survival of fungi is very suitable (figure 2) so that the development of mycorrhizal may increase 189.7% in rehabilitated area with more than 4 years. According to Hermawan [3], twelve-year revegetation process has improved the physical (texture, bulk density, porosity, water concentration, and penetration resistance) and chemical (pH, organic carbon, minerals, exchangeable cation, and exchange capacity) variables of soil.

Figure 2. The vegetation of trees and shrubs in nickel post-mining rehabilitation land in Sorowako (In Sumasang 1 (a); Sumasang 2 block A (b); and Sumasang 2 block B (c)).

The rhizosphere is suitable habitat for microbes due to the availability of organic matter. This condition allows microbes to grow and develop that are called as rhizosphere effect. This effect is centrifugal and tends to increase as the root branching system become more complicated. Besides, there are also several factors that determine the rhizosphere effect, namely soil type, soil humidity, pH, Temperature, plant's age, and plant's condition [17].

In general, there are two major types of the rhizosphere, namely endo rhizosphere including stele, cortex, endodermis, and root cap also recto rhizosphere that covers the area surrounding the root such as soil contact zone [17]. [18] Stated that rhizosphere includes space surrounding root, in which there are complex biological and ecological processes take place. It mediates biological interaction inside the soil [19].

Root secretes several organic matters in the form of exudate needed by microorganism as their energy resources and substrate for their metabolism such as sugar, amino acid, organic acid, fatty acid, sterol, nucleotides, flavonom, enzyme, and miscellaneous [20]. There are several factors that determine exudate composition and quantity such as plant species, plant age, environmental condition (temperature, radiation, soil humidity, soil type, plant nutrition, and plant stress), and microorganism existence [17]. Beside exudate, roots actively pump out several materials called as root secrets [21]. This can be physiological response toward external environment, such as organic acid that is secreted by rice as the response to Aluminum existence [22]. Another material that is secreted by root is root lysate such as protein, fat, and amino acid which are passively produced during autolysis of root cells [21]. Next material is musigel as the form of polysaccharide synthesized by golgi apparatus inside the cells of root cap, which moves through the cytoplasm to plasmalemma [23]. For instance, musigel can be formed as root secets, epidermis cell debris, and microbial cell fuse with root cap cell debris, a metabolic product, organic colloid, and an organic colloid [21]. At last, the main enzymes produced by root are oxidoreductase, hydrolase, lyase, and transferase. On the other hand, microbes secrete cellulose, dehydrogenase, urease, phosphatase, and sulphatase in the rhizosphere [24].

In succession process, vegetation is initially formed by the invasion of the new plant into the area through stages. This plant along with the other plants then adapting to the environmental condition that is called as aggregation stage. Following this, those plants competing for each other in order to reach the ecosystem balance or called as stabilization stage [25].

On the other side, the abundance of spore population on the land that has been rehabilitated for less than 2 years shows insignificant growth (figure 3). This can be expected as the effect of stress caused by backfilling activity, in which the spore has not been adapted to the new host. Sundari *et al.* [26] added that the existence of AM is determined by an environmental factor. Moreover, Delvian [27] deduced several theories about AM growth and development. The germination process of AM spore is divided into 4 phases, namely hydration, activation, sprout channels growth, and hyphae growth. In the first phase, water enters into spore for hydrating the organelles and macromolecules such as ribonucleic acid and enzymes to activate metabolism. Spore becomes active in the second until the tenth day of hydration. It follows by the formation of sprouts channel and hyphae growth.

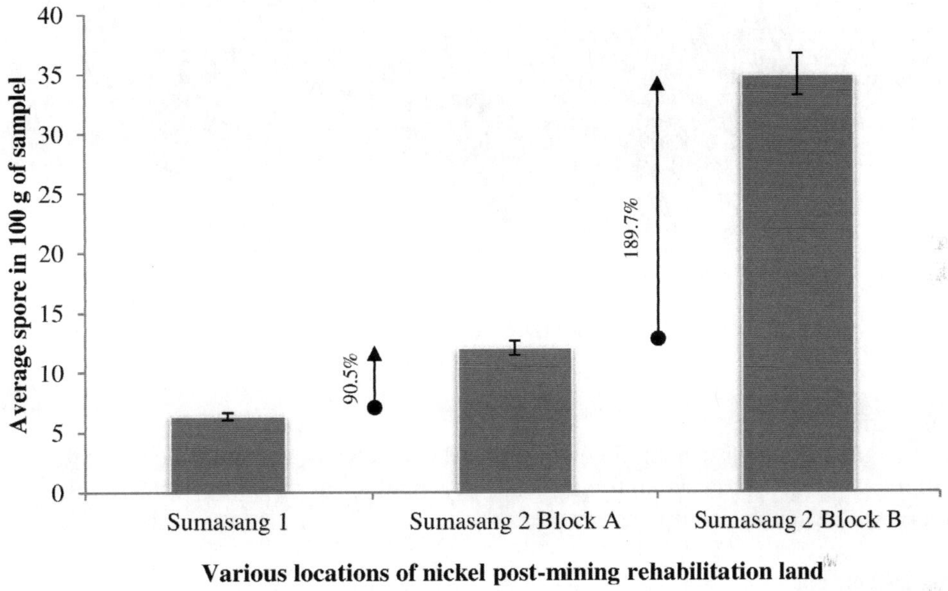

Figure 3. Increased population of AM Spores in various locations of nickel post-mining rehabilitation land in Sorowako.

There is several factors influence spore infection. Specifically, there are two steps of infection namely primary and secondary infection. The primary infection is mainly influenced by the other fungi spore growth, hyphae growth in soil. Besides, some AM also depends on sprout channel growth that is possibly influenced by root exudate, soil fertility, water supply, and enter spots in the root. These factors can be limitations for AM formation. In addition, secondary infection is entirely determined by host physiology due to it provides energy from photosynthesis for hyphae spread, which is transferred from plant to fungi through arbuscular or internal hyphae.

The characteristics of spores obtained from observation are varied. The classification was done based on their shape and colors. There are three genera identified in this study, *Glomus* sp, *Acalauspora* sp, and *Gigaspora* sp. Analysis of spore abundance shows that *Gigaspora* sp. least dominating the area. Moreover, each spore shows specific character. For instance, thick wall and ornamental spore are shown by *Acaulospora.* Following this, *Gigaspora* has bulbous suspenser in its hyphae base and has no sprouts layer. At last, *Glomus* spore has subtending hyphae (figure 4).

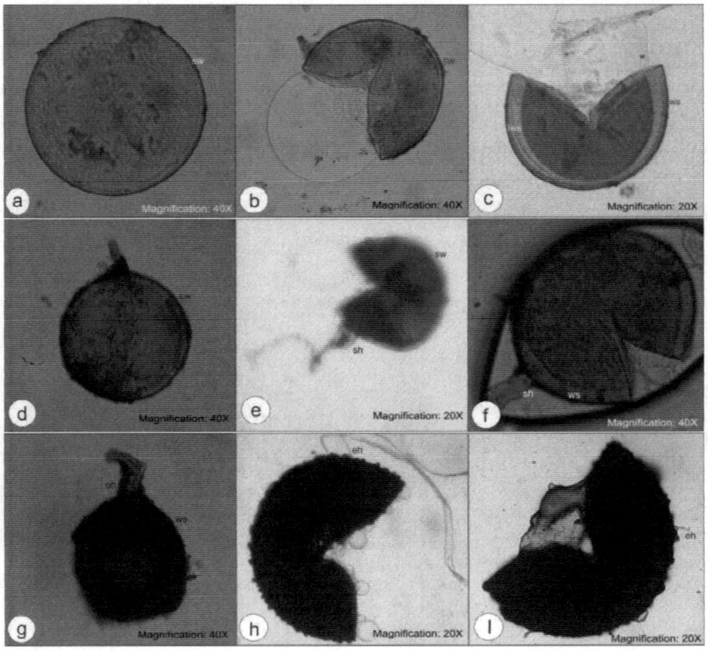

Figure 4. Spores of arbuscular mycorrhiza in the nickel post-mining rehabilitated land (*Acaulospora* sp. (a - c), *Gigaspora* sp. (d - f), *Glomus* sp. (g - i)).

4. Conclusion

There is a significant improvement in the biological characteristic of soil in rehabilitation area of nickel post-mining land in Sorowako through time. Arbuscular mycorrhizal form *Acauolospora* sp. and *Gigaspora* sp. are found dominating the area.

Acknowledgement

The author would like to thanks to Vale Indonesia for the support. Also thanks to Research and Development Center for Environment and Forestry of Makassar for the advice during the study.

References

[1] Maruto R 2013 *PT Vale Akan Lepas 13.000 Hektare Lahan* www.antarasulteng.com Accessed 8 March 2016

[2] Yasid M 2015 *Tahun ini Vale akan mereklamasi lahan 77 Ha* www.industri.kontan.co.id Accessed 8 March 2016

[3] Hermawan B 2011 *Peningkatan Kualitas Lahan Bekas Tambang melalui Revegetasi dan Kesesuaiannya Sebagai Lahan Pertanian Tanaman Pangan Pros. Seminar Nasional Budidaya Pertanian* Bengkulu 7 Juli 2011 pp 60-70. ISBN 978-602-19247-0-9

[4] Abubakar F 2009 *Evaluasi Tingkat Keberhasilan Revegetasi Lahan Bekas Tambang Nikel Di PT. INCO Tbk. Sorowako, Sulawesi Selatan* (Bogor: Institut Pertanian Bogor) http://repository.ipb.ac.id Accessed 9 March 2016

[5] Sufaati S, Suharno and Bone I H 2011 Endomikoriza yang berasosiasi dengan tanaman pertanian non-legum di lahan pertanian daerah transmigrasi Koya Barat, Kota Jayapura *J. Biologi Papua* **3** 1 – 8

[6] Cahyani N K M D, Nurhatika S dan Muhibuddin A 2014 Eksplorasi mikoriza vesikular arbuskular (mva) indigenous pada Tanah Aluvial di Kabupaten Pamekasan Madura *J. Sains dan Seni Pomits* **3** 2337-3520

[7] Setiadi Y and Setawan A 2011 Study of arbuscular mycorrhizal fungi status at rehabilitation post-nickel mining area (Case study at PT INCO TBK. Sorowako, South Sulawesi) *J. Silvikultur Tropika* **3** 88-95

[8] Estrada B, Hermosob M B, Palenzuela J, Iwase K, Lozano J M R, Barea J M, Oehl F 2013 Diversity of arbuscular mycorrhizal fungi in the rhizosphere of *Asteriscus maritimus* (L.) Less., a representative plant species in arid and saline Mediterranean ecosystems *J. of Arid Env.* **97** 170-175

[9] Krishnamoorthy R, Kim C G, Subramanian P, Kim K P, Selvakumar G and Sa T M 2015 Arbuscular mycorrhizal fungi community structure, abundance and species richness changes in soil by different levels of heavy metal and metalloid concentration PLOS ONE **2** 1-15

[10] Brundrett M, Bougher N, Dells B, Grove T and Malajozuk N 1996 *Working with Mycorrhizas in Forestry and Agriculture* (Canberra: Australian Ventre for International Agricultural Research)

[11] Muis R, Ghulamahdi M, Melati M, Purwono and Mansur I 2016 Diversity of arbuscular mycorrhiza fungi from trapping using different host plants *Int. J. of Sci.: Basic and App. Res. (IJSBAR)* **27**158-169

[12] Shamini S and Amutha D K 2014 Techniques for extraction of arbuscular mycorrhizal fungi spores. Int. J. of Front. in Sci. and Tech. **2** 1-6

[13] INVAM 2017 *International Culture Collection of (Vesicular) Arbuscular Mycorrhizal Fungi.* Available at: http://invam.caf.wvu.edu/Mycoinfo/Taxonomy/classification.htm Accessed 8 March 2017

[14] Pas A A, Sopandie D, Trikoesoemaningtyas and Santosa D A 2015 Application of Phyllosphere and Rhizosphere Microbial Consortium to Improve Rice rowth and Production *J. Pangan* **24** 15-24

[15] Santosa D A, Handayani N and Iswandi A 2003 Isolation and selection of growth promoting phyllosphere bacteria from leaf of rice *(Oryza sativa* L.*)* Cultivar IR·64 *J. Ilmu Tanah dan Lingkungan* **5** 7-12

[16] Widyati E 2013 Understanding on plants-microbes interaction *J. Tekno Hutan Tanaman* **6** (1)13-20

[17] Budiyanto G 2015 *Interaksi Biologi Nitrogen Dalam Tanah Makalah Kuliah Umum Program Studi Agroteknologi* (Yogyakarta: Fakultas Pertanian Universitas Muhammadiyah)

[18] Bais H P, Weir T L, Perry L G, Gilroy S and Vivanco J M 2006 The role of root exudates in rhizosphere interactions with plants and other organisms *Ann. Rev. of Plant Bio.* **57** 233 – 266

[19] Huang X F, Chaparro J M, Reardon K F, Zhang R, Shen Q and Vivanco J M 2014 Rhizosphere interactions: Root exudates, microbes, and microbial communities *Botany* **92** 267–275

[20] Wihardjaka A 2010 Emisi gas dinitrogen oksida dari tanah sawah tadah hujan yang diberi jerami padi dan bahan penghambat nitrifikasi *J. Biologi Indonesia* **6** 211-224

[21] Sari D R 2015 Isolation and identification soil bacteria around plant roots *Bio-Site* **1** 21-27

[22] Fajarwati I 2007 *Organic Acid Secretion on Al- Stressed Rice* (*Skripsi Departemen Biologi Fakultas Matematika Dan Ilmu Pengetahuan Alam*) (Bogor: Institut Pertanian Bogor) http://repository.ipb.ac.id Accessed 8 March 2017

[23] Takehisa H, Sato Y, Igarashi M, Abiko T, Antonio B A, Kamatsuki K, Minami H, Namiki N, Inukai Y, Nakazono M and Nagamura Y 2012 Genome-wide transcriptome dissection of the rice root system: implications for developmental and physiological functions *The Plant J.* **69** 126–140

[24] Sudana W 2005 Pemanfaatan mikrobia pelarut fosfat dan mikoriza untuk perbaikan fosfor tersedia, serapan fosfor tanah ultisol dan hasil jagung pada ultiso. *J. Ilmu-Ilmu Pertanian Indonesia* **6** 8-13

[25] Susilawati and Maryati A 2012 *Identifikasi Mikroba Rhizosfer Tumbuhan Pioner Di Lahan Eks Penambangan Batubara Sebagai Bahan Bioremedias* (Kalimantan Tengah: Balai Pengkajian Teknologi Pertanian) www.pkpp.ristek.go.id Accessed 13 September 2016

[26] Sundari S, Nurhindayati T and Trisnawati I 2011 *Isolasi dan Identifikasi Mikoriza Indigenous dari Perakaran Tembakau Sawah (Nicotiana tabacum L) di Area Persawahan Kabupaten Madura* (Surabaya: Fakultas matematika dan Ilmu Pengetahuan Alam, Institut Teknologi Sepuluh November)
[27] Delvian 2005 *Respon Pertumbuhan dan Perkembangan Cendawan Mikoriza Arbuskula dan Tanaman Terhadap Salinitas Tanah* www.library.usu.ac.id Accessed 13 September 2016

IOP Conf. Series: Earth and Environmental Science **157** (2018) 012023 doi:10.1088/1755-1315/157/1/012023

Biophysics and economic potential analysis of vertisols for maize in the humid tropics of Indonesia

R Neswati, C Lopulisa, A Ahmad and M Nathan

Department of Soil Science, Hasanuddin University, Jalan Perintis Kemerdekaan KM 10, Makassar, 90245, Indonesia.

E-mail: *riesma76@yahoo.com*

Abstract. The main objective of this study is to establish the potential of Vertisols both biophysical and economic found in the humid relatively dry tropics region in the southern part of Sulawesi to the development of maize. This study used a spatial approach for establishing research sites totalling in 3 observations units of soil profile and involved 30 farmers as respondents. Land potential analysis of was conducted using parametric approach while economic analysis using the benefit-cost (B-C) ratio. The results show that the growing period of the study sites was in November to June, or 240 days and classified as climate type E3. The locations potentially moderately suitable or S2s for maize with soil texture as limiting factors in the growing season from November to June. In the growing season from July to October, the locations are not suitable or Nc with limiting factors such as very limited rainfall in the period of crop ripening. Land suitability Index of study site during the growing season from November to June ranged from 52 to 72 with an average productivity of maize obtained by farmers ranged from 4.3 to 5.7 tons of dry grain per hectare. Analysis of B-C ratio indicates that cultivation of maize in the growing period at such study locations are feasible with the value of B-C ratio ranged from 1.8 to 2.5. These results show that Vertisols in the humid tropics of Indonesia are physically and economically potential for maize development.

1. Introduction

South Sulawesi, one of the major food-producing provinces in the eastern part of Indonesia produces a large amount of maize. This region has varied climate types. According to the criteria of Oldeman [1], the climate in South Sulawesi classified as B1, C1, C2, D1, D2, E1, E2, and E3. Average of maize production in South Sulawesi varies in the range of 3-5 tons/ha. The highest average productivity can be found in the southern region with relatively dry climate, while the lowest productivity is in the northern region with relatively wet (humid). Based on the exploratory map of Indonesia [2], one of the main soil types found in South Sulawesi is Vertisols (3.6 %) which has to swell and shrink properties. One of the maize-producing districts in South Sulawesi, Jeneponto District has the most extensive land area of Vertisols. The vertic property can be one of the major growth inhibiting factors of maize growth at the time of shrinking which results in the breaking of plant roots. Maize productivity is strongly influenced by the quality/characteristics of the land, including soil and climate. Agricultural development always needs soil suitability evaluation that can be expressed as "land suitability index" measured on different types of soil. The types of soil formed by a certain pedogenesis process may provide information on the potential productivity of the soil [3]. Soils also provide several ecological and social functions [4, 5]. The soil characteristics such as fertility, water availability, and land management have not been to date under consideration by farmers in agricultural development. Land evaluation is necessary to produce recommendations for appropriate land management according to land potential. To obtain production level according to land potential, crops should be grown under conditions of sufficient water and nutrients, while pest, weed, and disease should be controlled [6]. Therefore, land suitability classification needs to be taken into account in the development of high commodity because it is based on knowledge of the requirements of crops, soil characteristics, and soil management [7]. The result of land evaluation should reflect not only possible yield but also, more importantly, the ease or difficulty of management of the land parcels for a given purpose and ensure the land's sustained use for a long period of time [8]. The main objective of this study is to determine the potential of Vertisols found in the humid tropics relatively dry southern part of Sulawesi both biophysically and economically for the development of maize.

2. Methodology

2.1. Study Area and Data Collection

The study area located at 3 land units in District of Jeneponto (figure 1) which classified as Vertisols where maize is cultivated (existing land use). The three land units are characterized by high humidity range between 66.0% and 86.7%, mean annual rainfall distribution of 1.500 mm and mean temperature ranged between 26.5 °C and 27.1 °C. The study area is underlined by carbonate rocks, volcanic breccias. Ten-year meteorological data used in this study were collected from a climatology station to determine rainfall, temperature, relative humidity and daylength.

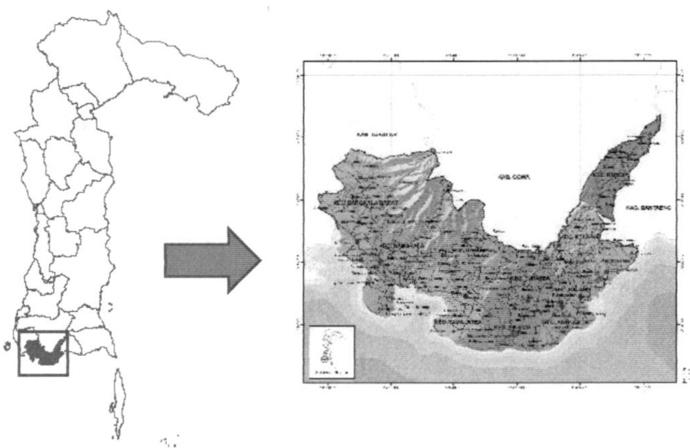

Figure 1. Location of study area, in the box, is Jeneponto Regency

2.2. Field Survey and Laboratory Analysis

Soil characteristics surveyed in 2016 was analyzed using purposive sampling method. Identification of land under traditional maize cultivation in the study area was undertaken during cultivation period (November to February and March to June). As many as 30 farmers (focusing on maize cultivation) were sampled and 3 representative soil profiles were also made to examine soil horizons. Soil samples were collected and analyzed using standard procedures, as follows, soil particle size distribution was determined using hydrometer method [9]; soil organic carbon was determined using dichromate oxidation procedure [10]; exchangeable bases (Ca, Mg, K, Na) and cation exchange capacity (CEC) were determined with 1 N NH_4OAc, pH 7.0. Soil reaction (pH) was measured in water (1 : 2.5, soil water ratio). Maize harvested at a randomly determined unit sample of 2.5 x 2.5 sq meters, with 3 replications. The harvested seeds of maize were dried, then weighted and their weights were calculated by using (1) [11]:

$$W \text{ (tons/ha)} = 1.6 \times W \text{ sample} \quad (1)$$

2.3. Data Analysis

The soils were classified in a subgroup group level according to Keys to Soil Taxonomy [12]. The suitability classification was carried out separately for each soil unit identified in the survey area. Land suitability evaluation was performed using the parametric method of [13]. The parameters used for land quality calculation include rainfall of growing cycle, mean temperature, relative humidity, length of radiation, slope, drainage, soil physics (texture, depth), soil fertility (pH, organic carbon, CEC, base saturation). In the evaluation, the soil characteristics were then compared with crop requirement of maize [14]. Land suitability index (I) calculated from the individual ratings was determined using Square Root method [15] with equation:

$$I = Rmin \times \sqrt{A/100 \times B/100 \times \ldots\ldots} \qquad (2)$$

where I= Land suitability index; R min = Minimum rating; A, B, ..= Other ratings beside the minimum rating.

An Economic analysis using the benefit-cost ratio can be defined as the ratio of the equivalent value of benefits to the equivalent value of costs. The equivalent values can be present values, annual values, or future values. The benefit-cost ratio (BCR) is formulated as:

$$BCR = B / C \qquad (3)$$

where B represents the equivalent value of the benefits associated with the project and C represents the project's net cost [16]. A B/C ratio greater than or equal to 1.0 indicates that the project evaluated is economically advantageous.

3. Results and Discussions

3.1. Soil Classification

The three soil profiles were classified at Subgroup levels. The two of pedons classified as *Typic Calciusterts* and *Typic Haplusterts* show Vertisols characteristics which have slickensides and have more than 30 percent clay in the fine-earth fraction either between the mineral soil surface and a depth of 18 cm or in an Ap horizon. In addition, the pedons have cracks that open and close periodically and high base saturation (vary from 52 to 60%). *Typic Calciusterts* has calcic horizons within 63 cm. *Typic Haplusterts, Typic Calciusterts,* occupy the gently undulating lower slope, shallow with soil depth ranging from 20 to 65 cm, and well drained. They are found in climatology zones of E2 according to Oldeman classification [1]. Their parent materials are limestone and volcanic breccias.

3.2. Distribution of Maize Yields and Suitability Class

The results of field study show that maize yield variation ranges from 4.3 to 5.7 tons ha^{-1} with land suitability index (I) ranges from 52 to 72. Based on [13], the index value (I) of land ranging from 75 to 100 is classified as highly suitable (S1); 50 to 75 as moderately suitable (S2), and 25 to 50 as marginal (S3). Their parent materials are mostly limestone and volcanic breccias which consist of a lot of basic cations such as calcium and magnesium. Therefore, soils formed from limestone minerals have high base saturation and soil pH tend to be around neutral [17]. Soil fertility characteristics like base saturation and soil pH as well as water availability influence maize yield [18]. In the period of November to June, 3 units of analysed land classified as an S2 class or sufficient according to soil texture limiting factors, whereas in the growing season July to October due to the problem of water availability, the land unit classified as unsuitable (N2).

3.3. Benefit-Cost Ratio of Land Unit at Study Area

Based on the analysis of the B-C ratio on 30 respondents in the research location, it can be seen that the feasibility of maize farming attempted by farmers in Vertisols soil was still economically feasible with B-C ratio range between 1.8 and 2.5. The B-C ratio indicates that the investment of maize in this area is economically viable and on an average 1 investment brings 1.8 to 2.5 returns.

4. Conclusion

The growing period of the study sites was in November to June, or 240 days and classified as climate type E3. The location potentially moderately suitable or S2 for maize with soil texture as limiting factor in the growing season from November to June. In the growing season from July to October, the location is not suitable or Nc with limiting factors such as limited rainfall at the period of fruit ripening. Land suitability index of the study site during the growing season from November to June ranged from 52 to 72 with an average productivity of maize was obtained by farmers ranges from 4.3 to 5.7 tons of dry seed per hectare. Analysis of B-C ratio shows that the cultivation of maize in the growing period at locations such studies are feasible with the value of B-C ratio ranges from 1.8 to 2.5.

Acknowledgments
The authors deeply thankful to the Ministry of Research & Technology, Higher Education of Indonesia for providing research fund through the "PTUPT" Scheme, and the Department of Soil Science, Faculty of Agriculture, Hasanuddin University, for providing data, and other supporting facilities for this study.

References

[1] Oldeman 1977 *The Agroclimatic Map of Sulawesi* (Bogor: Contr. Centr. Res. Ins. Agric.) No. 33. 30p+map.
[2] Subagyo H, Suharta N and Siswanto A B 2000 *Agriculture soils in Indonesia* (Bogor: Indonesian Center for Soil and Agroclimate Research and Development, Ministry of Agriculture of Indonesia)
[3] Mueller L, Schindler U, Mirschel W, Shepherd T G, Ball B C, Helming K, Rogasik J, Eulenstein F and Wiggering H 2010 Assessing the productivity function of soils A Review *Agron. Sustain. Dev.* **30** 601-614 INRA.EDP Sciences
[4] Lal R 2008 Soils and sustainable agriculture. A review *Agron. Sust. Dev.* **28** 57-64
[5] Jones A, Stolbovoy V, Rusco E, Gentile A R, Gardi C, Marechal B and Montanarella L 2009 Climate change in Europe. Impact on soil. A review *Agron. Sust. Dev.* **29** 423-432
[6] Verdoodt A, and van Ranst E 2003 *Land Evaluation for Agricultural Production in The Tropics* (Gent, Belgium: Laboratory of Soil Science, Ghent University)
[7] Ande O T 2011 Soil suitability evaluation and management for cassava production in the derived savanna area of South Western Nigeria *Int. J. of Soil Sci.* **6** (2) 142-149
[8] Baja S 2009 *Land Use Choice and Land Resources Assessment in Agriculture. A review* (*CAB Reviews: Perspectives in Agriculture, Veterinary Science, Nutrition and Natural Resources*)
[9] Bouyoucos G J 1962 Hydrometer method improved for making particle size analysis of soils *Agron. J.* **54** 464-465
[10] Walkley A and Black G A 1934 An examination of Degtjareff NHO for determining soil organic matter and a proposed modification of chronic acid titrate *Soil Sci. Battinose* **37** 29-38
[11] Baja S, Amrullah A, Ramli M, and Ramlan A 2011 *Spatial-based fuzzy classification of land suitability index for agriculture development: A model validation perspective* (*Proc. of 3rd International Joint Conference on Computational Intelligence-International Conference on Fuzzy Computation Theory & Applications IJCCI-FCTA*) Paris, France, 24-26 October 2011 p 435-440
[12] Soil Survey Staf 2010 *Keys to Soil Taxonomy. Eleventh Edition* (United States Department of Agriculture, Natural Resources Conservation Service)
[13] Sys C, van Ranst E and Debaveye J 1991 *Land Evaluation. Part I. Principles in Land Evaluation and Crop Production Calculations Agricultural Publications No. 7* (Brussel-Belgium: General Administration for Development Cooperation)
[14] Sys C, van Ranst E and Debaveye J 1993 *Land Evaluation. Part III. Crop Requirements. Agricultural Publications No. 7* (Brussel-Belgium: General Administration for Development Cooperation)
[15] Khiddir S M 1986 *A Statistical Approach in the Use of Parametric System Applied to the FAO Framework for Land Evaluation* (Belgium: State University of Ghent) p 144
[16] Ulukan Z and Ucuncuoglo C 2010 Economic analyses for the evaluation of is projects *J. of Inform. Sys. and Tech. Manag.* **7** (2) 233-260
[17] Sys C 1977 *Regional Pedology. Agricultural Publications No. 7* (Brussel-Belgium: General Administration for Development Cooperation)

[18] Calvino P A, Andrade F H and Sadras V O 2003 Maize yield as affected by water availability, soil depth, and crop management *Agron. J. 2750* **95** (2) 275-281

GIS-based agroecological assessment of land suitability for food crop development at a regional scale: A study case of Buton Island.

A Ramlan[1], S Baja[1], S Arif[2] and R Neswati[1]

[1] Department of Soil Science, Faculty of Agriculture, Hasanuddin University, Jl. Perintis Kemerdekaan KM 10, Makassar, 90245, Indonesia.
[2] Department of Physics, Hasanuddin University, Jl. Perintis Kemerdekaan KM 10, Makassar, 90245, South Sulawesi, Indonesia.

E-mail: *andi-ramlan@agri.unhas.ac.id, ramlanmartin74@gmail.com, sbja02@yahoo.com.au*

Abstract. Agriculture has long become a prime sector for regional development in Buton Island, although local government emphasis on perennial crops. Food crop have been developed in very limited land areas, mainly on transmigration areas, as parts of central government programs. Today, the central government has launched a national strategic program on food self-sufficiency and has imposed the local government to optimize available land for cultivating food crop. The primary aim of study is to develop rapid assessment on a spatial basis using GIS for agricultural land suitability evaluation of agriculture commodities, *i.e.,* rice (irrigated paddy field, rainfed rice) and corn *(Zea maize L.)*. The study was undertaken using the following procedures: (i) conducting reconnaissance soil survey based on land units; (ii) constructing soil database in a GIS; and (iii) classifying land suitability using the FAO method. Spatial data were generated from digital topographic map, soil survey, soil characteristics, as well as climate data. Preliminary results indicate that quite large area available for food crop cultivation both in the context of land suitability (mostly in S2 and S3 classes) and land availability. All data bases were managed in GIS, then it is amenable to various operations in GIS to accommodate possible additional assessment including socio-economic and policy assessment.

1. Introduction

Agricultural development program essentially needs availability of information on land resources which is clear and accurate [1, 2]. The lack of success in agricultural development is often caused by incomplete data and information on land resources. To support the agricultural development program in Buton island, it is necessary to provide reliable agroecological information. The use of land and commodity options in different areas appropriate for development can be established within a GIS. Through agro-ecological zone mapping activities for planned commodity zoning within a predetermined timeframe, it is hoped that a map of agroecological information and important commodity development alternatives can be generated in a more informative and interactive GIS form.

Along this frame, the local government of Buton regency has formulated development policies that emphasizing agriculture as the strategic sector, where the main activities are agriculture commodity zoning. The objective of commodity zoning, as this paper presents is to regionalize land suitability for the importance of allocation of activities to increase the volume of yields in every part of the region in accordance with local agroecology and agroeconomics.

2. Research objectives

The purposes of research are as follows:

- Detecting, monitoring, and analysing the development of the use of food crop lands on Buton island;
- Analysing the suitability and productivity of food crop commodities in Buton island;
- Building an evaluation system, through zoning of land management units based on agroecological character, based on spatially referenced data in GIS.

3. Methodology

3.1. Research sites

The project was conducted in Buton island, Southeast Sulawesi for 8 months (April-November 2017). Laboratory analysis was conducted at Department of Soil Science, Faculty of Agriculture, Hasanuddin University and soil research station, Agriculture Department of Republic of Indonesia, Maros.

3.2. The stages of research

3.2.1. Preparation of the base map. A base map was developed from a simplified 1: 50,000 topographic maps. The base map was used later to generate the land unit and the commodity zoning map. The base map was digitized and stored in a spatial database [1].

Soil characteristics were obtained from land systems map (RePPProT map) at scale 1:250,000 [3]. Land system map was being detailed by landform approach using topographic maps which result 58 land units. Field observation and soil sampling was established on the land unit.

3.2.2. Terrain analysis. Terrain analysis was carried out from topographic maps. Land units were developed by means of topographical data namely; lithology, reliefs and slopes, elevation, and land use. Relief and slope estimated based on contour line, elevation or altitude obtained from contour lines; landform and lithology are extracted from geological maps.

The results of terrain analysis were then depicted as the base map that used as a working map for ground observation. Terrain data was acquired from DEM (digital elevation model) of SRTM (shuttle radar topography mission) which has 30 m pixel resolution. Data processing was completed to generate slope in percent unit. The procedure was started by extracting data and converting coordinate system from geographic coordinate system to UTM (Universal Transverse Mercator) coordinate system. Determination of slope unit was generated by means of geographic information system module.

3.2.3. Observation of soil in the field. Land observation in the field was conducted by transect or toposequence (A distinctive soil characteristics related to topographic situation) [2, 4]. Soil classification was determined according to Soil Survey Staff to sub-group level [5]. Field observation and soil sampling were established on the land unit. The field survey included the identification of morphological parameters such as effective depth, soil drainage, slope, soil surface rock, outcrop rock, flood frequency, and others. This stage included analysis of climate data.

3.2.4. Analysis of soil and water samples. Soil sample analysis was conducted primarily for land evaluation. Soil properties analysed include soil physics and chemical. Properties of soil physics consist of: a). soil texture, b). drainage, c). soil depth, d). rock surface, and e). inundation (flooding). Properties of soil chemical are: a). soil acidity (pH), b). base saturation, c). organic matter, d). cation exchange capacity (CEC), and e). salinity.

3.2.5. Database and data processing. Spatial data were also stored in digital form and then processed for land suitability classification. This stage consisted of activities of developing, organizing, and managing digital databases to the mapping of results.

3.2.6. Land evaluation. Land suitability analysis was conducted based on land direction technical guidance [6, 7]. Evaluated crops namely; rice paddy, paddy fields, corn, peanuts, soybeans, sweet potatoes, cassava, green beans.

4. Results and discussion

4.1. Research sites characteristics

Buton county was located in the Buton islands (Buton island and Muna island), the south-eastern island of Sulawesi. This area includes parts of Muna island and Buton island. The most widespread districts are Pasarwajo District with 356.40 km^2, Lasalimu District 327.29 km^2 and Mawasangka District with an area of 271.55 km^2 or 14.31%, 13.14% and 10.89% respectively of the total area of Buton Regency. Meanwhile, the smallest area was Batu Atas District which consist of 7.18 km^2 or 0.29% of the total area of Buton Regency.

Climatic conditions in Buton according to Schmidt and Ferguson [8], was an area of varying rain type that include type B, C and D. According to Koppen classification, the climate type is tropical rain characterized by dry months (<60 mm/month). The dry months (less rainfall 60 mm/month) between 0 to 4 months and the number of wet (rainfall over 100 mm / month) between 6 to 10 months. Average rainfall varies from 1,233 mm to 3,020 mm, the highest is in Kaongkeongkea area of 3,020 mm and the lowest around Mawasangka is 1,403 mm.

Buton Island has a more complex geological, consisting of Mesozoic red mesoxic rocks, etersier sediment, river deposits and ultrabasic rocks [9]. The region also was divided toward 7 main landform units, namely: alluvium plains, coastal plains (marine), piedmont terrain, marine terraces, limestone limestones, hill systems and mountain systems. Soil order was divided into 6 classes, namely: Entisols, Inceptisols, Vertisols, Alfisols, Ultisols and Oxisols. Most areas of Buton (46.79%) were located at 0 - 115 m above the sea level, where the highest altitude is at 808 m above sea level.

4.2. Land unit as evaluation basis

Soil survey was conducted by means of existing map for sampling purposes, namely: soil map of Muna island and Buton island, Southeast Sulawesi [10] and land system of regional physical planning project for transmigration [11]. Map detailing is completed by conducting land survey on land units and recorded using GPS.

4.3. Land suitability classification

Land suitability class reflected degrees of suitability. The class of suitability consists of; S1 = highly suitable; S2 = moderately suitable; S3 = marginally suitable, N1 = currently not suitable and N2 = not suitable permanently. The most potential commodities to be developed were corn and beans, which has potentially more 6,000 ha (class S1) and 120,000 ha (class S2) for corn. While beans (peanuts and green beans) has potentially 79,000 ha (S1) (table 1). Soybeans potentially to be developed in 118,000 ha but has a significant limitation. The barrier varies different limiting factors for each land unit. Potential development of other commodities were cassava, sweet potatoes, and rice fields. Rice is only in the range of S2 and S3 classes.

IOP Conf. Series: Earth and Environmental Science **157** (2018) 012024 doi:10.1088/1755-1315/157/1/012024

Table 1. Land suitability and development area of food crops in Buton regency (ha)

Crops		Land Suitability			Potential	Existing	Potential
		S1	S2	S3	Area	Crop	Development
1.	Irrigated paddy	-	-	237,088	237,088	989	236,099
2.	Paddy field	-	10,580	228,299	238,879	3	238,876
3.	Corn	6,845	120,460	23,012	150,317	8	150,310
4.	Soybean	-	118,673	117,542	236,215	498	235,717
5.	Peanut	6,420	79,894	119,524	205,838	498	205,340
6.	Green bean	6,420	79,894	119,524	205,838	71	205,767
7.	Sweet potato	-	37,027	134,376	171,403	217	171,186
8.	Cassava	-	193,944	49,447	243,390	5	243,385

Soil acidity in this area is generally quite good, organic content (C-Organic) is generally medium, while nitrogen, cation exchange capacity and base saturation are generally moderate level. Potassium and nitrogen takes at medium to high level. The degree of limiting factor of soil chemical is relatively mild to be overcome, while the efforts of handling physical barriers such as soil depth, coral, drainage and water table are high in some area of land unit.

Micro-constraints are the limits to the development commodities such as low pH, exchangeable bases, phosphor, phosphate, and nitrogen, as well as organic matter in some land units. So that the improvement to overcome the limiting factor through fertilization and calcification [12, 13].

Potential development of food crops is quite good (generally S2 and S3) and there is still a wide development area. Corn is the most prospective commodity, including beans. The barrier that should be considered in land development is the existence of land use. Another thing is mainly in terms of the status of the area which in the middle of the island of Buton is generally occupied by forests with protected functions.

Land suitability analysis of paddy rice field indicated that this commodity could be developed in area of 238,879 ha, which consist of 10,580 ha in S2 and 228,298.9 ha in S3. Corn available at S1 located at 6,845 ha and potentially to be developed become 120.460 ha at S2 level. Soybean are suitable to be cultivated at 118,673 ha (S2) and 117,542 ha (S3). Peanut has potentially to be cultivated toward 205,808 ha which consist of 6,420 ha (S1), 79,894 ha (S2) and 119,524 ha (S3). The suitable area for green bean located class S1 is 6,420 ha, class S2 79,894 ha and class S3 119,524 ha, where dominantly located in Lapandewa sub district. Sweet potato is suitable at 37,027 ha (S2) and 134,376 ha (S3). There is no available location (class S1) for sweet potato in Buton region. Sampolawa sub-district has the extensive area for sweet potato, while Mawasang sub-district has the widest. The other location for sweet potato plant is 71,987 ha (N1) and 64,102 ha (N2). Land suitability analysis of cassava indicated that class S2 reached 193,944 ha and 49,447 ha in class S3, while N1 is 2,664 ha. The examples of land suitability distribution in Buton presented in the figures below (figure 1).

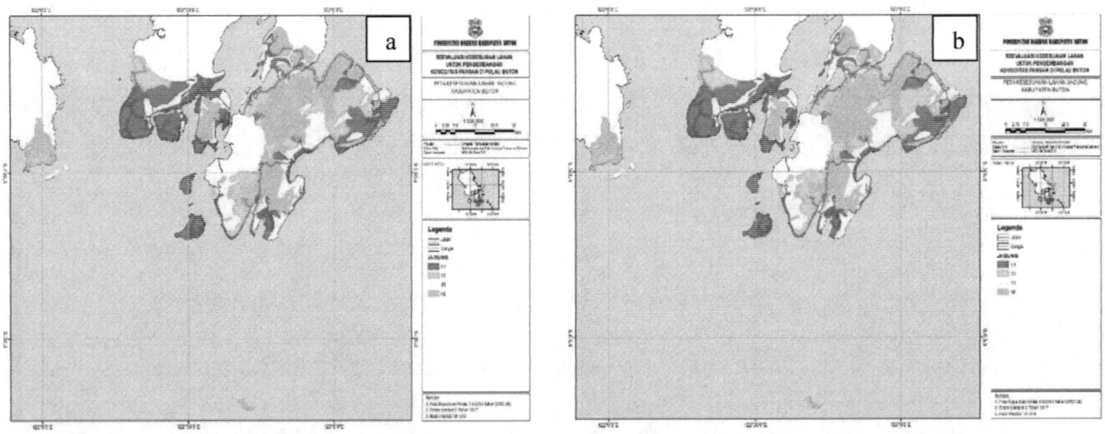

Figure 1 Land suitability map of corn (a) and peanut (b).

5. Conclusion

From this analysis, it was found that there are still available land areas for cultivating food crops in this region. Most potential food crop commodities with good suitability class include: corn and beans (peanut, corn), which may be cultivated in most parts of available land units. It was found that the main limiting factors to the development of food commodities in this region are rock outcrops, soil fertility, and topography. The results of spatial analysis in GIS confirms that Some suitable land units are found within the forest area with conservation status, according to local spatial planning map.

Acknowledgments

Partly of this research has been financed by The Ministry of Research, Technology, and Higher Education, Republic of Indonesia, through a competitive based PUPT research scheme.

References

[1] Baja S, Chapman D M and Dragovich D 2002 A conceptual model for defining and assessing land management units using a fuzzy modeling approach in GIS environment. *Environmental Management* **29** 647-61

[2] Davidson D A and Jones G E 1986 A land resource information system (LRIS) for land use planning *App. Geog* **6** (3) 255-265

[3] Regional Physical Planning Project For Transmigration (RePPProT) 1988 Land System Map (Indonesia: Department of Transmigration Republic of Indonesia)

[4] Steers C A and Hajek B F 1979 Determination of map unit composition by a random selection of transects. *Soil Sci. Soc. Am. J.* **43** 156-60 doi:10.2136/sssaj1979.03615995004300010030x

[5] Soil Survey Staff 2014 *Keys to Soil Taxonomy* (United States: United States Department of Agriculture, Natural Resources Conservation Service)

[6] Djaenudin D, Marwan H and Suharta 2000 *Kriteria Kesesuaian Lahan untuk Komoditas Pertanian. Badan Penelitian dan Pengembangan Pertanian* (Bogor: Pusat Penelitian dan Agroklimat)

[7] Soil Research Center and Agroclimate 2001 *Technical Guidelines for Preparation of Agricultural Commodity Map based on Agro Ecology Zone (ZAE) Scale 1: 50,000*) Bogor: Center for Soil and Agro-climate Research, Balitbang Pertanian, Departemen Pertanian)

[8] Schmidt F H and Ferguson J H A 1951 *Rainfall types based on wet and dry period rations for Indonesia with Western New Guinea* (Jakarta: Ministry of transportation of meteorology and geophysics)

[9] Sikumbang N and Sanyoto P 1981 *Peta Geologi Lembar Buton dan Muna, Sulawesi Tenggara, Sekala 1 : 250.000, S 81 – 4*

[10] Indonesia Soil Research Center 1985 *Soil map of Muna Island and Buton Island, Southeast Sulawesi*
[11] Regional Physical Planning Project For Transmigration (RePPProT) 1988 *Land System Map* (Indonesia: Department of Transmigration Republic of Indonesia)
[12] Mahanty B, Kim S and Kim C G 2013 Assessment of a biostimulated or bioaugmented calcification system with *Bacillus pasteurii* in a simulated soil environment *Microbial Eco.* **65** (3) p 679
[13] Sikirić B, Mrvić V, Stevanović, D, Maksimović S, Stajković O and Bogdanović D 2009 The effects of calcification, urea and Al salts on Fe, Mn and Al contents in the soil and raspberry leaves *Agrochimica* **53** (4) 250-259 ref.37

IOP Conf. Series: Earth and Environmental Science **157** (2018) 012025 doi:10.1088/1755-1315/157/1/012025

Land use and land suitability assessment within the context of spatial planning regulation

S Baja[1], R Neswati[1] and S Arif[2]

[1]Department of Soil Science, Hasanuddin University Jl. Perintis kemerdekaan KM. 10 90245 Makassar, South Sulawesi, Indonesia
[2]Department of Physics, Hasanuddin University Jl. Perintis kemerdekaan KM. 10 90245 Makassar, South Sulawesi, Indonesia

E-mail: *sbja02@yahoo.com.id*

Abstract. This paper presents the results of study aimed at assessing land suitability at a regional scale, and examining spatial matching between suitability classes and spatial planning regulations for Mamminasata, one of the strategic regions in Indonesia. We focussed on assessing areas for producing food crop, in Gowa Regency, using integrated techniques of ground surveys and geographic information systems (GIS) technology. We used available soil and climate information to generate land suitability for food crops. Ground survey was carried out to verify several soil data, and SPOT XS images (resolution 20 meters) were employed to map land use in the area of interest. A set of spatial planning data bases developed by local government was used to understand planned land use types. Spatial matching procedures within GIS were performed to see the best fit between suitability classes and planned land use according to spatial planning regulation of Mamminasata region. The same procedure was also performed between present and planned land use types. The results show that some suitable areas for food crops have been allocated for residential development. Similarly, areas that should be protected according to the spatial planning scheme, in fact, have been converted or practiced for long as, intensive agriculture and residential. The results of this study may give insights into the future anticipation of land use development in this region.

1. Introduction

Land use planning is a complex area where different approaches may be used to solve a variety of problems in land use and management. Land suitability assessment is one of important components of land use planning which is directed towards the evaluation of diverse phenomena and indicators related to the land in terms of potential use and change, value, suitability, or possible level of degradation, through a series of principles and methodologies. From a regional planning perspective, different steps can be distinguished in land resource assessment, notably, investigation of problems, identification of opportunities for change, selection of feasible land use alternatives, and choice of suitable scenarios before implementation as well as monitoring. In essence, these steps are *context dependent* and *site specific*. For a complex area like Mamminasata and within a such context of land suitability assessment, land use model is a useful tool that can be reproduced, to complement the mental capability to analyse changes in land use and to make more informed decisions on land use planning and implementation [1]. On the other hands, land use planning is needed to guide decision makers in choosing the right type of land use, determining the optimal spatial location of the planned activities, identify opportunities and formulate land use changes, and anticipate the consequences of changes in land use policy [2, 3].

This study was carried out in the Mamminasata (as an acronym for Makassar, Maros, Sungguminasa, and Takalar) region, one of the strategic regions in Indonesia according to Government Act number 26, 2008. The Mamminasata region covers four districts in the southern parts of South Sulawesi Province, including the whole area of Makassar City, and some parts of Maros, Gowa, and Takalar regencies. Spatial planning regulation was devised to cope with the rapid change in land use, and the fact that rapid land use changes has occurred within the last decade, due to intensive urban development. Therefore, as stated in [3], spatial model in GIS is useful to describe the complexity of the biophysical and spatial patterns of land use for now and the future. This study is aimed at assessing land suitability at a regional scale, and examining spatial matching between suitability classes and spatial planning regulations for Mamminasata region.

2. Materials and Methods

This study was carried out in a spatial context, using integrated techniques of ground surveys and geographic information systems (GIS) technology [4, 5]. We used available soil and climate information to generate land suitability for food crops. Ground survey was carried out to verify several soil data, and SPOT XS images (resolution 20 meters) were employed to map land use in the area of interest.

Field surveys were conducted using a GPS and base maps to find out actual condition on the ground that represents a wide range of physical environments within the study area. Such sets of information were used for refining soil and land use information (obtained from satellite images), and for assessing the actual conditions in conjunction with spatial planning regulation. A set of spatial planning data bases developed by local government was used to understand planned land use types.

In the present study, suitability classification is undertaken at a class and sub-class levels according to the FAO Framework for Land Evaluation [4]. The classification at class level adopted is as follows [5]:

- Class S1 (highly suitable): land having no significant limitation or only have minor limitations to sustain a given land utilization type without significant reduction in productivity or benefits and will not require major inputs above acceptable level.
- Class S2 (moderately suitable): land having limitations which in aggregate are moderately severe for sustained application of the given land utilization type; the limitations will reduce productivity or benefits and increase required inputs to the extent that the overall advantage to be gained from the use, although still attractive, will be appreciable compared to that expected from Class S1 land.
- Class S3 (marginally suitable): land having limita- tions which in aggregate are severe for sustained application of the given land utilization type and will so reduce productivity or benefits, or increase required inputs, that any expenditure will only be marginally justified.
- Class N (not suitable): This class consists of two categories: N1 for currently not suitable, and N2 for permanently not suitable.

Spatial matching procedures within GIS [6] were performed to see the best fit between suitability classes and planned land use according to spatial planning regulation of Mamminasata region. The same procedure was also performed between present and planned land use types. A procedure was developed in GIS to integrate all sets of the information obtained from field surveys, digitized GIS data, and local government vector-based spatial data bases related to spatial planning regulation for the Mamminasata region [7, 8].

3. Results and Discussion

3.1. Land suitability classification

Distribution of land suitability derived from applying the FAO method [4, 9] can be seen in figure 1. The result from land suitability analysis in Gowa Regency for maize cultivation indicates that around 9% of land areas have class of moderately suitable, ±51% fall within marginally suitable, and ±40%

are not suitable) (figure 1). The main limiting factor for maize cultivation in this region is land surface, especially slope gradient (s).

The figures indicate that nearly half of the study area is not suitable for the development of maize, and predominantly caused by slope gradient factors, although it was found that soil attributes are in a good condition. Furthermore, more than half of the study area is available (as appropriate) for the development of maize, by assuming that the decision makers will give the highest priority to areas like marginally and moderately suitable).

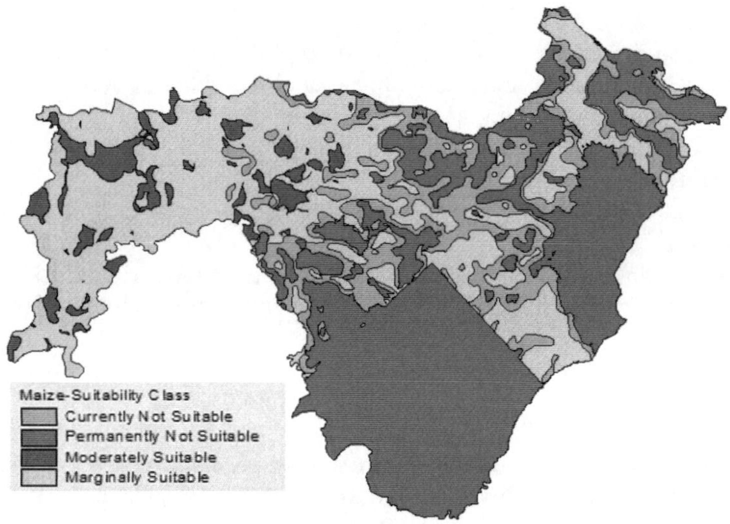

Figure 1. Map of land suitability for maize development

3.2. Land Use Classification

A basic procedures of image classification were undertaken to produce present land use/land cover of the study area, using a combination technique of visual interpretation and image analysis. The result of image analysis is presented in figure 2, and the statistics of the analysis for the image classification can be seen in table 1.

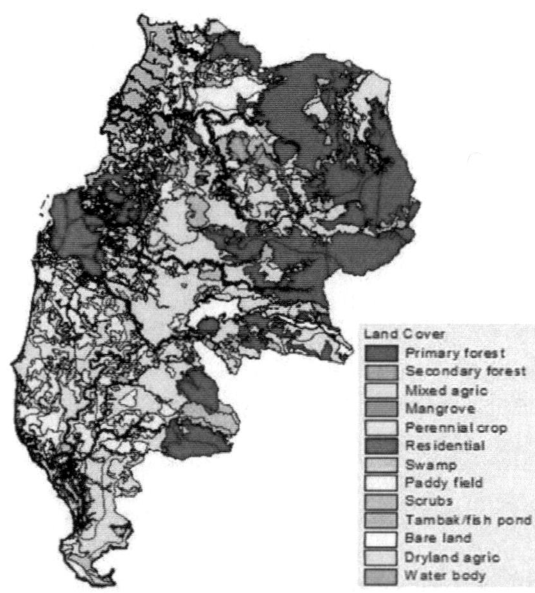

Figure 2. Land use/land cover map of Mamminasata

Table 1. Land use types and their coverage area derived from SPOT XS images

Land use type	Area (ha)	Percent
Mangrove (secondary)	1,099	0.4
Primary forest	133	0.1
Secondary forest	14,767	6.0
Planted forest	5,031	2.0
Residential	10,035	4.1
Dryand agriculture	7,180	2.9
Mixed dryland agriculture	72,886	29.6
Swamp	124	0.1
Grassland	302	0.1
Paddy field	88,015	35.7
Scrubs	27,076	11.0
Water body	1,945	0.8
Fishponds/tambak	15,948	6.5
Bare land and Extraction	501	0.2
Dam/reservoir	1,203	0.5
Total area	**246,244**	100.0

As the table shows, mixed dryland agriculture and paddy field cover the dominant part of the region approximately 72,886 ha and 88,015 ha, respectively, followed by scrubs (27,076 ha), fish pond/*tambak* (15,948 ha), and secondary forest (14,767 ha). Forest is dominantly found in the upland of Maros and Gowa regencies, and may become protected and conserved zones for the region. Land use conversion for the last five years is greatly found for residential, and other built-up areas. This is due to a rapid growing urban area in many parts of the region.

3.3. Spatial Planning Database

Based on the local government vector data base (see figure 3), an analysis was performed in GIS to calculate the distribution of land function/status that should be used by the local government in directing land use within a period of 20 years. According to the land zoning or status/function map, conservation forest, protected forest, and other protected areas cover a total area of 16,426 ha, 14,072 ha, and 14, 489 ha, respectively (table 2).

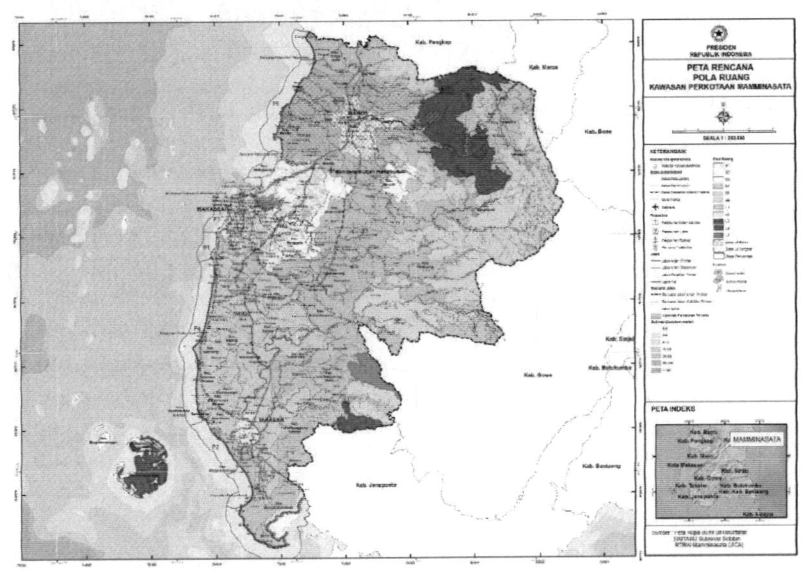

Figure 3. Land use zoning according to land use planning scheme (Presidential Decree No. 55, 2011 about Spatial Planning of Mamminasata Region)

Table 2. Land use protection zones (Code L) from government database

Land use zoning	Area (ha)
Conservation forest (L3=wildlife protection, mangrove, national park)	16,425.50
Protected forest (L1)	14,071.85
Other protected areas (L2=small island, coastal buffer zone, river buffer zone, L6=hunting park)	14,489.45

Spatial analysis was also underway to examine the spatial matching between land use types and the land status according to the spatial planning regulation stated by the government under the Government Act No. 26, 2007 about Spatial Planning. The same procedure was also applied to see the corresponding phenomenon between land suitability classes and the zoning map as stated in the spatial planning regulation of Government. It is important to note here that that according to spatial planning regulation of Mamminasata region, conservation forest consists of land zoning code L3=wildlife protection, mangrove, national park and other protected areas consists of land zoning code L2=small island, coastal buffer zone, river buffer zone, L6=hunting park (see table 2). It should be also pointed out that this study focused only on protected zone, while the statistics for management zones (B1, B2, B3, B4, B5, and B6) (see figure 2) are not presented here. These zones should be functioned for various land use activities for the sake of increasing economic values of the land.

The results show that, for Gowa Regency, land suitability classes mostly correspond to the spatial planning zones. Most of Not Suitable land classes fall within the protection zones with L1, L2, and L3 codes. This implies that most suitable land classes have been practiced for developing agricultural land like paddy fields, mixed agriculture, etc. within management zone (code B). In the contrary, present land use type of some proportions of the area do not match according to their function as stated in the spatial planning map produced by the government, particularly for those already been determined as protected forest and other types of conservation zones. For example, some areas of mixed dryland agriculture, in fact are found on the protected forest, conservation forest, and other protected zones. Based on this detailed analysis assisted by ground truth data, other types of protected areas (L2=small island, coastal buffer zone, river buffer zone, and L6=hunting park) were occupied by quite a large area of mixed agriculture, dryland agriculture, and rice field, with some proportion of residential areas. This phenomenon can be found in most developed region [10].

4. Conclusion

This study has resulted in two different phenomena in relation to the development of food crops particularly maize. First, the results for Gowa Regency shows that land suitability classes mostly correspond to the spatial planning zones, where most of Not Suitable land classes fall within the protection zones with L1, L2, and L3 codes. This implies that most suitable land classes have been practiced for developing agricultural land like paddy fields, mixed agriculture, etc. within management zone (code B). Second, present land use type of some proportions of the area do not match according to their function as stated in the spatial planning map produced by the government, particularly for those already been determined as protected forest and other types of conservation zones. This indicates that agriculture practices, especially food crops were not only developed in the management zone, but also has expanded to areas prone to degradation in the sloping zone hat by regulation assigned as protection zone. This study is thus useful for land resource management and monitoring in Mamminasata region, and may give insights into the future anticipation of land use development on the region on a spatial basis.

Acknowledgment

The author deeply thankful to the The Ministry for Research, Technology, and Higher Education (Kemenristekdikti), The Republic of Indonesia for providing research fund through the "MP3EI"

Scheme, and the Department of Soil Science, Faculty of Agriculture, Hasanuddin University, for providing data, and other supporting facilities for this study.

References

[1] Ruth M and Costanza R 1998 Using dynamic modeling to scope environmental problems and build consensus *Env. Management* ***22*** 183–195

[2] Baja S, Dragovich D and Chapman D 2007 Spatial Based Compromise Programming for Multiple Criteria Decision Making Modeling in Land Use Planning. *Env. Mod. & Ass.* ***12****:* 171-184

[3] Verburg P H, Schot P P, Dijst M J and Veldkamp A 2004 Land use change modelling: current practice and research priorities *GeoJournal* ***61*** 309–324

[4] FAO 1976 *A Framework for Land Evaluation* (Rome: Food and Agriculture Organitation of the United Nations)

[5] Djaenudin D, Marwan H and Hidayat A 2003 *Petunjuk Teknis Evaluasi Lahan untuk Komoditas Pertanian Versi 3* (Bogor: Balai Penelitian Tanah, Puslitbang Tanah dan Agroklimat)

[6] Baja S 2009 *Land Use Choice and Land Resources Assessment in Agriculture. A review.* (*CAB Reviews: Perspectives in Agriculture, Veterinary Science, Nutrition and Natural Resources*)

[7] Baja S, Chapman D M and Dragovich D 2002 Using GIS and remote sensing for assessing and mapping the present status of land use and land qualities in the lower Hawkesbury-Nepean Catchment, Australia *Geocarto Int.* **17** 15-24

[8] Baja S, Amrullah A, Ramli M and Ramlan A 2011 *Spatial-based fuzzy classification of land suitability index for agriculture development: A model validation perspective (Proc. of 3rd International Joint Conference on Computational Intelligence-International Conference on Fuzzy Computation Theory & Applications IJCCI-FCTA*) Paris, France, 24-26 October 2011 p 435-440

[9] El Baroudy A A 2016 Mapping and evaluating land suitability using a GIS-based model. *Catena* **140** 974-981

[10] Maxwell S K and Sylvester K M 2012 Identification of "ever-cropped" land (1984–2010) using Landsat annual maximum NDVI image composites: Southwestern Kansas case study *Remote Sensing of Environment* **121** 186–195

Contribution of urban farms to urban ecology of a developing city

H Iswoyo[1], T Dariati[1], B Vale[2] and M Bryant[2]

[1]Department of Agronomy, Faculty of Agriculture, University of Hasanuddin, Tamalanrea Campus, Jl. Perintis Kemerdekaan KM. 10, Makassar, 90245, Indonesia.
[2]School of Architecture, Faculty of Architecture and Design Victoria University of Wellington, 139 Vivian St., Wellington, 6140, New Zealand.

E-mail: *iswoyo@yahoo.com; hari.iswoyo@unhas.ac.id*

Abstract. Urban ecology has become a more popular concern as the awareness for mutual sharing between humans and other ecosystem members is increasing. This study aimed at assessing the value of urban farms in the city of Makassar, a fast-developing Indonesian city which according to its city council classification covered significant area of the city. The research employed Rapid Biodiversity Assessment (RBA) to assess quality of urban farms. The method assessed two important aspects of spaces in terms of ecology *i.e.* vegetation structures and vascular plants. Results showed the biodiversity of urban farms in Makassar compared to other typologies was high. Urban farms in Makassar in general have potential for ecological spots because despite they are more cultural than natural, their high plant biodiversity score, dominance of trees and less built areas make them always available for improvement to become more ecological spots.

1. Introduction

Urban areas are now main habitat for people in this planet. The fact that urban areas only covers around 3% of earth's space, does not limit the more than 50% world population to reside in them [1, 2], it equals to around 3 billion people of the 21st Century's first decade [3]. Urbanization has become the inevitable results of urban development, forming cities, peri urban and urbanized rural areas. To define a city, it is a district within urban areas with powers and corporate status of self-government. It is relatively large in size and become the home for a range of population size and background. On the other hand, urban region is a city plus its surrounding areas with all connections and interactions in between [3]. Other distinction character of a city is the population size, an area can be considered as urban when there are at least 150 people living within a square kilometre area [4].

Human occupancy has resulted in dynamic changes which include economy, society, culture and technology which all aimed at increasing life quality of the urban residents. However, development always has inevitable implications to the environment and other biotic components, hence affected ecosystem ecology [5] which tend to threaten the natural sites [6] through habitat fragmentation [7][8]. As a result, urban areas tend to become improving in economy but declining in ecological quality. Urban development relates with land use conversion, and among many types of lands that tend to be 'sacrificed' is agricultural land or in other term urban farms.

Several concepts for ecological improvement in urban areas have been introduced [9] all targeting at the improvement of spaces in urban area that favor the interests for creation of habitat for all possible species besides human [10]. The intention is not to stop urban development, but how to preserve the ecological function of urban area whilst developing [9].

Being a developing country, Indonesia has many growing cities and urbanizing area. One of big cities and the biggest in Eastern Indonesia is Makassar. This city needs preservation of its natural areas, to maintain their existence and function while the city is still developing [11]. The growing infrastructure and built areas in cities has unfavorably affected the existence of productive sites such as urban farms. It has been observed in may big cities in Indonesia such as Depok [12], Bandung [13], Yogyakarta [14], Tegal [15], Malang [16], Bali [17] and also observable in Makassar [18, 19]. It could be seen as a disadvantage as urban agriculture is one of strategic resources which are valuable in many terms such as economy, social and environmental terms [20].

Despite some thoughts regarding agricultural activities to be more cultural than natural, in environmental perspectives agriculture sites in general are more compatible with principle of environmental preservation, not to mention the intrinsic values of agriculture sites as preservation of biodiversity and for educational purposes [15]. Even agriculture sites are not really natural, their existence preservation within urban area at least will limit the expansion of built structures within.

Therefore, it is necessary to understand the state of urban farms within a city, and this paper select Makassar to study how is the existence of them and what values they have that might contribute to ecology of the city. One important parameter of ecology is the state of biodiversity, as many definitions of biodiversity emphasizes on existence living organisms as the main component with ecological complexity [21, 22].

In terms of application, biodiversity assessment in general comprises both animal and plant species [23], for urban context, biodiversity relates with the existence of natural sites (or its remnants), cultural and even artificial. Understanding the intensive human interaction in urban sites, it might not be significant to expect species of wild animal within, therefore urban biodiversity state could be assessed by regarding two factors: vegetation structures and vascular plants diversity. These two factors should be able to represent the general state of biodiversity since structure of vegetation determines habitat complexity and many studies have indicated the complexity and composition of habitat are acceptable indicator of biodiversity in general [24]. Additionally, vegetation structural composition of habitats in urban areas could be used as substitute in assessing their biodiversity [25] and coverage of vegetation is an important contributors to ecological health and hence is indicative of wildlife habitat and ecological value through their vegetation attributes [26]

This paper presents the study and assessment on biodiversity of urban farms in the city of Makassar, using a Rapid Biodiversity Assessment (RBA) developed by Tzoulas and James [24] in order to get general description of biodiversity state of urban farms and how they can be improved for their ecological contribution, because according to Iswoyo *et al.* [11] assessment of biodiversity could be taken as the first stage of assessment of urban ecology.

2. Research Methodology

2.1. Urban Farm Identification among Urban Spaces

One of common land use in urban areas is agriculture sites, it has been a common classification and be adapted with various terms including urban farm. This study adapted the classification of spaces which was developed through aerial and ground observation and defined in form of typologies [9]. The urban farm is classified as one of patches in Makassar and defined as '*Areas which are used for farming by people living nearby, mostly with seasonal plants, including paddy fields*'[9 p.9]. Based on that classification, all spaces in the city are digitized and defined with GIS which urban farms later are surveyed for field observation and assessment. Sampling locations are determined based on farming types as seen through aerial photograph and based on secondary references. The number of sampling points for each urban farm location is determined based on variabilities of land coverage as seen through aerial photograph and covered at least 10% of the area. This study identified 186 spots of urban farms within the city, and the distribution based on district can be in table 1 and figure 1.

Table 1. Identified urban farm in Makassar based on spatial observation and randomized ground confirmation

District and Type of Urban Farm	Area (hectare)
Biringkanaya	**770.31**
Farm field	215.21
Paddy field	288.50
Paddy field and farm	248.68
Plantation	17.92
Makassar	**3.44**
Farm field	3.44
Manggala	**837.97**
Farm field	229.08
Paddy field	578.67
Paddy field and farm	30.23
Panakkukang	**40.50**
Farm field	1.81
Fish pond, farm	5.78
Mixed Agriculture field	32.91
Rappocini	**14.80**
Farm field	3.54
Paddy field	11.26
Tallo	**18.43**
Farm field	18.43
Tamalanrea	**548.75**
Farm field	238.11
Paddy field	143.29
Paddy field and farm	167.35
Tamalate	**614.08**
Farm field	97.53
Paddy field	516.55
TOTAL	**2848.28**

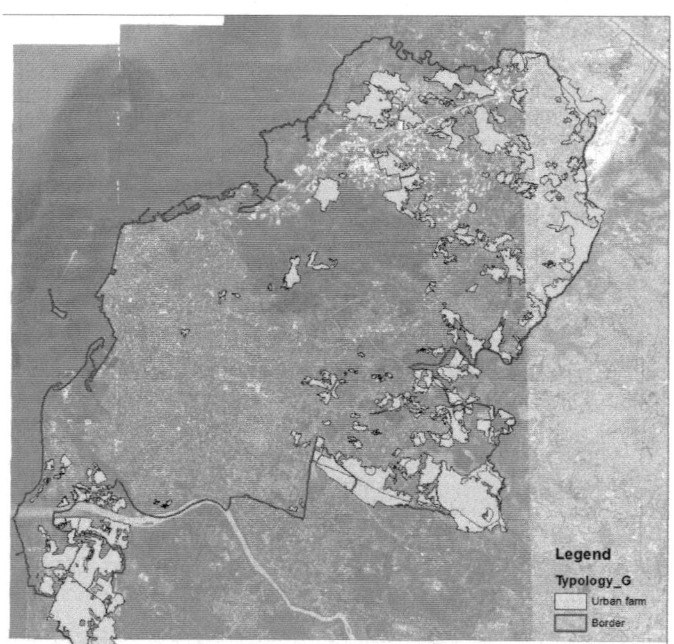

Figure 1. Distribution of urban farms in Makassar

Among all those identified urban farm spots, this study selected 11 spots that were taken as sampling points which represent few locations and 4 dominating districts. The selection was based on the type of identified farms, and pattern variation based on aerial observation.

2.2. Biodiversity Scoring with Urban Rapid Biodiversity Assessment (RBA) method
This method was developed by Tzoulas and James [24] based on a study in the UK. This method was adapted and adjusted to local developing and tropical context. It consists of several stages.

2.2.1. Development of Field Sheet Record. A sheet to guide the surveyor in performing vegetation structure assessment. All structures are prescribed based on the composition and height variability of trees, shrubs, forbs and grasses as well as built structures and water features in an area. The structures were assessed for their Domin value [27].

2.2.2. Vascular plant checklist development. The study could not obtain existing list of vascular plants in Makassar, therefore the study developed a list based on vascular plant species recorded at the survey locations.

2.2.3. *Assessing land coverage by vegetation structure and diversity of vascular plants*. This stage consists of three main activities: determining sample size, number of sampling points, and recording vegetation structure as well as vegetation diversity.

2.2.4. Recording of vegetation structures and diversity of vascular plants. This stage started by defining visual horizons of each sampling location, then from consistent exact position, structures of vegetation, built and water bodies were recorded for their dominance. After Domin value of each structure is recorded, vascular plants were identified and recorded into a vascular plants list.

2.2.5. Combining all indicators into a biodiversity score. Using approach developed in Tzoulas and James [24] a final score calculation produced scores of each sampled location of urban farms in Makassar.

3. Results and Discussion

3.1. Biodiversity Score.
As a city that keeps growing and developing, Makassar only has little space for agricultural fields. Most urban farms in Makassar are located in the urban fringe and only few are within the city area. Results of biodiversity assessment conducted in 6 locations and 11 sampling points is presented in table 2, locations of sampled urban farms are shown in figure 2.

As seen in figure 2, most urban farms are in the fringe of Makassar. There are some scattered patches of farms more towards the centre of the city although they are not significant in size. These farm spots are possible remnants of previous large agricultural fields. Those fields were encroached and fragmented due to construction of infrastructure development. Understanding the development trend, there is no guarantee that they will still be exist in the further future. Most of urban farms sampled in this study are privately owned, or under collective ownership. These statuses clearly posed a 'threat' for the farms for possible conversion in the future. It is a common thing due to civilization which tend to cause agricultural landscapes to contain only small isolated patches [28]. On the other hand, some mentioned agriculture sites to be the cause of natural sites fragmentation [29, 30]. This of course will need to be seen from the perspective of how the urban farms are formed and the direction of conversion. Creating farms out of natural forest would of course be negative, but preserving farms from conversion to a shopping mall will certainly be positive in ecological perspectives.

Table 2. Biodiversity score of each urban farm sampling point.

Location Name and District	Sampling point No.	Vegetation Structures Scores								Number of Vascular Plants	Bio-diversity score
		High trees	Low trees	Bushes	High grasses and forbs	Low grasses and forbs	Ground flora	Aquatic	Built		
Baddoka (Biringkanaya)	01	4	1	2	2	7	2	0	0	11	13
Baddoka (Biringkanaya)	02	7	2	2	3	5	2	0	1	15	15
Baddoka (Biringkanaya)	03	8	5	4	3	6	5	0	0	17	14
South of BTP (Tamalanrea)	04	1	1	2	0	7	1	2	1	18	15
Bukit Baruga (Manggala)	05	1	1	2	6	7	0	1	2	34	17
Antang (Manggala)	06	6	3	3	2	2	4	2	0	25	17
Antang (Manggala)	07	8	7	7	7	6	7	0	0	41	18
Antang (Manggala)	08	1	1	3	5	6	1	2	1	32	19
Borong (Panakkukang)	09	2	2	3	3	3	9	5	3	30	16
Urip Sumoharjo (behind Finance Bld) (Panakkukang)	10	3	2	4	2	3	3	1	6	42	14
Urip Sumoharjo (behind Finance Bld) (Panakkukang)	11	5	2	2	1	2	2	5	2	27	17
Total		46	27	34	34	54	36	18	16	292	**175**
Average		4.18	2.45	3.09	3.09	4.91	3.27	1.64	1.45	26.55	**15.91**

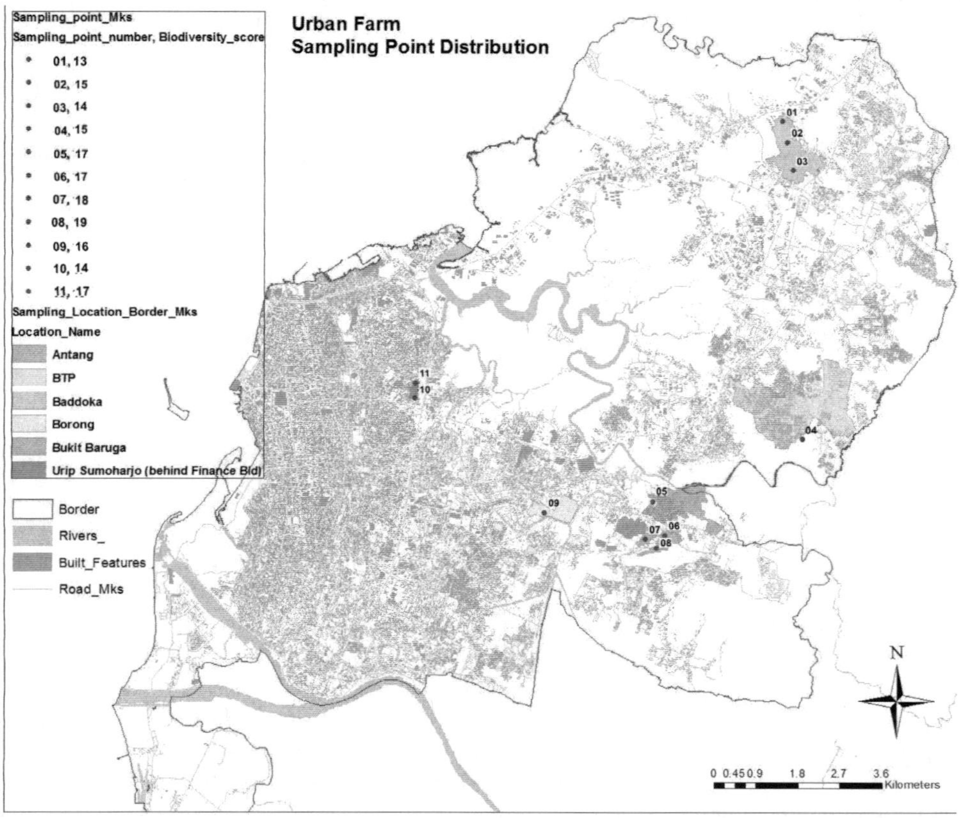

Figure 2. Sampled urban farms in the city of Makassar.

The following figures show the appearance of selected surveyed urban farms in Makassar and the structures of vegetation recorded in panoramic photographs.

Figure 3. Aerial view of two urban farms in Makassar

Figure 4. Panoramic views of a vegetable farms in Makassar

Regarding the biodiversity score, the average score of all sites are 15.91. Based on the original method of RBA, there is no range of 'high', 'medium' or 'low' class for score of individual site or typology. The scoring purpose is for comparing the results with other sites or typologies. In order to see how is the biodiversity of urban farm in Makassar, we compared it with scores of other typologies of spaces in this city from other study. Based on the results of scoring of other typologies [11], the score range between the lowest and the highest was 4 – 16. Therefore, using deviation standard for making 3 classes (high, medium and low) the score of urban farm in this study could be classified 'high'. As for vegetation structure, high trees is more dominant than other structures and average number of recorded vascular plants (26) can be considered good compared to the results found in [24].

Being one type of most fragile agriculture field, paddy fields are actually important ecosystem for some animal species. It is very common for most paddy fields in Indonesian also deliberately culture various species of fish such as 'Nila' (*Oreochromis niloticus*), 'Mas' (*Cyprinus carpio*), 'Bandeng' (*Chanos chanos*), and various edible crabs. Other common non consumable animals usually reside in paddy fields are some reptiles, frogs, some species of birds and small mammals [31]. As paddy fields normally connected to rivers through irrigation lines, they have potential to also harbor varieties of river fishes. Fish existence in paddy fields also help farmers in predating insect larvae of potential pests, and their wastes could increase soil fertility [32]. Additionally, the significance of paddy fields as habitat for various plants and animals also reported [32]. Some of them as also found in this study are *'kangkung'* (*Ipomoea aquatica* Conv.), '*Semanggi*' (*Marsilea crenata* Mars) and '*Genjer*' (*Limnocharis flava* Limn).

Figure 5. Panoramic views of two paddy fields in Makassar.

3.2. Further Improvement for the Sake of Ecology

For a city like Makassar where green spaces are limited, the existence of land that is still a working urban farm is a potential for ecology and hence need to be preserved.

Human activities whatsoever are dominant in agriculture, therefore, potentially pose disruption to the natural system. Apart from that, vegetation in urban farms are dominated by agricultural crops and cultivated plants, both are not favorable components of a natural ecological habitat. Consequently, as this study also observed, they have low potential in terms of habitat function. However, they still have value as green areas and functioning accordingly, if that they can be preserved from conversion. It is due to the fact that vegetation diversity is quite high as reflected from the recorded vascular plants, although they are rather cultural than natural.

At locations in Antang, tree coverage in the farms is quite good, at least of course it is justifiable to expect usual environmental services provided by trees such as mentioned in [33, 34, 22]. Another promising feature of urban farm for urban habitat is the absence of built structures.

Agricultural fields, especially paddy fields, are open poor in plant diversity other than the cultured varieties. The openness of paddy field makes them less feasible to serve as a site of refuge for wildlife. However, there is an opportunity as paddy fields have non-planted areas as well as the borders that might be optimized by planting vegetation with specific ecological function and uses. According to USDA guidelines [35], "*conservation buffers are strips of vegetation planted in the landscape of fields to effect ecological processes favorably and provide a variety of goods and services to people and the ecosystem*".

Figure 6. Conservation buffer in agricultural fields [35].

For agriculture fields around streams or rivers, setting up these buffers would act as barriers that protect the water bodies from spray drift which may harm non-target species as well as poisoning stream water. Moreover, conservation buffers could serve as homes for beneficial insects. In the event of overflow, these buffers can slow the run off and absorb excess water, hence farm fields are less threatened by flooding [35].

4. Conclusion

Despite the area is declining due to development of the city, urban farms in Makassar still exist and provide benefits in economy, culture and ecology. Although in general urban farms have been strongly influenced by human intervention which result in causing them to become more cultural and less natural, their high plant biodiversity score, status as green areas and less built structure around them giving them great potential for future development to become better sites for ecology, ecosystem and urban habitats.

References

[1] Murray K, Mohareb A and Ogbuagu C 2009 Sustainable Cities Enhancing the Seven Forms of Community Capital *A Resour. Guid. Pract. to Engag. Munic. Employees Mov. Towar. Sustain.* **47**

[2] Singh V S, Pandey D N and Chaudhry P 2010 *Urban Forests and Open Green Spaces: Lessons for Jaipur, Rajasthan India* vol No. 1/2010 (Rajasthan: Rajasthan State Pollution Control Board)

[3] Forman R T T 2008 *Urban Regions: Ecology and Planning Beyond The City* (Cambridge, New York, Melbourne, Madrid, Cape Town, Singapore, Sao Paulo, Delhi: Cambridge University Press)

[4] Niemelä J 1999 Is there a need for a theory of urban ecology? *Urban Ecosyst.* **3** 57–65

[5] Alberti M 2008 Urban Patterns And Ecosystem Function *Advances in Urban Ecology* (Springer US) pp 61–92

[6] Aminzadeh B and Khansefid M 2010 A case study of urban ecological networks and a sustainable city: Tehran's metropolitan area *Urban Ecosyst.* **13** 23–36

[7] McPherson G, Esparza A X and DeStefano S 2009 Wildlife Corridors and Developed Landscapes *The Planner¿s Guide to Natural Resource Conservation:* (Springer New York) pp 85–102

[8] Davenport J, Davenport J L, Huijser M and Clevenger A 2006 Habitat and corridor function of rights-of-way *The Ecology of Transportation: Managing Mobility for the Environment* vol 10 (Springer Netherlands) pp 233–54

[9] Iswoyo H, Vale B and Bryant M 2011 Assessment and optimization of spaces as the basis for an urban ecological network in an Indonesian city *The 1st ACIKITA International Confrence of Science and Technology* ed P Kahar, J Agus, D Akbas, K Tarman and I Iswaldi (Jakarta: ACIKITA Foundation) pp 64–77

[10] Jongman R and Pungetti G 2004 *Ecological networks and greenways: Concept, design, implementation* (Cambridge: Cambridge University Press)
[11] Iswoyo H, Vale B and Bryant M 2012 Scoring urban spaces in an Indonesian city based on biodiversity *Second Int. Conf. Environ. Energy Biotechnol.*
[12] Zain A F M, Syahban G and Ermyanyla M 2013 Analisis perubahan peruntukan lahan pertanian perkotaan (urban agriculture) menjadi non pertanian di Kota Depok (Analysis on the landuse shift of urban farms into non-farm in the city of Depok) *Lokakarya Nasional dan Seminar Forum Komunikasi Perguruan Tinggi Pertanian Indonesia* (Bogor, Indonesia: Fakultas Pertanian Institut Pertanian Bogor Forum Komunikasi Perguruan Tinggi Pertanian Indonesia)
[13] Iqbal D 2016 Penyusutan Lahan Pertanian Kota Bandung Capai 200 Hektar dalam Setahun (Depreciation of farms sites in Bandung reaching 200 hectares a year)
[14] Kharisma W 2016 Lahan Pertanian Kota Yogyakarta Tinggal 56 Hektare (Farm sites in Yogyakarta are only 56 hectares left)
[15] Setyaki D 2012 *Alih fungsi lahan dan dampaknya bagi produktivitas pertanian di Kota Tegal (Land conversion and its effect on agricultural productivity in the city of Tegal)* (Tegal)
[16] Wati D L 2015 *Pengalihan fungsi lahan pertanian di kota Malang (Land use diversion of urban farm in Malang)* (Malang)
[17] Tribun Bali 2017 Miris, Alih Fungsi Lahan Pertanian di Badung Tak Terbendung
[18] Nurhikmah 2016 Luas lahan pertanian produktif di Kota Makassar semakin berkurang (Productive farm sites in Makassar is declining) *Rakyatku.com*
[19] Tribun Makssar 2014 Lahan Petani Kota Makassar Sisa 2.636 Hektar, Danny: Jangan Jual Sawah (Farm sites in Makassar are only 2,636 hectares left, Danny (the Mayor):"Do not sell your paddy field") *Trib.Makassar*
[20] Knowd I, Mason D and Docking A 2006 Urban Agriculture: The New Frontier *City* **23** 1
[21] Magurran A E 2004 *Measuring Biological Diversity* (Victoria: Blackwell Science Ltd)
[22] Hubbell S P 2001 *The Unified Neutral Theory of Biodiversity and Biogeography* (Princeton University Press)
[23] van Jaarsveld A S, Freitag S, Chown S L, Muller C, Koch S, Hull H, Bellamy C, Krüger M, Endrödy-Younga S, Mansell M W and Scholtz C H 1998 Biodiversity Assessment and Conservation Strategies *Science (80-.).* **279** 2106–8
[24] Tzoulas K and James P 2010 Making biodiversity measures accessible to non-specialists: an innovative method for rapid assessment of urban biodiversity *Urban Ecosyst.* **13** 113–27
[25] Cornelis J and Hermy M 2004 Biodiversity relationships in urban and suburban parks in Flanders *Landsc. Urban Plan.* **69** 385–401
[26] Cook E A 2002 Landscape structure indices for assessing urban ecological networks *Landsc. Urban Plan.* **58** 269–80
[27] Kent M and Coker P 1992 *Vegetation Description and Analysis - A Practical Approach* (London: Belhaven Press)
[28] Hilty J A, Jr. W Z L and Merenlender A M 2006 *Corridor Ecology: The science and practice of linking landscapes for biodiversity conservation* (Washington: Island Press)
[29] Krisp J M 2002 GIS supported visualization of ecological networks *Int. Symp. GIS*
[30] Pattanavibool A and Dearden P 2004 Habitat fragmentation in north Thailand: a case study. *Bird Conserv. Int.* **14,** 13–22
[31] Suriapermana S and Bogor 1994 *Mina-padi: usahatani berwawasan lingkungan meningkatkan pendapatan* (Bogor: Pusat Penelitian dan Pengembangan Tanaman Pangan, Badan Penelitian dan Pengembangan Pertanian, Departemen Pertanian)
[32] Puspita L, Ratnawati E, Suryadiputra I N N and Meutia A A 2005 *Lahan Basah Buatan di Indonesia (Artificial Wetland in Indonesia)* (Bogor: Wetlands International - Indonesia Programme)
[33] Chiesura A 2004 The role of urban parks for the sustainable city *Landsc. Urban Plan.* **68** 129–38
[34] Nowak D J, Crane D E and Stevens J C 2006 Air pollution removal by urban trees and shrubs in

the United States *Urban For. Urban Green.* **4** 115–23
[35] Bentrup G, Center U N A and Southern Research Station F S 2008 *Conservation buffers: design guidelines for buffers, corridors, and greenways* (Lincoln: Department of Agriculture, USA)

Screening assays of termite gut microbes that potentially as probiotic for human to digest cellulose as new food source

R Abdullah, K R T Ananda and Wijanarka

Department of Biology, Faculty of Science and Mathematics, Diponegoro University, Jl. Prof H. Soedarto SH Tembalang Semarang, 50275, Indonesia.

E-mail: *ridwanabdulloh20@gmail.com*

Abstract. According to UN, earth population will increase approximately 7.3 billion people up to 11.2 billion from 2015 until 2100. On the other side, food needs are not balance with the availability of food on earth. People of the world need solution for a new food source. By cellulose digesting ability, people analyzed can consume cellulose as the new food source to get glucose. The aims of research is obtaining termite gut cellulase bacteria selected which is potential as probiotic to split cellulose. Method used was as follows; isolation of termite gut microbes, microbial cellulase purification by screening method and probiotic test includes microbial pathogenicity test and human stomach acid and salt osmotic concentration resistance test. The result shows, 3 pure isolates of termite gut microbes can break down cellulose in the medium 1% CMC and 0.1% congo red (indicator of cellulose degradation activity) and life at pH 2- 2.5 and osmotic salt condition. Two isolates show the activity of gamma hemolysis (non-pathogenic in terms of pathogenicity on human blood). In conclusion, there are isolated termite gut microbes can be used as probiotic candidate for human to digest cellulose of the new food source for global food scarcity era.

1. Introduction

Nowdays, the world's concern about the population and food needs are increasing over the year. It is based on data reported by FAO/Food and Agriculture Organization that consumption needs of the people in the world, especially sugar, will increase continuously until 2050 with are about 230 million tons [1]. The increasing of population and consumption of staple food needs is not equal with the global food availabilities. Decreasing of global food availabilities is caused by disappearing of the food supply and castaway as a waste. It will cause inefficiency of processing and managing of food materials [1, 2].

Global warming go on annualy, and reported to be responsible as factor of global food scarcity in the future [3]. Agricultural land degradation also needs to be concerned for the next agriculture [4]. Global people need a solution. Grasses can live everywhere even in extreme region [5]. As far as, grasses aren't intended for food content, it's intended for the production of bioethanol and the alternative of animal feed, especially families of *Cyperus*. However, grasses are known to have high cellulose (40-55%) [6], that is polimer of carbohydrate consist of glucose monomers binded by β $(1\rightarrow4)$-glycosidic bonds [7]. Cellulose cannot be digested by human because there is no cellulase enzymes in human gut digestion. It means, human do not have enzymes like termites [8]. Termites can digest cellulose as it is mutually symbiotic with particular microorganisms which produce specific enzymes [9]. Cellulase changes cellulose materials into fermentable mono- and oligo-saccharides monomer [8]. Recently, despite fuel and textile industry, the uses of cellulase bacterial are still only glued to improve the digestive efficiency of ruminant animals, [10]. On the other hand, bacteria that is useful for the host like animal or human in digestion is known as probiotic [11]. Probiotics are traditionally related with fermented foods and from bifidobacteria and lactobacilli. However, the developement of probiotics has been made recently [11].

The research efforts to make grasses become a new food source in the future needs be done. Using the termite gut microbes selected as cellulose probiotic as a natural splitter of cellulose for human's body, is the solution where it aims to increase the efficiency of human's digestive system in the future, in the global food scarcity.

2. Methods

2.1. Sample Collection and Isolation

The termite samples were obtained in the Gondang village of Semarang. *Coptotermes* sp were taken from decaying wood in the shade side. The preparation of isolation was carried out by preparing selectively modified NB media with 1% of CMC / Carboxymethylecellulase [12]. CMC is used as a substitute of cellulose polymer [13]. Isolation was carried out by washing the termites for 4 times aseptically in Alcohol 70% for 5 minutes, Aquadest and Alcohol 70% for 30 seconds. Termites were separated between the abdomen and the superior part with a sterile knife. The extract was transferred to the NB + CMC medium and incubated for 4 days in the 120 rpm 37^0C [12]. Then, bacterial culture was prepared to transferred in medium NA + CMC. The pouring was carried out by graded dilution technique using oxalate salt solvents [14]. Bacterial culture were incubated for 48 hours 37^0C.

2.2. Cellulolytic test

Isolated spesific bacteria had tested in the cellulase screening. The medium was made by regulating the acidity level for 2-2,5 (as an acid resistance test in stomach) [15]. Medium enriched with 1% of CMC and 0.1% of Congo red cellulolytic indicator reagent [14]. Isolate was transferred to the test medium using streak technique. The culture was incubated for 48 hours 37^0C in the incubator. The grown isolates with a clear zone indicates that the bacterial isolates passed the screening [15].

2.3. Pathogenicity Test

Isolated cellulase bacteria were inoculated into the new blood medium (Blood Agar base + 5% of type O of human blood (sterile) using the streak technique [16]. Human blood and blood agar base were obtained from Laboratory of Diponegoro Hospital. The isolated in the blood medium were incubated for 48 hours and 37 ° C. Isolates that didn't show the clear zones and didn't change the color of the blood medium are identified as non-pathogenic bacteria or gama hemolysis [16]. Probiotics bacteria must be non pathogens, so that were not harmful when consumed [17].

2.4. Human Stomach Acid and Salt Osmotic Concentration Resistance Test

The escaped isolates were inoculated into test medium before incubated for 48 hours by acid treatment on disk paper, pH 7 and pH 2.5. The isolated bacteria was also inoculated into 3.5% and 6.5% (w / v) of sodium cloride solution, then incubated for 2 -4 hour 37^0C [18] and poured into NA + CMC medium for 24 hours. Starch agar plates containing different sodium chloride concentrations were used to analyze the sodium chloride tolerance of the bacteria [19]. The bacteria resistance were showed by the grown of bacteria colonies in post test.

3. Findings and Results

The various stages of screening under take to obtained cellulase probiotic. Destruction of the abdomen was carried out by separating the superior parts, so the pure isolate was obtained without contamination from the other body parts [20]. Table 1 showed that bacterial isolates were selected from the results of the dilution, and screening assay: cellulolytic/cellulase enzyme activity test, pathogenicity test, human stomach acid and salt osmotic resistance test.

Table 1. Screening assays results: The selected bacteria isolates potentially probiotics in 10^{-4} dilution/ isolates 4.1, 4.1' dan 4.3

No	Isolates	Cellulolytic Test	Human stomach acid resistance test	Pathogenicity test	Salt osmotic concentration resistance test
1	4.1	+	+	+	+
2	4.1'	+	+	-	+
3	4.2	-	-		
4	4.3	+	+	+	+
5	4.5	-	-		
6	5.1	-	-		
7	5.2	-	-		
8	5.3	-	-		
9	5.4	-	-		

(+) : Positive : (qualified on test)
(-) : Negative : (not qualified on test)

3.1. Cellulolytic Test

The screening was carried out to test bacterial activity of cellulose. Figure 1 showed that isolate of 4.3, 4.1 and 4.1 '(isolates were taken in the same dilution) released the clear area faded around bacterial colonies. This result can be analyzed that the isolate selected can break down the cellulose medium into glucose by screening test using congo red medium and CMC. Congo red as an indicator for cellulolytic enzyme activity released by bacteria [21]. The ability of the cellulolytic activity that broke down into glucose, derived from the analysis that cellulase bacteria in termite colon was useful for obtained the glucose in the termite's body from consumed of wood [22]. The endurance of 4.1, 4.1 'and 4.3 bacteria in the acidic medium also showed that bacteria potentially qualified the stomach acid, that was as a probiotic candidate. Cellulolytic enzymes to broke down cellulose, must be in acidic conditions to be able to hydrolyze the bonding cellulose polymer between glucose [23].

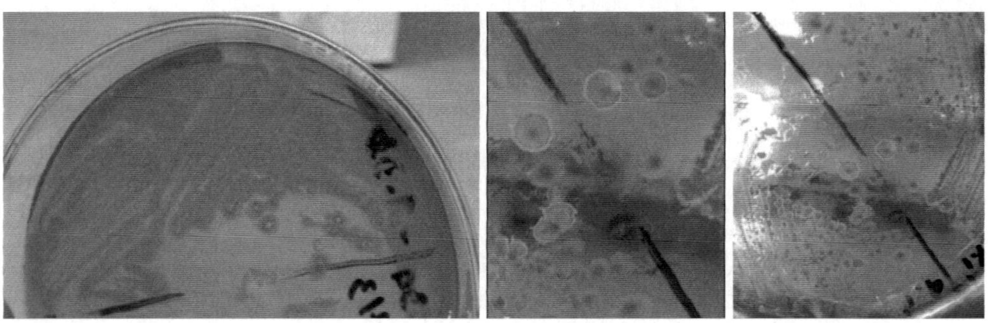

a. b. c.

Figure 1. Results of sceening bacterial cellulase and acid resistance: a-isolates 4.3; b-isolates 4.1; c-isolates 4.1‘ (on a 10^{-4} dilution).

3.2. Pathogenicity Test

Blood agar base medium and sterile human blood type of O are used to detect the pathogenicity activities of the bacteria by hemolysis assasement [24]. The cellulase bacteria strains were streaked on blood agar plates that containing 5% human blood and 2.5% sodium chloride (NaCl), the plates were incubated for 48 hours at 37°C [25]. It was carried out to determined the potential in terms of safety for the human body as probiotics. Figure 2 showed the results that isolate passed through cellulolytic screening 4.1' had the less of hemolysis, β hemolysis. But it had less of ability to lyse the red blood cells with the certain amount. On the other hand, cellulase bacteria of 4.3 and 4.1 showed the positive result with non pathogenic activity (γ hemolysis). Bacterial isolates of 4.3 and 4.1 showed that there were not the damage of the blood medium, there were not discoloration or the formation of the clear zones around the colonies. From the results, they were potentially non pathogenic for probiotic cellulase candidates, as in [17]. The γ hemolysis strains are the most promising probiotic candidates, and should not be either pathogenic or able of producing toxic substances that may harm for the body [25].

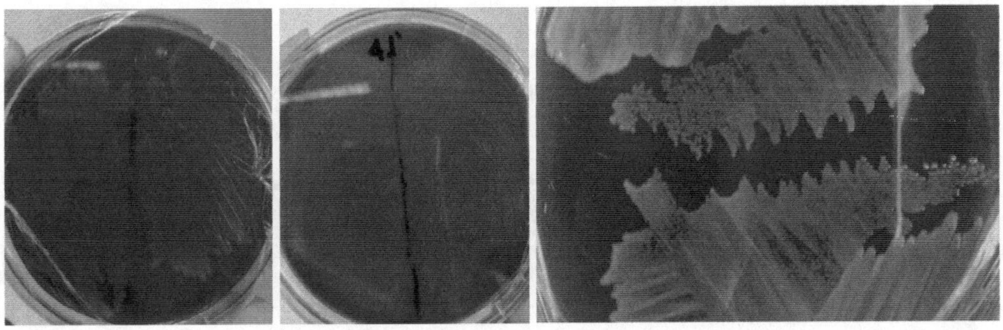

a. b. c.

Figure 2. Results of pathogenicity test: a-isolates 4.3(γ hemolysis); b-isolates 4.1' (β hemolysis); c-Isolates 4.1 (γ hemolysis).

3.3. Human Stomach Acid and Salt Osmotic Concentration Resistance Test

Probiotic should be able to lived in the digestive conditions [26]. Among the conditions in the digestion was the stomach acid pH 2-2.5 and the osmotic salt (6.5% NaCl) [27]. The resistance was necessary to bacteria kept working according to the function [28]. Table 2 showed the test results that Isolates of 4.1 and 4.3 can grow from the pH 2.5 treatment and with 3.5% and 6.5% salt osmotic concentrations for 2 hours (2H) and 4 hours (4H) digestive incubation period. Both of the isolates were tolerant with the high osmotic concentration of salt (NaCl 3.5% and 6.5%), after the treatment and incubation for 24 hours. Isolate grown in testing media by streaking, after incubation time in human gut condition. It can be analyzed that both of bacterial isolates can survive in the digestive conditions than other acids, such as NaCl osmotic concentration as Menconil *et al.* [18]. All of bacteria strains had a similar behaviour to different pH exhibiting no growth at pH 1-3, but grow at pH higher than 3. It has been estimated that the survival rate of traditional probiotics in the host's gut is only 20-40%, being gastric acidity one of the main obstacles. Other properties, to be considered are tolerance to salt and different pH values as well as the in vitro ability to colonize the host gastrointestinal tract [29]

The availability of human ability to digest polymer cellulose to glucose from termite gut microbes was the chance to process the grass as a new staple food, without concerning the agriculture result. Through cellulase probiotic, whatever the materials were high in cellulose, it can be directly consumed and digested by humans' body with the help of termite gut microbes.

Table 2. Results of human stomach acid and salt osmotic concentration resistance test in the digestive incubation period (2-4 hours)

No	Isolates	Stomach acid resistance test				Salt osmotic concentration resistance test			
		pH 7		pH 2- 2,5		3,5% NaCl		6,5% NaCl	
		2H	4H	2H	4H	2H	4H	2H	4H
1	4.1	+	+	+	+	+	+	+	+
2	4.1'	+	+	+	+	+	+	+	+
3	4.3	+	+	+	-	+	+	+	+

(+) : Positive : (tolerant on treatment)
(-) : Negative : (not tolerant on treatment)

4. Conclusion

There are isolated termite gut microbes (bacteria) analyzed can be used in human gut as probiotic candidate to digest cellulose to glucose by in vitro testing. It can be sustainability research for in vivo testing, before the termite gut microbes selected can be consumed by human in the future for probiotic in the global food scarcity to eat cellulose material, like grasses and sawdust with high cellulose.

References

[1] Alexandratos N and Bruinsma J 2012 *World agriculture towards 2030/2050: The 2012 Revision* ESA Working Paper (Rome: FAO)

[2] Gustavsson J, Caderberg C, Sonesson U, Otterdijk R V, Meybeck A 2011 *Global food losses and food waste – Extent, causes and prevention* (Rome: FAO)

[3] Campbell B M, Corner-Dolloff C, Girvetz E, Loboguerrero A M, Ramires-Villegas J 2016 Reducing risk to food security from climate change *Global Food Security* **11** 34-43

[4] Blum W E H 2013 Soil and land resources for agricultural production General trends and future scenarios-A worldwide perspective *Int. Soil and Water Conservation Res.* **1** 1-14

[5] Sanderson M A and Paul RA 2008 Perennial forages as second generation bioenergy crops *Int. J. Mol. Sci.* **9** 768-88

[6] Niemi P, Pihlajaniemi V, Rinne M, Siika-aho M 2017 Production of sugars from grassilage after stream explosion or soaking in aqueous ammonia *Industrial Crops and Products* **98** 93-99

[7] Han Y J and Chen H Z 2007 Synergism between corn stover protein and cellulose *J. Enzyme and Micro. Technol.* **41** 638-45

[8] Manhar A K, Bashir Y, Saikia D, Nath D, Gupta K, Konwar B K, Kumar R, Namsa N D, Mandal M 2016 Cellulolytic potential of probiotic Bacillus Subtilis AMS6 isolated from traditional fermented soybean (Churpi): An in-vitro study with regards to application as an animal feed additive *Microbiological Res.* **186–187** 62–70

[9] Muwawa E M, Budambula N L M, Osiemo Z L, Boga H I, Makonde H M 2016 Isolation and characterization of some gut microbial symbionts from fungus-cultivating termites (Macrotermes and Odotermes) *African J. Micro. Res.* **10** 994-1004

[10] Yang W, Meng F, Peng J, Han P, Fang F, Ma L, Cao B, 2014 Isolation and identification of a cellulolytic bacterium from the tibetan pig's intestine and investigation of its cellulase production *Electronic J. of Biotech.* **17** 262-7

[11] Halim M, Mustafa N A M, Othman M, Wasoh H, Kapri M R, Ariff A B 2017 Effect of encapsulant and cryoprotectant on the viability of probiotic Pediococcus acidilactici ATCC 8042 during freeze-drying and exposure to high acidity, bile salts and heat *LWT - Food Sci. and Technol.* **81** 210-06

[12] Huang S, Sheng P, and Zhang H 2012 Isolation and identification of cellulolytic bacteria from the gut of *Holotrichia parallela* larvae (Coleoptera: Scarabaeidae) *Int. J. Mol. Sci*, **13** 2563–577

[13] Li J, Lewis R B, and Dahn J R 2007 Sodium carboxymethyle cellulose: a potential binder for Si negatives electrodes for Li-ion batteries *Elechtrochemical and Solid-state Letters* **10** A17-20

[14] Davis C 2014 Enumeration of probiotic strains: Review of culture-dependent and alternative techniques to quantify viable bacteria *J. microbiological methods* **103** 9-17

[15] Plessas S, Nouska C, Karapetsas A, Kazakos S, Alexopoulos A, Mantzourani I, Chondrou P, Fournomiti M, Galanis A, Bezirtzoglou E 2017 Isolation, characterization and evaluation of the probiotic potential of a novel Lactobacillus strain isolated from feta-type cheese *Food Chemistry* **226** 102–8

[16] Ferbiyanto A, Rusmana I, Raffiudin R 2015 Characterization and identification of cellulolytic bacteria from gut of worker macrotermes gilvus *HAYATI J. Biosci.* **22** 197-200

[17] Yeh E, Pinsky B A, Banaei N, and Baron E J 2009 Hair sheep blood, citrated or defibrinated, fulfills all requirements of blood agar for diagnostic microbiology laboratory tests *PLoS ONE* **4** e6141

[18] MenconiI A, Kallapura G, Latorre J D 2014 Identification and characterization of lactic acid bacteria in a commercial probiotic culture *Biosci. of Microbiota, Food and Health* **33** 25-30

[19] García-Bernal M, Campa-Córdova A I, Saucedo P E, Casanova-González M, Medina –Marrero R and Mazón-Suástegui J M 2015 Isolation and in vitro selection of actinomycetes strains as potential probiotics for aquaculture *Veterinary World* **8** 170-6

[20] Suman S K, Dhawaria M, Raturi D T V, Adhikari D K, Kanaujia P K 2016 Investigation of lignin biodegradation by *Trabulsiella* sp isolated from termite gut *Int. Biodeterioration & Biodegradation* **112** 12-17

[21] Mamin´ska R T, Szymona K, Madej H, Wong W Z, Bala A, Brutkowski W, Krajewski K, S H'ng P, Mamin´ ski M 2015 Cellulolytic and electrogenic activity of Enterobacter cloacae in mediatorless microbial fuel cell *Appl. Energy* **160** 88-93.

[22] Ni J, Tokuda G 2013 Lignocellulose-degrading enzymes from termites and their symbiotic microbiota *Biotechnol Adv.* **31** 838–50

[23] Fatyasari N I, Chairul I, Primata M, Kang L C. 2015 Carbon-based strong solid acid for cornstarch hydrolysis, *J. Solid State Chemistry* **230** 163-8.

[24] Zomorodian K, Rahimi M J, Safaei A, Bazargani A, Motamadi M, Kharazi M, Mostaghni S, Pakshir K, Ghaedi H, Afsarian M H 2011 Analysis of beta-hemolysis in human blood agars by Streptococcus pyogenes *J. Microbiol Methods* **85** 233–4

[25] Heindl H, Thiel V, Wiese J and Imhoff J F 2012 Bacterial isolates from the bryozoan Membranipora membranacea: Influence of culture media on isolation and anti-microbial activity *Int. Microbiol* **15** 17-32

[26] Fernandez M F, Boris S and Barbes C 2003 Probiotic properties of human lactobacilli strains to be used in the gastrointestinal tract *J. Appl. Microbiol* **94** 449-55

[27] Huang Y, and Adams M C 2004 In vitro assessment of the upper gastrointestinal tolerance of potential probiotic dairy propionibacteria *Int. J. Food Microbiol* **91** 253-60

[28] Han Q, Kong B, Chen Q, Sun F, and Zhang H 2017 In vitro comparison of probiotic properties of lactic acid bacteria isolated from Harbin dry sausages and selected probiotics *J. Functional Foods* **32** 391–400

[29] García-Bernal M, Campa-Córdova A I, Saucedo PE, Casanova-González M, Medina-Marrero R and Mazón-Suástegui J M. 2015 Isolation and in vitro selection of actinomycetes strains as potential probiotics for aquaculture *Veterinary World* **8** 170-6

[30] Mokhtari S, Jafari S M, Khomeiri M, Maghsoudlou Y, Ghorbani M 2017 The cell wall compound of Saccharomyces cerevisiae as a novel wall material for encapsulation of probiotics *Food Res. Int.* **96** 19-26

[31] Hook C D, Samsonov V V, Ublinskaya A A, Kuvaeva T M, Andreeva E V, Gorbacheva L Y, Stoynova N V 2016 A novel approach for Escherichia coli genome editing combining in vivo cloning and targeted long-length cromosomal insertion *J. Microbial Methods* **130** 83-91

Enzymatic production of maltodextrins derived from sago flour using heat-stable alpha-amylase and pullulanase

A Laga, A Syarifuddin and A Dirpan

Department of Agricultural Technology, Hasanuddin University, Jalan Perintis Kemerdekaan KM 10, Makassar, 90245, Indonesia.

E-mail: *amranlaga@yahoo.co.id*

Abstract. Maltodextrins are produced by starch modification in a partial hydrolysis thus altered physical sago properties. Sago as one of starch resources has characteristic with high amylopectin that influences high viscosity during cooking. Partial hydrolysis or liquefaction will influences starch hydrolysis and the size of maltodextrin produced. The aim of this study was to analyze the degree of sago starch hydrolysis during the enzymatic process using single α-amylase and combination with pullulanase The starting solids content was 20% (w/v), with adjusted pH to 6.5, and calcium (Ca_2+ ions) addition as high as 50 ppm. The majority of starches used in the study contain 0.2 % (w/v), to combination of 0.2 % (w/w) and 0, 3 gram per kg of sago. The sago suspension temperatures were started from 105 °C lowered to 60 °C for 30 minutes, respectively. Optimum liquefied starch yields, which accounted for virtually all of the starch present, were obtained at temperatures of 80°C and above, for 120 minutes, with each sampling every 20 minutes. Observed parameters were levels of reducing sugars, degree of hydrolysis, and refined sago starch. The result showed that there was a significant increase in reducing sugars, degree of hydrolysis during 120 minutes until liquefaction process for both enzymatic treatments. The amount of reducing sugars was 95.76 g/L at 120 min for the single α-amylase and 98.84 g/L combination with pullulanase. The degree of hydrolysis was 37.93 % at 120 minutes for the single α-amylase and 37.32 % combination with pullulanase, whereas 0.035 % and 0.038 % for refined sago starch value respectively.

1. Introduction

Sago (*Metroxylon sagu*) is an Indonesian indigenous plant species which can produce sago starch. In part of Indonesia, sago has been consumed as a traditional staple food. For example, Kapurung in South Sulawesi and Papeda in Papua. Because of its cheap, food manufacturer has encouraged to create sago starch-based food industry such as maltodextrin. In food, maltodextrin can be used as an ingredients because it offers many advantages such as consistency, viscosity, mild texture and stability. In addition, as it was added into food the soluble solids increased allowing better inhibit crystallization and control the freezing point [1]. Maltodextrin are enzymatic and/or acid hydrolysis products of starch, consisting of a-(1,4) linked D-glucose oligomers and/or polymers, which are normally defined as having a dextrose equivalent (DE) value < 20 [2]. A number of maltodextrin have been produced from different starches. In the manufacture of maltodextrins, the long chain of starch molecules was hydrolyzed by thermal, acid and enzymes treatments. Various components have been used to account for the efficiency of starch hydrolysis to produce sugars and maltooligosaccharides which are α-amylase, glucoamylase, and pullulanase enzymes action [3].

Due to enzymatic modifications have improved the functional properties of starch, most of the maltodextrin production enzymatically was carried out by liquefaction of starch using α-amylase obtained from bacteria. With enzymatic method, sago starch can be modified by α-amylase which first attach to the granule surface and peel away section of granule [4] and then pullulanase, an important debranching enzyme, has been widely utilised to hydrolyse the α-1,6 glucosidic linkages in starch, amylopectin, pullulan, and related oligosaccharides, which enables a complete and efficient conversion of the branched polysaccharides into small fermentable sugars during saccharification process [5]. The size of maltodextrin molecule can be adjusted by adjusting the degree of starch hydrolysis to the duration of the required enzymatic reaction. Commercial maltodextrin generally ranges from DE 3 – 25 in which the DE value of the maltodextrin is dependent on its use [6].

In the present work, sago starch was hydrolyzed by commercial α-amylase and pullulanase. The aim of this work was to investigate the effect of α-amylase and α-amylase combined with pullulanase on the degree of hydrolysis of sago starch during liquefaction reaction.

2. Material and Method

2.1. Materials
Sago starch was supplied from local market, α-amylase and pullulanase (Novo Nordisk A/S) in the form of liquid concentrate were used as hydrolysis enzyme.

2.2. Experimental procedure
First of all, a suspension containing sago starch (20% w/v) was adjusted to pH 6.5 (Dakton pH 510 Series). $CaCl^{-2}$ 50 ppm was added into the suspension and the reaction was started by adding α-amylase 0.01%. The suspension was heated at 105 °C for 7 minutes. After gelatinization reached, the reaction temperature was adjusted at 60 °C for 30 min and then added pullulanase 0.3 g/kg of substrate and α-amylase of 0.1% w/w. Then, the process of liquefaction was started by increasing the reaction temperature at 80 °C. The liquefaction reaction was carried out in a stirred fermentor for 120 minutes with specified sampling time were 20 minutes.

2.3. Chemical analysis
Parameters observed in this work were reducing sugars (DNS method), degree of hydrolysis [7], remaining iodine content, viscosity modified from [8].

2.4. Data analysis
An experimental design with type of enzyme: α-amylase and combining α-amylase with pullulanase and time of liquefaction varying from 0 to 120 with specified sampling time 20 minutes were used. Data analyses were performed using SPSS 16. One way ANOVA was used to determine significant differences among parameters. Three replications were used. For all data analyses, the effects were considered significant when $p<0.05$.

3. Results and discussion

3.1. Reducing sugar
The reducing sugar is a sugar having the ability to reduce an electron receiving compound because it has an aldehyde and ketone free group. Except sucrose and starch, all monosaccharides (glucose, fructose, galactose) and disaccharides (lactose, maltose) were reducing sugar. In this study, the reducing sugar obtained by combine α-amylase and pullulanase was higher (64.49 g / L - 103.61 g / L) than reducing sugar obtained by α-amylase alone (58.74 g / L - 101.51 g / L). ANOVA showed that time liquefaction was significant effect on reducing sugar ($p<0.05$) or in the other words the reducing sugar increase with increasing time of liquefaction (Fig.1). It was caused by the length of contact time between the enzyme and the substrate allowing hydrolysis process in the α-1.4 glycosidic bond in

short chain formation is increasing. Jane and Chen [8] explains that reaction time of starch hydrolysis enzymatically will affect the duration of enzyme and substrate contact. This phenomenon is then further elaborated by Jane *et al.* [9]. The authors reported that the reaction time increases, the more hydrolyzed starch leading to increase enzyme activity to break the α-1.4 glycosidic bonds into simple oligosaccharide chains. The more glycosidic bonds, the amount reducing sugar is increasing. However, type of enzyme and interaction between type of enzymes and time liquefaction did not show significant.

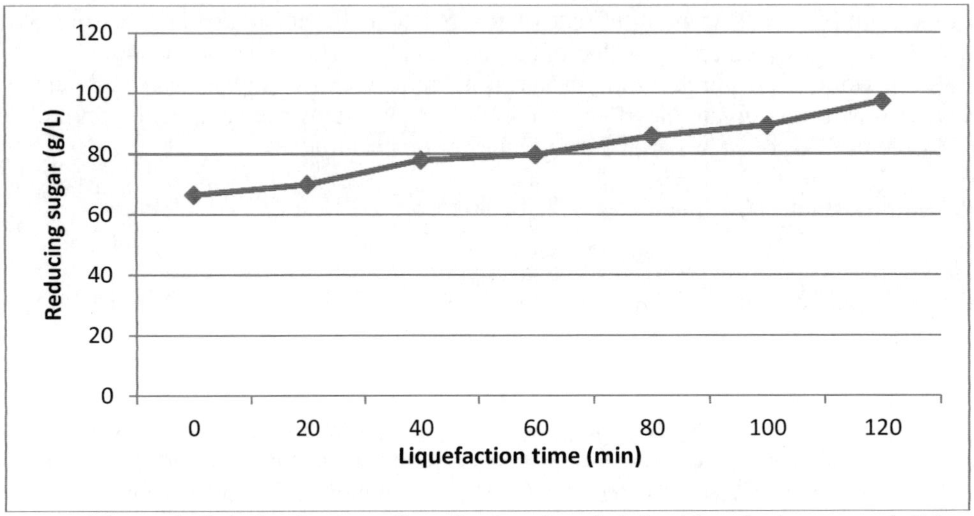

Figure 1. Relation between liquefaction time (min) with reducing sugar (g/L)

The use of single α-amylase enzyme and α-amylase and pullulanase combined were not significantly on reducing sugar. It was due to time for debranching process was only for 30 minutes, so that the cutting branching process α-1,6 glycosidic was ineffective resulting in the formation of reducing sugar not significant. Leong *et al.* [10] reported that the best debranching on sago substrate was 8 hours. This phenomenon also explained that the time to cut the α-1,6 glycosidic bond required 30 time more longer than to bypass α-1,4 glycosidic.

3.2. Degree of Hydrolysis (DH) Relative Humidity

Another property measured was degree of hydrolysis. The product of hydrolyzed starch is generally characterized by the degree of hydrolysis which is expressed by the DH value (hydrolysis degree), the percentage of pure dextrose in the total of solid of the hydrated substrate. In this study, the DH of α-amylase was 27.26% - 39.10%. and the DH of combined α-amylase and pullulanase was 27.99% - 38.92%.

ANOVA showed that time liquefaction was significant effect on the DH ($p<0.05$) or in the other words the degree of hydrolysis (DH) increased with increasing time of liquefaction (figure 2). According to Nakamura [11] the longer the hydrolysis time, the more hydrolyzed molecules leading to increase reducing sugar and DH. The same results was also reported by [12]. The authors reported that the longer hydrolysis time the more the hydrolyzed component.

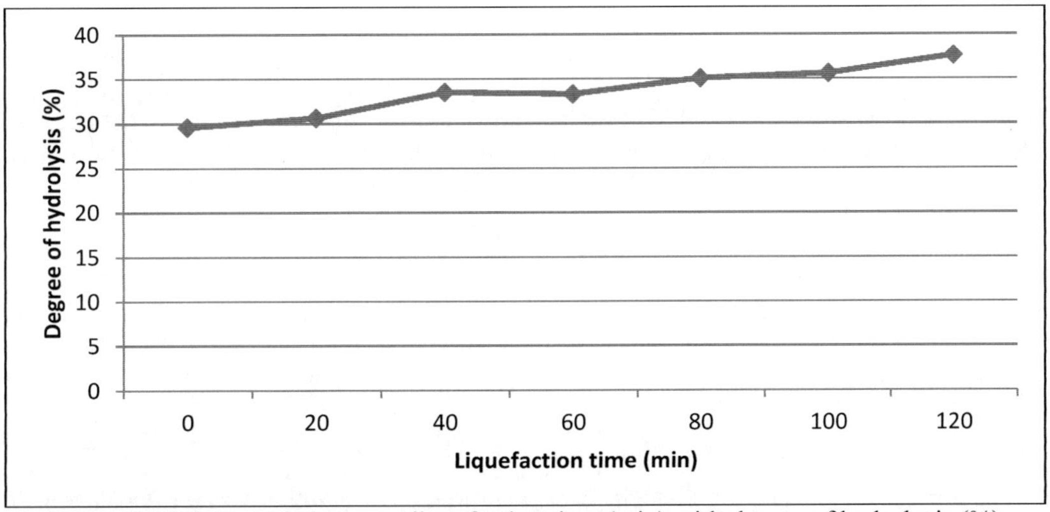

Figure 2. Relationship between liquefaction time (min) with degree of hydrolysis (%)

The use of single α-amylase enzyme and combined α-amylase and pullulanase did not significant on the DH. α-amylase activity cuts the α-1.4 glycosidic bond in a straight chain while the pullulanase cuts the α-1.6 glycosidic bond with each product in the form of a straight chain or oligosaccharide. This phenomenon shows that the cutting activity of α-1.6 glycosidic did not significant in the DH. It is caused by the condition or timing of the cutting of the unfavorable branching chain. According to Griffin and Brooks, pullulanase will work well under suitable conditions such as pH 5.2 [13].

3.3. Residual starch content

Residual starch represents un-hydrolyzed starch content into a simple form of starch. The residual starch test is to find out the content of enzymatic digestibility of raw starch granules due to the differences in amylose content and branch-chain length of the amylopectin. The Residual starch content of α-amylase enzymatic usage was 0,021% - 0,042%. Whereas the combination of α-amylase and pullulanase 0,029% - 0,044%. The starch content in of α-amylase enzymatic is lower than α-amylase and pullulanase combined (figure 3).

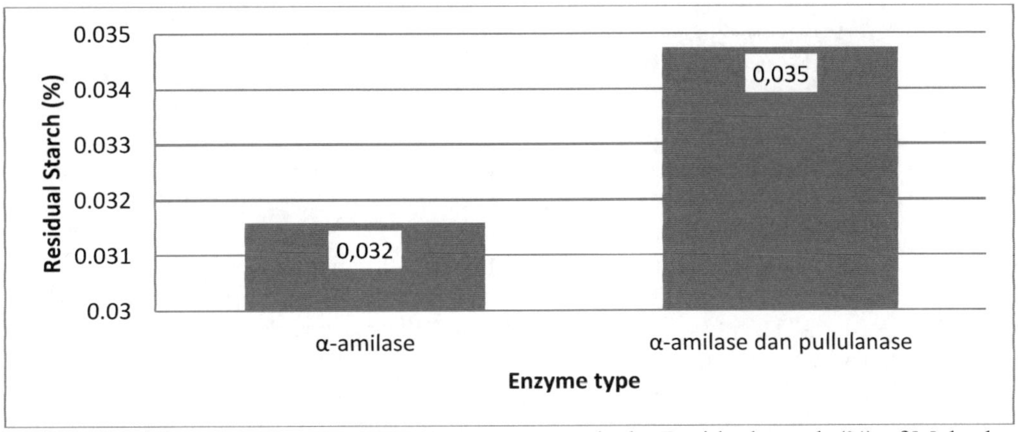

Figure 3. Correlation of different enzymatic usage towards the Residual starch (%) of Maltodextrins

The analysis of variance (ANOVA) for the Residual starch (%) of Maltodextrin showed that susceptibility to the enzyme production and liquefaction, or both interactions were not significant to the amount produced.

Susceptibility of the residual starch during enzyme hydrolysis for α-amylase and pullulanase combined showed the least amount of enzyme. The residue value was stable during increased of liquefaction time. The residual starch contributes the highest amount of enzyme units during liquefication 120 minutes was for 0,037%. The combination of different enzymatic usage towards the residual starch (%) showed that the basic gelatinization step at 105 °C for 7 min was not sufficient to convert crystalline sago starches formed. Therefore there are many of crystalline starches, but it's fine granular structure.

Starch crystallinity affects thermal properties of the starch granules due to the dense packing of cellulose chains due to its resistance to enzymatic hydrolysis. Breaking down the intermolecular bonds of starch crystallinity can be done under high temperature and pressure. The starch crystallinity of the starch granule is difficult to breakdown by the production of starch hydrolyzing enzymes, hence slow breakdown during hydrolysis. The stronger the starch crystallinity, the higher temperature and energy total needed [10].

3.4. Viscocity

The data for the production of viscosity test was presented quantitatively. The viscosity can be measured for a product thickness. Viscosity from maltodextrins are usually supplied with the dextrose equivalent value (DE) as the only information from the source of the starch or hydrolysis of starch.

Viscosity value obtained from α-amylase enzyme utilization was around 46 – 80 cP, for the lowest of 100 to 120 min reaction. While the viscosity number from a combination of α-amylase and pullulanase enzyme was between 56 – 74 cP, with the same time duration as a single enzyme. Figure 4 showed the relationship between liquefaction reaction (min) and viscosity (cP) of maltodextrin viscosity.

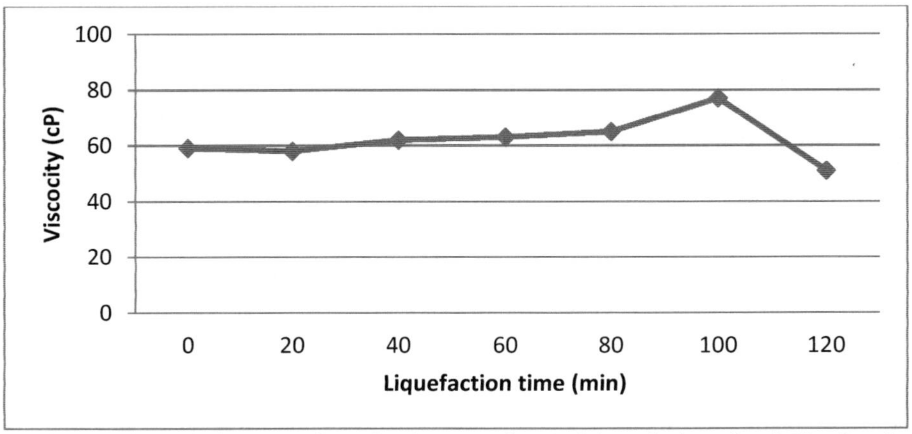

Figure 4. The relationship between liquefaction reaction (min) and viscosity (cP) of maltodextrin viscosity.

The viscosity value increased during the reaction time of 100 minutes due to the starch fraction, i.e. amylopectin and a portion of the amylose extracted out of the granules and suspended in water thereby increasing the viscosity. The fraction of amylopectin is faster out compared with amylose because it has higher molecular weight. Suspected amylopectin fractions have crystalline regions with tight structures, raising viscosity because of their ability to bind strongly with water through intermolecular hydrogen bonds.

The presence of water as a solvent and sufficient heat energy causes the starch granules to swell to the point where the starch granules can absorb water. Furthermore, the starch granules ruptured causing the amylose and amylopectin extracted out of the granules. It is clear that compounds that have high molecular weight would come out faster than compounds that have low molecular weight. Amylopectin has a higher molecular weight than amylose so that amylopectin will break faster than amylose. Nakamura [11] described the increase in the viscosity of the paste that increases the viscosity of the paste due to the development of starch granules that form molecular bonds through hydrogen bonds so that the starch soles form of an opaque and firm gel. The formation of the opaque gel is caused by the intermolecular hydrogen bonds that make up the aggregate in the water-insoluble crystalline region.

Furthermore, the viscosity decreases in the duration of the 120-minute reaction are due to the amylose fraction remaining in the outgoing starch granules and suspended in maltodextrin. The more amylose that is suspended will decrease the viscosity. Decreased viscosity also occurs quickly due to the hydrolysis of the starch fraction by the enzyme. The amylose fraction and some suspended amylopectin are hydrolyzed by the enzyme into fractions with lower molecular weights such as short-chain dextrins. The formation of the dextrin unit causes the DH value to rise, and the viscosity decreases at a reaction time of 120 min.

The effect of the amylose fraction on viscosity decrease can be influenced by the higher proportion of amylose and cause lower product density. Nakamura [11] further explained that the viscosity of the starch solution rapidly decreases in the event of hydrolysis by α-amylase (liquefaction of starch). The decrease in viscosity is a change in physical properties of natural starch when experiencing enzymatic hydrolysis. The enzyme will bypass the amylose and amylopectin fractions from the starch granules to the simpler sugars thus decreasing the viscosity of the starch suspension.

4. Conclusion

Hydrolysis degree evaluation was not significant to make a difference between maltodextrin obtained from the use α-amylase and combination α-amylase with pullulanase on sago starch. The highest yield of maltodextrin determined from degree of hydrolysis reveals that the yield depends on the viscocity at 120 min of liquefaction time.

References

[1] Console 1998 *Tese de Douturado* (Brasil: Universidade Estadual de Campinas)

[2] Lambri M, Dordoni R, Roda A and De Faveri D M 2014 Process development for maltodextrins and glucose syrup from cassava *Chem. Eng. Trans.* **38** 469–74

[3] Plácido M G R, Do Canto L R, Amante E R and Soldi V 2005 Cassava and corn starch in maltodextrin production *Quim. Nova* **28** 596–600

[4] Fujii M, Homma T and Taniguchi M 1988 Synergism of α-amylase and glucoamylase on hydrolysis of native starch granules *Biotechnol. Bioeng.* **32** 910–5

[5] Hii S L, Tan J S, Ling T C and Ariff A Bin 2012 Pullulanase: Role in starch hydrolysis and potential industrial applications *Enzyme Res.* **2012** Article ID 921362, 14 p doi:10.1155/2012/921362

[6] Gupta R, Gigras P, Mohapatra H, Goswami V K and Chauhan B 2003 Microbial α-amylases: A biotechnological perspective *Process Biochem.* **38** 1599–616

[7] Chiba S 1988 *Handbook of Amylases and Related Enzymes Their Sources, Isolation Methods, Properties and Application* (first edition) ed. The Amylase Research Society of Japan (Oxford: Pergamon Press) pp104–16

[8] Jane J and Chen J F 1992 Effect of amylose molecular size and amylopectin branch chain length on paste properties of starch *Cereal Chem.* **69** 60–5

[9] Jane J, Chen Y Y, Lee L F, McPherson A E, Wong K S, Radosavljevic M and Kasemsuwan T 1999 Effects of amylopectin branch chain length and amylose content on the gelatinization and pasting properties of starch *Cereal Chem.* **76** 629–37

[10] Leong Y H, Karim A A and Norziah M H 2007 Effect of pullulanase debranching of sago (metroxylon sagu) starch at subgelatinization temperature on the yield of resistant starch *Starch/Staerke* **59** 21–32

[11] Nakamura Y 1996 Some properties of starch debranching enzymes and their possible role in amylopectin biosynthesis *Plant Sci.* **121** 1–18

IOP Conf. Series: Earth and Environmental Science **157** (2018) 012028 doi:10.1088/1755-1315/157/1/012028

[12] Griffin V K and Brooks J R 1989 Production and size distribution of rice maltodextrins hydrolyzed from milled rice flour using heat-stable alpha-amylase *J. of Food Sci.* **54** 190-193

Expression of CYP2A6, KIF12, and SULT1C1 in liver of sheep with divergent sheepmeat flavour and odour

K Listyarini[1], Jakaria[1], A Furqon[1], C Sumantri[1], M J Uddin[2] and A Gunawan[1]

[1] Department of Animal Production and Technology, Faculty of Animal Science, Bogor Agricultural University, 16680, Bogor, Indonesia.
[2] School of Veterinary Science, The University of Queensland, Gatton Campus, Gatton Qld, 4343, Australia.

E-mail: *aagun4780@gmail.com*

Abstract. The aim of this study was to investigate the expression of some of the key enzymes involved in liver sample of sheep with high and low sheepmeat flavour and odour. The study was conducted with Indonesian Javanese fat tailed sheep. Sheep having a fat branched chain fatty acids 4-methylnonanoic (MNA) greater and less than 215 μg g^{-1} and 229 will be defined as low and sheep meat odour, respectively. For the flavour, sheep having a fat skatole level less than 0.25 μg g^{-1} and greater than 0.25 μg g^{-1} will be defined as low and high flavour samples, respectively. The enzymes investigated were cytochrome P450 2A13 (CYP2A6), kinesin-like protein KIF12 (KIF12), and sulfotransferase 1C1 (SULT1C1). Expression of CYP2A6 in liver had differ between animals with high and low sheep meat flavour. Expression of CYP2A6, which catalyses the first stage of oxidation degradation, was increased in high sheep meat flavour and odour ($P > 0.05$). Similar pattern, the expression of SULT1C1, which catalyse the second stage of conjugation steroid catabolism, was increase in high sheep meat flavour and odour ($P > 0.05$). In contrast, the expression of KIF12 was decreased in high sheep meat flavour and odour animals. It is suggested that accumulation sheep meat flavour and odour in liver tissue of Indonesian Javanese fat tailed might be related to a high rate of oxidation in metabolic stage I and conjugation degradation in metabolic stage II.

1. Introduction

Sheep meat consumption is uncommon in many Asian countries including Indonesia, whose consumers often complain about the unpleasant flavour or odour of the sheep meat [1]. The main causes of sheepmeat odour are the two compounds (branched chain fatty acid and skatole): branched chain fatty acid (BCFA), present in all fatty tissue, have been implicated as cause of such flavours as, has indole, which originated from pastoral diets [2, 3]. Branched chain fatty acids (BCFAs; 4-methyloctanoic (MOA), 4-ethyloctanoic (EOA) and 4-methylnonanoic acids) are the chemical compounds that are accepted as the main contributors for odour and flavour [3]. 3-methylindole (skatole) also involved with 'boar' taint in pigs, and to a lesser extent 4-methylphenol (p-cresol) are the main compounds implicated as contributors to 'pastoral' flavour [4].

A number of candidate genes have been identified for reducing sheepmeat flavour and odour compound with the final goal to utilize this genetic information in the breeding scheme [5]. A recent study using RNA deep sequencing technology identified a number of genes including cytochrome P450 2A13 (CYP2A6), kinesin-like protein KIF12 (KIF12), and sulfotransferase 1C1 (SULT1C1) that were expressed differentially in sheepmeat with divergent levels of flavour and odour compounds [5].

CYP2A6 is one of the enzymes involved in the hepatic metabolism of a naturally produced compound, skatole, in the pig. Low CYP2A6 activity has been linked to excessive accumulation of skatole in pig adipose tissue and development of the phenomenon "boar taint" [6]. KIF12 is the gene encoding a novel kinesin, underlie the Chr 4 main effect on the renal and biliary phenotypes. KIF12 is transcriptionally downregulated by chronic exposure to fatty acids treatment in beta cells [7]. Sulfotransferase enzymes including SULT2A1, SULT2B1 and SULT1A1 are associated with metabolic clearance of androsterone and skatole since they are known to conjugate steroid hormones and drugs into more water soluble compounds that facilitates excretion.

Notably, no study investigated the expression of these genes in sheepmeat flavour and odour. However, functional and positional studies suggested that these genes could be important candidate genes for sheepmeat flavour and odour compounds. Therefore, the present study was aimed to investigate the expression of some of the key enzymes (CYP2A6, KIF12, and SULT1C1) involved in liver sample of sheep with high and low sheepmeat flavour and odour in Indonesian Javanese fat tailed sheep. The results of this study might highlight the selected genes for further study with regard to the sheepmeat flavour and odour.

2. Material and Methods

2.1. Animals

Tissue samples were collected from the Indonesian Javanese fat tailed sheep animals. All the sheep (n=30) were slaughtered in a commercial abattoir. The carcass and meat quality data were collected according to guidelines of the Indonesian performance test. Tissue samples from liver were frozen in liquid nitrogen immediately after slaughter and stored at -20°C until used for isolation of total RNA.

2.2. Isolation of Total RNA and Quantitative Real Time PCR

Total RNA was isolated from the liver using RNeasy Mini Kit (QIAGEN, Hilden, Germany), according to the manufacturer's instructions. DNase treated total RNA was reverse-transcribed with the High-Capacity cDNA Reverse Transcriptase Kit (Applied Biosystems, Foster City, CA, USA) to produce cDNA. For qRT-PCR, total RNA and cDNA synthesis from liver tissues of two divergent groups of animals (sheepmeat flavour and odour) were done as described above. Gene specific primers for CYP2A6, KIF12, and SULT1C1 for the qRT-PCR were designed by using the Primer3 software [8] (table 1). In each run, the 96-well microtiter plate contained each cDNA sample and no-template control. The qRT-PCR was conducted with the following program: 95 °C for 1 min and 40 cycles ; 95 °C for 15 s; 55 and 62 °C for 30 s; and 72 °C for 30 sec on the Analytikjena qTower. For each PCR reaction, 2 µl iTaqTM SYBR® Green Supermix with Rox PCR core reagents (Bio-Rad), 2 µL of cDNA (50 ng/µL) and an optimized amount of primers were mixed with ddH_2O to a final reaction volume of 10 µL per well. All samples were analysed twice (technical replication) and the geometric mean of the Ct values was further used for mRNA expression profiling. The geometric mean of two housekeeping genes GAPDH and Beta Actin was used for the normalization of the target genes. The delta Ct (ΔCt) values were calculated as the difference between target gene and geometric mean of the reference genes: ($\Delta Ct = Ct_{target} - Ct_{housekeeping\ genes}$) as described previously [9]. The final results were reported as fold change calculated from delta Ct-values. The differences in CYP2A6, KIF12, and SULT1C1 gene expressions were analysed with the paired t tests. The values of $P < 0.05$ were considered to indicate statistically significant differences.

Table 1. GenBank accession numbers and primer sequences

Gene name	Accession number	Primer sequence	Size (bp)	Tm (°C)
CYP2A6	XP_004015316.1	F: 5'- CAAGGATACCAAGTTTCGAG -3' R: 5'- ATGATGGTGGTGAAGAAGAG -3	246	55
KIF12	XP_011967829.1	F: 5'- CTGGATGTTCCTGTCACTCT -3' R: 5'- TAAGACCCATTTGCAGAAGT -3	196	55
SULT1C1	XP_004005969.1	F: 5'- ATCATCCATCACACATCCTT -3' R: 5'- CAGCCATCTTCCTCTCATAG -3	190	55
GAPDH	NC_019460.2	F: 5'- GAGAAACCTGCCAAGTATGA -3' R: 5'- TACCAGGAAATGAGCTTGAC-3	203	62
Beta Actin	NC_019471.2	F: 5'- GAAAACGAGATGAGATTGGC -3' R: 5'- CCATCATAGAGTGGAGTTCG-3	194	62

3. Result and Discussion

3.1. mRNA Expression by Quantitative Real Time PCR

Quantitative real-time PCR analysis showed the abundance of CYP2A6, KIF12, and SULT1C1 transcript in animals with divergent sheepmeat flavour and odour. The results showed that the CYP2A6, KIF12, and SULT1C1 mRNA was differentially regulated between animals with high and low sheepmeat flavour and odour values ($P > 0.05$) in the liver. Expression of CYP2A6, which catalyses the first stage of oxidation degradation, was increased in high-sheepmeat flavour and odour ($P > 0.05$) (figure 1). The mRNA expression of CYP2A6 which were up-regulated in high sheepmeat flavour and odour (figure 1) might indicate its function with regard to a high rate of oxidation in metabolic stage I. The expression pattern of the CYP2A6 gene was consistent with the previous study of sheepmeat flavour [10]. Babol *et al.* [11] Diaz and Squires [12] reported that the first stage of hepatic skatole metabolism involves the P450 system, in particular the isoforms CYP2E1 and CYP2A6. Lin *et al.* [13] established that CYP2A6 mRNA was only present in liver and kidney. The expression of CYP2A6 mRNA was much higher in liver compared with the other tissues. Cytochrome P450 isoenzymes are the main enzymes playing roles in phase 1 skatole metabolism, where skatole is degraded to several intermediate products including such as indole-3-carbinol (I3C), 2-aminoacetophenone (2AAP) and 3-metyloxyindole (3MOI) (details reviewed by [14, 15]. The main enzymes of phase 2 skatole metabolism are UDPGT (uridine 5'-diphospho-glucuronosyltransferase) and SULTs (sulfotransferase) [16].

In contrast, the expression of KIF12 was decreased in high sheep meat flavour and odour animals (figure 2). It is suggested that accumulation sheep meat flavour and odour in liver tissue of Indonesian Javanese fat tailed might be related to a high rate of oxidation in metabolic stage I and conjugation degradation in metabolic stage II.

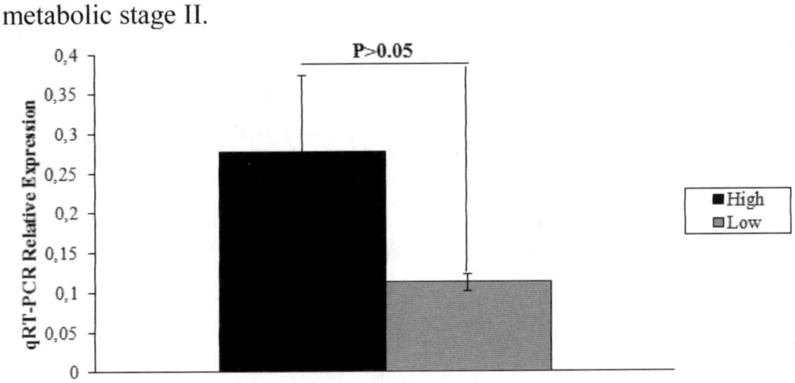

Figure 1. mRNA expression of CYP2A6 in liver tissues from high and low sheepmeat flavour and odour

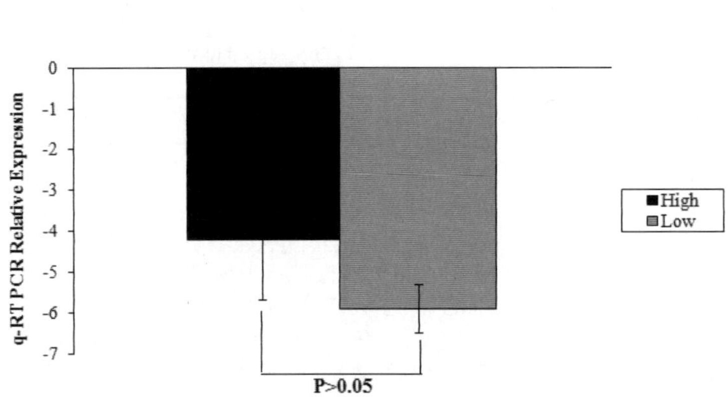

Figure 2. mRNA expression of KIF12 in liver tissues from high and low sheepmeat flavour and odour

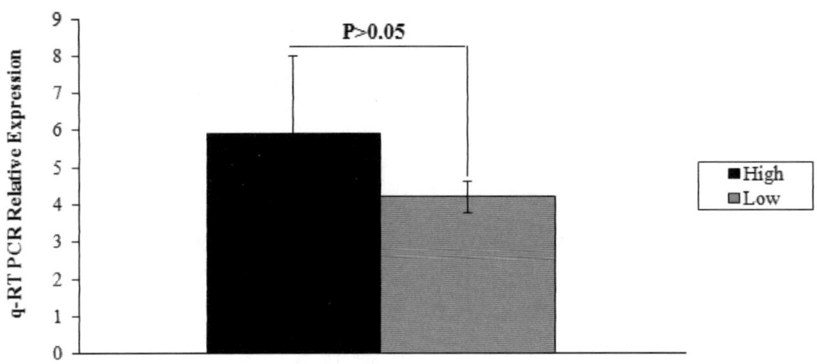

Figure 3. mRNA expression of SULT1C1 in liver tissues from high and low sheepmeat flavour and odour

The expression of KIF12 was decreased in high sheepmeat flavour and odour animals (figure 2). Yang *et al.* [7] reported that KIF12 is transcriptionally downregulated by chronic exposure to fatty acids, this antioxidant cascade involving KIF12 and Hsc70 is proposed to be a critical target of nutritional excess in beta cells in diabetes. Similar pattern with CYP2A6, the expression of SULT1C1, which catalyze the second stage of conjugation steroid catabolism, was increase in high sheep meat flavour and odour ($P > 0.05$) (figure 3). The expression pattern of the SULT1C1 gene was consistent with the previous study of rats [17]. Klaassen *et al.* [17] reported that SULT1C1 mRNA increase in males. Hypophysectomy abolished SULT1C1 expression in male rats. It is suggested that accumulation sheepmeat flavour and odour in liver of Indonesian Javanese fat tailed might be related to a high rate of oxidation in metabolic stage I and conjugation degradation in metabolic stage II that CYP2A6, KIF12 and SULT1C1 were involved.

4. Conclusion

Expression of CYP2A6 and SULT1C1 which catalyses the first stage of oxidation degradation and the second stage of conjugation steroid catabolism was increased in high sheepmeat flavour and odour ($P > 0.05$). In contrast the expression of KIF12 was decreased in high sheepmeat flavour and odour animals. This study proposed candidate genes such as CYP2A6, KIF12, and SULT1C1, that might be involved in the liver for sheepmeat flavour and odour metabolism.

Acknowledgment
This work was financially supported by Directorate of Research and Community Service Directorate General of Strengthening Research and Development Ministry of Research, Technology, and Higher Education Fiscal Year 2016 Number: 079/SP2H/LT/DRPM/II/2016 date 17 Februari 2016.

References
[1] Young A, Reid D H, Smith M E, Braggins T J 1994 In *Flavour of Meat and Meat Products* ed F Shahidi (Glasgow:Blackie Academic and Professional) p. 71
[2] Watkins P J, Kearney G, Rose G, Allen D, Ball A J, Pethick D W and Warner R D 2014 Effect of branched-chain fatty acids, 3-methylindole and 4-methylphenol on consumer sensory scores of grilled lamb meat *Meat Sci.* **96** 1088-1094
[3] Young O A, Berdagué J L, Viallon C, Rousset-Akrim S and Theriez M 1997 Fat-borne volatiles and sheep meat odour *Meat Sci.* **45** 183-200
[4] Young O A, Lane G A, Priolo A and Fraser K 2003 Pastoral and species flavour in lambs raised on pasture, lucerne or maize *J. Sci. Food and Agric.* **83** 93-104
[5] Gunawan A, Jakaria, Listyarini K, Furqon A, Sumantri C and Uddin M J 2016 *Transcriptome Analysis of Liver for Meat Odour and Flavour in Javanese Fat-Tailed by Using RNA Deep Sequencing (Innovation for Food Sovereignty) Proc. Conference on Research Results and Community Service* December 1, 2016 IPB Convention Center Bogor Indonesia
[6] Chen G, Cue R A, Lundstrom K, Wood J D and Doran O 2008 Regulation of CYP2A6 protein expression by skatole, indole, and testicular steroids in primary cultured pig hepatocytes. *DMDF* **36** 56-60
[7] Yang W, Tanaka Y, Bundo M and Hirokawa N 2014 Antioksidant signaling involving the microtubule motor KIF12 is an intracellular target of nutrition excess in beta cells *Dev. Cell* **31** 202-214
[8] Rozen S and Skaletsky H 2000 Primer3 on the www for general users and for biologist programmers *Methods Mol. Biol.* **132** 365-86
[9] Silver N, Best S, Jiang J and Thein S L 2006 Selection of housekeeping genes for gene expression studies in human reticulocytes using real-time PCR *BMC Mol. Biol.* **7** p 33
[10] Shaw A 2007 Characterisation of skatole metabolism in sheep in relation to animal health and meat quality. [Thesis]. Agriculture and Holticulture Development Board
[11] Babol J, Squires E J and Lundstro¨m K 1998 Hepatic metabolism of skatole in pigs by cytochrome P4502E1 *J. Anim. Sci.* **76** 822-828
[12] Diaz G J and Squires E J 2000 Metabolism of 3-methylindole by porcine liver microsomes: responsible cytochrome P450 enzymes *Toxicol. Sci.* **55** 284-292
[13] Lin Z, Lou Y and Squires E J 2004 Molecular cloning, expression and functional characterisation of the cytochrome P450 2A6 gene in pig liver *Anim. Genet.* **35** 314-316
[14] Robic A, Feve K, Larzul C, Billon Y, van Son M, et al 2011 Expression levels of 25 genes in liver and testis located in a QTL region for androstenone on SSC7q1.2 *Anim. Genet.* **42** 662-665
[15] Deslandes B, Gariepy C and Houde A 2001 Review of microbial and biochemical effects of skatole in animal production *Livest. Prod. Sci.* **71** 193–200
[16] Agergaard N and Laue A 1993 *Absorption from the gastrointestinal tract and liver turnover of skatole. In Measurement and Prevention of Boar Taint in Entire Male Pigs* ed Bonneau M (Paris:Institut National de la Recherche Agronomique)
[17] Klaassen C D, Liu L and Dunn R T 1998 Regulation of sulfotransferase mRNA expression in male and female rats of various ages *Chem. Biol. Interact.* **109** 299-313

IOP Conf. Series: Earth and Environmental Science **157** (2018) 012030 doi:10.1088/1755-1315/157/1/012030

Variant discovery in the sheepmeat odour and flavour in javanese fat tailed sheep using RNA sequencing

M A M Abuzahra[1], Jakaria[1], K Listyarini[1], A Furqon[1], C Sumantri[1], M J Uddin[2] and A Gunawan[1]

[1]Department of Animal Production and Technology, Faculty of Animal Science, Bogor Agricultural University, 16680 Bogor, Indonesia.
[2]School of Veterinary Science, The University of Queensland, Gatton Campus, Gatton Qld 4343, Australia.

E-mail: *aagun4780@gmail.com*

Abstract. High-throughput RNA sequencing (RNA-Seq) reveals new challenges for the detection of transcriptome variants (SNPs) in different tissues and species. The aims of this study was to characterize a SNP discovery analysis in the sheep meat odour and flavour transcriptome using RNA-Seq. Six liver samples from divergent sheep meat odour and flavour were analyzed using the Illumina Genome Hiseq 2500 Analyzer. The SNP detection analysis revealed 142 SNPs in sheep meat samples, and a large number of those corresponded to differences between high and low sheep meat odour and flavour ovis genome assembly OAR v4.0.Among them, about 90.4% of genes had multiple polymorphisms within 12 genes (JAML, ANGPTL8, LOC101103463, SEPW1, SCN5A, LOC101113036, DOCK6, GTSE1, KIF12, KCTD17, KANK2, CYP2A6). Several of the SNPs (JAML, CYP2A6, SEPW1, and KIF12) found in this study could be included as suitable markers in genotyping platforms to perform association analyses in commercial populations and apply genomic selection protocols in the sheep meat production.

1. Introduction

Next generation technologies of modern time for sequencing has been provided new opportunities for maximum output like genome annotation, functional genomic research dividing gene expression, detection and profiling of aberrant transcription and discovery of ncRNA [1]. It may be used in entire transcriptome to determine the changes in gene expression. RNA sequence also provides positive results to identify the splicing events, different transcripts of different family isoforms as well as polymorphism [2].

RNA-Seq has been widely used to detect single nucleotide polymorphisms (SNPs), alternative splicing (AS) events differentially expressed genes (DEGs) between two gene expression patterns and insertion or deletion (InDels). In addition, RNA-sequence helped to approach and describe previously inaccessible complexities in the transcriptome, such as isoforms, allele-specific expression and novel promoters [3]. Even in recent times the use of next-generation sequencing platforms to determine and to express the sheep transcriptomes is not too much developed and is very limited, and globally there is no authentic publication regarding analysis of sheep liver transcriptome by using RNA sequence. So to approach and for better understanding of fat deposition and metabolism in sheep, data of liver transcriptomes from Javanese fat tailed sheep was analyzed.

Among these different techniques, RNA sequencing (RNA-Seq) is a reliable and powerful new approach to quantify and for the mapping of transcriptomes used for the analysis of global gene expression in different tissues. In present, this approach efficiently works and also very cheap technique systematically used for the identification of SNPs in transcribed regions in different species [4]. RNA-Seq generates sequences on a very large scale at a fraction of the cost required for traditional Sanger sequencing, allowing the application of sequencing approaches to biological questions that were logically, logistically and economically difficult to practicalized [5].

The aim of this study was to characterize the SNPs discovery analysis in the sheep meat odour and flavour in Javanese fat tailed sheep using RNA-Seq. the transcriptome profile, polymorphisms and alternative splicing analysis using RNA deep sequencing could reveal potential candidate genes affecting odour and flavour of sheep meat in Javanese fat tailed sheep. It is exhibited that in future these polymorphisms could be used as markers for sheep meat related with odour and flavour trait.

2. MATERIAL AND METHODS

2.1 Animals and samples

A total of six liver samples from divergent of sheep meat flavour and odour in Javanese Fat Tailed Sheep were analyzed by using the Illumina Genome Hiseq 2500 Analyzer. All lambs were healthy and development was consistent with age. Management and feed rations were the same. Sheeps were slaughtered after 10 months following the "Guidelines on Ethical Treatment of Experimental Animals, which was formulated by the Ministry of agriculture Indonesia. Liver tissues were collected and frozen in liquid nitrogen within 10 min of slaughter. Subsequently, liver tissues were stored at -80°C until RNA isolation.

2.2 Library construction and sequencing

The ideal method for transcriptomics should be able to directly identify and quantify all small or large RNAs. RNA-Seq provides enough explanation and description of identification of splicing events and different family isoforms of transcripts, transcriptome profiling and polymorphisms [6]. Deep RNA sequencing approach was used to find polymorphism, differential expression and alternative splicing detection. To obtain this purpose, full-length cDNA was obtained from 1 μg of RNA, with the SMART cDNA Library Construction Kit (Clontech, USA), according to the manufacturer's instructions. Libraries of amplified RNA for each sample were prepared following the Illumina mRNA-Seq protocol. The library preparations were sequenced on an Illumina HiSeq 2500 as single- reads to 100 bp using 1 lane per sample on the same flow-cell (first sequencing run) at Macrogen, South Korea. All sequences were analysed using the CASAVA v1.7 (Illumina, USA).

2.3 RNA-Seq analysis and SNP detection

Short sequence reads (36-40 bp) were assembled and mapped to the annotated bovine reference genome B tau4.0 (http://www.ncbi.nlm.nih.gov/genome/guide/cow/index.html) using CLC Genomics Workbench software (CLC Bio, Aarhus, Denmark). Sequencing reads for each of the six samples were pooled to perform the RNA-Seq and SNP discovery analyses. Stringent criteria was applied in order to reduce the rate of detection of false-positive SNPs. For the assembly procedure, the sequences were mapped to the consensus genome accounting for a maximum of two gaps or mismatches in each sequence. SNP detection was performed using the following quality and significance filters: (1) the minimum average quality of surrounding bases and minimum quality of the central base were set as 15 and 20 quality score units, respectively; (2) minimum coverage was set at ten reads; (3) minimum variant frequency or count was set at 20% or two read counts per SNP; and SNPs located in read ends (last three bases) were not considered in the analysis due to possible sequencing errors.

2.4 Gene Variation Analysis

In this analysis, SNP calling was performed on the mapping files generated by TopHat algorithm using samtools mpileup command and associated algorithms. From the variants so generated, only those variants with a minimum Root Mean Square (RMS) mapping quality of 20 and a minimum read depth of 100 were selected for further analysis. In the final step, the selected variants were cross- checked against dbSNP database to identify mutations that are already studied. In the next step, to understand whether these identified polymorphisms segregate either in only one sample group (high odour and flavour or low odour and flavour group) or in both groups (high and low odour and flavour group), we calculated the read/coverage depth of these polymorphisms in all the samples . The identified SNPs were furthermore classified as synonymous or non-synonymous using the GeneWise software (http:// www.ebi.ac.uk/Tools/psa/genewise/last accessed 21.03.2013) by comparing between protein sequence and nucleotides incorporated SNP position.

3. RESULTS AND DISCUSSION

3.1 Variant Discovery in Javanese fat tailed Sheep

In addition to the quantification process of transcriptome, RNA-Seq approach provides valuable information regarding gene polymorphisms which could be directly correlated with the relevant phenotype [2] .The method of variant detection applied to sequence javanese fat tailed sheep was the pooled samples analyses, and it revealed 142 SNPs in 37 genes.

In the results observation that most of the SNPs were in intergenic region with the maximum number of 142 SNPs (figure 1), among them, about 90.4% of genes had multiple polymorphisms (figure 2 A). It should also be noted that the *O. aries* reference genome was used in the present study; therefore, development of a *O. aries* reference genome could greatly enhance results from SNP discovery studies.

Localization and predicted effect of polymorphisms discover in Javanese fat tailed

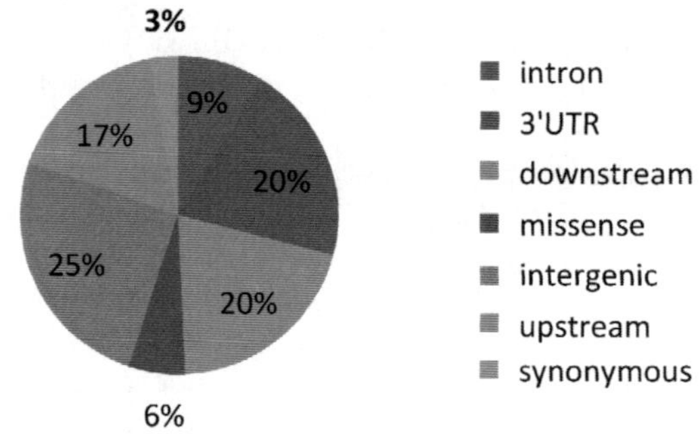

Figure 1. Descriptive statistics results from the Variant Effect Predict (VEP) tool f Ensembl using the 142 SNPs identified in Javanese Fat Tailed Sheep.

Because SNPs were detected in the RNA sequence that the current annotations described as being in introns, a transcript discovery analysis was conducted. Specifically, in Illumina Genome Hiseq 2500 assigned 37 SNPs to exons. Only 8 of these SNPs in the exons were non-synonymous (figure 1). Illumina Genome Hiseq 2500 classified some SNPs as splice variants. This means that polymorphisms occurred within the region of the splice site and potentially influenced the length of the RNA transcript [7]. SNPs located within the non-coding region (3'UTR) of the candidate gene may be located in a binding site for mRNA and/or the transcriptional machinery [8]. Moreover, SNPs in 3'UTR region can also influence mRNA stability [9]. RNA transcripts often contain a poly-A tail, which could have a role in gene expression as well as post-transcriptional stability [10]. Also have to take care that it should that SNPs located within an intron may also affect transcription factor binding sites or mRNA, transcription of other genes would be effected [8]. SNPs located within the upstream intergenic area of a gene could be within or can influence binding site for transcription factors [11].

Analysis of SNPs discovered in Javanese Fat Tailed Sheep samples revealed that 142 SNPs associated with sheepmeat odour and flavour within 37 genes (figure 1). The present study candidate 12 genes related with sheepmeat odour and flavour (JAML, LOC101119620, KIF12, KCTD17, ANGPTL8, LOC101103463, SEPW1, SCN5A, LOC101113036, DOCK6, GTSE1, KANK2, CYP2A6) (figure 2 B). SCN5A gene was the fewest numbers of SNPs with one polymorphism detection. The GTSE1 gene contained the largest number of mutations with 17 SNPs. The average number of SNPs per gene was 3 SNPs.

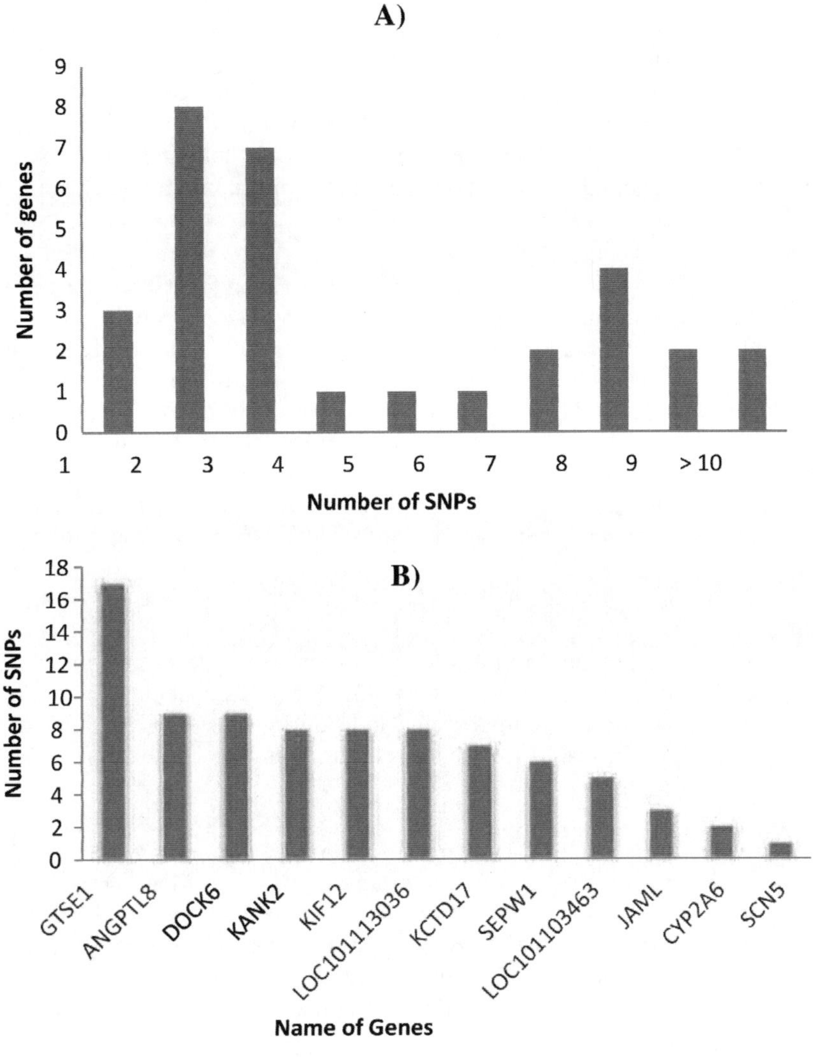

Figure 2. Distribution of the number of SNPs detected in the DEGs. The distribution of the number of SNPs occurred in each gene (A); numbers of SNPs in the genes selected for the association validation (B).

After SNP discovery executed in javanese fat tailed sheep, a comparative analysis was carried out Furthermore, the number of mutations could depend on the gene length, and the GTSE1 gene was the longest gene among this studies. Gene KIF12 is a gene encoding a novel kinesin, mapped beneath the 34 cM QTL peak and has expression level variants and strain-specific sequences that were associated with renal disease severity in affected mice [12]. ANGPTL8 (also called betatrophin) is a major and an important influencing factor in triglyceride levels and adipose tissue mass as well as in replenishing the

IOP Conf. Series: Earth and Environmental Science **157** (2018) 012030 doi:10.1088/1755-1315/157/1/012030

adipose tissue triglyceride store [13]. Dock6 gene, Expression and the cellular phenotype of fibroblasts of individual further confirms the role of DOCK6 in Adams-Oliver syndrome (AOS) pathogenesis, which appears to converge with that reported for ARHGAP31 in perturbation of the actin cytoskeleton through inactivation of Cdc42 and Rac1 [14]. Gadea and Blangy [15] had identified Kank2 proteins as talin activators that decrease the grip between the integrin–talin complex and actomyosin to regulate cell migration velocity.

Among the 12 genes that contained SNPs in Javanese Fat Tailed Sheep in the present study, CYP2A6 was associated with boar trait in a gene expression study by [16]. Nonetheless, the results of our study may be useful for the design of a focused SNP panel for use in genotype-to-reproductive phenotype association analyses in a multi-breed setting. The focused (i.e. functional) SNP panel is an approach that can reduce costs associated with genotyping [17]. Biologically relevant polymorphisms are used to design a focused or low-density panel of selected SNPs [17]. This is the first study used candidate genes to be determined as associated with sheepmeat odour and flavour in javanese fat tailed sheep, the SNPs discovered should be useful to reduce sheepmeat odour and flavour.

4. Conclusions

In this study for the very first time NGS technology has been used to analyze the expression profiles of sheepmeat odour and flavour in javanese fat tailed sheep by using RNA Sequencing. Several of the SNPs (JAML, CYP2A6, SEPW1, and KIF12) were founded in this study could be included as suitable markers in genotyping platforms to perform association analyses in commercial populations and apply genomic selection protocols in the sheepmeat production. This transcriptome, polymorphisms and alternative splicing analysis using RNA deep sequencing revealed potential candidate genes affecting odour and flavour of sheepmeat in Javanese fat tailed sheep. It is exhibited that in future these polymorphisms could be used as markers for sheepmeat related with odour and flavour traits. However, further research and evidences are required to confirm the effect of these genetic markers in other sheepmeat populations.

Acknowledgment

I would like to pay my special thanks to my supervisor honorable Dr Asep Gunawan for his guidance, encouragement, availability and support to perform this research and achieve main objectives. I'm also thankful to my lab fellows for their support and help during research, Authors greatly thankful to development ministry of Research, Technology, and higher Education year 2016 number: 079/SP2H/LT/DPRM/II/2016.

References

[1] Cánovas A, Rincon G, Islas-Trejo A, Wickramasinghe S and Medrano J F 2010 SNP discovery in the bovine milk transcriptome using RNA-Seq technology *Mamm. Genome.* **21** 592

[2] Gunawan A, Sahadevan S, Cinar M U, Neuhoff C, Große-Brinkhaus C, Frieden L, Tesfaye D, Tholen E, Looft C, Wondim D S and Hölker M 2013 Identification of the novel candidate genes and variants in boar liver tissues with divergent skatole levels using RNA deep sequencing *PloS One*. **8** e72298.

[3] Li C, Cai W, Zhou C, Yin H, Zhang Z, Loor J J, Sun D, Zhang Q, Liu J and Zhang S 2016 RNA-Seq reveals 10 novel promising candidate genes affecting milk protein concentration in the Chinese Holstein population *Sci. Rep.* **6** 26813

[4] Cirulli E T, Singh A, Shianna K V, Ge D, Smith J P, Maia J M, Heinzen E L, Goedert J J and Goldstein D B 2010 Screening the human exome: a comparison of whole genome and whole transcriptome sequencing *Genome Biol.* **11** R57

[5] Marguerat S, Wilhelm B T and Bähler J 2008 Next-generation sequencing: applications beyond genomes *Biochem. Soc. Trans.* **36** 1091-6

[6] Marguerat S and Bähler J 2010. RNA-seq: from technology to biology *Cell Mol Life Sci.* **67** 569-79

IOP Conf. Series: Earth and Environmental Science **157** (2018) 012030 doi:10.1088/1755-1315/157/1/012030

[7] Cunningham F, Amode MR, Barrell D, Beal K, Billis K, Brent S, Carvalho-Silva D, Clapham P, Coates G, Fitzgerald S and Gil L 2014 Ensembl 2015 *Nucleic Acids Res.* **43** 662-9

[8] Jin Y and Lee C G 2013. Single nucleotide polymorphisms associated with microRNA regulation *Biomolecules* **3** 287-302

[9] Matoulkova E, Michalova E, Vojtesek B and Hrstka R 2012 The role of the 3'untranslated region in post-transcriptional regulation of protein expression in mammalian cells *RNA Biol.* **9** 563-76

[10] Kuersten S and Goodwin E B 2003 The power of the 3 [variant prime] UTR: translational control and development Nature reviews *Genetics* **1** 626

[11] Vinsky M, Islam K, Chen L and Li C 2013 Association analyses of a single nucleotide polymorphism in the promoter of OLR1 with growth, feed efficiency, fat deposition, and carcass merit traits in hybrid, Angus and Charolais beef cattle *Can. J. Anim. Sci.* **93** 193-7

[12] Mrug M, Li R, Cui X, Schoeb T R, Churchill G A and Guay-Woodford LM 2005 Kinesin family member 12 is a candidate polycystic kidney disease modifier in the cpk mouse *J. Am. Soc. Nephrol.* **16** 905-16

[13] Li Y and Teng C 2014 Angiopoietin-like proteins 3, 4 and 8: regulating lipid metabolism and providing new hope for metabolic syndrome *J. Drug Target.* **22** 679-87

[14] Shaheen R, Faqeih E, Sunker A, Morsy H, Al-Sheddi T, Shamseldin H E, Adly N, Hashem M and Alkuraya F S 2011 Recessive mutations in DOCK6, encoding the guanidine nucleotide exchange factor DOCK6, lead to abnormal actin cytoskeleton organization and Adams-Oliver syndrome *Am J Hum Genet.* **89** 328-33

[15] Gadea G and Blangy A 2014 Dock-family exchange factors in cell migration and disease *Eur. J. Cell Biol.* **93** 466-77

[16] Habier D, Fernando RL, Dekkers JC 2009 Genomic selection using low-density marker panels *Genetics* **182** 343-53

[17] Fortes M R, Porto-Neto L R, DeAtley K L, Reverter A, Thomas M G, Moore S S, Lehnert S A and Snelling W M 2014 *Genetic Markers in Transcription Factors of Differentially Expressed Genes Associated With Post-Partum Anoestrus Predict Pregnancy Outcome in An Independent Population Of Beef Cattle* 10th World Congr. Genet. Appl. Livest. Prod., Vancouver, Canada

The use of colour indicator as a smart packaging system for evaluating mangoes Arummanis (*Mangifera indica* L. var. Arummanisa) freshness

A Dirpan[1], R Latief[1], A Syarifuddin[1], A N F Rahman[1], R P Putra[2] and S H Hidayat[1]

[1] Department of Agricultural Technology, Faculty of Agriculture, Hasanuddin University, Jalan Perintis Kemerdekaan KM 10, Makassar, 90245, Indonesia.
[2] Department of Agriculture Technology Education, Faculty of Engineering, Universitas Negeri Makassar, Jalan A. P. Pettarani, Makassar, 90222, Indonesia.

E-mail: *dirpan@unhas.ac.id*

Abstract. The high demand and public sensitivity of packaging products need the emergence of various packaging for fresh fruit and vegetables. One of the packaging that can be applied is smart packaging. The aimed of this study was to make smart packaging indicator that can be applied to mangoes. The benefits of this research were to facilitate the consumer to know the quality of the fruit and to promote in choosing the fruit according to freshness or desired maturity without damaging the packaging. The research was carried out in the following two stages: preparation of solution as bacterial growth medium *Acetobacter xylinum*, (incubated for 14 days) and making colour indicator and application to smart packaging. The results showed that long immersion of bacterial cellulose with bromophenol blue solution was 24 hours and temperature of ± 30 °C. The colour indicator that has been applied to the packaging changed from blue to green colour for over-ripe product, which can be visible to the naked eye. The colour change of the indicators reflects the pH of the mango packaging. It was also in similar trends to the change of several parameters (total acid, and soluble solids content (SSC)) that was typically used to characterize the freshness of mango. In general, this indicator can be used as smart packaging.

1. Introduction

There have been more demands from the consumers to have food products with high quality in developed countries. This led to the emergence of various innovations in the field of packaging technology that allows consumers to choose fresh fruit and vegetable products that have been packaged[1, 2]. One of the packagings that can be applied is smart packaging [3, 4]. Smart packaging is the package function switches on and off in response to changing external/internal conditions and can include communication with the customers or end users as to the status of the product. Therefore, intelligent packaging can be stated as a system that monitors the condition of the packaged food to provide information about the quality of transport and distribution [4, 5].

Some studies on constructing and applying smart packaging have been conducted. Kuswandi and Nurfawidi [6] have developed the label that can be used as a simple and practical indicator for freshness monitoring in packaged beef. In their study, the label responses accurately to the beef freshness, in term of its pH change due to the deterioration as shown by both colour change at room and chiller temperatures. Similarly with the other parameters, such as the total volatile basic amine (TVBN), total viable count (TVC) and sensory evaluation. These levels were reached at eight hour and seven days as the threshold of spoilage at room and chiller temperature respectively. Cavallo *et al.* [7] investigated a colourimetric sensor of milk spoilage which was built from adsorption of methylene blue (MB) into a modified polypropylene film (PP). The results showed that devices were suitable sensors to be used in the development of smart packaging. Kuswandi *et al.* [8] created a real-time monitoring of shrimp spoilage using on-package sticker sensor based on natural dye of curcumin. The result shows that the curcumin/bacterial cellulose membrane was successfully used as a sticker sensor for real-time monitoring of shrimp spoilage in ambient and chiller conditions. Hong and Park [3, 9] constructed colour indicators for kimchi (fermented vegetable product) packaging. These colour indicators were developed to monitor the quality of such commercial products during storage and distribution. Kuswandi *et al.* [10] also have created colour indicator based on bromophenol blue and tests have been conducted to assess the freshness of guava (*Psidium guajava L.*). It can be seen that the indicator can be used for real-time visual monitoring of freshness state of packaged guavas. From the above studies, it can be concluded that smart packaging can be applied to the food product including fruit.

The application of smart packaging can also be applied to mango fruit. This is because of the mango fruit including climacteric fruit group. Climacteric fruits do not need to be harvested when the ripe fruit is full in the tree because the fruit picked at the level of ages before ripening in the trees can be perfectly ripe after being stored. At the time of transportation with the distance, far enough climacteric fruit can be damaged. Due to the increase in temperature inside the crate, this comes from heat released by the fruit plus heat from the outside environment which can cause more rapid damage.

To the best of the author's knowledge, this is the first study to create smart packaging on mango arummanis, particularly in South Sulawesi, Indonesia. Therefore the purpose of this study were: (1) To create and study the indicator of freshness as an element of smart packaging (smart packaging) on mango arummanis fruit (2) To know the profile of colour change on freshness indicator of mango arummanis fruit.

2. Material and Method

This experiment was conducted in January - May 2017 in Microbiology and Food Biotechnology Laboratory and of Food Quality Analysis Laboratory, Food Science and Technology study Program, Department of Agricultural Technology, Hasanuddin University, Indonesia.

2.1. Preparation of Cellulose Membrane

As previously described [8, 10, 11], membrane sheets were prepared with some modification from bacterial cellulose (10 g), blended until homogeneous, then cast into the glass plate, pressed and left overnight (12 h). Finally, the white membrane sheet dried at 60 and ready to be used.

2.2. Immersion of Bacterial Cellulose

Bromophenol blue (BPB) was prepared by dissolving 10 ml BPB in 10 ml of ethanol (70%). After obtaining the cellulose membrane, the bacteria are then added with a 5% NaOH solution, for 60 min at 100 °C to remove the bacterial cell and substrate from the layer in the bacterial cellulose film. Then rinsed with tap water until reach neutral condition or pH 7.0. The bacterial cellulose membrane layer was stored in a glass plate, pressed and allowed to overnight (12 hours). A white membrane sheet was dried at 60 °C and ready for use [10].

2.3. Preparation of the Colour Indicator Labels

Colour indicator labels were produced by immersing the membrane sheet into 10 ml of BPB solution (1 mg/ml) at ambient condition (28 ± 2^{0}C) for overnight (12 h). Practically, the successful immobilization of BPB into bacterial cellulose membrane was shown by the dark blue colour of the membrane as the original membrane colour was white. Then, the BPB/cellulose was washed with tap water to remove excessive and unbound BPB solution. Afterward, the membrane was conditioned with acetate buffer

(0.01 M) at pH 4.8 (prepared with sodium acetate and acetic acid to give the dark blue colour). The membrane was then dried with an electrical drier (Philips) and cut into the desired shape [10].

2.4. Smart Packaging Application on Mango
Mango weighing about 200 g was used for analysis, then placed on a styrofoam and covered with polyethene plastic film. Samples were stored in room condition with normal light exposure. pH analysis, total soluble solid content (SSC), total acid. The colour change index was performed every 48 hours.

2.5. Measurement of SSC, pH and Total Acid
The determination of soluble solids content (SSC) was performed by homogenized the mango (three replications) and the homogenate filtered through several layers of cheese cloth to obtain a clear juice. The SSC (%) was recorded with a refractometer (Model: PAL-1, Otago co, Ltd, Tokyo, Japan). The pH values were recorded by a pH meter.

Total acid was measured acid by a titration method with 10 ml suspension plus 2-3 drops of phenolphthalein indicator, then titrated with 0.1N NaOH solution until the colour turns to light colour and the colour does not change back for 30 seconds. At the end of the titration, the amount of NaOH used was calculated. Finally, total acid was calculated using this equation:

$$\%\ \text{total acid} = \frac{V\ NaOH\ x\ N\ NaOH\ x\ FP\ x\ Grek}{1000\ x\ sample\ weight}$$

2.6. Measurement of the Colour Indicator Using Colour Analysis
The colour indicator was placed inside the package of the mango samples attached to the plastic wrap of the package using double transparent tape, where the indirect indicator contact with the atmosphere inside the package and covered with the label for viewing the colour of the indicator along with the reference colour for the state of freshness. Then the samples were stored in ambient condition. Besides, the colour indicator can be viewed by the naked eye in term of the colour change during mango deterioration state. For quantification of colour measurement of the indicator, a simple method was used using a digital camera [10]. A microscope digital (model: digital microscope 5-500 x) was used to record the colour of the indicator. The graphics software CorelDrawx6 was then used to analyze the colour of the indicator. The term ''measure'' means that the digital camera is used to obtain the colour values of the pixels on the colour indicator membrane. The term ''analyze'' means that CorelDraw is used to measure those colour values to obtain colour distribution, averages, and so on. Here, we used the 'Eyedropper Tool' for the colour value of a selected area in the membrane image of the colour indicator. The blue value was used as the colour value for all membrane colour measurements, since the membrane colour change from blue to green. By using this method, it is more suitable and simpler as compared to the visual inspection by the naked eye as the main purpose for this indicator rather than using reflectance measurement [10, 12]. The principles of colour measurement can be found elsewhere [10, 13].

2.7. Statistical Analysis
SSC, pH and total acid values were analyzed by one-way analysis of variance (ANOVA) while the significance of differences among means was determined using Tukey's test with a level of significance of $P < 0.05$. SPSS software for Windows version 16.0 (SPSS Inc., IL, USA) and Microsoft Excel 2010 was used to analyze the data.

3. Results and discussion

3.1. pH
The aim of the pH determination was to determine the acidity of mango fruit and to study the effect of pH on the freshness indicator of mango fruit.

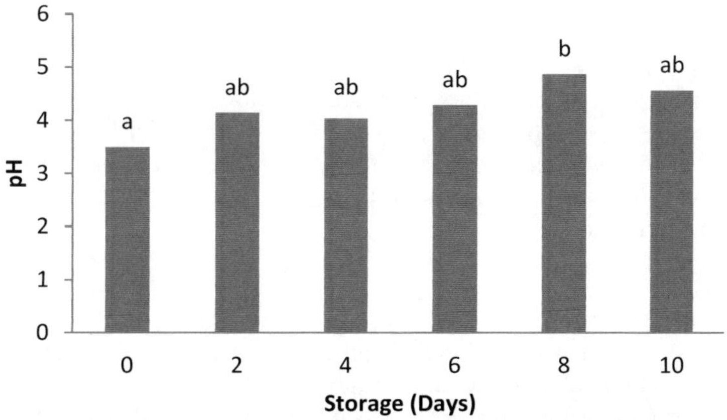

Figure 1. Changes in pH for Mango fruit at different storage

Figure 1 explains the typical diurnal changes of pH from 0 to 10 days storage. It can be seen that there is an increase in pH every day during storage. The results of storage on day 0 showed the pH value of mango around 3.5 while on day 8 showed pH value of 4.7. The occurrence of pH increase in mango fruit is caused by the influence of long-time storage, enzymatic reactions, and microbiological changes. Also, Arumanis Mango stored at room temperature showed increase in pH value or decrease in acidity. Furthermore, higher acid content is present in maturation fruits than in ripe and decaying fruit. This is by the statement Muchtadi *et al.* [14] which states that maturation fruits contain lots of organic acids and will decrease during the fruit ripening process.

3.2. Total Acid

Total acid is the overall amount of acid contained in a material. Total acid associated with pH but different principles. The pH level measures the total strength of the acid, while the total acid measures the amount of acid.

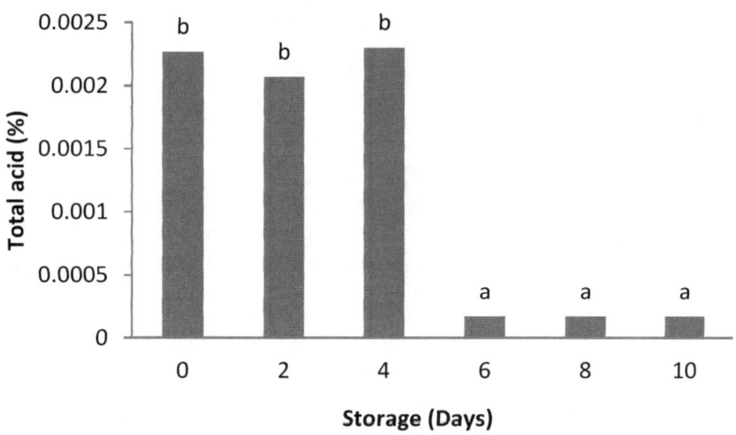

Figure 2. Changes in total acid for Mango fruit at different storage

Figure 2 shows the total acid on mango fruit during storage. The difference of total acid values in ten days storage was detected significantly by Tukey's test at 5% level ($P < 0.05$). The highest total acid occurred at 0 days, the average value of total acid was about 1.2%. This value was decreased significantly until ten-day storage. Differences in total acid could be due to the use of organic acids in the mango fruit by process of respiration that requires energy. The energy required is derived from overhauling the nutrients contained in the food. Also, a decrease in the total acid value of the fruit is due to the absence of an enzymatic reaction (alcohol dehydrogenase), the ethanol is oxidized to acetaldehyde, then oxidized to acetic acid during ripening. Thus, regardless of the influence of fruit metabolism, the fruit maturity level also affects the acid level. This is in line with the statement of Muchtadi *et al.* [14] which states that maturation fruits contain lots of organic acids and will decrease during the fruit ripening process.

3.3. Soluble Solids Content (SSC)

The figure 3 showed that mango fruit has total soluble solid at 0 days storage is 5.4%, and 13.0% at ten days storage. Based on these data it is known that during the storage, the total soluble solid in the mango fruit increased slightly. This is due to the change of starch in the mango into sugar so that the sugar content in the fruit increases during storage. This is consistent with the statement Pandarinathan *et al.* [15] that increased sugar content in fruit during the ripening process will increase the total value of the SSC, the formed sugars will be used as energy for respiration. Also, the packed mango fruit will undergo an anaerobic respiration process that also will increase the SSC.

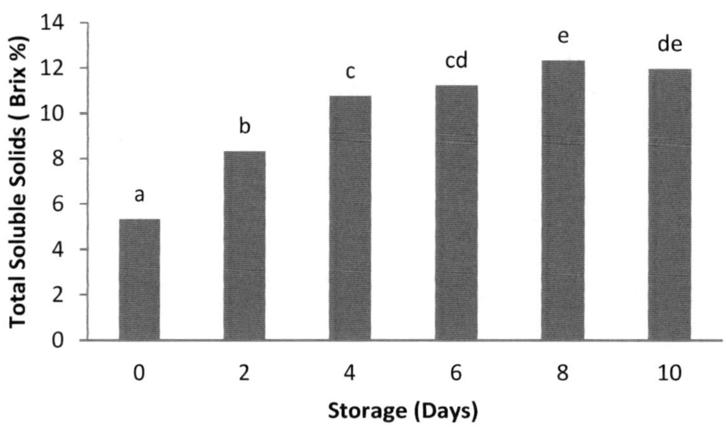

Figure 3. Changes in total soluble solid for Mango fruit at different storage

3.4. Colour Changes Index

Figure 4 showed the rate of colour changes of the indicator. It can be seen that the colour indicator response decrease steadily (as the indicator colour change to green) within ten days of the experiment was observed. Here, the colour indicator gradually changed colour from blue to green after six days. The colour indicator response as the colour change was similar with deterioration state of the climacteric nature of mango that was well reported in the literature [10]. The onset of overripe was detected after six days of storage. This indicated that the packaged mango released volatile acids (e.g. acetic acid) at a relatively slow rate in comparison with normal mango since its freshness lasted longer within six days. On the other hand, the packaging influence on other acid compounds in the mango, such as ascorbic acid, was found positive and effective in preventing the losses of this acid in fruit during ripening stage [10, 16, 17].

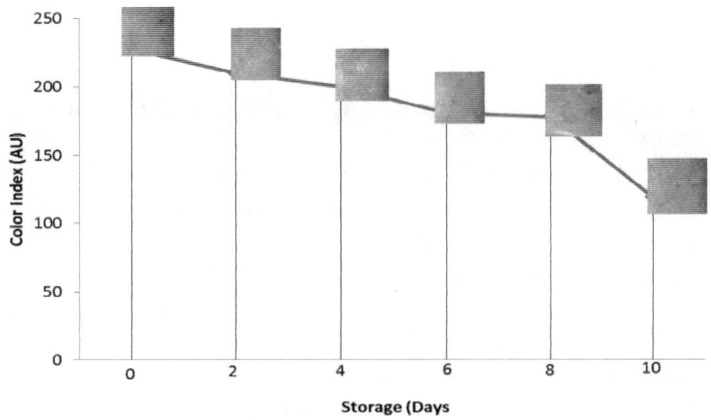

Figure 4. The rate of colour changes of the indicator response towards mango freshness state

4. Conclusion

1. Freshness indicator of mango fruit can be produced by utilizing bacterial cellulose from *Acetobakter xilinum* which then soaked with bromophenol blue solution.
2. Profile of colour change on the freshness indicator of mango arummanis showed the change of colour from dark blue indicating fresh fruit, light blue is firm and green colour indicates the fruit has been rotten

References

[1] Syarifuddin A, Hasmiyani, Dirpan A and Mahendradatta M 2017 Physical, mechanical, and barrier properties of sodium alginate/gelatin emulsion based-films incorporated with canola oil *IOP Conf. Ser. Earth Environ. Sci.* **73** 12018

[2] Dirpan A, Hikida Y, Chiba H, Aji G K and Rahman A N 2017 Improving the accuracy of intercellular space volume of citrus prediction by artificial neural networks *Acta Hortic.* 1179 37–44

[3] Hong S I and Park W S 2000 Use of color indicators as an active packaging system for evaluating kimchi fermentation *J. Food Eng.* **46** 67–72

[4] Yam K L, Takhistov P T and Miltz J 2005 Intelligent packaging: concepts and applications *J. Food Sci* **70** 1–10

[5] Kuswandi B, Wicaksono Y and Abdullah A 2011 Smart packaging: sensors for monitoring of food quality and safety *Sens. and Instrument. for Food Quality and Safety* **5** (3-4)137-146

[6] Kuswandi B and Nurfawaidi A 2017 On-package dual sensors label based on pH indicators for real-time monitoring of beef freshness *Food Control* **82** e123–e123

[7] Cavallo J A, Strumia M C and Gomez C G 2014 Preparation of a milk spoilage indicator adsorbed to a modified polypropylene film as an attempt to build a smart packaging *J. Food Eng.* **136** 48–55

[8] Kuswandi B, Jayus, Larasati T S, Abdullah A and Heng L Y 2012 Real-Time Monitoring of Shrimp Spoilage Using On-Package Sticker Sensor Based on Natural Dye of Curcumin *Food Anal. Methods* **5** 881–9

[9] Hong S-I and Park W-S 1999 Development of Color indicators for kimchi packaging *J. Food Sci.* **64** 255–7

[10] Kuswandi B and Maryska C 2013 Real time on-package freshness indicator for guavas packaging *Food Meas.* **7** 29–39

[11] Surma-Ślusarska B, Presler S and Danielewicz D 2008 Characteristics of bacterial cellulose obtained from Acetobacter Xylinum culture for application in papermaking *Fibres Text. East. Eur.* **16** 108–11

[12] Byrne L, Lau K T and Diamond D 2002 Monitoring of headspace total volatile basic nitrogen from selected fish species using reflectance spectroscopic measurements of pH sensitive films *Analyst* **127** 1338–41
[13] Gunasekaran S 1996 Computer vision technology for food quality assurance *Trends Food Sci. Technol.* **7** 245–56
[14] Muchtadi 2010 *Teknologi Proses Pengolahan Pangan* (Bogor: CV. Alfabeta IPB)
[15] Pandarinathan S, Sivakumar S, Nadu T and Arts S S 2010 Studies on biochemical changes in mangoes due to artificial ripening *Int. J. Agric. Sci.* **1** 347–55
[16] Lee S K and Kader A A 2000 Preharvest and postharvest factors influencing vitamin C content of horticultural crops *Postharvest Biol. Technol.* **20** 207–20
[17] Singh S P and Pal R K 2009 Ionizing radiation treatment to improve postharvest life and maintain quality of fresh guava fruit *Radiat. Phys. Chem.* **78** 135–40

Process optimization of emergency food originated from denatured whey protein concentrate and dried sweet potato puree

S Khairunnisa, R Andoyo, H Marta and G L U Saripudin

Department of Food Industrial Technology, Faculty of Agro-industrial Technology, Universitas Padjadjaran, Sumedang, 40600, Indonesia.

E-mail: *r.andoyo@unpad.ac.id*

Abstract. Denatured whey protein concentrate can be added into food products to increase protein content without changing the texture considerably. One food product that requires high protein content is emergency foods. The aim of this study was to determine the optimum parameters of emergency food processing based on denatured whey protein concentrate combined with dried sweet potato puree with high lethality of microbes and low physical change. Optimizing process has being conducted using Response Surface Method with Box-Behnken scheme. Emergency foods was prepared with containing 6.1-11.6% (w/w) protein, placed in three packaging materials *e.g* glass jar, nylon bag, and metallized bag and treated with 5.0 to 12.0 (D, decimal reduction time) thermal processing level. The optimum parameters of emergency food produced were 10.14% (w/w) protein content, placed in metallized pouch, and sterilized in 11.11 thermal processing level.

1. Introduction

Application of whey protein as food fortification had attracted a great attention. The purpose of addition of whey protein into various food products, such as sport drinks, baby foods, baked products, etc. is not merely for increasing the protein content of such products, but also may affect the physical characteristics. The change of the food characteristics is caused by functional properties of whey protein in native state, such as high solubility, high water holding capacity, and gel forming capability [1]. However, application of whey protein would usually generate negative sensorial for some food products. To prevent this change, the whey in native state is modified by some heat treatment. Different from the whey in native state, the whey in denatured state can increase protein content of food without changing the texture considerably [2].

Denatured whey protein can be added into emergency food specified processed for those whose impacted by disaster. Based on their function, the emergency foods must have high protein content to fulfill nutrition needs for certain victim category, such as children age of less than 5. The emergency foods must be ready-to-eat, have high energy and other important nutrition content, and can fulfill victim's need for 15 days in emergency situation [3].

The emergency foods are also designed to have long storage period so can always be available in unpredicted situations. To prolonging such storage period, the emergency foods can be treated with thermal process. Heat treatment applied to foods can inactivate even kill spoilage microbes [4]. However, uncontrolled thermal process can degrade product characteristics. Therefore, in application of the thermal processing we must consider its effect on food characteristics. It is expected that the change of characteristic of foods after thermal treatment should be as small as possible.

Different kind and ratio of raw materials will produce different thermal characteristics of products. Protein-based and carbohydrate-based foods have slight difference in thermal characteristic. For example, whey protein based food system has higher thermal conductivity and specific heat capacity compared to mashed potato [5]. In other words, food with high protein to carbohydrate ratio will have larger thermal penetration.

The thermal characteristics of food are also affected by packaging material. Different packaging material as well as different packaging geometry may generate different heat penetration into the foods [4]. Other than retort pouch that usually, foods that have been treated at high temperature can be packaged in glass jars.

Fortunately, the thermal process can be controlled through processing time. Shorter processing time is able to prevent overheating. In addition, the thermal processing level, *i.e.* longer shelf life due to reduction of microorganism, can be modified by varying the processing time.

The objective of this study was to investigate the optimum thermal processing parameters, namely protein content, packaging material, and thermal processing level for producing high protein emergency food. Response Surface Method (RSM) with Box-Behnken design was used to obtain optimum processing parameters for emergency foods with high lethality of microbes and low physical change.

2. Material and Methods

2.1. Raw Material Preparation

Denatured Whey Protein Concentrate (WPC-d) made of fresh cow milk was obtained from Laboratory of Dairy Cattle, Faculty of Animal Husbandry, Padjadjaran University. The material has been processed as reported reference [6]. Dried sweet potato puree made of sweet potato (*Ipomoea batatas,* var. Ace Putih) has been processed according to reference [7]. Each raw material was packaged in airtight container and placed in dry area at ambient temperature.

2.2. Emergency Food Preparation

Emergency food was made by mixing dry materials (WPC-d and dried sweet potato puree, at protein contents ranging from 6.1 to 11.6% (w/w)) with warm water (70 °C). The ratio of dry material and water was kept at 1:2 (w/w). The mixture was then packaged in nylon pouch, metallized pouch, or glass jar that obtained from local market.

2.3. Thermal Processing

The samples were then placed in autoclave (Astell Scientific), set at the coldest point of the autoclave. The sample temperature was recorded by a data logger (Lutron). The processing time was trial adjusted to reach 5.0-12.0 thermal processing level at 110 °C.

2.4. Experimental Design and Process Optimization

Experimental design and process optimization were conducted by Response Surface Method with Box-Behnken design using Design Expert 7.0.0 (Stat-Ease Inc.). Three factors were tested, namely protein content (6.11-11.6% (w/w)), packaging material (nylon pouch, metallized pouch, and glass jar) and thermal processing level (5.0-12.0). The experiment was conducted with 15 running includes 3 repetition of center point. The sample run are listed in table 1.

Measured responses were consists of total microbial numbers, water activity, color, texture, and energy content. The total microbes number was analysis according to reference [8]. The water activity was measured by Aw meter (AquaLab LITE). The texture was measured by texture analyzer (Stable Micro System TA.XT) installed by P1Ks probe with marmalade project. The texture analysis was conducted with speed of 2 mm/min using adhesiveness, gel strength, and brittleness indicators. The color response was analyzed by spectrometer (Minolta CM-5) using Hunter color system.

We also estimated the response model equation. From the equation, we can predict value of response when inserting the desired variables. We used the model equation as following

$$y = \beta_0 + \sum_{i=1}^{k} \beta_i x_i + \sum_{i=1}^{k} \beta_{ii} x_i^2 + \sum_{i=1}^{k-1} \sum_{j=2}^{k} \beta_{ij} x_i x_j + \varepsilon \quad (1)$$

with y is the response, x_i is the controlled parameter, β_o is the independent response, β_i is the coefficient contributed by linear response , β_{ii} (i = 1, 2,..., k) is the coefficient generated by quadratic response, and β_{ij} (j = 1, 2,..., k) is the coefficient contributed by interaction between different parameters, and ε is the random error [9].

Table 1. Experimental design

Run	Variables			Run	Variables		
	Protein content (%)	Packaging	Thermal processing level (D)		Protein content (%)	Packaging	Thermal processing level
1	6.1	Glass jar	8.5	9	6.1	Metallized pouch	12.0
2	8.9	Nylon pouch	5.0	10	8.9	Nylon pouch	12.0
3	8.9	Metallized pouch	8.5	11	6.1	Metallized pouch	5.0
4	8.9	Metallized pouch	8.5	12	8.9	Glass jar	5.0
5	6.1	Nylon pouch	8.5	13	11.6	Metallized pouch	5.0
6	11.6	Nylon pouch	8.5	14	11.6	Glass jar	8.5
7	8.9	Metallized pouch	8.5	15	8.9	Glass jar	12.0
8	11.6	Metallized pouch	12.0				

3. Results and Discussion

Generally, the objective of this work was to identify the optimum process for producing emergency food originated from WPC-d with dried sweet potato puree. The factors investigated at the present work include the protein content (6.1%, 8.9%, and 11.6% (w/w)), packing material (nylon pouch, metallized pouch, and glass jar), and thermal processing level (5.0, 8.5, and 12.0).

3.1. Thermal Processing Time

Prior to identification of the optimum process, trial and error thermal processing steps were conducted to obtain the optimum holding time because of unavailability of data found in references. The thermal processing was kept at temperature of below standard commercial sterilization (110°C) to prevent appreciable damage to product. Modification of the processing temperature was also expected to modify processing time. When using the standard process at 121 °C, the holding times at thermal levels of 5.0, 8.5, and 12.0 were 1.05, 1.79, and 2.52 minutes, respectively. The F values that processed at 110 °C was calculated by a trapezoidal method. This method measures the total surface area under curve of thermal processing time to sample temperature [10]. At arbitrary temperature T, the F value satisfies

$$F = \int 10^{(T - T_{ref})/z} \, dt \quad (2)$$

with F is the thermal processing time, T is temperature, T_{ref} is standard temperature, and z is the characteristic temperature for reducing the microbe by one log. It was observed that the thermal process at 110 °C at different thermal levels of 5.0, 8.5, and 12.0 required the holding times of 15, 25, and 35 minutes, respectively.

3.2. Process Optimization

Optimization of process was conducted by using parameters as listed in table 1. It was obtained the relationship between the measured responses and the corresponding controlled parameters are given by equation (2) with the best factors were listed in table 3. Y, A, B, and C represent, respectively, the response, the protein content, the thermal processing level, and the packaging types.

Based on optimum constants listed in table 2 figure 1 showed the examples of 3D plot of total microbes' response in sample packaged with (a) nylon pouch (b) metallized pouch (c) glass jar. Using ANOVA analysis, it was identified 4 of 7 tested responses, i.e. total microbes number, color, brittleness, and energy content, exhibited significant equation model. By the significant equation model, it meant 3 responses affected 3 tested factors. To the contrary, the water activity, adhesiveness, and gel strength provided only minor effect.

Table 2. The best constants for estimating various responses based on equation (1)

Response	Model Equations
Total microbe number	$Y = 3{,}245\times10^{3} + 1{,}111\times10^{2}A - 6{,}780\times10^{2}B + 8{,}147\times10^{2}C + 1{,}405\times10^{1}AB - 2{,}459\times10^{1}AC - 5{,}714\times10^{1}BC - \ldots$ $\ldots - 1{,}405\times10^{1}A^{2} + 2{,}483\times10^{1}B^{2} + 1{,}417C^{2}$
Water activity	$Y = 1{,}173 - 7{,}159\times10^{-2}\,A + 1{,}346\times10^{-2}\,B + 1{,}278\times10^{-3}\,C - 1{,}223\times10^{-3}\,AB + 8{,}197\times10^{-4}\,AC + 8{,}197\times10^{-4}\,BC + \ldots$ $\ldots + 4{,}418\times10^{-3}\,A^{2} - 2{,}211\ 10^{-4}\,B^{2} + 0{,}019\,C^{2}$
Color (*L)	$Y = 51{,}544 + 0{,}928\,A - 0{,}797\,B + 0{,}378\,C$
Color (*a)	$Y = 5{,}211 + 0{,}202\,A + 0{,}253\,B - 0{,}334\,C$
Color (*b)	$Y = 18{,}029 + 0{,}211\,A - 0{,}078\,B - 0{,}031\,C$
Adhesiveness	$Y = -38{,}637 - 0{,}330\,A - 1{,}146\,B - 4{,}932\,C + 0{,}118AB - 0{,}144\,AB + 0{,}683\,BC$
Gel strength	$Y = 34{,}896 - 5{,}835\,A - 1{,}451\,B - 2{,}223\,C + 0{,}226\,AB + 0{,}219\,AC + 0{,}142\,BC + 0{,}212\,A^{2} - 0{,}034\,B^{2} + 1{,}193\,C^{2}$
Brittleness	$Y = 18{,}469 - 0{,}954\,A - 1{,}804\,B + 1{,}644\,C + 0{,}207\,AB - 0{,}492\,AC + 0{,}360\,BC$
Energy content	$Y = 3210{,}537 + 95{,}036\,A + 18{,}821\,B - 137{,}250\,C$

It was identified from figure 1, that the dominant factor affecting the total microbes number in emergency food was the thermal processing level. By increasing the thermal processing level, the total microbes number in the emergency food decreased. Heat treatment applied to the food product was able to kill microbes through different mechanism, such as destruction of cell membrane and substantial organs in cells such as ribosome, RNA, and DNA [11]. High temperature resistant and spore forming bacteria could remain exist after thermal process and continue to grow in necessary condition [12].

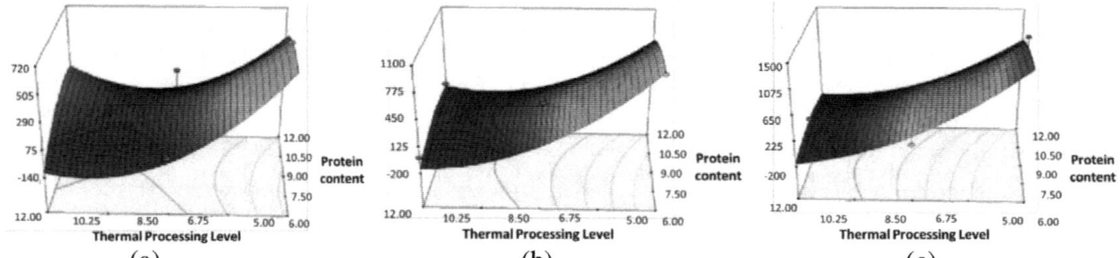

Figure 1. 3D plots of total microbes response in sample packaged with (a) nylon pouch (b) metallized pouch (c) glass jar

High Aw value on emergency foods containing WPC-d at higher ratio compared to dried sweet potato puree has been observed. This WPC-d was made by initially heating the whey in liquid state at high temperature prior to centrifugal separation and drying process. Applying high temperature heating at protein caused the water holding capacity of the protein decreases. This circumstance was similar to that reported in reference [13], where heating treatment to the whey has lowering the water holding capacity on yoghurt compared to yoghurt made by untreated whey. The thermal process will form hydrophobic chain to imply the whey having lower water holding capacity and solubility [14].

The emergency food that has been treated by thermal process became darken, more reddish, and yellowish. The change of color was caused by non-enzymatic browning as a result of interaction between reducing sugars with amino acids. The use of high temperature accelerated reaction of the two compounds. Furthermore, presence of reducing sugar at high content will conduct high non-enzymatic browning [15].

After thermal processing, the brittleness values increased and the adhesiveness decreased, while gel strength decreased only slightly. Adhesiveness determines the force needed to pull the probe from the food surface, while brittleness measures the fracture of the food [16]. Decreasing of adhesiveness caused by the forming of whey aggregates implied to lowering food's ability to stick other object. The adhesiveness of emergency foods decreased with protein content. This circumstance is similar to yoghurt substituted by heat-denatured whey and buttermilk, where the adhesiveness decreases when increasing whey substitutions. It caused by decline of sample's capability to adhere to probe by increase of denatured whey [17]. Brittleness increased because of the presence of interaction in between food components such as protein, carbohydrate and water. Such interaction was also observed to cause hardening of snack bar produced added with whey [18]. The energy was measured to range between 3762 – 4625 cal/g in dry matter. Different energy content in each samples mainly caused by different nutrient content in each treatment, where carbohydrates could present higher Gross Energy Value than protein [19].

3.3. Model Validation

Based on observed responses, Design Expert 7.0.0 suggested the optimum process for producing the emergency foods was protein content of 10.14% (w/w), packaged in metallized pouch, and processed at 11.11 thermal level. Validation of the proposed optimum value was conducted by experiment. The result can be seen at table 3. As seen in table 4, the responses having very high reliability are water activity, color changes, brittleness and energy content. However, the adhesiveness has only high reliability, while then total microbe number and gel strength has low reliability.

Table 3. Validation of optimum process

Response	Prediction	Result	Validation
Total microbes number	42.56 cfu / mL	10 cfu / mL	11.75
Water activity	0.88	0.81	94.32
Color (*L)	52.11	18.32	97.49
Color (*a)	10.06	9.67	96.07
Color (*b)	19.30	17.63	91.32
Adhesiveness	-41.41 gsec	-51.49 gsec	77.11
Gel strength	2.74 g	4.85 g	34.80
Brittleness	12.08 g	10.32 g	97.76
Energy content	4381.82 cal/g	4239 cal/g	96.78

4. Conclusion

The present study showed that the optimum process of emergency food producing from denatured whey protein concentrate and dried sweet potato puree performed on product's formulation in 9.84% (w/w) protein content which packaged in metallized pouch and processed under 11.11 log cycle of thermal process. Among responses that measured in this study, protein content, package material, and thermal processing level affected significantly toward total microbes, color, brittleness, and energy content.

References

[1] Fox P F and McSweeney P L H 1998 Dairy chemistry and biochemistry (New York: Springer Science & Business Media)
[2] Tsakali E, Petrotos K, D'Alessandro A G and Goulas P 2010 *A Review on Whey Composition and the Methods Used for Its Utilization for Food and Pharmaceutical Products* (*6th International Conference On Simulation And Modelling In The Food And Bio-Industry (Foodsim 2010) (Eurosis)*) pp 195–201
[3] IOM 2002 *High-Energy, Nutrient-Dense Emergency Relief Food Product* (Washington DC: National Academies Press)
[4] Toledo R T 2007 *Fundamentals of Food Process Engineering* (New York: Springer)
[5] Chen J, Pitchai K, Birla S, Gonzalez R, Jones D and Subbiah J 2013 Temperature-dependent Dielectric and Thermal Properties of Whey Protein Gel and Mashed Potato *Biol. Syst. Eng. Pap. Publ.* **56** 1456–7
[6] Sitindaon A 2017 *Produksi dan Kajian Sifat Fisikokimia Whey Protein Concentrate pada Beberapa Waktu Pemanasan* (Sumedang: Universitas Padjadjaran)
[7] Sunyoto M, Andoyo R, A H R and Nurmalinda R 2017 Kajian karakteristik pure kering ubi jalar dengan perlakuan suhu dan lama annealing sebagai sediaan pangan darurat *JST J. Sains Dan Teknol.* **6** 1–10
[8] AOAC 1995 *Official Methods of Analysis of AOAC International* (Washington DC: AOAC International)
[9] Aksu Z and Gönen F 2006 Binary biosorption of phenol and chromium(VI) onto immobilized activated sludge in a packed bed: Prediction of kinetic parameters and breakthrough curves *Sep. Purif. Technol.* **49** 205–16
[10] Valentas K J, Rotstein E and Singh R P 1997 *Handbook of Food Engineering Practice* (New York: CRC Press)
[11] Russell A D 2003 Lethal effects of heat on bacterial physiology and structure *Sci. Prog.* **86** 115–37
[12] Keir L, Wise B, Krebs C and Kelley-Arney C 2007 *Medical Assisting: Administrative and Clinical Competencies* (New York: Delmar Cengage Learning)
[13] Sodini I, Mattas J and Tong P S 2006 Influence of pH and heat treatment of whey on the functional properties of whey protein concentrates in yoghurt *Int. Dairy J.* **16** 1464–9
[14] Walstra P, Walstra P, Wouters J T M and Geurts T J 2005 *Dairy Science and Technology, Second Edition* (Boca Raton: CRC Press)
[15] Echavarría A P, Pagán J and Ibarz A 2012 Melanoidins Formed by Maillard Reaction in Food and Their Biological Activity *Food Eng. Rev.* **4** 203–23
[16] Ahmed J, Ptaszek P and Basu S 2016 *Advances in Food Rheology and Its Applications* (Abington: Woodhead Publishing)
[17] Saffon M, Richard V, Jiménez-Flores R, Gauthier S F, Britten M and Pouliot Y 2013 Behavior of Heat-Denatured Whey: Buttermilk Protein Aggregates during the Yogurt-Making Process and Their Influence on Set-Type Yogurt Properties *Foods Basel Switz.* **2** 444–59
[18] McMahon D J, Adams S L and McManus W R 2009 Hardening of high-protein nutrition bars and sugar/polyol-protein phase separation *J. Food Sci.* **74** E312-321
[19] McArdle W D, Katch F I and Katch V L 2010 *Exercise Physiology: Nutrition, Energy, and Human Performance* (Philadelphia: Lippincott Williams & Wilkins)

Premix formulation for making the Indonesian otak-otak

A B Tawali[1], N Wakiah[1], A R Ramli[2], M Mahendradatta[1], S Tawali[3] and S Made[4]

[1]Department of Food Science and Technology, Hasanuddin University, Jl. Perintis Kemerdekaan KM 10, Makassar, 90245, Indonesia.
[2]Centre of Information, Dissemination and Property Rights, Hasanuddin University, Jl. Perintis Kemerdekaan KM 10, Makassar, 90245, Indonesia.
[3]Department of Medicine, Hasanuddin University, Jl. Perintis Kemerdekaan KM 10, Makassar, 90245, Indonesia.
[4]Department of Marine Science and Fisheries, Hasanuddin University, Jl. Perintis Kemerdekaan KM 10, Makassar, 90245, Indonesia.

E-mail: *andi.rahmayanti28@gmail.com*

Abstract. Otak-otak is one indigenous foods from Indonesia which made from fish paste, tapioca flour and spices. The aim of this research was to produce premix flour for making otak-otak using mackerel surimi, tapioca flour and spices and to evaluate their chemical properties and sensory acceptability of the product. Mackerel surimi was first dried, miled into powder form then mixed with tapioca flour and spices. The result showed that, otak-otak which made from Premix-B formula (40% mackerel surimi powder and 60% tapioca flour) was the best result for all sensory attributes. It presented "like moderately" for hedonic score. Protein, ash, moisture, carbohydrate, and fat contents of the premix-B formula were 23.51%, 2.44%, 7.31%, 71.02%, 4.28%, respectively, which met the SNI requirements. Overall, this study clearly showed that it was possible to make otak-otak with less complicated process and the product did not have any negative results on sensory perception.

1. Introduction

Fish is one of the most abundant natural resources in South Sulawesi, Indonesia. South Sulawesi has a favorable geographical conditions because it is surrounded by the sea. According to the 2015 Statistics Indonesia, the fish production form marine reached 295 239.20 tonnes [1]. These potential resources must be managed properly so that the protein needs of Indonesian people can be fulfilled.

Fish-based diversification product is done continuosly, one of the most popular fish product in Indonesia is otak-otak. Otak-otak is an indigenous food from Indonesia which made from fish paste, tapioca flour and spices. This product can be served as snack or as part of meal. In Malaysia, otak-otak is a popular street food among tourist and locals [2]. The type of fish used to make otak-otak vary, but the most commonly used in Indonesia is mackerel because it contains low fat and has a good capability for gel forming.

The preparation of otak-otak from fresh mackerel takes quite a long time. It takes several steps from preparing fish paste until otak-otak is ready to eat. In addition, nowadays, the food processing are required to follow the trend and consumer taste who tend to prefer instant process and easy to serve product. On the other hand, the mackerel is not always available in the market and the quality is often the problems. Therefore, premix technology can be one of the alternative solutions.

Premix is a mixture of several different types of flour. This product is widely practiced in baking industries [3]. The purpose of the premix technology itself is to make a product consisting a balanced

IOP Conf. Series: Earth and Environmental Science **157** (2018) 012033 doi:10.1088/1755-1315/157/1/012033

formula that is easy to use and has a long shelf life. Moreover, premix technology can reduce the costs, quality and product standardization, optimization of labor, simply sanitization, and reduce the space of the raw material storage room [4]. In the present work, we described the premix production using mackerel surimi, tapioca flour and spices for making otak-otak and evaluated their chemical properties and sensory acceptability of the product.

2. Materials and Methods

2.1 Surimi Powder Preparation

The mackerel fish was cleaned from skin, bones, fins, and gill. The meat fish was washed with running water then minced into fish paste. The fish paste was leached with 0.3% NaCl solution with ratio of 1:4 (w/v) for 15 minutes at 5-10 °C. The leaching process was repeated for four times then filtered using filtered cloth. Surimi was dried using blower for 12 hours at 65 °C until the water content reach 4%. The dried surimi then grinded using grinder and sieved using a 100-mesh sieve.

2.2 Premixes preparation

There are three premixes were formulated by mixing surimi powder, tapioca flour and spices. Surimi powder was mixed with tapioca flour with the following ratio displayed in table 1. The spices then added into the mixture as much as 10% of the total volume of the mixture. The premix formulas and spices compositions could be seen in table 1 and table 2, respectively.

Table 1. Ratio of surimi powder and tapioca flour of premix formulas

Formula	Surimi Powder (%)	Tapioca Starch (%)
Premix-A	50	50
Premix-B	40	60
Premix-C	60	40

Table 2. Spices composition of the premix

Spices	Quantity (%)
Garlic powder	2.6
Onion powder	2.48
Coconut milk powder	64.00
Refined sugar	13.38
Refined salt	6.69
Flavor enhancer	10.51

2.3 Preparation of Otak-otak

Otak-otak was made by kneading each formulated premix and cold water (1:1) into dough. The dough was wrapped in banana leaves then steamed for 20-25 minutes at 90-100 °C. Freshly prepared otak-otak was analyzed for sensory attributes.

2.4 Sensory Analysis

Otak-otak made by premix formulas were subjected to sensory analysis for attributes of color, aroma, texture and taste using Hedonic Scoring Scale [5]. The scoring scale used was between 1-5 with the scores representing the hedonic attributes of 5,4,3,2,1 were "like very much"; "like"; "like moderately"; "dislike"; "dislike very much", respectively. The samples were tested by 25 panelist.

2.5 Chemical Analysis

The chemical compounds of the premix best formula and otak-otak product were measured using AOAC methods [6]. Oven drying and weighing methods (926.12, 41.1.02) were used to measure the moisture content. Ash content was measured by weighing and furnace methods at 600°C for 3-5 h (942.05, 4.1.10). Fat extraction using sohxlet distillation and chloroform as a solvent was used to measure the fat content (948.22, 40.1.05). The protein content was measured using kjeldahl distillation and the nitrogent value was converted to protein value using conversion factors (960.52, 12.1.07). The carbohydrate content was measured by difference method.

3. Result and Discussion

3.1 *Sensory* analysis

Sensory analysis was an important test in the product development. Because, consumer point of view was always a predominant determinant of acceptability of a new product [7]. The best formula of the premix was evaluated by making otak-otak from each formula then subjected to sensory acceptability. The results was showed in figure 1.

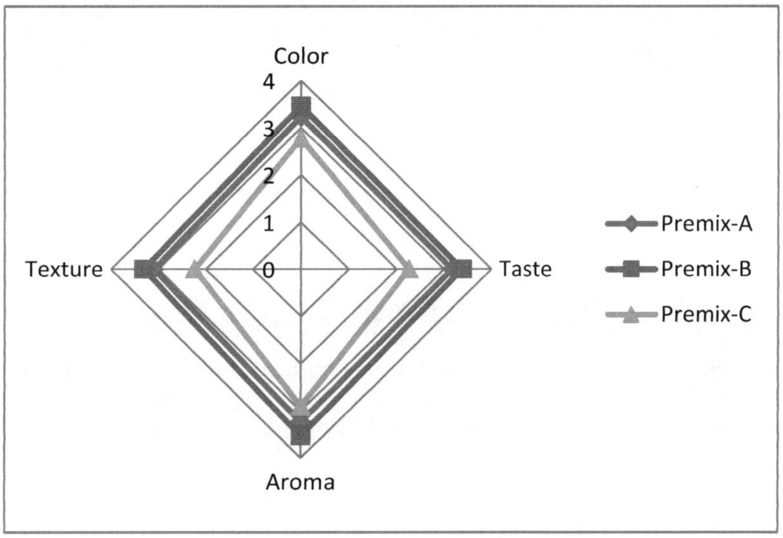

Figure 1. Sensory analysis of otak-otak

Figure 1 showed that otak-otak which made from Premix-B formula has the highest score for all sensory attributes while Premix-C has the lowest. Product with Premix-B formula was scored 3.5 for color attributes. On a scoring scale, these corresponded to near "like" or "like moderately". Similar trend was seen for the attribute of aroma wherein the score also 3.5 for Premix-B formula while Premix-C formula given the lowest score 2.9. This may be due to the ingredients used. The aroma was mostly influenced by mackerel surimi powder and spices. The spices could slightly hide the fishy aroma from fish but Premix-C formula contained the highest level of mackerel surimi powder resulting in strong fishy aroma. Meanwhile, color attributes contributed by the mackerel surimi powder. The color of otak-otak turned into slightly brown affected by Maillard reaction which occurred in drying process of mackerel surimi. Those indicated that higher level of mackerel surimi powder resulted more brown color of the product.

The score given for attribute taste and texture of Premix-B formula were 3.4 and 3.3, respectively. The texture of otak-otak was mostly contributed by tapioca flour. Tapioca flour gave a chewy mouth-feel to the products. This chewy texture was determined by the ratio of amylose and amylopectin which contained in starch. The taste of the otak-otak was mostly contributed by mackerel surimi powder and spices.

3.2 Chemical composition of premix

The chemical composition of premix-B formula and otak-otak product were presented in table 3. The moisture content was indicated the amount of water per unit weight of material and it was the most fundamental parameter because it affected the shelf life of the product. The moisture content of the premix-B formula was 7.27%. Those indicated that our products meet the SNI requirements. The standard of the premix flour referred to the standard of the fish flour. According to the SNI number 01-3709-1995 [8], the water content of the fish flour was maximum of 10%. Drying process of the mackerel surimi and other dried materials which used were the determinant factors of the low moisture content of the premix flour. In this study, the mackerel surimi was dried until the moisture content reached 4.03%. Moisture content from different premix have been reported. The moisture content of the premix for preparing flat breads and noodles were 4.05% [7]. The wide range of moisture content of the premix flour was affected by the moisture content of raw materials.

Table 3. Chemical composition of Premix-B and otak-otak products

Constituents	Premix-B (%)	Otak-otak (%)
Moisture	7.27	59.92
Fat	4.67	2.29
Protein	22.23	16.07
Ash	2.92	0.98
Carbohydrate	62.81	20.22

The fat content of premix was low (4.67%) due to leaching process in surimi production. The leaching process in surimi production increase the moisture content and reduce the contents of protein, fat and minerals [9]. The fat content of the premix was mostly contributed from coconut milk powder. Otherwise, the protein content of the premix was high at 22.23% which would have been contributed for almost all ingredients, particularly mackerel surimi which has high protein content. Fish was widely known as source of protein. Mackerel fish has high protein content and low fat. The protein and fat contents of mackerel fish were 27.6 % and 6.5%, respectively [10]. Similarly the carbohydrate content was also high at 62.81% which influenced by tapioca starch. On the other, the chemical composition of otak-otak which made by premix-B formula also meet the SNI requirements of otak-otak number 7757.2013 [8]. The standard of protein, ash, moisture and fat contents of otak-otak were 16%, 1.48%, 60%, and 5%, respectively.

4. Conclusion

It was concluded that the best formula of the premix for making otak-otak was Premix-B formula which consisted of 40% mackerel surimi powder and 60% tapioca flour. It was presented the hedonic score "like moderately" for all sensory attributes. Premix-B formula contained protein, ash, moisture, carbohydrate, and fat as much as 22.23%, 2.92%, 7.27%, 62.8%, and 4.67% of fat, respectively. Overall, this study clearly underlined that it was possible to make otak-otak with less complicated process and the product did not have any negative results on sensory perception.

References

[1] Statistics Indonesia 2016 Fish production data https://www.bps.go.id/linkTabelStatis/view/id/1705 Accessed March 14 2017

[2] Tang J Y H, Mat-Sa'ad S H, Banerjee S K, Ho L H, and Son R 2017 Survivability of Vibrio parahaemolyticus in satar and otak-otak, Malaysian fish-based street food. *Int. Food Res. J.* **24** (3) 1000-1005

[3] Mahalingam P, Veluppillai S, and Ekanayake S 2014 Study on preparation of rice-wheat bread using premix *J. Agr. Sci.* **9** (1) 31-36

[4] Valduga E, Sfredo M A, and Luccio M D 2004. Assessment of flow and technological behaviour of french bread premixes *Braz. J. of Food Nutri.* **15** (1) 1-6

[5] Larmond E 1977 *Laboratory Methods for Sensory Evaluation of Food.* (Ottawa: Research Institute, Canada Department of Agriculture)

[6] AOAC 2005 *Determination of Moisture, Ash, Protein and Fat* (*Official Methods of Analysis*) 18th edn (Washington DC: Association of Official Analytical Chemist)

[7] Faiza S, Shyamala B N, Oghbaei M, and Prakash J 2015 Formulation of nutritious premixes based on natural ingredients and evaluating their efficacy for value addition *Int. Food Res. J.* **22** (2) 546-555

[8] SNI National Standardization Agency of Indonesia http://sisni.bsn.go.id/

[9] Karthikeyan M, Dileep A O, and Shamasundar B A 2006 Effect of water washing on the functional and rheological properties of proteins from threadfin bream (*Nemipterus japonicus*) meat. *Int. J. Food Sci. Tech.* **41** 1002–1010

[10] Guizani S E O and Moujahed N 2015 Seasonal variation of chemical and fatty acids composition in atlantic mackerel from the Tunisian Northern-East Coast *J Food Process Technol* **6** 9 doi: 10.4172/2157-7110.1000487

IOP Conf. Series: Earth and Environmental Science **157** (2018) 012034 doi:10.1088/1755-1315/157/1/012034

Potential hazards from hygiene, sanitation and bacterium of refill drinking water at Barrang Lompo island (water and food safety perspective)

A B Birawida, M Selomo and A Mallongi

Department of Environmental Health, Publich Health Faculty, Hasanuddin University Jalan Perintis Kemerdekaan km 10, Makassar, 90245, Indonesia.

E-mail: *agusbirawida@gmail.com*

Abstract. Coliform bacteria are suspected to come from faeces. As a result, their presence in various places ranging from drinking water, foodstuffs or other ingredients to human needs are not expected. This research aimed to describe bacterial contaminations, depot sanitations, equipment sanitations, worker hygiene, raw water quality, and refill drinking water in Barrang Lompo island in 2017. The observational research applied descriptive approach. Water samples collected grab from the drinking water depot taps, then examined using Most Probable Number (MPN) method. The results showed that three of six depots have qualified drinking water quality. Raw water samples from all depots indicated that none were eligible. The samples examined contained gram-negative bacteria. The types of bacteria that grew on the sample were *Klebsiella pneumonia* and *Pseudomonas aerogenosa*. In the environmental sanitation depots and worker hygiene, there was no one eligible. Sanitary appliances were all eligible and there were depots that used reserve osmosis methods and used combination methods between reserve osmosis and ultraviolet light. It was concluded that almost all samples of drinking water were contaminated by bacteria. Owners and depot workers were advised to improve and implement better hygiene and sanitation.

1. Introduction

According to Global Water Supply and Sanitation Assessment 2000 Report claimed by WHO/UNICEF around 1.1 billion world populations still found difficulties to access clean water and almost 2.5 billion inhabitants has no accessed to proper sanitation [1]. In addition, about 10 thousands dwellers that lived in developing countries were dead every day because of the diseases caused by unavailability of clean water and low quality of environmental sanitation. In spite of having means to access clean water easier, most of the inhabitants in developing countries prefer to use water without any process first [2].

More than 100 million Indonesia's citizens lack the access to clean water and at 150 million water sources contaminated by pollutants [3]. One of the Indonesia's government strategies to fulfill water needs was through a policy to create local water company or Perusahaan Daerah Air Minum (PDAM) that enganged in distribution and water supply facilities for general public. Current drinking water standard refers to the Regulation of Health Minister No. 492 of 2010.

Clean water supply for drinking water should comply the existing regulation. Drinking water was categorized safety for societies just if it met the physical, microbiological, chemical, and radioactive parameter. For microbiological parameter, it included bacterial *Coliform* and *Escherichia coli*. The higher contamination of *Coliform* bacteria, the higher attendance risk of other pathogens, such as bacteria, virus, and parasite [4]. One example of the pathogens was *Escherichia coli* which caused symptomps of diarrhea, fever, abdominal cramps, and vomiting [5]. *Escherichia coli* was usually made as indicator of contamination of human feces and it also showed the appearance of other potential pathogens in drinking water [6]

Diarrhea was in the second position that led to the mortality of children under five years old in the world. In Indonesia, diarrhea was the main cause of death in the children under five years. Riskesdas reports showed that diarrhea was behind the 31% mortality of children in age from a month to a year, and 25% mortality in children from one to four years old [7]. Based on the data on Barrang Lompo Primary Healthcare in Singkarang Sub-district in Makassar City (2015) diarrheal diseases was one of the notable public health problems and was classified on ten major diseases in this primary health care scope. The number of people with diarrhea in 2013, 2014, and 2015 was 192, 187, and 199 inhabitants, consecutively [8]. Till this time, the needs for clean water and drinking water for the residents of Barrang Lompo was fulfilled through dug well, boreholes, and clean water supply bussiness [9].

Drinking water was the main source consumption for the family as well, such refill water. Residents could feel economic benefits for this water, especially in lower middle class families. However, not all depot gave guarantee to consumer towards the good quality of product that they produce. According to research in risk analysis in drinking water in Semarang (2015) there was a risk of bacterial contamination in clean water sources and water treatment process, and even the high risk bacterial contamination still found in processed water [10]. Another research found that there was connection between existence of *Escherichia coli* in drinking water and diarreal for consumers in refill drinking water in Ciputat Timur [11] and that about 55.6% samples of refill drinking water in Padang did not meet the specified requirements the regulation of Health Minister [12]. A research result also concluded that there was a significant correlation between hygine santitation and microbiological quality of drinking water refills [13].

The number of refill drinking water depots in Barrang Lompo Island in 2016 was six depots which used source of raw water from boreholes that most likely made *Coliform* bacteria easy to contaminate. Oppositely, it was different from raw water in refill drinking water in Makassar that used raw water from PDAM. Many numbers of DAMIU scattered in Barrang Lompo Islands were not guarantee the quality of drinking water in this region since the management and filling of the depots could not always been supervised. As a consequence, lots of public health issues could be emerged distinctively.

2. Materials and Methods

This research used observational method with descriptive approaches. This research conducted in refill drinking water depots in Sangkarang Sub-district in Barrang Lompo Island in Makassar City and also in Microbiology Laboratoryy in Medical Faculty in Hasanuddin University. This research held from 20 February 2017 to 6 March 2017. Population in this study was all refill drinking water depots in Sangkarang Sub-district in Barrang Lompo Island in Makassar City and the sample was six depots of refill drinking water in Sangkarang Sub-district in Barrang Lompo Island in Makassar City. In six depots, AMIU was taken and 150 ml raw water per sample, with total 12 samples. Type of sampling was non probability sampling which sampling techniques that do not provide the same opportunity for each member of the selected population to be sampled because of the relatively small population.

3. Results

The number of *Coliform* bacteria, total colony, types of the bacteria that was found in raw water and refill drinking water in DAMIU in Barrang Lompo Island shown in the table 1. Table 1 illustrates that there were three depots which did not meet the requirements both in raw water and refill drinking water and therewere three depots that did not fulfill the criteria of raw water, yet they still met the qualifications for refill drinking water. Depot that has the highest MPN was AB D3 at 460 and the lowest one were AM D3at 28, respectively. Also, table 1 shows that there was no one *Escherichia Coli* that found in both raw water and refill drinking water, yet there were found two types of bacteria called *Klebsiella pneumonia* dan *Pseudomonas aerogenosa.*

Table 1. *Coliform* bacteria contaminations in refill drinking water and raw water in dam in Sangkarang sub-district in Barrang Lompo island in Makassar City 2017.

No.	Sample code	MPN/100ml Index	Total Colony	Culture and Bacterial Identification
1	AB D1	4	12	*Pseudomonas aerogenosa*
2	AM D1	0	0	uncultured and unidentified
3	AB D2	15	52	*Klebsiella pneumonia*
4	AM D2	0	0	uncultured and unidentified
5	AB D3	460	88	*Pseudomonas aerogenosa*
6	AM D3	28	80	*Pseudomonas aerogenosa*
7	AB D4	93	130	*Klebsiella pneumonia*
8	AM D4	4	6	*Klebsiella pneumonia*
9	AB D5	8	59	*Klebsiella pneumonia*
10	AM D5	7	32	*Klebsiella pneumonia*
11	AB D6	4	15	*Klebsiella pneumonia*
12	AM D6	0	0	uncultured and unidentified

Observational result for environmental sanitation, sanitary equipments, and workers hygiene in DAMIU that existed in Sangkarang in Barrang Lompo Island showed on the in table 2. Table 2 illustrates that in environmental sanitation aspect, there were two depots which met the standard and others still did not meet the defined criteria according to regulation of the Indonesia's Minister of Industry and Trade No. 651/MPP/Kep/10/2004. Based on building aspects of the six drinking water depots, none of them fulfilled 100% criteria according to regulation of the Indonesia's Minister of Industry and Trade No. 651/MPP/Kep/10/2004. Moreover, sanitary equipments in the six drinking water depots in Barrang Limpo Island has qualified with the requirements according to regulation of the Indonesia's Minister of Industry and Trade No. 651/MPP/Kep/10/2004. Furthermore, no one drinking water depots qualified the defined standard according to regulation of the Indonesia's Minister of Industry and Trade No. 651/MPP/Kep/10/2004. Unfulfilled criteria was in using medical mask which showed that 100% workers did not use this mask in workhours.

4. Discussion

The laboratory results on raw water and refill drinking water depots in Sangkarang Sub-district in Barrang Lompo island showed that raw water in all depots have no qualified to standard, however, there were three depots that could meet the standard on drinking water according to Indonesia's Health Ministry regulation No. 492 year 2010.

All raw water samples were stated positive contained *Escherichia coli* with MPN/100 ml water, in which the largest one found in D3 samples at 460 MPN/100 ml water, while the smallest one found in D1 and D6 samples. The presence of Coliform bacteria in drinking water refills indicated the possibility of contamination and the existence of enteropatogenic microbes that can interfere with heath. The possible factors that could cause positive results and persumtive MPN test was the contamination of drinking water refills on processing method, specifically on raw water reservoir, desinfection, and filtration. Other factor that influenced water quality of production was raw water, the type of equipment used, the maintenance of equipments and handling of water treatment and distribution [14].

Based on the following research, not one of *Coliform* bacteria was found on samples, yet *Klebsiella pneumonia* and *Pseudomonas aerogenosa* were discovered. The results of bacterial examination samples that grew much in samples was *Klebsiella pneumonia* bacteria. This bacteria could be easily found on skin, throat, digestive tract, and even in sterile wounds which also led to pneumonia and bronchitis [15, 16].

Table2. Environmental Sanitation, Sanitary Equipments, and Workers Hygiene in DAMIU in Sangkarang Sub-district in Barrang Lompo Island in Makassar City 2017

Environmental Sanitation	**N**	**%**
Site		
a. Qualified	2	33
b. Unqualified	4	67
Building		
a. Qualified	0	0
b. Unqualified	6	100
Total	6	100
Sanitary Equipments	**N**	**%**
Made of food tare, corrosion-resistant, do not react with chemicals	6	100
a. Qualified	0	0
b. Unqualified		
The tool used is still in its lifetime		
a. Qualified	6	100
b. Unqualified	0	0
Materials not made of heavy metals that are soluble in water		
a. Qualified	6	100
b. Unqualified	0	0
Total	6	100
Workers Hygiene	**n**	**%**
Must be free from infectious diseases		
a. Qualified	6	100
b. Unqualified	0	0
Free from wounds, boils, and other injuries		
a. Qualified	6	100
b. Unqualified	0	0
Wash hands before and after serving the consumers		
a. Qualified	5	83
b. Unqualified	1	17
While serving the consumers, not eating, not having long nails, not scratching, and not prying ear/nose/teeth	6	100
a. Qualified	0	0
b. Unqualified		
Use medical mask		
a. Qualified	0	0
b. Unqualified	6	100
Must be free from infectious diseases		
c. Qualified	6	100
d. Unqualified	0	0
Total	6	100

Furthermore, sanitation and hygiene of DAMIU itself could affect MPN results. Poor sanitation and low hygiene induce contamination. Based on observational research in, there were some depots that have poor sanitation and low hygiene and desinfection process of less qualified containers.Contamination was not only caused by the high contain of microbial pollution from raw water used but also inadequate filtration process, sterilization used ultra violet or ozonization that were usually done in drinking water refill depots, and sanitation along with the process of filling the water to gallons of drinking water [17]. Based on location criteria that involves free from pollutants, such dust, garbage dump, breeding place of insects, poor sewer system, the results showed that just two depots that were eligible to the drinking water requirements. The observational output for building criteria such as clinelines of the floor, wall, ceiling, door, lighting, and ventilation pointed that there was no one depot to fulfill 100% standard issued. If properly observed that it could contaminate processed water.

Observation on adequate lighting on production signified that all depots have utilized adequate lighting on production area, washing place, rinsing, sterilization, filling gallons. These places should be bright enough to find out physical contamination, thus workers has a sight to contamination products [18]. Results of research conducted on drinking water depots in Sangkarang Sub-district in Barrang Lompo in Makassar City showed that every depot has met the criteria needed. Observational results on sanitation equipments with tools and fittings criteria should be made by food tare, not corrosive, not react to with chemical material, tools still in a lifetime, and not made from heavy metal that is soluble in the water. This represented that depot owners has understood sanitation equipments of qualified drinking water depots.

Equipment was highly important to turn raw water become drinking water. The great condition and qualified criteria on equipment condition on drinking water treatment would produce high quality of drinking water as well. A significant correlation was found between condition of drinking water refills processing and its bacteriological quality [19] but on the contrary, in other study no correllation between drinking water treatment and biological quality of refill drinking water was found [20]. Observational results and direct interview with workers in six drinking water depots in Sangkarang Sub-district in Barrang Lompo in Makassar City showed that the use of masks was still not eligible, in which no one workers used it in their working hours. This happened because most of them consider consumers would feel offended when seeing workers used masks. Moreover, criteria of workers hygiene that was not fulfilled in one depot was wash hands before and after serving consumers. This would lead to drinking water contamination by bacteria on hands with the result that could generate the certain diseases for those who consume the water later. Workers who did not behave cleanly and healthily as if not wash their hands or smoking when serving the consumers undeniably would affect drinking water.

Observational results for healthiness of workers such as not having infectious diseases, skin diseases, ulcers, and other diseases showed that all workers were in good health. In terms of use of clean clothes, all laborers used net clothes when in workhours or when doing production process of refill drinking water [21]. Good habits were found in all laborers in drinking water depots in Sangkarang Sub-district Barrang Lompo island in Makassar City such as do not eating, not having long nails, no scratching, and scraping ears/nose/teeth when they were in production process.

5. Conclusion

The bacteriological quality of raw water and refill drinking water in six depots showed that none raw water in depots that fulfilled the proper standard, while in refill drinkng water, three depots has been eligible to reach the certain parameter of drinking water refills. *Klebsiella pneumoniae* dan *Pseudomonas aerogenosa* was found on refill drinking water depots. Environmental sanitation for drinking water depots based on location has qualified criteria set by Ministry of Home Affairs. Yet, in building aspect, all depots did no meet the defined criteria. On the contrary, equipment sanitation of drinking water depots represent that all depots have fulfilled criteria issued by Ministry of Home Affairs.

References

[1] UNICEF Indonesia 2012 *Air Bersih, Sanitasi dan Kebersihan (Ringkasan Kajian UNICEF)* (Jakarta: UNICEF Indonesia)
[2] Dey N C, Parvez M, Dey D, Saha R, Ghose L, Barua M K, Islam A and Chowdury M R 2016 *Int. J. of Hyg. and Env. Health* **220** 621-636
[3] Parera M J, Supit W, Rumampuk, J F 2013 *J. e-Biomedik* **1** (Manado: Universitas Sam Ratulangi)

p 466
[4] Divya A H and Solomon P A 2016 *Int. Conf. on Emerg. Trends In Eng., Sci. and Tech.* **24** p 631
[5] Bambang A G, Fatimawali and Kojong N S 2014 *Pharmacon J. Ilmiah Farmasi* **3** (Manado: Universitas Sam Ratulangi) p 325
[6] Messner M J, Berger P and Javier J 2017 *Int. J. of Hyg. and Env. Health* **220** p 736
[7] Diallo F A, Cong X, Henderson W A and Mc Grath J 2016 *Int. J. of Nursing Studies* **66** p 82
[8] Puskesmas Pulau Barang Lompo. Data. 2015
[9] Selintung M, Zubair A and Rakhman D 2013 *Jurnal Teknik Sipil* (Makassar: Universitas Hasanuddin) p 1
[10] Budiyono, Ginandjar P, Saraswati L D, Pangestuti D R, Martini, Jati S P and Rahfiludin Z 2015 *Proc. Env. Sci.* **23** p 93
[11] Jayadisatra Y S 2013 *Thesis* (Jakarta: UIN Syarif Hidayatullah)
[12] Wandrivel R, Suharti N and Lestari Y 2012 *J. Kesehatan Andalas* **1** (Padang: Universitas Andalas) p 129
[13] Indirawati S M 2009 *Thesis* (Medan: Universitas Sumatera Utara)
[14] Mirza M N 2014 *J. Kesehatan Masyarakat* **9** (Semarang: Universitas Negeri Semarang) p 167
[15] Warganegara E, Apriliana E and Ardiansyah R 2012 *Pros. SNSMAIP* **3** (Bandar Lampung: Universitas Lampung) p 344
[16] Elfidasari D, Noriko N, Mirasaraswati A, Feroza A and Canadianti S F 2013 *Al-Azhar Indonesia Seri Sains Dan Teknologi* **2** (Jakarta: Universitas Al Azhar Indonesia) p 41
[17] Kemenperindag 2004 *Keputusan Menteri Perindustrian Dan Perdagangan RI Nomor: 651/Mpp/Kep/10/2004 Tentang Persyaratan Teknis Depot Air Minum Dan Perdagangannya.*
[18] Asfawi S, Jazuli N and Sulistiyani 2004 *J. Kesehatan Lingkungan Indonesia* **3** (Semarang: Universitas Diponegoro) p 50
[19] Maharani A D 2007 *Thesis* (Semarang: Universitas Diponegoro)
[20] Khoeriyah *et al* 2013 *Pros. Seminar Nasional Pengelolaan Sumberdaya Alam dan Lingkungan* vol 1 (Semarang: Universitas Diponegoro) p 70
[21] Kemenkes RI 2014 *Permenkes Nomor 43 Tahun 2014 Tentang Higiene Sanitasi Depot Air Minum*

IOP Conf. Series: Earth and Environmental Science **157** (2018) 012035 doi:10.1088/1755-1315/157/1/012035

The application status of Good Food Production Method (GFPM) production of corn crackers in SME Mawar Merah Luwu Utara

R Latief[1], A Dirpan[1], M M Tahir[1] and F V Albanjar[2]

[1] Department of Agricultural Technology, Faculty of Agriculture, Hasanuddin University, Jalan Perintis kemerdekaan KM 10, Makassar, 90245, Indonesia.
[2] Postgraduate Student at Department of Agricultural Technology, Hasanuddin University, Jalan Perintis Kemerdekaan KM 10, Makassar, 90245, Indonesia.

E-mail: *rindamias04@yahoo.com*

Abstract. Good Food Production Method (GFPM) requires several requirements that must be fulfilled by all industries involved in the whole production chain from raw materials to the fonal product. This study aims to identify the application of GFPM in the production of corn chips, and assess the status of GFPM implementation which was conducted through observations, interwiews, and documentation of all activities related to the production process. The assessment used a reference adapted from the Decree of Indonesian Minister of Industry Affairs (Permenperin) number 75/M-Ind/Per/7/2010 and the regulation of Head of National Drug and Food Control Agency (KBPOM) number HK.03.1.23.04.12.22007 in 2012. The result showed several inappropriate conditios including: (1) absence of continuing maintanace of the wall and ceiling of the production room: (2) unstandardized ventilation in the production room; (3) absence of net weight and production labels on the packages; and (4) the absence of health label informing health claims and nutrition claims. The status of GFPM implementation in UKM Mawar Merah is at D (poor) level with a rate of IV.

1. Introduction

Good Food Production Method (GFPM) is a guideline for production which aims to ensure that food producer meets the prescribed requirements to produce quality[1]. It is started from raw materials to processing stage and to go with until the final product for sure. The purpose of the study was to identify and to assess the applicability status of GFPM in the Mawar Merah Small and Medium Enterprises (SME), in North Luwu. The method used was descriptive method, in which that Small and Medium Enterprises (SMEs) was one of the activities of the community which would become a driving force in the country's economic development. Mainly it could improve economic resilience at the household level. Also, it became an important part that was influential in enhancing the competitiveness of a country. The movement of the SME sector was to create economic growth and employment; it encouraged equitable distribution of income and distribution of development outcomes, and it was adaptable to the ups and downs and direction of market demand. It contributed to the provision of food products for public consumption. However, in its operations, there were some disadvantages such as safe, vulnerable, and unattractive packaging [2]. Without any strategic effort, SMEs would have a high risk of losing competitiveness. One of the initiatives that could be done was to apply Good Food Production Method (GFPM). It is a guideline for production strategy which is aimed at ensuring that

IOP Conf. Series: Earth and Environmental Science **157** (2018) 012035 doi:10.1088/1755-1315/157/1/012035

food producer. It meets the prescribed requirements to produce quality and safe food products by consumer demands [3]. The purpose of this study is to identify and assess the applicability status of GFPM in the Mawar Merah SME, North Luwu.

2. Material and Method

This research was conducted at SME Mawar Merah, North Luwu which produced the most popular maize crackers. The method used was descriptive method, which described the phenomenon, application and business conditions observed. Working procedure: (1) Observation, interview, and documentation of all activities related to the production process were performed. (2) Observation to perceive precisely the movement and behaviour, environment, and general picture of SME location. Interviews were conducted in depth to the business owner. Any reference to the status of GFPM-IRT application used is adapted from Permenperin RI No.75 / M-Ind / Per / 7/2010 and KBPOM regulation no. HK 03.1.23.04.12.22007 year 2012 shown in table 1.

Table 1. References to the appraisal of GFPM-IRT

Rating levels	Amount of Deviation Value				Category
	Unfavorably	Serious	Mayor	Minor	
I	0	0	1	1	A (Very Good)
II	0	0	2-3	1	B (Good)
III	0	1-4	5 or more	Not used	C (Average)
IV	Yes	Not used	Not used	Not used	D (Less)

Source: Agency for Drug and Food Control (2012)

3. Results and discussion

The scope of the GFPM included production goods since raw materials enter the plant until the product was produced, including other requirements that must be met. Stages of corn cracker production process were; (1) sorting of raw materials, (2) immersion (3) boiling, (4) milling, (5) mixing, (6) stirring, (7) printing, (8) drying, (9) frying pan, (10) refrigeration and (11) packaging. According to Susianawati [4],there were 8 GFPM keys that must be applied by industry. By the results of observations at the study site, the application of GFPM in the Red Rose Mawar based on eight key GFPM was as follows.

3.1. Water Protection

The role of water in the industry was significant, ranging from washing production space, washing materials, washing equipment and variety of other means. Use of clean water should be considered to avoid contamination of dirty water. The water used in the Mawar Merah SME came from a local water company (PDAM), Luwu Utara, which was piped, then collected and deposited in a particular water reservoir. The protections were clean storage container and water in colourless condition, tasteless and odourless. The Regulation of the Minister of Health of the Republic of Indonesia (2011) was that the minimum quality of clean water must meet the physical requirements of colourless, tasteless and odourless. Also, the water used in the SME was available in sufficient quantities for production activities. According to [5], water used in the processing must be processed such as, firstly to reduce the number of microbes as low as possible, secondly treating water which was undergone chemical and physical treatment.

3.2. Conditions of Hygiene Surface in Contact With Foodstuffs

The expected objective was to assure that surfaces which direct contacted with food was in a clean state included the status of the contact surface with the food. In the SME Mawar Merah, equipment and containers conditions used have to be clean enough. Such as plastic basket, tray, bucket and frying pan. Before the activity begins and after completion of production activities, the equipment was cleaned by brushing and washed with soap and then all of them must be dried carefully. It was expected to reduce the occurrence of contamination of the product by hanging above the room and stored in a cupboard. It was explained that all material which was contacted directly with the product must be washed with proper washing technique and use sanitizer [3].

3.3. Prevention of Cross-Contamination

There was no place for raw material in SME Mawar Merah, but they kept it in a processing location. The cleanliness and sanitation of the processing area and the walls were poorly maintained, and it was not supported continuously. The practice of worker hygiene, clothing and hand washing as well were in good shape, even in the modest circumstances.

3.4. Handwashing, Sanitation and Toilet Facilities

The importance of maintaining hand-washing facilities, sanitation and toilets to avoid contamination was usually fatal because the possibility of pathogenic bacteria being carried could be contaminated in foodstuffs. Therefore, every worker or employee who performed the processing must ensure that the hand condition was clean constantly. State of hand-washing facilities, they were well-maintained toilets and located away from production sites, as well as the concentration of hand sanitizing materials which were available although not in accordance with the requirements. Each employee has a small container for washing their hands so that before starting the processing, each employee should wash his hands as well as after finishing the processing.

3.5. Protection from Contaminant Materials

The purpose of protection of contamination-causing products was to ensure that food product, packaging materials and direct contact surfaces were protected from microbial, chemical and physical contamination. The storage container was located separately from the processing room, and it was closed. The condition of ventilation in the production room was not yet suitable because pests such as bird that became carriers of bacteria could easily enter the production room.

3.6. Labeling and Storage

The containers used in food products should have label that included the kind of raw material and the name or address of the manufacturer. Food grade material and non-food grade one must be considered primarily. The labelling of corn crackers in SME Mawar Merah has been done quite well, but for marketing strategy, the position of the logo and writing brand "Sar Intan" should be designed bigger than the product name (Corn Chips product name). Therefore, the brand “Sar Intan” would be better known by consumers. Beside that, the critical information which not included in the packaging label was the net weight, the production code and the health claims as well.

3.7. The Supervision of the Personal Health Conditions

Owners should pay attention to the health conditions of workers. It was not justified when one of the employees performed the production process suffered from illness or they have wounds that could be as source of contamination. Employees worked in processing room, sorting, and packaging should always wear work clothes, headgear, masks and gloves. This was done to maintain cleanliness by paying attention to the aspects of sanitation and hygiene because these aspects were potential sources of contaminants.

3.8. Eliminating Pests from Processing Units

SME Mawar Merah was located close to the residential area, where no fence, and easy for animals such as chickens and birds to enter the processing area. In this case, it could be a source of contamination from a) flies and cockroaches: source of *Salmonella, Streptococcus, C. Botulinum, Staphylococcus*, *C.*

perfringens, Shigella., b) rodents: sources of Salmonella and parasite, c) birds: source of *Salmonella and Listeria*

Based on the assessment at the study site, and the guidance of the appraisal that were adapted from Permenperin RI No.75 / M-Ind / Per / 7/2010 and KBPOM regulation no. HK 03.1.23.04.12.22007 year 2012, showed that SME Mawar Merah has D status (unacceptable) with rating IV because some deviations occured in processing area (table 2).

Tabel 2. Irregularities in SME Mawar Merah

No	Serious deviation	Unfavorably deviation
1.	The walls and ceiling in the production room were not maintained continuously.	Packaging labels did not include net weight and production code
2.	Ventilation in the production room did not meet the standard.	There was no health labels that included health claims and nutritional claims (nutritional content)
3.	-	The existence of a pet in the production room

Severe and unfavorably deviations were irregularities which were not adequately corrected and would affect the food safety as well [6]. The application status of GFPM in various food industries varied. For example the SME Deni Deva in Surabaya which produced sea cucumber crackers. It got poor status and needed to be improved from design aspect, facility of production room, maintenance of production facility, sanitation activity and administrative record [7]. Another study stated that there were still many SME crackers in Padang City which have not applied GFPM optimally, either from the aspects of quality control, employee and infrastructure as well [8]. It has been reported that 38.78% of food households in Cianjur were considered to be lacking in water supply and treatment parameters, pest control and sanitation practices [9]. GFPM processing of corn crackers was essential to produce good quality products, guaranteed security and by consumer expectations. Ensuring products were manufactured and controlled consistently according to standards [9]. Thus the stages of improvement against deviations that occurred become the main thing in the process of development in the future [10]. This research provided some managerial implications, namely:

a. Fixing the walls and ceiling by the Regulation of the Minister of Health of the Republic of Indonesia 2011[11]to facilitate the maintenance of production space. (1) The inner wall surface should be flat, smooth, and easy to clean. If the surface of the wall was exposed to splashes, it should be coated with waterproof material and easy to clean as porcelain as high as 2 meters from the floor. Part of the waterproof wall was made smooth, flat and bright colored, (2) The ceiling surface should be flat, light-coloured, and easy to clean. Also, the ceiling should not be hollow, and the height of the ceiling was at least 2.4 meters from the floor.
b. Fix the vent in the production room and the packing room, and complete it by installing the insect barrier insertion. Ventilation was necessary to maintain comfort by lowering indoor heat, preventing condensation (humidity), and removing smells, smoke, and dust in the room. The ventilation was divided into natural and artificial ventilation. Natural ventilation should be sufficient (10% of the floor area) and capable of ensuring proper air circulation and should be able to remove steam, gas, smoke, odours, and indoor dust. Artificial ventilation was required when natural ventilation could not meet the requirements.

Implications of improvements to critical deviations were as follows [12]:

- Label include. According to government regulation No. 18 of 2012 where the provisions on food product label are as follows: (a) Product name according to IRT food type in Regulation of Head of

POM HK.03.1.23.04.12.2205 Year 2012 on Granting of Certificate (b) Name and address of PIRT, (c) Date, month and year of expiration, (d) Production code of the Household Industry, (e) List of materials or composition used;) P-IRT Number.b. Include health labels such as health clones and nutritional claims. Based on the Regulation of the Head of the POM Number HK.03.1.23.11.11.09909 In 2011, the claim was any form of description which states, suggests or declares the specific characteristics of a food concerning the content, nutritional content, nature, production, processing, or other quality factors. Claims on processed foods included dietary claims and health claims.

- Nutrition claims included: (1) claims of nutrient content i.e. claims that described the nutrient content in food; and (2) claims for the comparison of nutrients,i.e.claimsto compare the content of nutrients and energy content between two or more foods.
- Health claims included: (1) a nutritional function claim that was a claim that described the physiological role of nutrients for growth, development and normal function of the body; (2) another claim of function that was claim related to the beneficial effects of food or food components in the total diet to normal biological function or activity in the body, which was associated with positive effects to improve body function or maintain health; and (3) disease risk reduction claim was a claim that connected food consumption or food components in the total diet with a reduced risk of a particular disease or health condition. However, it should be understood also that by the reduction of the major risk factors for a disease that caused multifactor, it was not necessarily beneficial for health by reducing one risk factor for the occurrence of the disease.
- Protection from pets
  Based on the Regulation of the Minister of Health of the Republic of Indonesia (2011), food processing places should be spared from pets such as chickens, birds, rodents and insects as they could cause health problems such as dengue fever, malaria and dysentery. Each hole in the building must be installed with a wire of cassa measuring 32 eyes per inches in the vents to prevent the entry of animals and made trellises with a distance of 2 cm from the door to prevent the entry of rats and other pets.

4. Conclusion

The status of GFPM application in SME Mawar Merah, North Luwu producing corn cracker was not in good (status D) with rating IV.

References

[1] Latif R, Dirpan A and Indriani S 2017 The status of implementation of Good Manufacturing Practices (GMP) shredded fish production in UMKM Az-Zahrah, Makassar *IOP Conf. Ser. Earth Environ. Sci.* **101** 12040

[2] Rodmanee S and Huang W 2013 Hygiene and manufacturing practices, interagency collaboration, and a proposal for improvement : A case study of community food enterprise in thailand *Int. J. Soc. Sci. Humanit.* **3**

[3] Rotaru G 2005 Food Quality and Safety Management Systems : A Brief Analysis Of The Individual and Integrated **XI** 229–36

[4] Susianawati R 2006 *Kajian Penerapan GMP dan SSOP pad Produk Ikan Asin Kering dalam Upaya Peningkatan Keamanan pangan di Kabupaten Kendal* (Semarang: Universitas Dipanogoro)

[5] Winarno F G 2004 *keamanan pangan* (Bogor: M-Brio Press)

[6] Masrifah E, Noorachmat and Sukkmawati 2015 Kesesuaian penerapan manajemen mutu ikan pindang bandeng (*Chanos chanos*) terhadap standar nasional indonesia *Manaj. IKM.* **2** 163–72

[7] Angriani T and Yudhastuti R 2014 Penerapan Good Manufacturing Practices pada industri rumah tangga kerupuk teripang di Sukolilo Surabaya *J. Kesehat. Lingkung.* **2** 148–58

[8] Shoffiyahti P 2014 *Analisis Penerapan Good Manufacturing Practices (GMP) Pada Industri*

Kecil Menengah Makanan (Studi Kasus : Industri Kerupuk Keripik Peyek dan Sejenisnya di Kota Padang) (Padang: Universitas Andalas)
[9] Patel, K. T., dan Chotai N P 2011 Good Manufacturing Practices *J. Young Pharm.* **3** 138–50
[10] Soeprapto F and Adriyani R 2003 Penilaian GMP dan SSOP pada Bagian Pengolahan Makanan di Katering X Surabaya dengan Metode Skoring sebagai Prasyarat Penerapan HACCP *Indones. J. Public Heal.*
[11] Peraturan Menteri Kesehatan Republik I 2011 Higiene Sanitasi Jasa Boga *Jakarta Menteri Kesehat. Republik Indones.*
[12] Badan Pengawasan Obat dan Makanan 2012 Cara Produksi Pangan yang Baik untuk Industri Rumah Tangga *Jakarta Badan Pengawas Obat dan Makanan.*

IOP Conf. Series: Earth and Environmental Science **157** (2018) 012036 doi:10.1088/1755-1315/157/1/012036

Modification of dry grain processing for rice nutrition produced

A N F Rahman, J Genisa, A Dirpan and A A Badani

Department of Food Science and Technology, Faculty of Agriculture, Hasanuddin University, Makassar 90245, South Sulawesi, Indonesia.

E-mail: *faidah83@yahoo.com*

Abstract. Rice is a staple food for people in Indonesia that provides high energy and nutrients of up to 360 calories per 100 g. Based on the research it was known that the nutrient content in rice will increased by soaking. This is suspected because the nutrient content in the aleurone layer adsorbed to the endosperm. The purpose of this research was to know the effect of dry grain immersion on the nutrition of rice produced. The method of this research was conducted through some stages: 1. Preparation of raw materials, 2. Grain immersion, 3. Grain drying, 4. Peeling chaff, 5. Testing the nutritional value of rice. The research was processed by using factorial randomized complete random design (RCRD) with three replications. The result showed that soaking the grain for 12 hours has the highest nutritional value increases compared to the control. Proximate test resulted from the best treatment were: protein content of 8.26%, ash content of 0.42% and thiamine content of 0.023%.

1. Introduction

High quality and nutrition of rice produced by several combination of processing to convert paddy into well milled silky-white rice [1]. Well milled silky-white rice (WMR) without bran and aleuron layer is prefer by the majority of consumers due to the fine colour of rice. Processing, variety of rice, methods of cultivation and cooking conditions are some factors that affect rice nutrition produced [2]. The aleurone layer is a layer of rice that contains lots of protein, fat, vitamins, and minerals [3], however these are removed by milling processing. Nowdays, consumers are more selective to consumption of rice milled to lower degrees (partially-milled rice: PMR) or brown rice (BR). In addition, many nutritional rice have also been developed such as germinated brown rice (GBR) and rice bread (RB). Hence, in this research we try to produce nutritious rice by soaking the dry grain in the water for various time. We expected the nutrient which dissolved in water such as thiamine (vitamine B1) and mineral in germ and outer layer could adsorbed into the rice through endosperm.

2. Materials and Methods

2.1 Materials

The fresh rice grain from Ciherang variety was purchased from Sidrap, South Sulawesi, Indonesia. Reagents used were obtained from chemical market in Makassar, Indonesia.

IOP Conf. Series: Earth and Environmental Science **157** (2018) 012036 doi:10.1088/1755-1315/157/1/012036

2.2 Methods

2.2.1 Sample preparation. Dry grain 10 kg (water content 10.55%) were soaked in water with different level treatment of water (5, 10, 20 L) and time spans (4, 8, 12 h). The grain was mixed every 2 hours. After drying (water content 13.70%), the grain was then milled and analyzed for the proximate content.

2.2.2 Determination of Protein. Protein was analyzed by using Kjeldahl method [4]. A 0.2 mg of sample was mixed with 1 g of N catalyst (mixture of Na_2SO_4 : HgO (20:1)). Three mL of concentrated sulfuric acid (93-98% free N) was added to the destruction of the protein and heated in the acid room until the solution to be clear and colorless, then cooled. After cool the Kjedahl flask was washed with distilled water and boiled for 30 minutes. Further distillation was carried out with NaOH to alkalis, the distillate was placed in an Erlenmeyer flask containing 5 ml of 4% boric acid solution and methyl red indicator. The distillates are then titrated with HCl 0.02 N. Protein content was determined by the Kjeldahl method using the conversion factor 6.25. Protein was expressed as the percentage of total protein.

2.2.3 Determination of Ash. Before the measurement of ash content, the cup was dried for 1 hour at 750°C. The cup was then cooled for 10 minutes using desiccator and weighed. Five g of sample were then added into cup and heated for 3 hours. After 3 hours, the sample was then cooled in desiccator for 10 minutes then weighed. Ash content was expressed as the percentage of ash [4].

2.2.4 Determination of Thiamine. Thiamine was analyzed by using HPLC method using C18 column, UV-Vis detection at 254 nm and using CH_3OH-0.05 M CH_3COONH_4 as mobile phase [5].

2.3 Statistical analysis
The experimental was designed by using factorial randomized complete random (RAL) with three replications and the data was analysed by using SPSS ver. 22

3. Results and Discussion

3.1 Protein Content
Protein content of rice was analyzed. Figure 1 showed protein content for all treatments increased after 12 hours. The 10 kg dry grain which were soaked in the 5 liters of water;10 liters of water; and 20 liters of water have protein content of 8.30%; 7.91%; and 8.26%, respectively. Two of the results obtained were higher than protein content of grain without soaking treatment (control) that was 8.01%. Based on statistical analysis of variance showed that the treatment of soaking time and water volume ratio at the time of immersion had no significant effect on the 5% level.

The long treatment of immersion may increase the protein content caused by the diffusion of proteins and amino acids from aleurons layers into the endosperm of rice. The types of proteins present in aleurons are distinguished by their solubility, namely albumin (soluble in water), globulin (soluble in salt), prolamin (soluble in alcohol), and glutellin (soluble in alkali). The research results of Paiva *et al.* [6] showed that, an increase of protein levels from red rice were due to the migration of protein to the endosperm. Hasbullah and Riskia [7] showed that, the longer time immersion, the higher increasing of protein levels of rice. And based on statistical analysis, time immersion had significant effect on increase of protein levels.

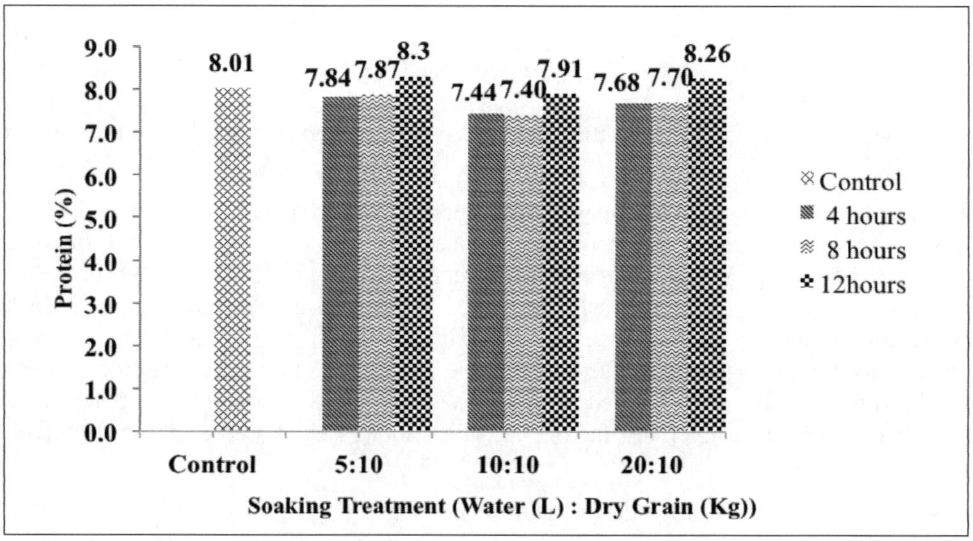

Figure 1. Protein content of dry grain in different treatments

3.2 Ash Content

As shown in figure 2, ash content for all treatments increased after 12 hours. The 10 kg dry grain which were soaked in the 5 liters of water, 10 liters of water and 20 liters of water have ash content 0.39%, 0.36% and 0.42%, respectively. Two of the results obtained were higher than ash content of grain without soaking treatment (control) that is 0.37%. Based on statistical analysis of variance showed that the treatment of soaking time and water volume ratio at the time of immersion had significant effect on the 5% level.

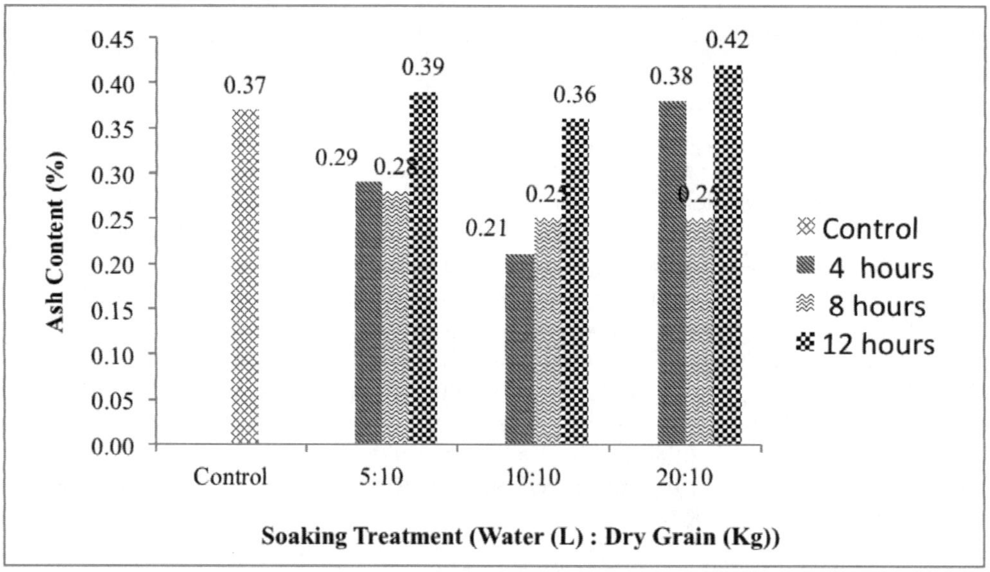

Figure 2. Ash content of dry grain in different treatments

Based on the results of research, ash content decreased after 8 hours of soaking. This occurred due to the dissolution of mineral content from husk into the water during soaking process. But after 12 hours of soaking, the ash level increased due to the opening of the bonds of rice cells and mineral adsorbed into the rice. Widowati *et al.* [8] stated that the increase of ash content in the process of soaking caused by dissolving of material components including minerals from bran and husk and adsorbed to endosperm of rice.

3.3 Thiamine Content

As shown in figure 3, thiamine content for all treatments increased after 12 hours. The 10 kg dry grain which were soaked in the 5 liters of water, 10 liters of water and 20 liters of water have thiamine content 0.023%, 0.024% and 0.023%, respectively. The results obtained were higher than thiamine content of grain without soaking treatment (control) that was 0.022%. Based on statistical analysis of variance showed that the treatment of soaking time and water volume ratio at the time of immersion had no significant effect on the 5% level.

The effect of immersion on grain could increase thiamine content of rice, because in the process of soaking, water soluble component such as thiamine adsorbed into the endosperm of rice. The aleuron layer is a layer of rice that contains lots of protein, fat, vitamins, and minerals [3].

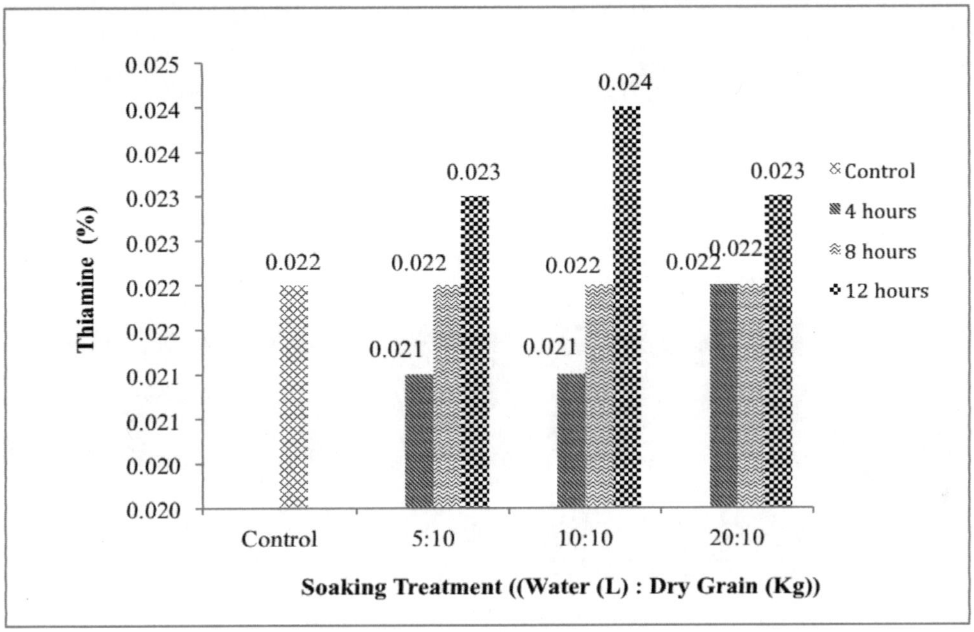

Figure 3. Thiamine content of dry grain in different treatments

4. Conclusions

The grain from Ciherang variety has been soaking in various time and several amounts of water. The results showed that the dry grain 10 kg with 12 hours immersion in 20 L water has the highest increase in nutritional value compared to the control and other treatments.

References

[1] Roberts L R 1979 Composition and taste evaluation of rice milled to different degrees. *J. Food Sci.* **44** 127-129

[2] Roy P, Orikasa T, Okadome H. Nakamura N and Shiina T 2011 Processing conditions, rice properties, health and environment. *Int. J. Environ. Res. Pub. Health* **8** 1957-1976

[3] Juliano B O and Bechtel D B 1985 *The rice grain and its gross composition* (*Rice Chemistry and Technology* 2nd ed.) ed B O Juliano (Eagan, MN, USA: American Association of Cereal Chemists) p 17-57

[4] Sudarmaji S, Haryono, B and Suhardi 1996. *Analisa Bahan Makanan dan Pertanian* [in Bahasa] (Yogyakarta: Liberty)

[5] Moreno P and Salvadó V 2000. Determination of eight water-and fat-soluble vitamins in multi-vitamin pharmaceutical formulations by high-performance liquid chromatography. *J. Chromatogr. A.* **870** 207-215

[6] Paiva F F, Vanier N L, Berrios Jde.J, Pinto V Z, Wood D, Williams T, Pan J and Elias M C 2015 Polishing and parboiling effect on the nutritional and technological properties of pigmented rice *Food Chemistry* **191** 105-112

[7] Hasbullah R and Riskia P D P 2013 The effects of soaking duration on parboiled rice quality of paddy cv IR 64 *J. Keteknikan Pertanian* **27** (1) 53-60

[8] Widowati S, Santosa B A S, Astawan M and Akhyar 2009 Penurunan indeks glikemik berbagai varietas beras melalui proses pratanak [in bahasa] *J. Pascapanen* **6** (1) 1-9

IOP Conf. Series: Earth and Environmental Science **157** (2018) 012037 doi:10.1088/1755-1315/157/1/012037

Time domain features in combination with a support vector machine classifier for constructing the termite detection system

M A Nanda[1], K B Seminar[1], D Nandika[2] and A Maddu[3]

[1]Department of Mechanical and Bio-system Engineering, Bogor Agricultural University, Bogor 16680, West Java, Indonesia
[2]Department of Forest Products, Bogor Agricultural University, Bogor 16680, West Java, Indonesia
[3]Department of Physics, Bogor Agricultural University, Bogor 16680, West Java, Indonesia

E-mail: *m.achirulnanda@gmail.com*

Abstract. Over the last decade, it has been broadly reported that wooden buildings have been massively degraded due to termite attacks. The termite detection system based on the acoustic signal has been proposed to overcome such termite attacks. In this study, we investigate the implementation of the support vector machine (SVM) at the termite detection system. In this work, the pine wood as the medium for termite infestation was divided into two groups, *i.e.*, the wood infested by termites *Coptotermes curvignathus* ('infested') and the normal wood ('uninfested'). The acoustic signal from each group was analyzed to produce the acoustic features, i.e., energy (*E*) and entropy (*H*). Subsequently, the acoustic features were included to construct the SVM Classifier. According to the numerical results, the SVM classifier achieved an accuracy of 93.21 ± 2.58%.

1. Introduction

Temperature and humidity (i.e., 28-30 ^{0}C and 70-90 %) in Indonesia are ideal for termite life, no surprise that there are 200 termite species that spread in various regions [1]. This certainly has a significant adverse effect on the damage of wooden buildings. Over the last decade, it is broadly reported that wooden buildings have experienced the degradation of their shelf life due to termite attacks. Arinana et al. [2] indicated that termites were able to destroy the wooden buildings up to the 33rd floor in Jakarta, Indonesia. As a result of this attack, Nandika et al. [1] estimated that the economic loss to house buildings in Indonesia reached IDR. 8.7 trillion in 2015, which is expected to increase continuously along with the enhancement in community settlements.

Among the 200 termite species, there is one that is considered to inflict the most significant impact loss, called *Coptotermes curvignathus* [3]. This termite is categorized as a subterranean termite species, the *Rhinotermitidae* family. This termite is widely reported to have damaged the apartments, books and buildings in Indonesia University [4]. One way to overcome the termite attack is with initial detection. Due to their cryptic behavior, termite attack on wood or wood products, it is difficult to detect visually.

Various approaches based on termite behavior when attacking the wood have been developed to detect termites including acoustic, temperature, moisture and gas related approaches. Termites produce the acoustic signal during foraging and moving. Therefore, this approach has been used to design termite detection systems by many researchers [5, 6]. However, Manzoor [7] reported that the termite detection systems still faces challenges to separate the acoustic signals generated by termites or noise from the environment. One way to solve this problem is to build a classification model. Therefore, we propose a support vector machine (SVM) classifier to construct a termite detection system.

2. Materials and Method

2.1. Timber Test

Pine wood, used as a medium for the feeding activity test of termites *Coptotermes curvignathus*. This wood has a size of 20 (l) x 9.5 (w) x 2.5 (h) cm and the inside has a hole with size 12 (l) x 6 (w) x 0.5 (h). Firstly, we group the wood into two classes, i.e., positive class (+1), the wood infested by 220 termites *C. curvignathus* ('infested'), and the negative class (-1), the normal wood that is not infested by termites ('uninfested'). Each class consists of four wood samples. In addition, the termite infestation is carried out for seven days.

2.2. Data Acquisition and Time Domain Features

The equipment used for the data acquisition is two electret microphone sensors by ITead Studio, with a sensitivity of -56 dB (frequency of 0.1-10 kHz), which is connected to the microcontroller. Furthermore, the acoustic signal will be displayed directly on the personal computer. In figure 1, a diagram of the acoustic signal to measure the acoustic signals in both classes is shown. The data acquisition is carried out for 3 seconds (repeated 10 times) on each wood.

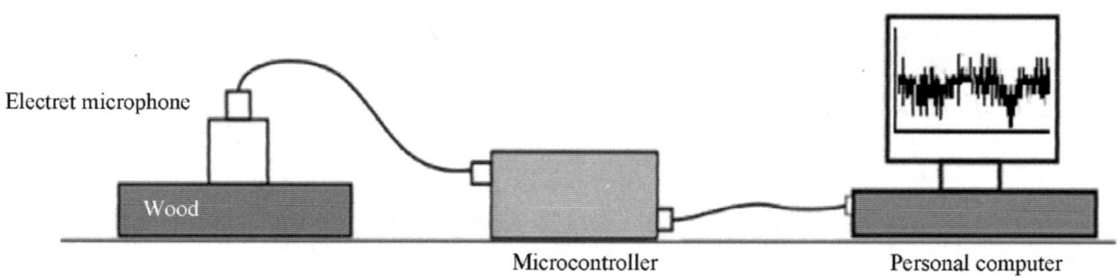

Figure 1. Diagram of acoustic signal sensing in the wood

The time domain feature is generated without requiring any a signal transformation first. This feature is easier and faster than features in other domains, in terms of computation, which is the frequency domain [8], because we extract its feature from the acoustic signal sample directly [9]. In this study, we propose the two time domain features, energy and entropy.

Furthermore, the energy (E_i) and entropy (H) can be calculated following equations 1 and 2 respectively [10]. Where, E_i is energy, H is entropy, W_L is length of frame size, e_j is the energy ratio of sub-frames with the total energy total and X_i is value of signal, n =1, . . ., W_L.

$$E_i = \frac{1}{W_L}\sum_{n=1}^{W_L}|X_i(n)|^2 \quad (1)$$

$$H = -e_j.\log_2.(e_j) \quad (2)$$

2.3. Support Vector Machine (SVM)

SVM is a binary classification method developed by Vladimir Vapnik in 1995 [11, 12]. The basic concept of SVM is to find the best decision boundary (hyperplane) that can separate perfectly the *d*-dimensional data into two classes (class +1 and class -1). As can be seen in figure 2, the hyperplane should be able to maximize the geometric distance from the hyperplane to its support vector (*SV*); this distance is well-known as margin (r). In addition, β and β_0 are a vector going through the center and the offset parameter sequentially.

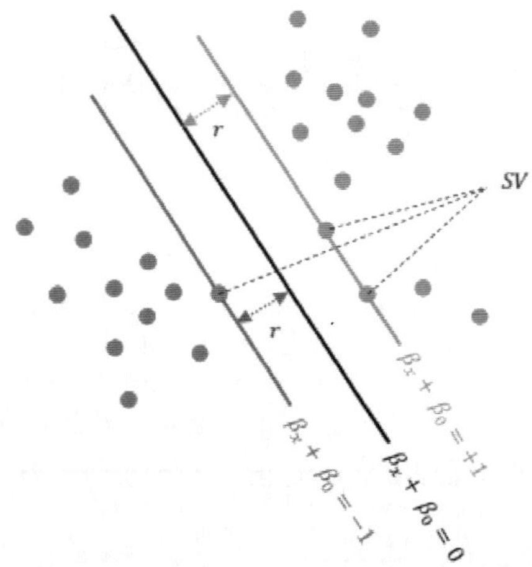

Figure 2. Generalized optimal separating using linearly separable

One of the problems in classification is variety of the data distribution, which can impact it, cannot be separated linearly. In this case, SVM introduces a kernel that can represent and transform the data to a higher dimension. Thus, the data already in higher dimensions can be easily separated by hyperplane [13]. In this study, we use the radial basis function (RBF) kernel. This kernel is formulated as follows:

$$f(x_d) = \sum_{i=1}^{n} \alpha_i y_i \exp(-\gamma \|x_i - x_d\|^2) + b \tag{3}$$

Where, we classify the data x_d into the class (i.e., +1 or -1) based on the calculation of several parameters, namely: $\propto_i$ which is a lagrange multiplier, y_i which is the membership of the class (+1, -1), x_i which represents the support vector data, γ as the gamma and b is intercept, with $i = 1, 2, 3, \ldots, n$.

Grid search is the step to take the best parameter value which is needed to construct the SVM classifier [14]. The required parameters in the RBF kernel are the cost parameter (C) and the gamma parameter (γ). The cost parameter can be viewed as a way to control the over-fitting of each support vector [15, 16]. In table 1, we set the lower and upper limit to find the best parameter value (i.e., C and γ).

Table 1. Interval determination to obtain the best parameter value (i.e., LL and UL)

Kernel function	Cost (C)		Gamma (γ)	
	LL[a]	UL[b]	LL[a]	UL[b]
RBF	2^{-20}	2^{2}	2^{-15}	2^{2}

[a] LL is the lower limit.
[b] UL is the upper limit.

2.4. Classifier Performance

Measurement of the classification performance was performed by taking 40 observations randomly from the training data set, and was repeated seven times. Furthermore, this performance produces 2 x 2 confusion matrix, including TP, TN, FP and FN, where, true positives (TP) were the actual infested class that was predicted correctly. True negatives (TN) were the uninfested class that was predicted correctly. False positives (FP) were the infested class that was predicted as an uninfested class. False negatives

IOP Conf. Series: Earth and Environmental Science **157** (2018) 012037 doi:10.1088/1755-1315/157/1/012037

(FN) were the uninfested class that was incorrectly classified. With a simple formula, we can measure the accuracy (AC) of the proposed classifier as follows [17]:

$$AC = \left(\frac{TP + TN}{TP + FP + FN + TN} \right) x100(\%) \tag{4}$$

3. Results and Discussion

3.1. Time Domain Features Distribution

Two proposed features in the time domain are energy (E_i) and entropy (H). As can be seen in figure 3, based on numerical results, we can plot the two dimensional data distribution between E_i and H in each class. Clearly, the features in both class (i.e., infested and uninfested) overlap; thus, they are difficult to separate with linear hyperplane. Therefore, kernel function is required to overcome this problem.

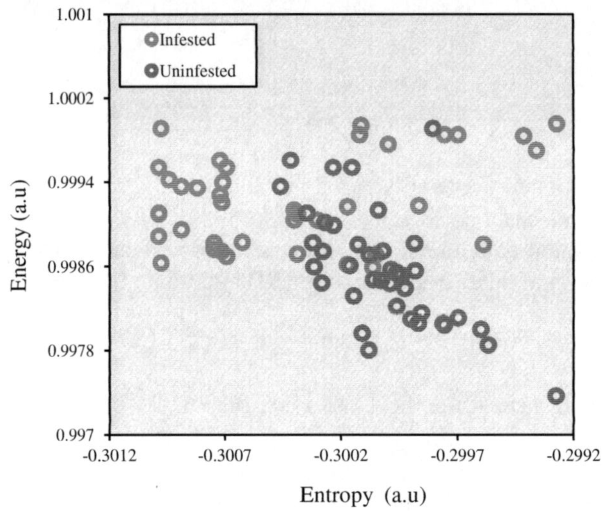

Figure 3. Time domain features distribution in the both groups

3.2. RBF Kernel Implementation

A common kernel is the radial basis function (RBF); this kernel is implemented into our termite detection system. Two important parameters need to be optimized to maximize the margin distance between two classes, i.e. cost (C) and gamma (γ). As a result, table 2 shows the best parameters for C and γ are 4 and 4 sequentially.

Table 2. The best parameters value using the grid search method

RBF kernel	
Cost (C)	4
Gamma (γ)	4

$$f(x_d) = \sum_{i=1}^{49} \alpha_i y_i \exp(-4\|x_i - x_d\|^2) + 0.162 \tag{5}$$

Based on the numerical calculation, this SVM classifier needs 49 support vectors. Furthermore, we can classify the new observation (x_d) into the class (i.e., infested or uninfested) following equation 5. The rule of this classifier is, if $f(x_d) > 0$, then x_d will be classified into the infested class. However, if $f(x_d) < 0$, then x_d will be classified into the uninfested class.

3.3. Performance of SVM Classifier

Based on the procedure described in Chapter 2.4, it produces the *TP*, *TN*, *FP* and *FN*. The performance of the SVM classifier can be seen in Table 3, where our proposed classifier that is implemented into the termite detection system has an accuracy of 93.21 ± 2.58%. This classifier is built using two acoustic features, namely energy (E_i) and entropy (*H*).in the time domain.

Table 3. Performance assessment of the proposed classifier

Repetition	TP	TN	FP	FN	AC (%)
1	17	21	0	2	95
2	16	22	1	1	95
3	18	17	2	3	87.5
4	17	20	2	1	92.5
5	17	21	1	1	95
6	16	22	2	0	95
7	18	19	0	3	92.5
mean ± stdev					93.21 ± 2.58

In previous research, Lewis et al. [18] tested the performance of the termite detection tool; this has been tested on different treatments, *i.e.*, various wood and the location of the sensor. Based on numerical results, that tool has an accuracy of 80%. Therefore, the opportunity of the further research is to test this classifier on various types of wood.

4. Conclusion

Detecting termites in wood is a real challenge in the context of integrated pest management, which aims to avoid higher damage of wood. In this paper, we have proposed an effective framework of a termite detection system based on acoustic signal by application of the SVM classifier. In this classifier, we implemented the radial basis function (RBF) kernel and optimized the parameters (*i.e.*, cost and gamma) to achieve the maximum accuracy. According to the numerical results, our termite detection system in combination with the SVM classifier achieved an accuracy of 93.21 ± 2.58%.

Acknowledgment

This work is supported by the accelerated master's program leading to doctoral research grants (PMDSU), Indonesia. In addition, the first author thanks Prof. Kudang Boro Seminar and Prof. Dodi Nandika, who provided the facility during the research at the Bioinformatics Laboratory and Termite Laboratory, Bogor Agricultural University.

References

[1] Nandika D, Rismayadi Y and Diba F 2015 *Rayap: Biologi dan Pengendaliannya*, edisi ke-2 (*Termite: Biology and its Control, 2nd Edition*) (Surakarta: Universitas Muhammadiyah Surakarta)

[2] Arinana, Philippines I, Koesmaryono Y, Nandika D, Aunu R, Harahap I S, Sumertajaya I M and Bahtiar E T 2016 *Coptotermes curvignathus* Holmgren (Isoptera: Rhinotermitidae) capability to maintain the temperature inside its nests *J. of Entomology* **13** (5-6) 199-202

[3] Sukartana P, Sumarni G and Broadbent S 2009 Evaluation of chlorfluazuron in controlling the subterranean termite *Coptotermes curvignathus* (Isoptera: Rhinotermitidae) *Indonesia J. of Tropic. For. Sci.* **21** 13-8

[4] Primanda A, Ischak T M and Basukriadi A 2003 Termite species richness on the campus of Universitas Indonesia, Depok Makara *J. of Sci.* **7** 9-14

[5] Mankin R, Hagstrum D, Smith M, Roda A and Kairo M 2011 Perspective and promise: A century of insect acoustic detection and monitoring *American Entomologist* **57** 30-44

[6] Lewis V, Leighton S, Tabuchi R and Haverty M 2011 Seasonal and daily patterns in activity of the western drywood termite, Incisitermes minor (Hagen) *Insects* **2** 555-63

[7] Manzoor F 2013 Biosensors for termite control. In: IOP Conference Series: Materials Science and Engineering, (Indus University, Pakistan: IOP Publishing) pp 1-3

[8] Griffin J 1989 An alternating frequency/time domain method for calculating the steady-state response of nonlinear dynamic systems *J. of App. Mech.* **56** 149

[9] Rivai M and Tasripan 2015 Fuel qualification using quartz sensors ARPN *J. of Eng. and App. Sci.* **10** 6737 – 43

[10] Giannakopoulos T and Pikrakis A 2014 *Introduction to Audio Analysis: A MATLAB® Approach* (USA: Academic Press)

[11] Bakheet S 2017 An SVM Framework for Malignant Melanoma Detection Based on Optimized HOG Features Computation 5 4

[12] Karatzoglou A, Meyer D and Hornik K 2005 Support vector machines in R *J. of Stat. Software* **15** 1-28

[13] Sakthivel N, Saravanamurugan S, Nair B B, Elangovan M and Sugumaran V 2016 Effect of kernel function in support vector machine for the fault diagnosis of pump *J. of Eng. Sci. and Tech. (JESTEC)* **11** 826-38

[14] Santos P, Villa L F, Reñones A, Bustillo A and Maudes J 2015 An SVM-based solution for fault detection in wind turbines *Sensors* **15** 5627-48

[15] Arun C and Durairaj D C 2017 Identifying medicinal plant leaves using textures and optimal colour spaces channel *J. Ilmu Komp. dan Informasi* **10** 19-28

[16] Nanda M A, Seminar K B, Nandika D and Maddu A 2018 A comparison study of kernel functions in the support vector machine and its application for termite detection *Information* **9** 5

[17] Ciptohadijoyo S, Litananda W, Rivai M and Purnomo M 2016 Electronic nose based on partition column integrated with gas sensor for fruit identification and classification *Comp. and Electronics in Agric.* **121** 429-35

[18] Lewis V R, Power A B and Haverty M I 2004 Surface and subsurface sensor performance in acoustically detecting western drywood termites in naturally infested boards *Forest Products J.* **54** 57-62

IOP Conf. Series: Earth and Environmental Science **157** (2018) 012038 doi:10.1088/1755-1315/157/1/012038

Detection of fungi from rice black bug *Paraeucosmetus pallicornis* Dallas (Hemiptera: Lygaeidae) and inhibition with crude extract of *Calatropis gigantea* (Asclepiadaceae)

S Sjam, U Surapati, Adiwena, A Syatri, V S Dewi and A Rosmana

Departemen Plant and Pest Diseases, Faculty of Agriculture, Hasanuddin University, Jl. Perintis Kemerdekaan KM 10, Makassar, 90245, Indonesia.

E-mail: *Sylviasjam@yahoo.com*

Abstract. Rice black bug (*P. palicornis*) is one of the pests that attack the rice plants in the generative phase that causes the rice easily destroyed when milled and tasted bitter after cooking hence reduces the quality and quantity of rice. The bitter taste in rice may be due to the fungus associated with rice black bug. The aimed of this research was to detect the associated fungi with rice black bug *P. pallicornis* using some sterilization methods and inhibition with of leaf crude extract of *C. gigatea*. Detection of fungi from *P. pallicornis* was conducted using three sterilization methods and control (without sterilization) namely: (1) sterilization with aquades + alcohol 70% (5, 10, 15 and 20 times dipping) + aquades; (2) aquades + alcohol 70% (10 and 20 times dipping); (3) Aquades + alcohol 90 %+ NaCl 0.5 % +alcohol 90 % + aquades . Inhibition of fungi from *P. pallicornis* with crude extract of *C. gigantea* obtained by maceration method and then made some concentration to see the effect of its inhibition on the fungi associated with the *P. pallicornis*. The results showed that without sterilization, four microbe were obtained: *Gliocladium* sp., *Aspergillus* sp., black hyphae fungus and white hyphae fungus, sterilization method of Aquades + alcohol 70% with 5 times dipping in alcohol obtained *Gliocladium* sp., 10 and 20 times dipping found *Aspergillus* sp. and *Gliocladium* sp and 15 times dipping found *Aspergillus* sp. Sterilization with 10 and 20 times dipping in alcohol 70% then washing 2 times with aquades found *Gliogladium* sp. and *Aspergillus* sp. Sterilization with Aquades + alcohol 90 % + NaCl 0.5% + alcohol 90% + aquades found *Gliocladium* sp. Crude extract of *C. gigantea* had the potential to inhibit fungi *Aspergillus* sp. and *Gliocladium* sp. from rice black bug.

1. Introduction

Rice black bug *P. pallicornis* (Dallas) (Hemiptera: Lygaeidae) is a new pest in rice plants belonging to quarantine pest with A2 categories and its currently a major problem for farmers and has spread in several rice production centers in South Sulawesi and West Sulawesi with the intensity moderate to severe thereby decreasing production. In the vegetative phase the pest attacks the rice by sucking the stem liquid with symptoms of leaves of yellowish-colored rice and generally occur in young leaves and the generative phase the pest attacks by sucking the base of the stem and the rice grain and causing bitter taste in rice due to the association with the fungus [1] this affecting the quality of the rice. The results of the laboratory studies show that on the grain and rice that has been attacked by *Aspergillus sp.* And *Gliocladium sp*. The fungus contained in rice plants attacked by *P. pallicornis* is pathogenic because it causes death in young rice leaves [2]. The previous results also showed that by sterilization method on the surface of the body of *P. pallicornis* was found many fungi [3]. Therefore, to reassure the assumption of an association between fungus and *P. pallicornis*, it is necessary to detect this by using different sterilization method from previous sterilization. Alleged existence of this association which says that the fungal origin of *P. pallicornis* can be pathogenic to rice crops then the necessity is also to a control method which can only control *P. pallicornis* but aslo can inhibit the development of the fungus associated with *P. pallicornis* by using plant extract *C. gigantea*. The antifungal activity of *C. gigantea* was also reported in some studies and it provides an important option for the biological control a plant pathogenic fungus [4] and can be used for the remedy of infectious diseases caused by pathogenic bacteria and fungi as well as for the control of insect [5]. Extract of Calatropis are harmless and nonphytotoxic, inhibitory effects on germination and on the viability of fungal spores as well and moderate to good activity against *A. niger* as it is a saprophyte in soil causing black mould of onion, garlic and shallot, stem rot of Dracaena, root stalk rot of Sansevieria, and boll rot of cotton; spoilage of cashew kernels, dates, figs, vanilla pods and dried prune [6]

2. Material and methods

2.1. Isolation and Identification of Fungi Emerged from P. pallicornis

Samples of *P. pallicornis* were obtained from Gowa Regencies at South Sulawesi. The adults were kept in a cage containing rice seedlings during the trip to the laboratory. *P. pallicornis* was randomly selected and stored in an airtight glass container which placed in a freezer with a temperature of ± -5°C for 1 hour before the fungus isolation process. Isolation of fungi associated with the separated insects done in three successive stages as follows: detection of fungi from *P. pallicornis* with three method sterilization was modified based on [7] and without sterilization. The process detection of fungi from *P. pallicornis*: (1). Adults were washed with aquades and Sterilized with alcohol 70% (5, 10, 15 and 20 times dipping) and washed again with aquades, (2). The adults were washed with Aquades and strelized with alcohol 70% (10 and 20 times dipping), (3). The adults were washed with Aquades and sterilized with alcohol 90% and subsequently with NaCl 0.5% + 90% alcohol and the finally washed with aquades. The sterilized insects were put in petri dish containing of 13 ml PDA medium. The dishes were checked for fungal growth after 7 days. The emerged fungi were picked-up and transferred to new PDA medium and identified by characterizing the morphology were described according to their macroscopic features such as colour, shape and growth of cultured colonies, as well as microscopic characteristics like structure of hyphae, conidia and conidiophores. Obtained data then compared with the descriptions of fungi species present in the literature [8].

2.2. Preparation of Calotropis gigantea Extract

Fresh leaves of *C. gigantea* were collected from areas in and around Takalar Regency of South Sulawesi, Indonesia and shade-dried for 2–3 days, macerated with methanol solvent for 7 days and filtered through filter paper. The solvent was removed by rotary-evaporator for 6–8 h to obtain crude extract in the form of paste. It was then placed in a reagent bottle, and stored in the refrigerator as a stock solution prior to its use. The extract were prepared 3 concentration that is 1%, 1.5%, and 3% by diluting the stock solution with acetone and then suspended with 1% ml of tween-80 for each concentration.

2.3. Preparation Liquid Media for Inhibition Extract to Fungi from P. pallicornis

The extract potato containing sugar 20 grams and 1 Chloromphenicol grain were homogenized in the erlenmeyer. Furthermore, the erlenmeyer was wrapped with aluminum foil and plastic wrapping and then sterilized in an autoclave for 2 hours at 121°C.

2.4. Method of Treatment

The liquid media was prepared as much as 100 ml and add the extract in accordance with concentration and then grow fungi from Rice black bug (*Aspergillus* sp. and *Gliocladium* sp.). Observations were made 2 times each at the age of fungi 7 HST and 14 HST were to calculate the weight of dried mycelium of fungi after being treated with *C. gigantea* leaf extract by filtering the fungus until no water dripped from the sieve. Then the dried mycelium is weighed using a digital scale.

3. Results and Discussion

The assessment of fungal detection in *P. pallicornis* revealed that the presence of *Aspergilus* sp. dan *Gliocladium* sp. was the most frequent with the process where adults were washed with aquades and sterilized with alcohol 70% with 5, 10, 15 and 20 times dipping and washed again with aquades. In NaCl treatment only presence Gliocladium. *Aspergilus sp., Gliocladium sp.*, white fungi and black fungi was detected in not sterilized as control. Macroscopic and microscopic observations of *Gliocladium* sp and *Aspergilus* are shown in table 1 and table 2.

These results show that *Aspergillus* sp. and *Gliocladium* sp. were consistently found in various sterilization methods. Interpretation of these data that the fungi are associated with *P. pallicornis.* The results also show that the presence of Aspergilus is not only found in adult insects but also on the nymph of *P. pallicornis* from rearing in the laboratory which is F1. From the results of previous tests showed that fungus was found in the left side body of the *P. pallicornis.* This may be because these fungi accumulate in the left gastrointestinal tract. The similarity between *Aspergillus* sp. isolated from insect and from rice grain, suggested that *Aspergillus* sp. from rice grain may come from *P. pallicornis* insect and that the fungi probably reside in the digestive tract including the mouth parts. From these sites, the fungi could be transferred to the rice grain through their stylets when they feed [1].

Association between *Aspergillus* sp. and insect attacking plant have been reported that *A. flavus* to be associated grew from spiracle, recta and mouth parts on surface insects, internal organs of Lygus and stink bug and colonies of *A. flavus* developed from intestinal tissue [9].

Table 1. Characterization of fungi in adults *P. pallicornis*

No.	Genera	Macroscopis	Microscopis
1	*Aspergillus* sp.	Colonies on agar culture are spread with pale dark brown colour. Growth with irregular shape.	Conidiophores, simple, bearing conidial head.
2	*Gliocladium* sp.	Colonies on agar culture are brownish green colony color. Growth with circular thick shape and entire margin.	Conidiophores hyaline, erect, simple or branched oppositely or verticillately especially in metula bearing spore masses at phialides on the apical branches.
3	White hyphae fungi	Colonies on agar culture are white. Growth like cotton with circular thick shape and entire margin.	Have no conidiophore and unbranched.
4	Black hyphae fungi	Colonies on agar culture are dark cotton with white edge. Growth with circular thick shape and entire margin.	Have no conidiophore and branched.

Aspergillus flavus has been widely accepted occurs in the field were damaging preharvest maize. Insect pest *M. nigrivenella* collected feeding on maize ears was found to be significantly correlated to *A. flavus* infection and aflatoxin contamination causes damage before and after harvest [10]. The presence of aflatoxin content may cause damage to plants in the vegetative phase causing yellow color while in the rice grain cause bitter taste. *P. pallicornis* can carry a fungus that produces a toxic substance and *Aspergillus* sp. isolated from rice grains was similar to that isolated from the adults *P. pallicornis* and from rice seed was just isolated *Aspergillus* sp., while from adult insect Isolated *Aspergilus sp.* and *Gliocladium* sp [1]. Further, this research confirmed the occurrence of an association between *Aspergillus* fungus and insect pest of *P. Pallicornis* in causing toxicity symptoms in rice.

IOP Conf. Series: Earth and Environmental Science **157** (2018) 012038 doi:10.1088/1755-1315/157/1/012038

Gliocladium sp is an antagonist fungus used to control some plant diseases so it is likely to cause damage to very small rice. *Gliocladium sp.* isolated from *P. pallicornis* does not give effect to the damage of seeds and rice plants [11]. The existence of *Gliocladium sp* may be derived from planting because of the habit of *P. pallicornis* who likes to hide in the fractures of the soil and most species of *Gliocladium* are saprophytes in various environments and can be associated with healthy plant roots to protect them from pathogenic [12] and *Gliocladium* can be carried by adults *P. pallicornis*. Sometimes many saprophytic fungi in the soil can be found in the area of the digestive system of insects [13]. The ability of various species of the genus *Gliocladium* is to produce hydrolytic enzymes and antifungal, and antibacterial compounds [7]. In addition, some species of the genus *Gliocladium* have the ability to encapsulate foreign substances. This is what causes the application of various insecticides against *P. pallicornis* have no real effect.

Table 2. Macrosopic and microscopic identification of fungi in *P. pallicornis*

Num.	Genera	Macroscopic	Microskopic
1	*Aspergillus* sp		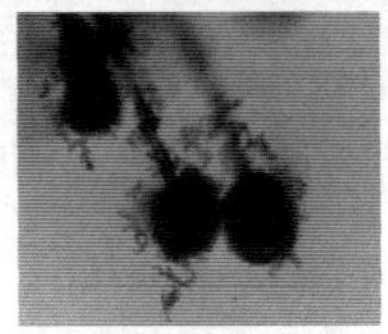
2	*Gliocladium* sp		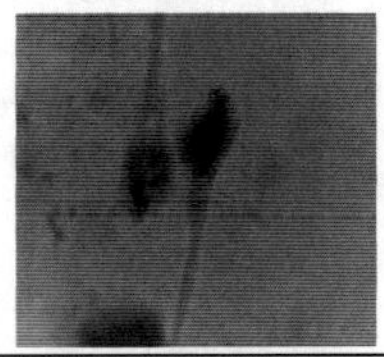
3	White fungi	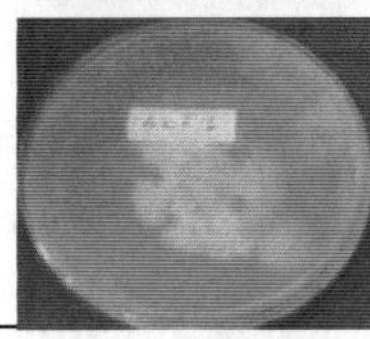	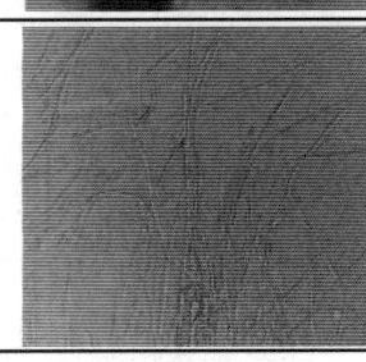
4	Black fungi		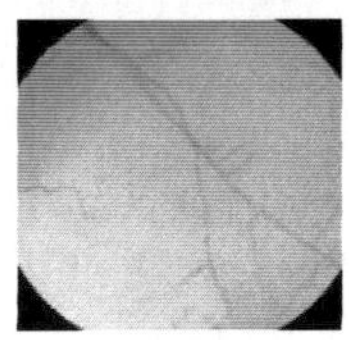

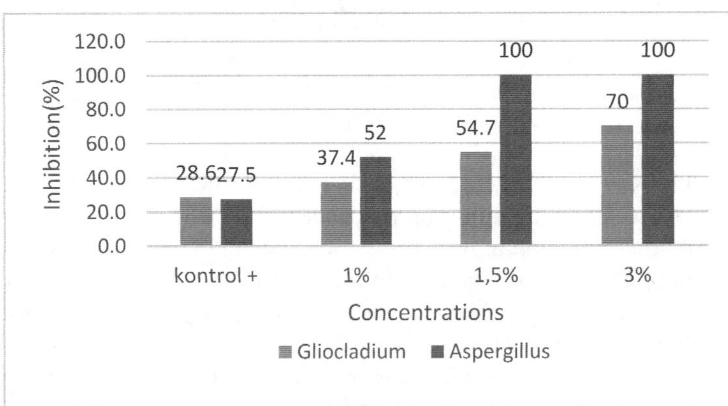

Figure 1. Percentage of inhibition fungi from *P. pallicornis* by *C. gigantea* leaf extract

Inhibition of fungi *Aspergilus* and *Gliocladium* from *P. pallicornis* with crude extract *C. gigantea* indicated that all concentrations of *C. gigantea* crude extract (1%, 1.5%, and 3%) were able to inhibit the development of *Aspergillus* sp. and *Gliocladium* sp (figure1). It can be seen that the weight of *Aspergillus* sp. and *Gliocladium* sp. after application of concentration was lower than control. This clearly shows that 3% concentrations inhibit than concentrations of 1.5% and 1%. *Calotropis* (*Asclepiadaceae*) commonly known as is a useful medicinal plant. The two species i.e. *C. gigantea* and *C. procera* are to a great extent having a very similar chemical properties like of tannins, saponin, flavonoids, terpenoids and cardiac glycosides [14, 15] cardenolide, glucosides, a non-protein, amino acid, flavonoids and steroids [16] and aqueous extracts of leaves have demonstrated strong inhibitory effect microorganisms with the mechanism of action could be by inhibition of fungal cell wall, protein and amino acid, sphingolipid biosynthesis and electron transport chain [17].

Acknowledgments
We would like to thank the State Minister of Research and Technology, Republic of Indonesia for the financial support provided for the study. We also expand our thanks to the head of Hasanuddin University Research Institute for his valuable advice during the study.

References

[1] Rosmana A, Sylvia S, Dian E S, Uvan N, Akbar A H, Ala A and Nasruddin G 2014 Fungi associated with *P. pallicornis* causing apparent symptoms of toxicity in rice grains and rice seedlings *Int. J. Curr. Microbiol. App. Sci.* **3** (2) 407-414

[2] Rosmana A, Sylvia S, Dian E S, Uvan N, Ala A and Nasaruddin 2012 Identification of fungi associated with *P. pallicornis* (Dallas) (Hemiptera: Lygaeidae) and study of its impact on incidence of rice seedlings with toxicity symptoms.

[3] Amin N, Daha L, Agus N, Rosmana A and Fadhlan M 2015 Diversity of some endophytic fungi associated with rice black bug *P. pallicornis* on rice plant. *J. of Chemic. and Pharma. Res.* **7** (4) 1246-1253

[4] Gaurav K, Loganathan K and Kokati V B R 2010 Antibacterial activity of aqueous extract *C. gigantea* leaves– an in vitro study *Int. J. of Pharmaceu. Sci. Rev. and Res.* **4** (2)

[5] Shumaia P, Kader Md A, Aktar U C, Shuaib Md A R and Ekramul Md H 2013 Antibacterial, antifungal and insecticidal activities of the n-hexane and ethyl-acetate fractions of methanolic extract of the leaves of *C. gigantea* Linn. *J. of Pharmacog. and Phytochem.* **2** (5) 47-51

[6] Varahalarao V, Mohan B, Kaladhar D S V G K, Kumar S S V N, Seshagiri B and. John P M 2012 Antimicrobial profile of crude extracts *Calotropis procera* and *Centella asiatica* against some important pathogens. *Indian J. of Sci and Tech.* **5** (8) ISSN: 0974- 6846

[7] Hallmann J, Berg G and Schulz B 2006 *Isolation Procedures for Endophytic Microorganisms* (*Microbial Root Endophytes (Soil biology)*) ed. Schulz B J E *et al.* (Heidelberg: Springer) pp 300–319

[8] Barnett H L and Hunter B B 1972 *Illustrated Genera of Imperfect Fungi* 3rd edition (USA: Burgess Publishing Co.)

[9] Stephenson L W and Russel T E 1974 *The Association of A. flavus with Hemipterous and other Insects Infesting Cotton Bract and Foliage* (Publication No .2298) (USA: Agriculture experiment station University of Arizona)

[10] Sétamou M, Cardwell K F, Schulthess F and Hell K 1997 *Aspergillus flavus* infection and aflatoxin contamination of preharvest maize in the Republic of Benin *Plant Disease* **81**1323-1327

[11] Sari D E 2012 Pengaruh Cendawan yang Diisolasi dari Kepik Hitam *Paraeucosmetus pallicornis* Dallas (Hemiptera : Lygaeidae) Terhadap Kerusakan Bibit Tanaman Padi (*Oryza sativa*)

(Makassar: Program Studi Ilmu Hama dan Penyakit Tumbuhan, Fakultas Pertanian, Universitas Hasanuddin)

[12] Heyler 2014 *Biological Control In Plant Protection: A Colous Handbook* Second Edition (London: CRC Press)

[13] Tanada Y and Kaya H K 1993 *Insect Pathology* (New York: Academic Press Inc.)

[14] Verma V N 2014 The chemical study of Calotropis *Int. Letters of Chem. Phy. and Astronomy* **20** 74-90

[15] Singh S, Sanchita Singh, Mishra R M and Shrivastava M P 2014 Preliminary phytochemical screening of *C. gigantea* leaf *Int. J. of Sci. and Res.* **4** (2) 1-3

[16] Kori P and Alawa P 2014 Antimicrobial activity and phytochemical analysis *C. gigantea* root latex extract *IOSR J. of Pharmacy* **4** (6) 7-11

[17] Hassan S W, Bilbis F L, Ladan M J, Umar R A, Dangoggo S M, Saidu Y, *et al* 2006 Evaluation of antifungal activity and phytochemical analysis of leaves, roots, and stem barks extracts of *C. procera* (Asclepiadaceae) *Pak. J. Biol. Sci* **9** (14) 2624-9

Endophytic fungi associated with cacao branch and their potential for biocontrol vascular streak dieback disease on cacao seedling

A Asman, N Amin, A Rosmana and T Abdullah

Department of Plant Pests and Diseases, Hasanuddin University, Jl. Perintis Kemerdekaan KM 10, Makassar, 90245, South Sulawesi, Indonesia.

E-mail: *asman_adi81@yahoo.com, asman@agri.unhas.ac.id*

Abstract. Endophytic fungi are one of the agents to control phytopathogen both bacterial or fungi including dieback disease on cacao such as Vascular Streak Dieback (VSD). Ten fungal isolates that associated with a healthy branch of VSD resistant clone was applied to cacao seedling, the fungi were a Curvularia-like colony, a Fusarium-like colony, a Geotrichum-like colony, an Aspergillus-like colony, a Gliocladium-like colony, a Colletotrichum-like colony and four isolates indicates as mycelia sterilia fungi. Those fungi were applied at 2.5-old months cacao seedling of moderate to susceptible clone against VSD, and 30 days after inoculation of those fungi, the seedling was applied with a bunch of mycelia emerged from infected branch that mix with water using mixer appliance and then the seedling was exposed at open area surrounded by infested cacao tree. The disease incidence was recorded started at 30 and then 60, 90 and 120 days after inoculation of the pathogen and exposed. The result of study shows that the disease appeared increasing gradually and 120 days fungus Curvularia-like colony exhibited lower incidence followed by Copper fungicide, Mycelia sterilia 1, Gliocladium-like colony, Fusarium-like colony, Geotrichum-like colony, Mycelia sterilia 3, Colletotrichum-like colony, Aspergillus-like colony, control (no application), Mycelia sterilia 2, Mycelia sterilia 4 with percentage of 3.0, 8.3, 9.5, 11.3, 15.8, 18.0, 19.0, 19.3 , 21.3, 26.5, 30.5, respectively. The result indicates that some fungi associated with VSD resistant clone were able to prevent VSD to develop and causing symptoms.

1. Introduction

Vascular streak dieback (VSD) is an important disease affecting Southeast Asian cocoa production, including Indonesia. Indonesia is the world's third largest producer of cacao [1], where mostly produced in Sulawesi Island. Estimates of production losses due to VSD is about 774 million US $ [2]. The disease has devastated some of the cacao plantations and then threaten cacao sustainability, particularly plants belong to smallholder farmer. The disease caused by the basidiomycete fungus *Ceratobasidium theobromae* (formerly *Oncobasidium theobromae*) [3, 4].

The disease has been spread out dramatically at cacao plantation area and seems difficult to control. The pathogen is wind-borne, leaf-penetrating, and xylem-infecting [5]. Basidiospores of the pathogen occurs mainly after midnight into the early morning [3]. Basidiospores germinate and hyphae penetrate unhardened leaves at branch termini by growing directly through the cuticle, above leaf veins [6]. Infected leaves do not show symptoms for 3 to 5 months, by which time the pathogen has ramified through the xylem in the adjacent stem [3]. Due to its systemic infection characteristic, the VSD disease is difficult to control. Control tools that needed are able to protect young leaves from germination of basidiospores and capable to ramify through the xylem vessel. Even though few fungicides had high fungitoxic activity against the pathogen under experimental condition such as the triazole compounds triadimefon, triadimenol, and propiconazole [7]. However, these fungicides only a few applied by farmers due to its high cost. Currently, cultural practices including regular pruning, fertilizing, and pruning of diseased plant material still remain the most prevalent applied in the diseased tree.

Biocontrol is an option to encounter losses caused by VSD. Endophytic fungi is a good alternative to suppress the disease, they proved to be effective against cacao pathogen such as black pod rot, frosty pod rot, witches' broom, and vascular streak dieback [8, 9, 10, 11, 12, 13, 14, 15,16, 17]. Endophytic fungi can colonize their host without causing disease symptoms [18], and could isolated both stem and leaves [15, 8, 18, 11, 12, 19]. In the present study, we used endophytic fungi from stems of VSD resistant cacao plants to further identify possible biocontrol agent for vascular streak dieback disease and the possibilities of preventing VSD to develop and penetrate on young leaves using endophytic fungi.

2. Materials and methods

2.1. Endophytic fungi application

Isolates of endophytic fungi were isolated from VSD resistant clone [19]. The Seedling was developed from VSD susceptible to moderate clone. The seedling was 2.5 months old, and before the seedling was treated, it maintained in a nursery building with UV plastic as roof and the seedling growth assisted with fertilizer. Twelve plants for each isolate and the controls were used. The treatments are no application (Control), Copper fungicide, *Curvularia*-like colony, *Fusarium*-like colony, *Geotrichum*-like colony, *Aspergillus*-like colony, *Gliocladium*-like colony, *Colletotrichum*-like colony, Mycelia Sterilia Isolate 1, Mycelia Sterilia Isolate 2, Mycelia Sterilia Isolate 3, Mycelia Sterilia Isolate 4

The fungi were formulated in powder form use rice medium, after colony of fungi was well distributed throughly in rice medium, the medium grinded until turn to powder form. Two grams of fungi powder applied to each seedling that already sprayed with 0.7% CMC before. Application of fungi powder to seedling using hand spayer and the seedling was wounded on stem use needle. For Copper fungicide is using Cuprous Oxide fungicide with concentrae 0.2 g/100 ml.

During 30 days after application of endophyte fungi powder, the seedling was kept under a nursery building with UV plastic as roof, after 30 days the seedling was applied with the disease inoculum and placed at open area surrounded by older high infested cacao tree, seedling placed near infested older cacao.

2.2. Evaluation of Endophytism status

This activity conducted through isolation of treated seedling before and after the seedling applied by Disease inoculum and placed at the open area. the isolation was carried out 30 days after inoculation (DAI) and 150 DAI. Part of seedling that isolated were leaf, stem, and root.

Initially, samples of leaf, stem, and root were rinsed with tap water, and then surface sterilized with NaOCl solution (2.5%) for three minutes and Ethanol (70%) for two minutes and sterilized water for one minute. and then the samples were dried above sterilized filter paper. Approximately 4 mm diametre pieces of plant materials were placed on Potato-Dextrose Agar (PDA) supplemented with Chloramphenicol and incubated at 25 °C.

2.3. Disease Record and Statistical Analysis

The incidence of the disease was recorded visually on leaf of the seedling. The data regarding the incidence of the VSD was collected 30, 60, 90, 120 after the last application. Aggressivity of fungi was determined using analysis of variance (ANOVA) and standard error. When significant difference detected, means were separated using Duncan test at 5% probability level.

3. Results and Discussion

3.1. Results

3.1.1. Incidence of Vascular Streak Dieback. Evaluation of ten different fungal isolates on seedling under field condition on VSD incidence are presented in table 1 and figure 1. The result of the effect of ten different fungal isolate and controls on 60 DAI observation, where *Curvularia*-like colony, Mycelia Sterilia Isolate 1, *Geotrichum*-like colony, Copper fungicide, *Fusarium*-like colony, *Aspergillus*-like colony, *Colletotrichum*-like colony, Mycelia Sterilia Isolate 3 shows the best result in suppressing the VSD incidence and significantly different with control. Meanwhile, on 90 DAI recording incidence of the disease was lower on *Curvularia*-like colony, Mycelia Sterilia Isolate 1, *Geotrichum*-like colony, Copper fungicide, *Aspergillus*-like colony, those treatments shows significantly different with control.

Table 1. Incidence of Vascular Streak Dieback on 60 and 90 days after inoculation and exposed on cacao seedling.

Treatments	Days After Inoculation (DAI) and Exposes	
	60	90
Control	15.00 ^{a}	17.50^{ab}
Copper fungicide	2.00 ^{cd}	8.25^{bcd}
Curvularia-like colony	0.00 ^{d}	0.00 ^{d}
Fusarium-like colony	1.50 ^{cd}	14.75 ^{abc}
Geotrichum-like colony	0.00 ^{d}	8.25 ^{bcd}
Aspergillus-like colony	5.00 ^{bcd}	9.75 ^{bcd}
Gliocladium-like colony	6.50 ^{bc}	10.50 ^{abc}
Colletotrichum-like colony	4.50 ^{bcd}	18.25 ^{ab}
Mycelia Sterilia Isolate 1	0.00 ^{d}	6.50 ^{cd}
Mycelia Sterilia Isolate 2	15.75 ^{a}	21.00 ^{a}
Mycelia Sterilia Isolate 3	10.50 ^{bcd}	18.00 ^{ab}
Mycelia Sterilia Isolate 4	0.41 ^{ab}	18.75 ^{ab}

Numbers in same column followed by same letters are not significantly different (P=0.05, Duncan test).

Incidence of Vascular Streak Dieback on 120 DAI (figure 1) was consistently suppressed by a number of fungal viz *Curvularia*-like colony, Copper fungicide, *Gliocladium*-like colony, Mycelia Sterilia Isolate 1 with significantly different with control.

3.1.2. Endophytism status of fungi isolated. Evaluation of endophyte fungi presence on seedling are presented in table 2. the result shows that fungi were reisolated only on the leaf, whereas on stem and root were not found. All fungi except *Colletotrichum*-like colony was found only on 30 DAI while on 150 DAI was not found.

3.2. Discussion

Two months after exposes and inoculation of disease inoculum of cocoa seedling a number of endophytic fungi viz *Curvularia*-like colony, Mycelia Sterilia Isolate 1, *Geotrichum*-like colony, Copper fungicide, *Fusarium*-like colony, *Aspergillus*-like colony, *Colletotrichum*-like colony, Mycelia Sterilia Isolate 3 able to inhibit VSD to develop, but after three months number of fungi were reduced remained four fungi including *Curvularia*-like colony, Mycelia Sterilia Isolate 1, *Geotrichum*-like colony,

Aspergillus-like colony as effective fungi, However, after four months only three fungi were effective that was *Curvularia*-like colony, *Gliocladium*-like colony, Mycelia Sterilia Isolate 1.

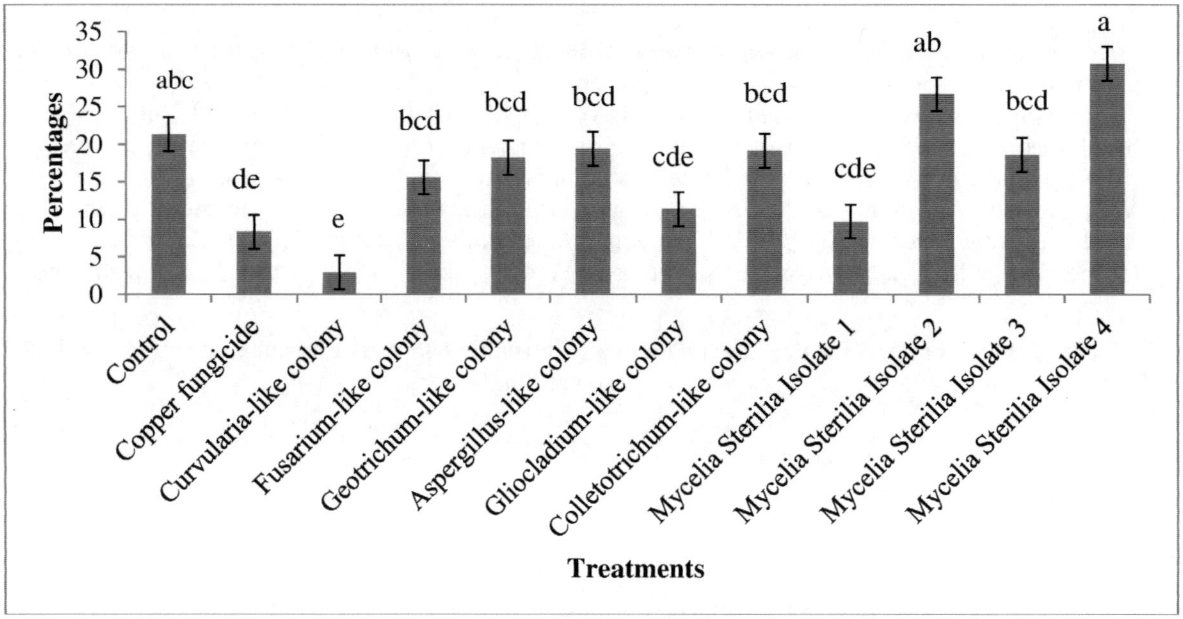

Figure 1. The incidence of Vascular Streak Dieback on 120 days after inoculation and exposed on cacao seedling. Numbers in the same color bar followed by same letters are not significantly different (P=0.05, Duncan test.

Table 2. Re-isolation of fungi from treated seedling on 30 DAI and 150 DAI.

Treatments	Leaf		Stem		Root	
	30 DAI	150 DAI	30 DAI	150 DAI	30 DAI	150 DAI
Curvularia-like colony	+	-	-	-	-	-
Fusarium-like colony	+	-	-	-	-	-
Geotrichum-like colony	+	-	-	-	-	-
Aspergillus-like colony	+	-	-	-	-	-
Gliocladium-like colony	+	-	-	-	-	-
Colletotrichum-like colony	-	-	-	-	-	-
Mycelia Sterilia Isolate 1	+	-	-	-	-	-
Mycelia Sterilia Isolate 2	+	-	-	-	-	-
Mycelia Sterilia Isolate 3	+	-	-	-	-	-
Mycelia Sterilia Isolate 4	+	-	-	-	-	-

* (+) : Found; (-) : Not Found;

These findings revealed that fungi associated with healthy tissue of cacao provide a potential role in suppressing VSD disease and could be used as biocontrol agents. The result supported by previously studies regarding endophytic fungi were used for helping tree from the infestation of pathogen in cacao [8, 12, 15, 16, 13, 14].

Curvularia-like colony and mycelia Sterilia Isolate 1 appeared consistently has low incidence of the disease. The fungi were able to protect seedling from VSD infestation. Obviously, the fungi should not be considered risk-free because some strains of the fungi found as pathogen at other crops [20, 21, 22]. Some of the fungi were used in this study found as pathogen at cacao viz *Fusarium* [23]), *Colletotrichum* [24, 25]. Further, even though they did not induce disease symptom we should be careful when we decide to use them as biocontrol agents.

The fungi were able colonized of seedling leaves, then the pathogen could not penetrate the leaves tissue, this verified by re-isolation of the fungi. The endophytic fungi protect plants from pathogen through the mechanisms of competition, induced resistance, antagonisms, and mycoparasite [9, 12]. The result of re-isolation of the fungi indicates that the fungi only found in leaf at 30 days after inoculation of endophytic fungi, this result suggested that the fungi should be re-inoculation for longer protection from VSD disease. Application frequency of antagonist fungi has significant role reduce cacao disease incidence [26].

Colonized and protecting of cacao leaves, especially young leaves were one of an important step to prevent VSD pathogen landed and germinate on leaf surface due to VSD pathogen, *Ceratobasidium theobromae*, where it basidiospores germinate and hyphae penetrate unhardened leaves [3].

Copper fungicide showing as one of the best treatment, cuprous oxide applied on leaf thoroughly. the fungicide is contact fungi and provides longer protection against fungal infection [27]. Copper-based treatments are effective against a wide spectrum of fungal pathogen. Their action is fungistatic and prevents spore germination [28]. However, application of this compound should be accurate since copper is toxic to plants, it must be used at low levels or in the insoluble form [27] and cuprous oxide is classified as FSC highly hazardous due to its acute toxicity to aquatic organisms.

4. Conclusion

Endophytic fungi was able to prevent VSD penetrating young leaves and slow down the pathogen to develop. However, some of endophytic fungi isolated found as pathogen at other crops. Therefore, those of endophytic fungi should not be considered risk-free. In addition, for longer protection, the endophytic fungi should be apply repetitive.

Acknowledgments

This study was supported by Hasanuddin University. The authors also would like to thank Bapak Arman Amir for providing his farm as research location and for assistance in managing the farm, Bapak Amiruddin for his technical support and providing research materials.

References

[1] International Cocoa O 2017 *ICCO Annual Report 2014/2015* (International Cocoa Organization (ICCO))

[2] Abdurrahman 2017 *Penerapan Budidaya Tanaman Sehat dalam Rangka Peningkatan Produksi Kakao di Sulawesi Selatan Pertem. Entomol. Indones. dan Fitopatol. Indonesia* Sulawesi Selatan di Balitsereal Maros, 27 July 2017

[3] Guest D and Keane P 2007 Vascular-Streak Dieback: A new encounter disease of cacao in Papua New Guinea and southeast asia caused by the obligate basidiomycete oncobasidium theobromae. *Phytopathology* **97** 1654–7

[4] Samuels G J, Ismaiel A, Rosmana A, Junaid M, Guest D, Mcmahon P, Keane P, Purwantara A, Lambert S, Rodriguez-Carres M and Cubeta M A 2012 Vascular Streak Dieback of cacao in Southeast Asia and Melanesia: In planta detection of the pathogen and a new taxonomy *Fungal Biol.* **116** 11–23

[5] Keane, J P 1992 *Diseases and Pests of cocoa : An overview FAO - Cocoa pest and disease management in southeast Asia and Australasia* ed J A . Keane, J. Philip ; Putter pp 1–11

[6] Prior C 1979 Resistance of cocoa to vascular-streak dieback disease *Ann. Appl. Biol.* **92** 369–76

[7] Varghese G, Zainal Abidin, M A, Lam H C 1992 Prospects for chemical control of vascular-

streak dieback of cocoa *FAO - Cocoa pest and disease management in southeast Asia and Australasia* pp 185–9
[8] Arnold A E, Mejıa L C, Kyllo D, Rojas E I, Maynard Z, Robbins N and Herre E A 2003 Fungal endophytes limit pathogen damage in a tropical tree *Proc Nat Acad Sci.* **100**
[9] Bailey B A, Bae H, Strem M D, Crozier J, Thomas S E, Samuels G J, Vinyard B T and Holmes K A 2008 Antibiosis, mycoparasitism, and colonization success for endophytic Trichoderma isolates with biological control potential in Theobroma cacao *Biol. Control* **46** 24–35
[10] Souza J T De, Bailey B A, Pomella A W V and Erbe E F 2008 Colonization of cacao seedlings by Trichoderma stromaticum , a mycoparasite of the witches ' broom pathogen , and its influence on plant growth and resistance *Biol. Control* **46** 36–45
[11] Hanada R E, Pomella A W V., Costa H S, Bezerra J L, Loguercio L L and Pereira J O 2010 Endophytic fungal diversity in *Theobroma cacao* (cacao) and *T. grandiflorum* (cupuaçu) trees and their potential for growth promotion and biocontrol of black-pod disease *Fungal Biol.* **114** 901–10
[12] Mejía L C, Rojas E I, Maynard Z, Bael S Van, Arnold A E, Hebbar P, Samuels G J, Robbins N and Herre E A 2008 Endophytic fungi as biocontrol agents of Theobroma cacao pathogens *Biol. Control* **46** 4–14
[13] Amin N, Salam M, Asman, Ryan, Darma M R 2015 Investigation of Endophytic Fungi Towards Vascular Streick Dieback Oncobasidium theobromae on Seedling of Cocoa Plant *J. Appl. Biol. Sci.* **9** 86–9
[14] Rosmana A, Samuels G J, Ismaiel A, Ibrahim E S, Chaverri P, Herawati Y and Asman A 2015 Trichoderma asperellum: A Dominant Endophyte Species in Cacao Grown in Sulawesi with Potential for Controlling Vascular Streak Dieback Disease *Trop. Plant Pathol.* **40**
[15] Rubini M R, Silva-ribeiro R T, Pomella A W V and Maki C S 2005 Diversity of endophytic fungal community of cacao (Theobroma cacao L .) and biological control of Crinipellis perniciosa , causal agent of Witches Broom Disease *Int. J. Biol. Sci.* 24–33
[16] Tondje P R, Roberts D P, Bon M C, Widmer T, Samuels G J, Ismaiel A, Begoude A D, Tchana T, Nyemb-Tshomb E, Ndoumbe-Nkeng M, Bateman R, Fontem D and Hebbar K P 2007 Isolation and identification of mycoparasitic isolates of Trichoderma asperellum with potential for suppression of black pod disease of cacao in Cameroon *Biol. Control* **43** 202–12
[17] Backman P A and Sikora R A 2008 Endophytes : An emerging tool for biological control *Biol. Control* **46** 1–3
[18] Carrol G 2013 Fungal endophytes in stems and leaves : From latent pathogen to mutualistic symbiont *Ecology* **69** 2–9
[19] Amin N, Salam M, Junaid M and Baco M S 2014 Original Research Article Isolation and identification of endophytic fungi from cocoa plant resistante VSD M . 05 and cocoa plant Susceptible VSD M . 01 in South Sulawesi , Indonesia *Int. J. Curr. Microbioliology Appl. Sci.* **3** 459–67
[20] Brecht M O, Stiles C M and Datnoff L E 2007 Evaluation of Pathogenicity of Bipolaris and Curvularia spp . on Dwarf and Ultradwarf Bermudagrasses in Florida Plant Health Progress Plant Health Progress *Plant Heal. Prog.*
[21] Girach R D, Aminuddin and Khan S A 1992 Ethnomedicinal uses of Achyranthes ospera L. in Orissa, India *Int. J. Pharmacogn.* **30** 113–5
[22] Kamaluddeen, Sobita, Simon and Abhilasha A L 2013 A new blight disease of rice caused by Curvularia lunata from Uttar Pradesh *Int. J. Agric. Sci. Res.* **3** 13–6
[23] Rosmana A, Iren, P, Papalangi, Sepsriyanti, Kannapadang, Rahim M D, Asman, Nasaruddin N 2014 Cultural and pathogenic characterization of Fusarium fungi isolated from dieback branches of cacao *Int. J. Curr. Res. Aca. Rev.* **2** 1–6
[24] Junianto J Y S-S 1992 Colletotrichum outbreak on cocoa in east Java *FAO - Cocoa pest and disease management in southeast Asia and Australasia* ed C A J Keane, J. Philip; Putter pp 157–62

[25] Peter P K and Chandramohanan R 2011 Occurence and distribution of cocoa (*Theobromae cocoa* L.) diseases in India *J. Res. ANGRAU* **39** 44–50
[26] Krauss U and Soberanis W 2002 Effect of fertilization and biocontrol application frequency on cocoa pod diseases *Biol. Control* **24** 82–9
[27] Waxman M F 1998 *The Agrochemical and Pesticides Safety Handbook* (Boca Raton: Lewis Publisher)
[28] Isaac S 1992 *Fungal-Plant Interactions* (London: Chapman and Hall)

IOP Conf. Series: Earth and Environmental Science **157** (2018) 012040 doi:10.1088/1755-1315/157/1/012040

Oceanographic conditions and sediment dynamic of the Barrang Caddi Island (Spermonde Archipelago, Indonesia)

M Lanuru[1], W Samad[1], K Amri[1] and D Priosambodo[2]

[1] Department of Marine Science, Faculty of Marine Science and Fisheries, Hasanuddin University, Jalan Perintis Kemerdekaan KM 10, Makassar, 90425, Indonesia.
[2]Department of Biology, Faculty of Mathematics and Natural Science, Hasanuddin University, jalan Perintis Kemerdekaan KM 10, Makassar, 90425, Indonesia.

E-mail: *mahat70@gmail.com*

Abstract. Small islands are vulnerable to long-term natural disasters like coastal erosion due to their size and topography. Barrang Caddi is one the small island in the Spermonde Archipelago (South Sulawesi) that encountered serious coastal erosion. Several attempts have been done by the relevant parties like by building a wave breaker to prevent erosion. But in fact some parts of the island are still eroded. A comprehensive oceanographic study of the wave climate and coastal processes at work to delineate the factors responsible for shoreline chance and to identify the location that need protection is needed. In this study, physical oceanographic data including waves, currents, tide, bathymetry, sediment characteristics and sediment transport were collected in the Barrang Caddi Island to analyze the factors responsible for shoreline chance (erosion) in the island. Results of the study showed that tide in the study site is mixed tide, predominantly semidiurnal with tidal range of 118 cm. Current measurements using a electromagnetic current meter revealed that current velocities at the study site were relatively low and vary spatially and temporally with magnitude of 0.02 – 0.58 m/s. Under normal conditions (no storms) the significant wave height (H 1/3) varied from 0.04 to 0.20 m. The wave height decreases from the fore reef to the reef flat due to the presence of coral reefs that reduce wave energy (wave height). Sediments were dominated by biogenic sand with grain diameter of 0.38 – 1.04 mm. Island erosion analysis showed that wave action was a main factor that responsible for shoreline chance (erosion) at the island. Current velocity alone with average of 0.19 m/s was not strong enough to move (erode) sediments at the island.

1. Introduction

Small island areas are vulnerable to long-term natural disasters like coastal erosion due to their size and topography. The erosion of the small island coast is a topic that is of concern throughout the world because it can cause huge losses of coastal land (residential), habitat, and the island. Coastal erosion is also threatening activities which can provide economic benefits such as aquaculture activities, tourism, mining, and transportation.

A wide variety of process and features can contribute to erosion in the small island coastal zone [1]. The obvious ones are storms and their associated waves, strong current, wind and gravity [2]. Commonly these factors act in combination. Important but small in scale is bioerosion, where organisms directly cause erosion. Some other factors in erosion are raising sea level. The rise of sea level makes it possible for processes such as waves and currents to continue attack sediments and rocks along the advancing shoreline. The presence of human modifications, such as structure, along the coast ia also may contribute to erosion in the small island. Structures of rock, steel, and concrete that are built along the coast can be a major factor in erosion, both directly and indirectly.

Barrang Caddi is one the small island in the Spermonde Archipelago (South Sulawesi, Indonesia) that encountered serious coastal erosion. Several attempts have been done by the relevant parties like by building a wave breaker to prevent erosion. But in fact some parts of the coast of the island are still eroded. A comprehensive oceanographic study of the wave climate and coastal processes at work to delineate the factors responsible for shoreline chance and to identify the location that need protection is needed. In this study, physical oceanographic data including waves, currents, tide, bathymetry, and bed sediment were collected in the Barrang Caddi Island to analyze the factors responsible for shoreline chance (erosion) in the island.

2. Material and Methods

2.1. Study Site

This study was conducted from July to October 2017 at Barrang Caddi Island. Measurements of coastal oceanographic parameters including waves, currents, tides, and bathymetry, and sediment characteristics were conducted to determine the main parameters or factors causing coastal morphological changes (erosion/accretion) at the study site.

2.2. Data Collection

Water level was measured every hour for seven days (25 July to 1 August 2017) using a tide staff at a pier of the east side of the Barrang Caddi Island to determine type of the tide and tidal range of the island. Water depth contour (bathymetry) of the island was produced from Indonesian Coastal Environment Map. Sediment characteristics of the island were determined by collecting bed sediment samples at the west, east, north, and south of the island using a sediment corer (PVC pipe with diameter of 10 cm and length of 30 cm). Sediment corers were pushed into the sediments to extract sediments that are approximately 15-20 cm in length. Only the top 3 cm of the sediments were used for grain size analysis in the laboratory. Sediment samples were dry sieved using standard laboratory test sieves of mesh sizes 0.063 mm, 0.125 mm, 0.250 mm, 0.500 mm, 1.0 mm, and 2.0 mm.

Current velocity measurements at three different depths, i.e. water surface, water column (middle), and near the bed were carried out for 3 days at the west, east, south and north side of the Island using an Electromagnetic Current Meter (Kenek LP 1100). At the east side, measurements were made for 13 hours at one hour intervals to cover one tidal cycle. While at the other three sides, measurements were made for 4 hours, *i.e.* two hours during shift period of flood tide to ebb tides, and two hours during shift period of ebb tides to flood tides.

Wave height and wave period measurements were conducted at four transects along west side of the island. At each transect, wave heights were measured at 3 points, i.e. fore reef, reef crest, and reef flat by observing wave crest and wave trough of the 51 incoming successive waves using a 2m length scaled pole. Wave period was measured using a stopwatch by counting the number of times the wave reaches its peak in a certain period of time for 51 incoming successive waves.

2.3. Coastal Erosion Analysis

Hjulstrom's Curve was used to analyze sediment movement (erosion) due to currents. The graph takes sediment particle size and water velocity into account (Figure 1). The upper curve shows the critical erosion velocity in cm/s as a function of particle size in mm, while the lower curve shows the deposition velocity as a function of particle size. To analyze sediment movement (erosion) due to waves, a formula of [3] was used to predict whether the island sediment is eroded or accretion based on wave height, wave period and sediment particle size at the study site:

$$\text{Erosion if: } \frac{H}{W_s\,T} \geq 3.2$$

Accretion if : $\dfrac{H}{W_s T} < 3.2$

Where, H is significant wave heights, Ws is sediment settling velocity, and T is wave period

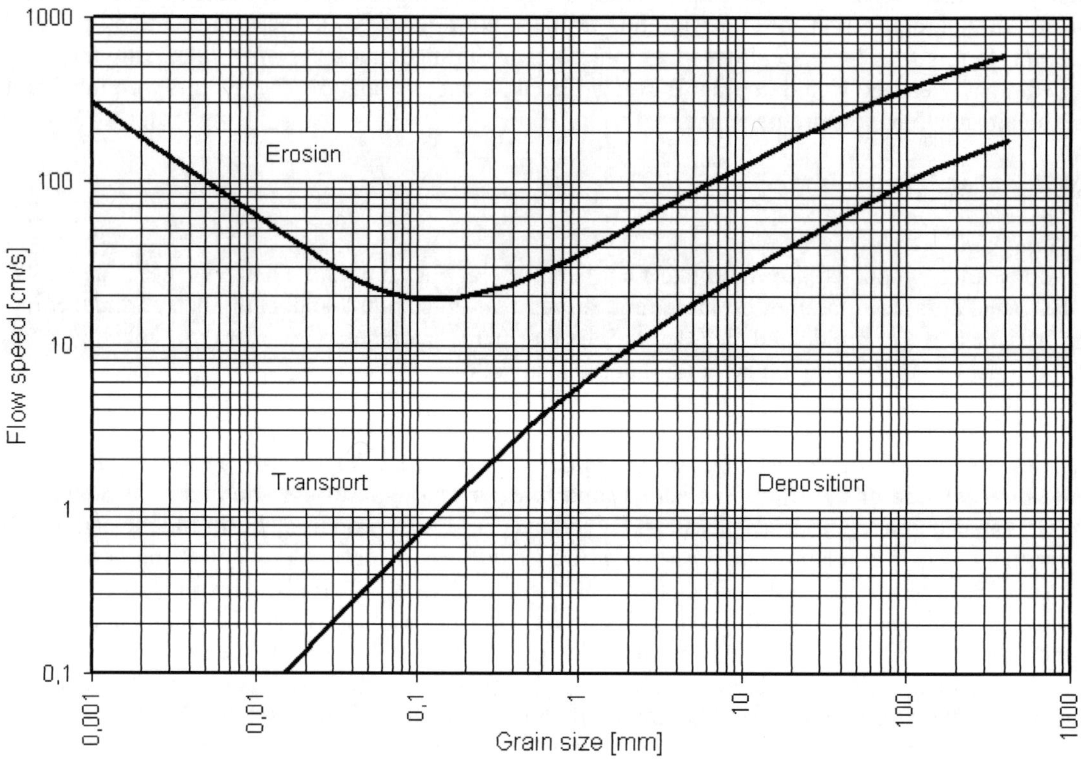

Figure 1. The Hjulström curve (https://en.wikipedia.org/wiki/Hjulström_curve)

3. Results and Discussion

3.1. Tide

Water level measurements indicated that tide at study site is Mixed Tide Prevailing Semidiurnal, i.e. two high tides and two low tides occur during a 24-hour period but exhibit two highs and two lows of significantly unequal height. Tidal range at the study site was also determined. Tidal range is the difference in water level between high tide and low tide. Tidal range at the study site was 118 cm and can be classified as microtidal coasts [4].

3.2. Bathymetry

The depth of Barrang Caddi Island water varied from less than 1 m to 34 m. At the west and south sides of the island, the coastal morphology was gentler with depth of the reef flats from less than 1m. While at the east and north sides of the island, the coastal morphology was steeper where the depth of the reef flats varied from 2 m to 6 m.

3.3. Sediment Characteristics

Sediments at the study site were dominated by biogenic sediment consisting of sand mixed with coral rubble, shells, fish bones and other marine organisms. Input of terrigeneous sediment from erosion of

IOP Conf. Series: Earth and Environmental Science **157** (2018) 012040 doi:10.1088/1755-1315/157/1/012040

rocks on land is relatively small due to the location of the study far from the mainland river mouth. Sand in the study sites is non-cohesive with very small mud content (<1%). The grain size of sediments at the study sites is presented in table 1. As shown in table 1, sediment particle size at north side of the island was slightly finer than those at the west, south, and east side of the island.

Table 1. The Grain size distribution of bed sediment in Barrang Caddi Island.

Location	Grain size (mm)	Type of sediment
East side	0.72	Coarse sand
North side	0.38	Medium sand
West side	1.04	Very coarse sand
South side	0.77	Coarse sand

3.4. Current Velocity

The result of current measurement using Drifter showed that the current velocity was relatively weak at the study site (0.04 - 0.21 m/s). Current measurements using a electromagnetic current meter (better accuracy) on four sides of the island at three different depths (surface, water column, near bed) indicated that current velocity varied spatially and temporally with a range and mean velocity of 0.02 - 0.58 m/s and 0.19 m/s, respectively.

Figure 2 shows the relationship between tide and tidal currents measured for 13 hours at east side of the island. Theoretically, magnitude of tidal current velocity will follow the tidal cycle, i.e. minimum or effective zero current velocity occurs during high or low water (slack waters) while maximum velocity occurs during phase between high water (flood tide) and low water (ebb tide) [5].

As shown in figure 2, the current velocity variation did not fully follow the rising and falling pattern of the water level (tide) indicating that variation of current velocity at the study site was not only caused by the tide. Other factors such as winds and waves may influence current velocity at the study site. Wind that blows over the water surface can generate surface currents that have a speed of about 3% of the wind speed at 10 m height for winds between 5 and 30 m/s [6]. This surface current velocity will decrease as the depth increases. In addition to wind, measured current velocity can also be influenced by the presence of longshore current generated by breaking waves after arriving at the beach [7].

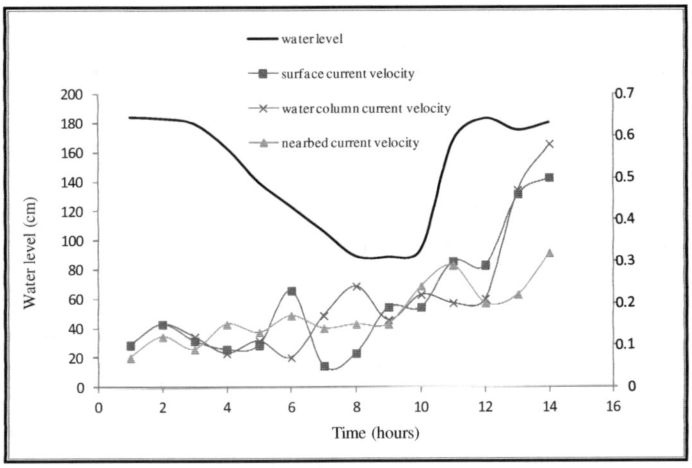

Figure 2. The relationship between the tide (water level) with the tidal current. The secondary Y-axis represents the current velocity in units of m/s.

Hjulstrom's Diagram can be used to analyze sediment movement (erosion) due to current. The size of sediment particles in the study sites was 0.38 - 1.04 mm. Based on Hjulstrom's Diagram, it takes a current strength of 0.25 to 0.37 m/s to erode the sediments of 0.38 to 1.04mm. The mean velocity of current at the study site was only 0.19 m/s which means that the current velocity alone was not strong enough to erode sediment at the study site.

3.5. Wave

Results of direct wave height measurements conducted at four transects along west side of the island is presented in Figure 3. Under normal conditions (no storms) the significant wave height ($H_{1/3}$) varied from 4 cm to 20 cm. The wave height decreased from the fore reef to the reef flat due to the presence of coral reefs that reduce wave energy (wave height).

Meta-analyses conducted by [8] revealed that coral reefs provide substantial protection against natural hazards by reducing wave energy by an average of 97%. Similarly, [9] reported that live corals together with seagrasses, and mangroves supply more protection services than any individual habitat or any combination of two habitats. Specifically, they found that mangroves are the most effective at protecting the coast under non-storm and storm conditions, live corals and seagrasses also moderate the impact of waves and storms, thereby further reducing the vulnerability of coastal regions.

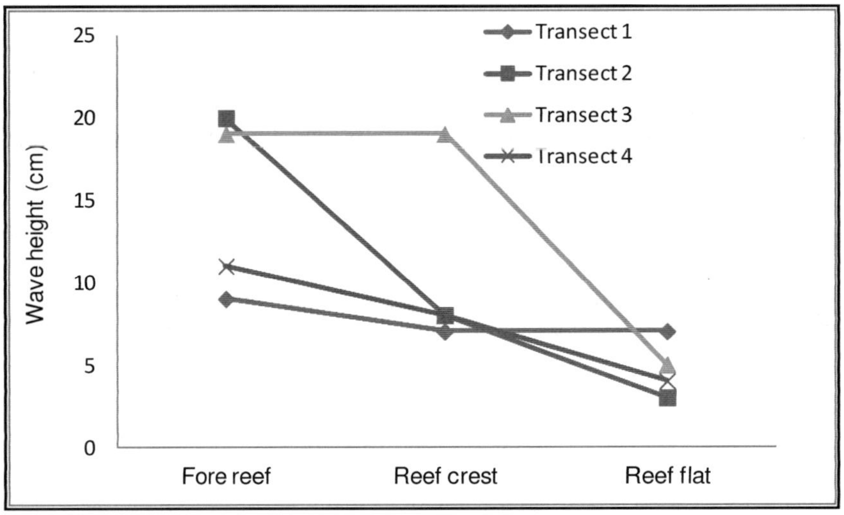

Figure 3. Wave heights at the fore reef, reef crest, and reef flats on four wave measurement transects

Wave conditions in a coastal waters determine the occurrence of erosion or accretion. The most important wave parameters in this case are the significant wave heights ($H_{1/3}$) and the wave period (T). Large waves with small wave periods that usually occur in extreme conditions or storms will cause coastal erosion. Conversely, small waves with larger periods that occur in normal conditions (calm) will cause the coast to experience accretion.

In addition to wave parameters, settling velocity of sediment particles also determines whether the coast is eroded or accreted. Settling velocity depends on the size of the sediment particles where the larger the particle the greater the settling velocity. In this study, [3] formula was used to predict whether the island sediment is eroded or accretion based on wave height, wave period and sediment particle size data. Large wave energy that has the potential to cause coastal erosion occurs when the wind blows from the west at the study site because the fetch length is greater and the wind frequency from the west is greater than the other direction. The result of erosion/accretion analysis showed that waves coming from west with wave height > 1.0 m can cause erosion in north side of the island (table 2).

Table 2. Result of erosion/accretion determination for wave coming from west direction.

Location	$H_{1/3}$ (m)	T (seconds)	Ws (m/s)	$H_{1/3}$ / (Ws x T)	Remarks
East side	0.4	4	0.125	0.8	Accretion
North side	1.0	4	0.05	5.0	Erosion
West side	1.2	4	0.17	1.8	Accretion
South side	1.1	4	0.13	2.1	Accretion

4. Conclusion

Tide in the study site is mixed tide, predominantly semidiurnal with tidal range of 118 cm. Sediments were dominated by biogenic sand with grain diameter of 0.38 – 1.04 mm. Current velocities at the study site were relatively low and vary spatially and temporally with magnitude of 0.02 – 0.58 m/s. Current velocity alone with average of 0.19 m/s was not strong enough to move (erode) sediments at the study site. Under normal conditions (no storms) the significant wave height ($H_{1/3}$) varied from 0.04 to 0.20 m. The wave height decreases from the fore reef to the reef flat due to the presence of coral reefs that reduce wave energy (wave height). Island erosion analysis showed that wave action was a main factor that responsible for shoreline change (erosion) at the island. Waves coming from west with wave height > 1.0 m can cause erosion in north side of the island

References

[1] Van Rijn L C 2011 Coastal erosion and control *Ocean & Coastal Management* **54** 12 867 – 887
[2] Davis R A and Fitzgerald 2004 *Beaches and Coast* (Malde, USA: Blackwell Publishing)
[3] Kraus N C, Larson M and Kreibel D L 1991 *Evaluation of Beach Erosion and Accretion Predictors Proc. Coastal Sediments '91*, ASCE 572-587
[4] Pinet P R 1999 *Oceanography, an introduction to planet oceanus*. (St. Paul. USA: West Publishing Company)
[5] Poerbandono and Djunasjah E 2005 *Survei hidrografi* (Bandung: PT. Refika Aditama)
[6] Weber J E 1983 Steady wind and wave-induced currents in the open ocean. *J. Phys. Oceanogr.* **13** 524–530.
[7] Tang J, Lyu Y and Shen Y 2014 Numerical study on water waves and wave-induced longshore currents in Obaköy coastal water *Acta Oceanol. Sin* **33** 40 – 46.
[8] Ferrario F, Beck M W, Micheli F, Shepard C C and Airoldi L 2014 The effectiveness of coral reefs for coastal hazard risk reduction and adaptation *Nat. Commun.* **5** 3794 doi: 10.1038/ncomms4794
[9] Guannel G, Arkema K, Ruggiero P and Verutes G 2016 The power of three: Cora reefs, seagrasses and mangroves protect coastal regions and increase their resilience *PLoS ONE* **11** 7 : e0158094. doi:10.1371/journal.pone.0158094.

IOP Conf. Series: Earth and Environmental Science **157** (2018) 012041 doi:10.1088/1755-1315/157/1/012041

Prospect of seaweed developement in South Sulawesi through a mapping study approach

S Yusuf[1], M Arsyad[2] and A Nuddin[1]

[1]Departement of Management and Accounting, Faculty of Economics, Muhammadiyah University of Parepare, Jl. Jend. Ahmad Yani Km. 6, Bukit Harapan, Soreang, 91131, South Sulawesi, Indonesia.
[2]Department Agricultural Socio Economic, Faculty of Agriculture, Hasanudddin University, Jl. Perintis Kemerdekaan KM 10, Makassar, 90245, Indonesia.

E-mail: *aryus_umpar@yahoo.com*

Abstract: Seaweed is one of fishery commodities which is extremely potential for foreign exchange income. Indonesia known as a nation which has many areas which are potential to develop seaweed, above all in Province South Sulawesi, encourages the researchers to know the prospect of seaweed production development. Therefore, this paper aims to present the potency of seaweed development in South Sulawesi in order to reduce proverty. The study was carried out in two regencies, namely Bantaeng and Luwu. The study identified the seawed development potency by using mapping study approach with analysis method. The data was obtained by using structured interview. The research result showed that both regencies developed some species of seaweed, they were *Eucheuma cottonii, Eucheuma spimosum* and *Gracilaria* sp. However, the farmers in both areas still used traditional ways. In Bantaeng, the farmer cultivated *Eucheuma cottonii* and *Eucheuma spimosum* utilizing long line method. Meanwhile, in Luwu, the farmer cultivated *Eucheuma cottonii* and *Gracilaria* sp with different method. *Eucheuma cottonii* was cultivated by using long line method, whilst *Gracilaria* sp was by bottom method. Generally, *Gracilaria* sp was cultivated in earthen dam. Consequently, the amount of seaweed production was still low. In addition, in this research founded that some of farmers applied an islamic-based system in cultivating seaweed but it still needs in-depth understanding.

1. Introduction

Indonesia is known as maritime nation with 13,667 islands and becomes major producer of marine cultured seafood products, specially seaweed [1]. Seaweed is one of fishery commodities which is extremely potential for foreign exchange income. In fact, it has been utilized as the main source income [2]. Seaweed is widely used as food, medicine, and important materials in the food industry, cosmetics, and pharmaceuticals [3]. Increasing consumption of seaweed as a food extract, medical use, and bio-refining is expected to drive the demand for seaweed, above all the species *Eucheuma cottonii* and

Glacilaria sp. Both of these seaweed species have a high economic value because it contains compounds used as raw materials for industries and food like agar and carrageenan [3]. Thus, both of these seaweed species are export thing which is used for meeting the global market share, particulary for some big countries, such China, Germany, America, Japan and other countries in Europe. The demand of seaweed for meeting the global market share has achieved 80.64 % and almost increase 15% per annum.

South Sulawesi is the largest producer of seaweed in Indonesia [4]. In South Sulawesi Province, seaweed has become the main means of livehood for fishers' communities. As we known that the production of seaweed in South Sulawesi is improved every year, particularly in two regencies, Luwu and Bantaeng. However, the important thing to be considered is that the development of seaweed farming is extremely affected by various factors, including the availability of socio-economic, resources, public policy, and technology [4]. Therefore, this research aims to investigate the potency and excellent of seaweed development in South Sulawesi in order to reduce proverty.

2. Research Method: Mapping Study Approach

This research was carried out in two districts in South Sulawesi Selatan, Bantaeng and Luwu, in 2016. Both of these districs are known as the largest seaweed producer in South Sulawesi. The researcher choose 115 respondents consisting of 15 respondents from bureaucracy, nine (9) collectors or traders, and 90 farmers. The study identified the seawed development potency by using mapping study approach with analysis method. The data was obtained by using *structured interview*.

3. Results and Discussion

3.1. The Potential of Seaweed Cultivation Development

Based on the research findings in two areas, founded that the seaweed farmers cultivated three kinds of seaweeds, namely *Eucheuma Cottonii, Euheuma Spinosum* and *Gracilaria sp*). In cultivating the seaweed, they used two different methods presented in table 1.

Table 1. The species of seaweed and their cultivation method

The cultivation area	Species	The method used
1. Bantaeng Regency	- *Eucheuma cottonii*	*Long line*
	- *Eucheuma spinosum*	*Long line*
2. Luwu Regency	- *Eucheuma cottonii*	*Long line*
	- *Gracilaria sp.*	*Bottom* or *basic method*

The research result showed that each area cultivated two species of seaweed using traditional ways. In Bantaeng Regency, the farmer cultivated two different species, namely *Eucheuma cottonii* and *Eucheuma spinosum* but both species were cultured by using same method, that was long line method. While, the species cultured in Luwu by farmer were two species, *Eucheuma cottonii* and *Gracilaria sp.* Both species were cultivated differently which *Eucheuma cottonii* was cultivated by using *long line method. and* was cultured using long line method, a seaweed cultivation method using raft (figure 1). Meanwhile, *Gracilaria sp* was cultured using bottom method (figure 2), a seaweed cultivation method by spreading the seaweed from the bottom [5]. Of both methods, the farmer of both regencies tended to cultivate seaweed by applying long line method. The farmers expressed that the usage of long line method was wear-resistant, practical and simple. The usage of long line method was tended to choice the seaweed farmers because simple and wear-resistant [6]. The difference of seaweed cultivation can be seen as follows:

IOP Conf. Series: Earth and Environmental Science **157** (2018) 012041 doi:10.1088/1755-1315/157/1/012041

Figure 1. Seaweed cultivation using long line method for *E. cottoni and E. Spinosum*

Figure 2. Seaweed cultivation using bottom method for *Glacilaria* sp.

In cultivating seawed, the researcher founded out that the farmers applied cultivation sytem which had similar procedures with Islamic economy system.

3.2. The Potential of Seaweed Production

Globally, the demand for seaweed production is increasing because of an increasing world population. The result of study indicated that seaweed farming has become the main income source for small-scale fishermen, particularly for both areas in South Sulawesi [2]. Therefore, the seaweed cultivation needs strategic management system to produce raw material with innovation and intervention [7]. Since 2010 utill 2014, seaweed production in Indonesia increase 32% per annum [1]. The research result indicated that seaweed production was potential to meet the sustainability of farmer's livelihood. The third of seaweed species developed have different characteristics, as shown in figure 3 below.

E cottonii *E spinnesum* *Glacilaria sp*

Figure 3. The kinds of seaweed cultivated in Bantaeng and Luwu

Although the farmers cultered the seaweed traditionally, the production of seaweed was very promising. The mapping study result of seaweed production for both areas showed different income. The production of *E. cottonii* was better than other species. The reason expressed by the farmers is suitable with Sukiman, et.al's finding that *E. cottonii* is one of seaweed species that can be harvested faster and has high economic value [3]. The production of developed seaweed can be seen in table 2.

Table 2. Mapping result of Seaweed Production shown by both areas

District	Type of Seaweed Species	Wide range (m^2/ha)	Production (kg)	Net Income (IDR)
Bantaeng	- *Eucheuma cottonii*	337,07 m	1.750	2.837.000
	- *Eucheuma spinosum*	600 m	2.300	1.900.000
Luwu	- *Eucheuma cottonii*	350,00 m	1.661	780.700
	- *Gracilaria sp.*	1.24 ha	1.030	1.986.400

Table 2 presents the production of seaweed species cultured in Bantaeng and Luwu District. Clearly, we can see that there is difference regarding net income for both areas. In Bantaeng, *E.cottonii* has better net income (IDR. 2, 837, 000.- with wide range of 337,07 m^2) than *E. spinosum* (IDR. 1,900,000.- with wide range of 600 m^2. While, in Luwu *Gracilaria sp.* has better net income (IDR. 1,986,400.- with wide range 1.24 ha) than *E. cottonii* (IDR. 780,700.- with wide range of 350.00 m).

4. Conclusion

Seaweed is one of the promising fishery communities that have high economic value. Both Districs in South Sulawesi, Bantaeng and Luwu cultivated three kinds of promising seaweed species namely *Eucheuma cottonii, Eucheuma sp.* and *Gracillaria sp.* However, the farmers still used traditional ways, namely long line and bottom (basic) *method* so that the seaweed cultivation needs strategic management system to produce raw material with innovation and intervention in order to improve the production volume.

Acknowledgment

The authors gratefully acknowledge the Department of Accounting, Faculty of Economy, Muhammadiyah University of Parepare. We would also thank the enemurators who have helped this research.

References

[1] Rimmer M A 2010. Mariculture development in indonesia; prosepect and constraints *Indonesian Agr. J.* **5** 2.

[2] Da Costa D K, Mohammad M and Bambang S 2015 Enviromental and area support capability analysis for seaweed mariculture development in hading bay of East Florest regency *Int. Journal of Sci. and Tech. Res.* **4** (11) 143-147

[3] Sukiman, Faturrahman, Rohyani I S and Ahyadi H 2014 Short communication: Growth of seaweed eucheuma cottonii in multitropic sea farming systems at gerupuk bay, central Lombok, Indonesia *Nusantara Bioscience* **6** (1) 82-85

[4] Zamroni A, Laoubi K, and Yamao M 2011 The development of seaweed farming as a sustainable coastal management method in Indonesia: An opportunities and constraints assessment *WIT Transaction on Ecology and the Environment* **150** 506-516

[5] Tatang 2011 Seaweed cultivation gracilaria verrucosa. https://suksesmina.wordpress.com/2011/05/03/budidaya-rumput-laut-gracilaria-verrucosa/. Accessed on 01 August 2016

[6] Al'Amin 2011 *Strategi Pengembangan Berkelanjutan Budidaya Rumput Laut Eucheuma cottonii di Kabupaten Barru* (Makassar: Prodi Ilmu Perikanan, Program Pascasarjana Universitas Hasanuddin)

[7] Saldyansah E, Hasanuddin A, Tantu F Y and Laapo A 2016 Determining potential volume of raw dried seaweed *Kappaphycus alvarezii* based on land suitability as agribusiness management plan *J. of Econ. and Sust. Dev.* **7** (20) 58-63

IOP Conf. Series: Earth and Environmental Science **157** (2018) 012042 doi:10.1088/1755-1315/157/1/012042

Microsatellite and SNAP markers used for evaluating pollen dispersal on Pati tall coconuts and Xenia effect on the production of 'Kopyor' fruits

S H Larekeng[1,2], A Purwito[2], N A Mattjik[2] and S Sudarsono[2]

[1]Biotechnology and Tree Breeding Lab., Faculty of Forestry, Hasanuddin University, Jalan Perintis Kemerdekaan KM 10, Makassar, 90245, Indonesia.
[2]Plant Molecular Biology Lab., Department of Agronomy, Faculty of Agriculture, Bogor Agriculture University, Jalan Meranti, Babakan, Dramaga, Bogor, 16680, Indonesia.

E-mail: *sitih5h.82@gmail.com, sittihalimah@unhas.ac.id*

Abstract Kopyor coconut is one of the many unique coconut types existed in Indonesia. To overcome the problem of low kopyor fruit yield, it is necessary to study xenia effect on the fruit yield of this coconut. The combination of kopyor coconut and normal coconut population selected at infarmers' coconut plantations consisted of 33 normal coconut trees, and 9 kopyor coconut trees. All adult trees surrounding the 9 kopyor heterozygous (Kk) palms were evaluated as potential male candidate parents (pollen donors). All samples genotypes were determined using four SNAP markers and six microsatellite marker loci, parentage analysis using CERVUS software version 2.0. Results of the analysis indicate that xenia effect reduced kopyor fruit yields. Kopyor heterozygous (Kk) female parents produced low number of kopyor fruits when they were surrounded by many normal homozygous (KK) pollen donors. Out of 99 harvested progeny arrays from the kopyor heterozygous (Kk) female parents, none exhibited kopyor phenotype. The results also indicate that the pollen dispersal from normal homozygous (KK) donor palms range from 0 m (self pollination) to 54 m (outcrossing). The occurence of outcrossing frequency was at least 95% and the selfing frequency is 5%.

1. Introduction

Kopyor coconut is one of the many exotic coconut types existing in Indonesia [1, 2]. The fruit was presumed from coconut plantation which is genetic mutation naturally. As a result of natural mutation, the number of kopyor coconut trees were very slight compared to the normal coconut. Accordingly, this mutant coconut is not develop, but it turns out these coconuts were found in several centers of coconut production in Indonesia. Kopyor properties were controlled by a mutant gene at the locus of K. The fruit couldn't germinate naturally while homozygous (KK) or heterozygous (Kk) normal fruit could germinate and produce coconut seedlings. One of the problems farmers faced is the production of kopyor fruit has not been optimum. Kopyor coconut plant in Pati could produce as much amount of kopyor fruit type about 1-4 grains per cluster of the total number of of 7-15 pieces of grains [1]. Theoretically, based on the pattern of segregation Mendel's Law, then the chances kopyor fruit formed should reach 25% of the amount of fruit harvested.

One of efforts to overcome the problem of low kopyor fruit yield is to study the xenia effect on the kopyor fruit yield. Xenia is a genetic phenomenon in the form of a direct effect of pollen to the phenotype of fruits yielded by the female parents. In the study of heredity, the expression of genes who carried from the male parent and a female parent are expressed in subsequent generations. With the xenia effect, gene expression brought from the male parent can be expressed in female parents (fruit) [3]. The phenomenon of xenia on coconut trees has been known to have influence on fruit size, weight copra and obstinacy of hybrid trees [4]. According to [5], the xenia effect in cross-pollinated plants not only influences the fruit set, but also determines the shape, maturity, size, color, flavor, and composition of fruits, and research of [6] to further understand xenia phenomenon, was the investigation of the xenia effect in *Castaneahenryi* using the cultivars of 'Huali 1', 'Huali 2', 'Huali 3', and 'Huangzhen' as materials.

Xenia related to phenotypes are controlled by a recessive gene. The formation of the kopyor coconut depends on the genotype of pollen to pollinate the female flowers of kopyor coconut plantations, so it is an example of xenia. This study to evaluate the effect of xenia on the kopyor coconut in Pati and determine the size range of self-pollination and cross-pollination that occurs in a population Coconut Pati using microsatellite and SNAP marker.

2. Materials And Methods

Samples were collected in kopyor coconut plantation in Dukuhseti Village, Pati Regency, Central Java with GPS location S6 27.649 E111 02.747. The research activities were conducted at the Laboratory of Plant Molecular Biology (PMB Lab) Department of Agronomy and Horticulture Faculty of Agriculture, Bogor. The population is a combination of kopyor heterozygous (Kk) coconut trees with normal coconut tree (KK) or coconut trees that never produce kopyor fruit with a total of 42 mature trees. Coconut trees in the planting area consists of coconut trees as much as 33 normal fruiting trees and 9 kopyor fruiting trees. Selection of the trees used as the female parent was done using purposive random sampling method. The female parents used were 7 palm kopyor trees. From each female parent trees, three palm fruit bunches (progeny) were harvested with a range the amount of progeny per female parent ranged from 2 to 12 pieces.

Total plant DNA sample was isolated with CTAB method [7] with modification [8], SSR primer pairs were selected from 36 SSR loci [9] and the most polymorphic six primer pairs [10] were used for genotyping parents and progeny were evaluated. Four loci markers were developed based on the SNAP gene nucleotide diversity SUS and WRKY also used genotyping parent and progeny. The development of gene based markers SNAP SUS and WRKY performed following phases of activities that have been conducted on various plants [11, 12]. Gene SUS and WRKY can be used as polymorphic markers for coconut [13, 14].

PCR amplification was conducted using KAPA Bio-PCR kit protocol. Visualization of PCR amplification products for each locus SSR markers was carried out by 6% polyacrylamide gel electrophoresis using Buffer SB [15] and staining gel with silver nitrate. Phases of staining gel with silver nitrate was carried out based on the modified method of [16, 17]. Visualization of PCR amplification products for each locus markers SNAP was carried out using agarose gel electrophoresis (1%) with TBE buffer solution. Visualization of DNA was conducted using red staining gel.

3. Data Analysis

3.1. Identification of the Candidate Males Parents

Identification of pollen donor parent conducted by analyzing the genotype of each progeny and comparing it with all of the mature plant genotypes that were evaluated, potentially as pollen donors. Genotype data of all progeny and mature plants were used to determine the male parents candidates sought. Determination of parents male candidates elected for each progeny was carried out with paternity analysis with software of Cervus version 2.0 [18].

3.2. Pattern Analysis of Pollen Dispersal

Known female parent and assigned males parents identified as pollen donor from any progeny were mapped for the position based on mature plant dispersal maps that have been done previously. The distance between each parent and the pollen donor parent who identified calculated using Map Source software. Self-pollination (selfing) was defined when the male parents who identified closely with their parent. Otherwise, outcrossing is occurred when a female parent and males parents shortly become donors pollen consists of two distinct adult plants. The frequency of occurrence of selfing, outcrossing for various categories of identity crosses are determined by the female parent and the male parents of the results of paternity analysis.

4. Result and Discussions

PIC value of the loci were using SSR markers indicate a higher value than the loci with SNP markers (table 1). Degree of variation decreased with the level of polymorphism. This was due to the amount of loci on SSR markers that being used more than the amount of loci, on SNP markers. PIC value is a measure of polymorphism between genotypes in a locus which used information on the number of alleles [19]. Total population of homozygous loci using SSR markers was relatively lower compared to markers SNP loci. Similarly, He value of the locus using SSR markers was also higher than that of SNP markers. Heterozygous alleles often found in SSR markers cause the value (He) is higher than the marker SNP. From the Cervus analysis, percentage of normal donor and Kopyor donor, which pollinate the Kopyor female tree could also be obtained which is shown in table 2.

Table 1. Number of alleles, observed (O) and expected (E) heterozygosity and polymorphic information content (PIC) at 10 loci molecular markers, the analysis results were in a population of coconut Cervus in Pati.

Name of loci	Number of alleles	Total		Heterozygosity		PIC
		Het	Hom	O	E	
CnCir_87	2	64	80	0.444	0.496	0.372
CnCir_86	4	89	54	0.622	0.687	0.635
CnZ-18	4	116	28	0.806	0.741	0.691
CnZ_51	5	89	55	0.618	0.707	0.66
CnCir_B12	5	107	37	0.743	0.745	0.699
CnCir_56	5	81	63	0.563	0.695	0.647
CnSus1#14	2	39	100	0.281	0.252	0.22
CnSus1#3	2	96	43	0.691	0.483	0.365
WRKY19#1	2	116	24	0.829	0.5	0.374
WRKY 6#3	2	123	21	0.854	0.495	0.371

Note: PIC= *Polymorphic Information Content.* Ho= *Observed heterozigosity.* He= *Expected*

Table 2. Scheme of cross pollination and pollination type were identified based on the analysis of the dispersal pollen from the progeny test populations in coconut Dukuhseti, Pati

Scheme of pollination	Type of pollination	Number of occurrences	Percentage
Kopyor x Kopyor	Self	3	3%
Kopyor x Kopyor	Outcross	25	25%
Kopyor x Normal	Outcross	74	73%
Total of progeny		102	100.00%

This study also shows that none of kopyor fruit could be harvested by farmers. It indicates that heterozygous (Kk) kopyor coconuts were pollinated by normal (KK) coconuts, thus no He of kopyor coconut was observed in this study and the trees only produced normal fruits having homozygous KK (50%) and heterozygous Kk (50%).

The existence of normal tree (KK) around the heterozygous (Kk) kopyor tree could directly influence the phenotype of the fruits harvested thus categorized as a xenia effect. Xenia effect could also be studied by pollen analysis was identified by the best parents pollen donors to increased fruit production. This was also done by [20] with an indication of xenia effect found on cucumber fruits and seeds.

Kopyor trait was controlled by a single locus with the recessive allele (k). Meanwhile, normal trait was controlled by allele (K). Coconut trees kopyor of the study sites have the heterozygous genotype (Kk), if it was pollinated by trees kopyor heterozygous (Kk) others, which would produce the expected value of 25% of fruits kopyor (genotype zygotic embryos (kk)) and 75% of normal fruits [21]. Figure 1 shows the class with the shortest distance pollen distance 0 m (selfing) and furthest <60 m, is 54 m. The highest frequencies were at a distance <20 m by 30 times pollination occurred. In *Diosphyros celebica* Bakh indicated that seeds were produced predominatly by outcrossing and pollen dispersal reached up to 166 m [22].

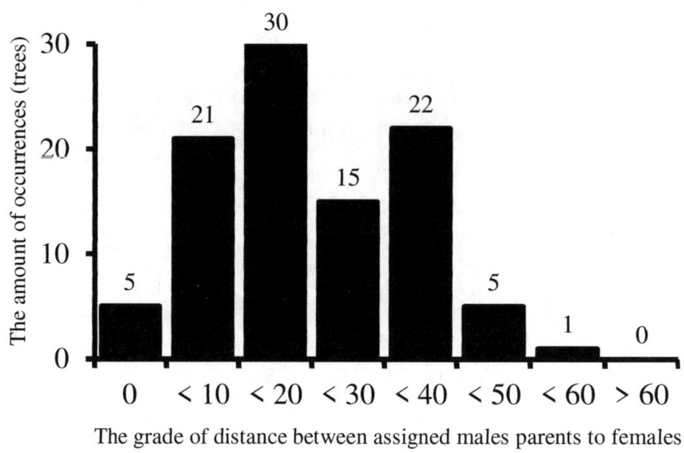

Figure 1. Numbers of pollination events for each distance class from the assigned male to the female kopyor coconut population in the Dukuhseti, Pati, Central Java.

Effect of the position and distances of the assigned male to the female parent recipient on the occurrence of kopyor fruits is presented in figure 1. Xenia negative influence was seen in the productivity of kopyor fruit where the percentage of all the fruits harvested became normal fruit. It is caused by normal palm tree population (KK) is greater in those populations. In mixed cropping conditions with a normal ratio of coconut trees only produce fruit kopyor average as much as 15.8% [23]. So in the research recommendations of [21], in a project Hi Links that most logging in normal coconut (KK) in between the coconuts of Pati's Genjah Kopyor (Kk) caused an increase in the percentage of kopyor harvested up to 10%. Increasing the percentage of kopyor harvested allegedly associated with the decreasing influence of xenia in cropping due to the number of normal trees (KK), which lesser at the sites.

In a population of coconut plantations, which tend also seen 7 times occur crosses (self pollination) based on the results of the analysis of Cervus. This possibility could occur because coconut having open morphological flower type. Xenia is a symptom of a genetic form of direct effect on the phenotype of pollen seeds and fruits was generated by the female parent. In the study of heredity, expression of genes that carried the male parent and a female parent was assumed, then expressed on the next generation.

With the xenia, the expression of genes that carried the male parent early has been expressed in female parents organs (fruits) embryo or endosperm. Symptoms xenia didn't only affect the color but also the form of sugar content, oil content, form and timing of ripening fruit. Research [24] about Xenia in kopyor coconut explain if homozygous KK normal is replaced with homozygous kk Kopyor coconuts in the field, negative xenia effects on kopyor fruit yield could probably be reversed. Mix coconut plantation consisting of heterozygous Kk and homozygous kk kopyor coconut provenances results in increased kopyor fruit yield because the homozygous kk kopyor coconut could become pollen contributors donating the k allele.

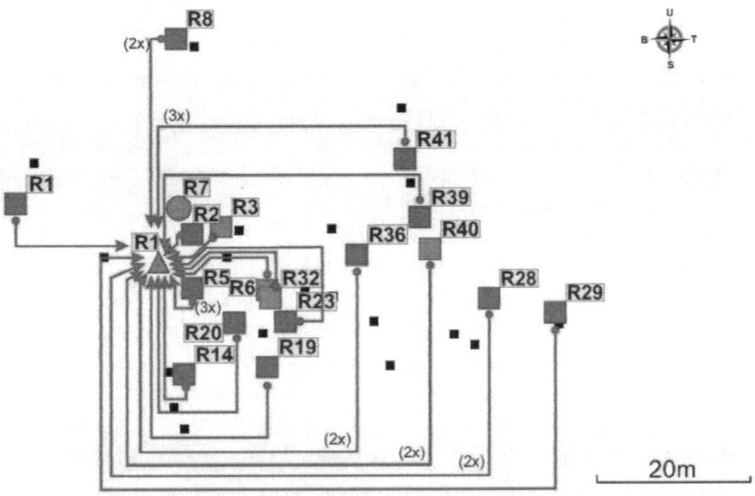

Figure 2. Dispersal pattern of kopyor coconut pollen Dukuhseti with female parent number R1 from parentage analysis. The symbol identified the position (▲) the female parent in kopyor, (■) assigned male normal (pollen donor), (■) the assigned male kopyor and (●) the asisgned male in genjah kopyor.

Xenia was not deviation from Mendelian inheritance, but a direct consequence of the multiple fertilization (double fertilization) occurring at flowering, plant embryo development process until the maturity of the seeds. At this stage of embryonic development, a number of genes in the embryo and endosperm expression affect the appearance of seeds, grains or fruits [25]. Unlike the xenia effect, metaxenia is inexplicable with the elements of heredity (chromosomes) carried in the pollen because unlike the chromosomes found in tissues that showed the direct influence of the pollen parent [26]. Metaxenia explained the effect of pollen on the network such as the pericarp fruit of maternal origin of, and other fruit components was not affected by pollen. On the other hand, xenia explained the effect of pollen in the tissues that contains at least one unit genes from the male parents of the embryo and endosperm [27].

5. Conclusion

Coconut plantation in Dukuhseti, Pati, tended to have negative xenia effect with the presence of normal coconuts surrounded by more kopyor fruits.. The distance of dispersal pollen range from 0 m (self pollination) to 54 m (the furthest). The frequency of distance the dispersal pollen was highest at a distance <20 m with a total of 30 times pollination. The occurence of outcrossing frequency was at least 95% and the selfing frequency is 5%.

References

[1] Maskromo I, Novarianto H and Mashud N 2007 Potensi pengembangan kelapa kopyor di Indonesia *Warta Penelitian dan Pengembangan Tanaman Industri* **13** 1

[2] Sudarsono, Novarianto H, Hossang M, Dinarti D, Yuliasti, Sukma D, and Maskromo I 2013 *Produksi bibit kelapa kopyor true to type dengan persilangan terkontrol dan peningkatan produksi buah kelapa kopyor dengan polinator lebah madu. Plant Molecular Biology Laboratory Annual Report* (Bogor: Institut Pertanian Bogor) p 34

[3] Bulant C, Gallais A, Matthys-Rochon E and Prioul J L 2000 Xenia effects in maize with normal endosperm *J. Crop Sci.* **40** 182

[4] Satyabalan K 1995 Xenia effect and hybrid vigor in coconuts *J. Coconut R&D.* **11** 41

[5] Alexander C M, Stuart A C and Maria A B 2012 Effects of natural and artificial pollination on fruit and offspring quality *Basic Appl. Ecol.* **13** 524

[6] Xuhui Z, Yuan D, Zou F, Fan X, Tang J, and Zhu Z 2016 A Study on the Xenia Effect in *Castaneahenryi Hort. Plant J.* **2** 301

[7] Rohde W, Kullaya A, Rodriguez J and Ritter E 1995 Genome analysis of Cocos nucifera L by PCR amplification of spacer sequences separating a subset of copis-like EcoR1 repetitive elements *J. Genet and Breed.* **49** 170

[8] Larekeng, SH, Maskromo I, Purwito A, Matjik NA and Sudarsono S 2015 Pollen dispersal and pollination pattern studies in pati kopyor coconut using molecular markers *CORD.* **31** 46

[9] Lebrun P, Baudouin L, Bopurdeix T, Konan JL, Barker JHA, Aldam C, Herran A and Ritter E 2001 Construction of linkage map of the Rennell Island Tall coconut type *(Cocos nucifera L.)* and QTL analysis for yield characters *Genome* **44** 962

[10] Larekeng, SH, Maskromo I, Purwito A, Matjik NA and Sudarsono S 2015 *Penyebaran polen berdasarkan analisis ssr membuktikan penyerbukan kelapa dalam kalianda normal ke kopyor Bul. Palma.* **16** 77

[11] Esteras C, Pedro G, Antonio JM, José B, Nelly V D, Cristina R, Fernando N and Belén P 2012 High-throughput SNP genotyping in *Cucurbita pepo* for map construction and quantitative trait loci mapping *Bio. Med. Central.* **13** 1471

[12] Sutanto A, Sukma D, Hermanto C, and Sudarsono 2014 Isolation and characterization of Resistance Gene Analogue (RGA) from Fusarium resistant banana cultivars *Emirates J. of Food and Agri.* **26** 508

[13] Herrera M M, Alan W M, James W B, David N K and Raymond J S 2007 Usefulness of WRKY gene-derived markers for assessing genetic population structure: An example with Florida coconut cultivars *Sci Hort* **15** 19

[14] Sukendah, Volkaert H and Sudarsono 2009 Isolation and analysis of DNA fragment of genes related to 'Kopyor' trait in coconut plant *Indon. J. Biotechnol.* **14** 1169

[15] Brody J R and Kern S E 2004 History and principles of conductive media for standard DNA electrophoresis *Anal. Biochem.* **333** 1

[16] Creste S, Tulmann A N and Figueira A 2001 Detection of single sequence repeat polymorphism in denaturing polyacrylamide sequencing gels by silver staining *Plant Mol. Biol Report* **19** 299

[17] Tinche, Asmono D, Dinarti D and Sudarsono 2014 Keragaman genetik kelapa sawit *(Elaeis guineensis* Jacq.*)* Populasi Nigeria Berdasarkan AnalisisMarka SSR *Bul. Palma.* **15** 14

[18] Marshal T C, Slate J, Krilek L E B and Pemberton J M 1998 Statistical confidence for likehood based paternity inference in nature populations *Mol. Ecol.* **7** 639

[19] Sajib A M, Hossain M M, Mosnaz, Hossain, Islam, Ali and Prodhan 2012 SSR marker-based molecular characterization and genetic diversity analysis of aromatic landreces of rice *(Oryza sativa L.). J. Bio. Sci. Biotech* **1** 107

[20] Olfati J, Sheykhtaher Z, Qamgosar R, Sabet A, Peyvast G, Samizadeh H, and Rabiee B 2010 Xenia and metaxenia on cucumber fruit and seed characteristics *Int. J. Veg. Sci.* **16** 243

[21] Sudarsono, Sudrajad, Novarianto, Hosang M, Dinarti D, Rahayu and Maskromo I 2014 *Pendekatan Genomik dan Molekuler untuk Pengembangan Kultivar Unggul Kelapa Eksotik*

Asal Indonesia, Penyediaan Bibit dan Pengendalian Hamanya Final Report of KKP3N (Bogor: IPB bekerjasama dengan Badan Penelitian dan Pengembangan Pertanian) p 46

[22] Restu Muh, Gusmiaty G and Siti Halimah Larekeng 2017 High Outcrossing Rate And Pollen Dispersal Distance Of *Diospyros Celebica* Bakh. (Ebenaceae), An Endemic Tree Species In Sulawesi Island, Indonesia *24 3* p 173

[23] Novarianto H and Lolong A A 2012 Peningkatan persentase buah kopyor melalui penyerbukan sendiri *Bul Palma.* **13** 7

[24] Maskromo I, Larekeng S H, Novarianto H and Sudarsono S 2016 Xenia negatively affecting kopyor nut yield in Kalianda Tall kopyor and Pati dwarf kopyor coconuts *Emirates J. of Food and Agri.* **28** 644

[25] Denney J O 1992 Xenia includes metaxenia *Hort. Sci.* **27** 722

[26] Bodor P, Gaal M and Toth M 2008 Metaxenia in apples cv. Rewena, Relinda, Baujade as influenced by scab resistant pollinizers *Int.J. Hort. Sci.* **14** 11

[27] Ehlenfeldt MK2003 Investigations of metaxenia in northern highbush blueberry *(Vaccinium corymbosum L.)* cultivars *J. of the America Pomological Society* **57**1 56

IOP Conf. Series: Earth and Environmental Science **157** (2018) 012043 doi:10.1088/1755-1315/157/1/012043

In vitro sterilization technique on embryo of black Toraja rice

F Haring[1], M Riadi[1], Rafiuddin[1], R Sjahril[1] and A R Muchlis[2]

[1]Faculty of Agricultural, Hasanuddin University, Jalan Perintis Kemerdekaan KM 10, Makassar, 90245, Indonesia.
[2]Master Student, Agro-technology, Faculty of Agricultural, Hasanuddin University, Jalan Perintis Kemerdekaan KM 10, Makassar, 90245, Indonesia.

E-mail: *feranita_haring@yahoo.co.id*

Abstract. Toraja black rice has a high anthocyanin content, a water-soluble pigments, with antioxidant activity. Toraja black rice has a variety of seeds colour in one panicles such as full black (the outside and inside the rice), medium black (the outside and slightly inside rice) and a little black (only the outside of rice). Embryo culture in vitro is one way to grow plants in sterile conditions. The presence of contamination and the death of the embryo require *in vitro* embryo culture. The sterilization technique is a very important first step to eliminate contamination and the death of embryos. This research aims to determine the right material composition for sterilization of black rice's embryo. The experiment was done by growing black rice on half strength MS media with the treatment of three method of sterilization, *i.e.*: S1 (70% alcohol for 5 minutes, 3% and 2% Chlorox each for 10 minutes,), S2 (70% alcohol for 3 minutes, 2% Clorox for 10 minutes) and S3 (70% alcohol for 3 minutes and 1% Clorox for 15 minutes). The materials used are rice seedlings that have been cut in two and opened the pericarp of paddy grain, leaving a piece of rice that has a complete embryo. The best sterilization for Toraja black rice embryo culture was using the S3 composition. Best germination was seen on the seeds with full and medium black color.

1. Introduction

Black rice from Toraja District / North Toraja region, has a unique phenotype character. A dark purple color that makes this rice looks black at a glance, caused by high anthocyanin content. Anthocyanins are water-soluble pigments, which have antioxidant activity. Black rice is known to increase the body's resistance to disease, repair damaged liver cells, prevent renal function, prevent cancer / tumor, slow aging, as an antioxidant, clean cholesterol in the blood, and prevent anemia [1, 2].

Black rice in Indonesia is predominantly derived from local rice from various regions. One of the areas in Indonesia that has a variety of local rice is South Sulawesi. Local rice is a major source of genetic diversity in rice breeding efforts, especially black rice. One of the local rice varieties used for generations as part of the tradition, ritual and culture of the community is black rice (Pare Ambo) from Toraja District / North Toraja District.

The productivity of local paddy especially Toraja black rice is still low. BPS data [3] noted that local rice productivity in the country of 3.99 tons ha^{-1} is much lower than that of paddy field productivity of 5.26 tons ha^{-1}. Other shortcomings due to black Toraja rice have a deep age, ranging from 5-6 months [4]. This cause the black rice is less developed in a broad scale.

In addition, Toraja black rice has a very low level of purity, to be used as a source of genetic diversity. Preliminary studies have been done, indicating that at one panicle there were different colors of seeds.

Some black seeds are black in whole (the outside and in the rice), others are medium black (the outside and slightly inside the rice) and are black in color (only the outside of the rice). In vitro culture has been widely used for the development of plant food cultivation. In plant breeding, embryo rescue culture applications are expected to maintain their integrity and grow into new plants [5]. One of the most widely used in vitro culture techniques is embryo culture, because embryo culture is one of the earliest studies to obtain plants that are free from contamination of microorganisms.

Through embryo culture, early embryo development can be studied. In the field of plant breeding, embryo culture can speed up the hybridization cycle [6]. The choice of embryos as explants due to the availability of seeds, has a high physiological uniformity and can be carried over time and at considerable distance [7].

One obstacle in the process of embryo culture in vitro is contamination. The sterilization technique is a very important first step to eliminate contamination and the death of embryos in *in vitro* cultures. In addition it is expected that sterilization does not change the color of black rice rice.

2. Materials and methods

2.1. Place and Time
The research was conducted at the Laboratory of Plant Biosciences and Reproduction Biotechnology, Department of Agronomy, Faculty of Agriculture, Hasanuddin Universit, Makassar. The study took place from April to November 2017.

2.2. Materials
Black rice seed of Pare Ambo variety was obtained from farmers survey in Ao 'Gading Toraja Utara, inorganic and organic material components for the manufacture of Murashige and Skoog (MS) media, clorox and alcohol.

2.3. Implementation of Research
There were two goals in this experiment. First experiment was carried out by testing kinds of sterilization technique for Toraja's black seed and the second experiment was to see germination performance of black seeds of Toraja.

2.3.1. Experiment 1. Preparation of materials. The seeds used were peeled Toraja black rice seeds cut crosswise into two parts to see the color within the endosperm. The part used was the embryo half pieces. Peeling and cutting the seed were conducted to select and classify seeds based on the same color criteria. The pieces are grouped into three sections of the rice color level: the full black (HP) (black on the surface and inside of the rice), the medium black (HM) (black on the surface and slightly inside the rice) and the black one (HS) (black only the surface of the rice) (figure 1). The seeds of embryo half pieces are grown in vitro on ½ MS germination medium after a sterilization process.

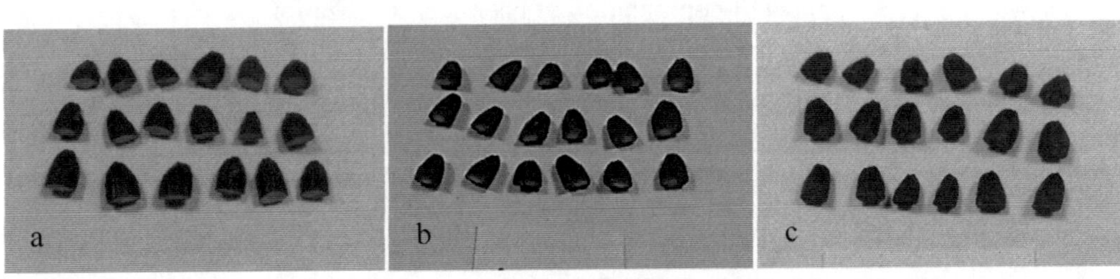

Figure 1. Cross-section of black rice seeds of Toraja (a) Black rice seeds with black color (HS), (b) Black rice seeds with medium black color (HM), and (c) Black rice seeds with full black color HM).

Sterilization was performed with three sterilization sequence treatments based on the materials used. Treatment of S1 was done by sterilizing rice explants consecutively using 70% alcohol for 5 minutes, rinsed with sterile distilled water 3 times, then soak in 3% Chlorox solution for 10 minutes, rinsed again with sterile distilled water for 3 times, continued with immersion in 2% Chlorox for 5 minutes and finally rinsed with again with sterile distilled water 3 times. Sterilization S2 treatment with 70% alcohol for 1 minute, rinsed with sterile distilled water 3 times, followed by 2% Chlorox solution for 10 minutes, and finally rinsed with sterile aquades 3 times. S3 sterilization treatment using 70% alcohol for 3 minutes, then rinsed with sterile distilled water 3 times, followed by 1% clorox for 15 minutes, and finally rinsed with sterile distilled water 3 times. Quantitative and qualitative observations were made on the color of cut seeds with embryo part after sterilization process.

2.3.2. Experiment 2. Preparation of materials. The second experiment was based on the first experiment. Seed materials used were black rice seeds of Toraja black rice which have the embryo part and are full black rice (surface and inside of part rice), medium black (surface and slightly inside of part rice) and thin black (only the surface of rice). Half-cut seeds were further planted in vitro on germination medium ½ MS after going through the sterilization process. The sterilization process is based on the best results of the first experiment. Quantitative observations include observing the number of days required to begin germination (day), and the length of the plumula (cm).

2.4. Methods and Analysis

The embryo culture in the laboratory consists of three treatment of sterilization sequence based on the materials used are S1, S2 and S3. Treatment S1 is a sequence of sterilization using alcohol 70% for 5 minutes, Chlorox 3% for 10 minutes and Chlorox 2% for 5 minutes. The treatment of S2 is sterilization with alcohol 70% for 1 minute and Chlorox 2% for 10 minutes. S3 treatment using alcohol 70% for 3 minutes and Chlorox 1% for 15 minutes. To know the difference of treatment, data was analysis by test of two middle value in pairs.

3. Result and discussion

3.1. Result

3.1.1. Sterilization experiment of black rice embryo. Table 1 shows germination of Toraja black rice embryo in combination treatment of sterilization composition. The results of the experiment showed that each sterilization sequence technique can be used for sterilization of black Toraja rice with different germination rates. This can be seen from the absence of contaminants in the culture. The best sterilization composition for black rice embryo culture is S3 sterilization composition with 93% germination.

Table 1. Germination of black rice embryo of Toraja on treatment of combination of chemical sterilization composition.

Sterilization composition	S1 (73)	S2 (80)	S3 (93)
S1 (73)	0.0	7.0 ^{ns}	20.0 ^{ns}
S2 (80)	7.0^{ns}	0.0	13.0 ^{ns}
S3 (93)	20.0 ^{ns}	13.0 ^{ns}	0.0

( ) = Presentation of embryo germination (%)
ns = Not significantly different

3.1.2. Black rice rice seed culture. Rice seeds of black rice Toraja result of embryo culture that has germinated, will grow and develop plantlet. The germination of black rice seeds of Toraja with HM and HP color character is better than HS (figure 2). The time it takes seeds to start germinating gives a difference to the average length of the planlet plumula on the three colors of the seed. The average length of plumula from a small black seed (HS) is 1.1 cm, medium black average length of 1.8 cm and full black average length of 1.7 cm (table 2).

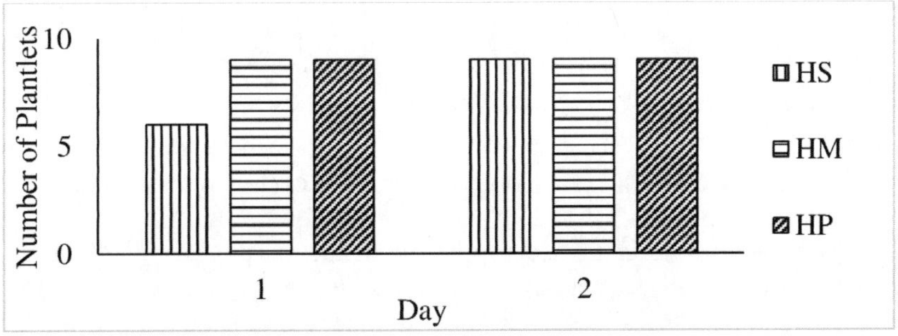

Figure 2. The germination of black rice seeds of Toraja rice is the result of embryo culture.

Table 2. The average length of black rice plumula of Toraja resulting from embryo culture at 7 days after culture.

The color of embyo	HS (1.1)	HM (1.8)	HP (1.7)
HS (1.1)	0.0	0.7 ns	0.6 ns
HM (1.8)	0.7ns	0.0	0.1 ns
HP (1.7)	0.6 ns	0.1 ns	0.0

( ) = The average length of plumula (cm)
ns = Not significantly different

3.2. Discussion

The first experiment was conducted by determining the exact order of sterilization materials for the sterilization of the black rice embryo of Toraja. The composition of the sterilizing agent is one of the main keys in tissue culture. The materials often used for wet sterilization are alcohol and chlorox. Alcohol is a suitable sterilizing agent but also very toxic to plants. Therefore, explant is usually exposed for several seconds or minutes [8].

Chlorox compounds are commonly used as a disinfectant or bleach. This compound is very effective in killing bacteria and viruses, cleaning microorganisms that are embedded in plant material, removing soil particles, dust and others. In plant tissue culture techniques, this compound is commonly used as a sterilizing agent of plant tissue surfaces [9, 10]. The use of Chlorox as a surface sterilizing agent from various sources of plant explants has been widely reported [11, 12, 13, 14, 15, 16].

The concentration and time of using Chlorox solution per explant vary. So with the right concentration and time of exposure, because the less Chlorox concentration the explants are more susceptible to the pathogen, but if the higher concentration of Chlorox the development of tissue explant become obstructed [17]. It is seen that by using S3 composition, the color of black rice seed did not fade Unlike the sterilization using material composition S1 and S2 (figure 3).

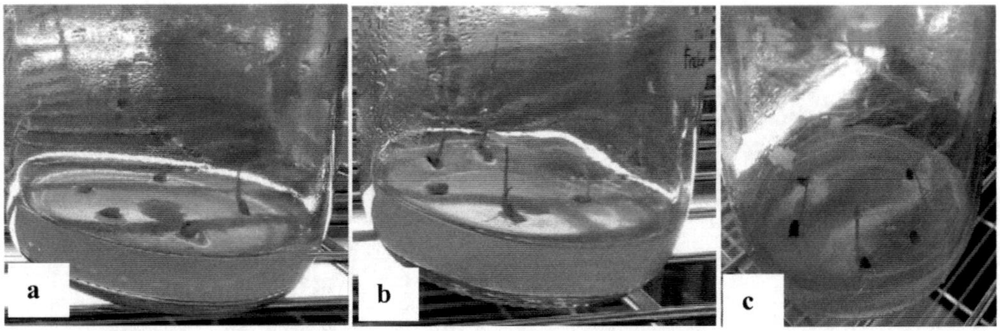

Figure 3. Culture of black rice embryo with different sterilization composition. (a) Composition S1, (b) S2 Composition, and (c) S3 Composition.

The process of seed germination is a complex sequence of physiological, morphological and biochemical changes. In the case of germination of black rice seeds of Toraja, the seeds that have a darker black color (black medium and black full) more quickly experience the germination process. We suspected thickness of the aleouron, cause faster enzyme activity, resulting in the hormone to work faster. This growth hormone activity that triggers the germination of black rice seeds Toraja faster.

The concentration and duration of seed immersion in the Chlorox solution, caused the black color of the rice seed to fade. In addition, the Chlorox solution and the composition of the sterilizing agent can also cause damaged embryos, browning, unable to grow and thrive in a culture bottle and eventually die.

4. Conclusion

The black rice seeds of Toraja have three color characteristics: black full (outside and inside of rice) (HP), medium black (outside and slightly inside rice) (HM), and thin black (only the outside of rice) (HS). The best sterilization for black rice embryo embryo culture Toraja using S3 sterilization sequence (Alcohol 70% for 10 minutes and Chlorox 1% for 5 minutes). Toraja black rice rice embryo culture on MS medium has succeeded in growing a sterile plantlet, with best growth in plantlet from seeds with medium black and black color.

References

[1] Suardi D and Ridwan I 2009 Beras hitam, pangan berkhasiat yang belum populer *Warta penelitian dan pengembangan pertanian.* **31** 9-10

[2] Suhartini T and Suardi D 2010 Potensi Beras Hitam Lokal Indonesia *Warta Penelitian dan Pengembangan Pertanian.* **32** 1

[3] Badan Pusat Statistik 2014 Data Sensus Produksi Padi (https://bps.go.id/Brs/view/id/1122)

[4] Limbongan Y and Fadjry J 2015 Karakterisasi dan Observasi Lima Aksesi Padi Lokal Dataran Tinggi Toraja Sulawesi Selatan *Buletin Plasma Nutfah.* **21** 2

[5] Gunawan and Livy W 1987 *Teknik Kultur Jaringan* (Bogor: Pusat Antar Universitas Bioteknologi Institut Pertanian Bogor)

[6] Monnier M 1990 *Zygotic embryo culture*, ed S S Bhojwani (Amsterdam: Elsevier) p 336-390

[7] Teixeira J, Sondahl M, and Kirby E 1993 Somatic embryogenesis from immature zygotic embryos of palm oil. Plant Cell Tissue and Organ Culture (Netherlands Kluwer Academic Pulishers) p 227-233

[8] Oyebanji O, Nweke O, Odebunmi O, Galadima NB, Idris MS, Nnodi UN, Afolabi AS, and Oghadu GH 2009 Simple, effective and economical explant-surface sterilization protocol for cowpea, rica, and sorghum seeds *African J. of Biotech.* **8** 5395-5399

[9] Sawant R and Tawar P 2011 Use of Sodium Hypochlorite as Media Sterilant in Sugarcane Micropropagation at Commercial Scale *Sugar Tech.* **13** 27-35

[10] Santoso U and Nursandi F 2003 *Kultur Jaringan Tanaman* (Malang: Universitas Muhammadiyah Malang)

[11] Miche L and Balandreau L 2001 Effects of rice seed surface sterilization with hypochlorite on inoculated Burkholderia vietnamiensis *Appl. Environ. Microbiol.* **67** 3046–3052

[12] Vejsadova H 2006 Factors affecting seed germination and seedling growth of terrestrial orchids cultured in vitro *Acta Biol. Cracoviensia Ser. Bot.* **48** 109–113

[13] Badoni A and Chauhan J S 2010 In vitro sterilization protocol for micropropagation of *Solanum tuberosum* cv. (Kufri Himalini Academia Arena) p 24–27

[14] Maina S M, Emongor Q, Sharma K K, Gichuki S T, Gathaara M, and Villiers S M. 2010. Surface sterilant effect on the regeneration efficiency from cotyledon explants of groundnut (*Arachis hypogea* L.) varieties adapted to eastern and Southern Africa *African J. of Biotech.* **9** 2866-2871

[15] Colgecen H, Koca U and Toker G 2011 Influence of different sterilization methods on callus initiation and production of pigmented callus in Arnebia densiflora Ledeb *Turkish J. Biol.* **35** 513–520

[16] Morla S, Rao C S V R and Chakrapani R 2010 Factors affecting seed germination and seedling growth of tomato plants cultured in vitro conditions *J. Chem. Biol. Phys. Sci.* **1** 328–334

[17] Rismayani and Hamzah F 2010 *Pengaruh Pemberian Chlorox (NAOCL) pada Sterilisasi Permukaan untuk Perkembangan Bibit Aglaonema (Donna carmen) Secara In Vitro Proc. Seminar Ilmiah dan Pertemuan Tahunan PEJ dan PFJ XX Komisariat Daerah Sulawesi Selatan* (Indonesia: Sulawesi Selatan)

IOP Conf. Series: Earth and Environmental Science **157** (2018) 012044 doi:10.1088/1755-1315/157/1/012044

Effect of selection agents to Chrysanthemum (*Chrysanthemum morifolium*) callus growth after *Agrobacterium*-mediated genetic transformation

R Sjahril[1], I Jamaluddin[2], M Nadir[3], Asman[4] and N E Dungga[5]

[1] Laboratory of Plant Biosciences and Reproduction Biotechnology, Faculty of Agriculture, Hasanuddin University, Jl. Perintis Kemerdekaan KM 10, Makassar, 90245, Indonesia.
[2] Agro-technology, Graduate School of Agriculture, Faculty of Agriculture, Hasanuddin University, Jl. Perintis Kemerdekaan KM 10, Makassar, 90245, Indonesia.
[3] Animal Feed Chemistry Laboratory, Faculty of Animal Husbandry, Hasanuddin University, Jl. Perintis Kemerdekaan KM 10, Makassar, 90245, Indonesia.
[4] Plant Pest and Disease Laboratory, Faculty of Agriculture, Hasanuddin University
[5] Department of Agronomy, Faculty of Agriculture, Hasanuddin University, Jl. Perintis Kemerdekaan KM 10, Makassar, 90245, Indonesia.

E-mail: *rinaldi.sjahril@gmail.com*

Abstract. Genetic transformation mediated by *Agrobacterium tumefaciens* requires an efficient selection method for successful progress of transformation. This study aims to determine the concentration and kind of antibiotics and selection agents used during transformation to formulate standard protocol of chrysanthemum in the process of propagating disease resistant Chrysanthemum mediated by *Agrobacterium tumefaciens* EHA105 (pEKB-WD). The experiments were performed by planting chrysanthemum explants leaf cutting (5 mm diameter on NAA medium 2 mg L^{-1} BAP 2 mg L^{-1}) with addition of Kanamycin: 25, 50, 100, 150 and 200 (mg L^{-1}); Hygromycin: 5, 10, 25, 50 and 75 (mg L^{-1}); Paromomycin: 10, 25, 50, 75 and 100 (mg L^{-1}). Experiment was arranged in a Completely Randomized Design (CRD). Each treatment was repeated five times thus 75 bottles of culture were used; each bottle consists of 5 pieces of leaf cuttings, resulted in total of 375 pieces. The results showed that selection agent had a critical value for Hygromycin 25 mg L^{-1} and Kanamycin 100 mg L^{-1} which can make explant experienced necrosis better than Paromomycin. Paromomycin at 100 mg L^{-1} was only able to kill explant's periphery. Remained callus stayed fresh more than 50% so that when used as the selection agent could produce more escape cell. The optimum transformation with concentration of 10% Agrobacterium (vol/vol) with 30 minutes co-cultivation can produce more efficient transformed callus. Considering the high price of Hygromycin, it was best to use Kanamycin as selective agents.

1. Introduction

Chrysanthemums occupy the largest floriculture commodity export to become a large foreign exchange producer as well but the development of chrysanthemums experiencing some obstacles and challenges. The most important obstacle is plant pests and diseases. Plants attacked by pests or diseases will decrease its quality, making it difficult to market. Plants that are free from pests and diseases are a key requirement to meet global export markets. One potential alternative is to develop resistant plants by utilizing biotechnology, especially in the field of genetic engineering. Genetic engineering technology allows genes from other organisms to be inserted into the genomes of target plants. Genetic engineering to produce disease-resistant chrysanthemums has been reported by [1]. Kim *et al.* [2] used the *chitinase* gene, but was limited to the resistance of pathogenic fungi. While resistance to viral disease was reported by Toguri *et al.* [3], who used the *pac*1 gene (RNA double ribbon from the specific ribonuclease from *Schizosaccaharomyces pombe*). Resistance to the fungal plaque *Alternaria* leaf spot (ALS) has also been reported by Xu *et al.* [4] using the Xpaa hoGo gene from *Xanthomonas oryzae*. It has also been reported the resistance of Chrysanthemum plants to pests using the cry1AB gene from *Bacillus thuringiensis* [5]. So far there has been no report on the use of genes that can provide resistance to leaf rust disease in chrysanthemums caused by fungi.

One of the anti-microbial protein coding genes that has been proven to encode antifungal proteins and bacteria is the wasabi defensin gene. Wasabi is a Japanese horseradish plant used as foodstuff and has anti-bacterial substance. Secondary metabolites from wasabi such as wasalexin and 6-methylsulfonylhexyl isothiocyanate have been reported to have anti-fungal and anti-bacterial activity [6]. The antibacterial protein coding gene (WjAMP-1) has been isolated from wasabi [7, 8] and is called the wasabi defensin gene by Kanzaki *et al.* [9]. Anti-microbial proteins encoded by this gene exhibit expression and inhibitory effects on the growth of fungi and bacteria in transgenic *Nicotiana benthamiana* [7]. Furthermore, Kanzaki *et al.* [9] also reported that the growth of fungi causing blast disease in rice was inhibited by over expression of the wasabi defensin gene. While on orchids can inhibit the growth of leaf blight caused by *Erwinia carotovora* [10].

Transformation with Agrobacterium mediation involves the use of soil bacteria known as *Agrobacterium tumefaciens* which have the ability to infect plant cells with a piece of DNA. There are several requirements that need to be considered in making genetic transformation such as efficient selection method [11]. Selection of transformant cells is a key factor in the success of methods developed for genetic transformation. The aim of selection for separating transformed and non-transformed cells, in vitro selection was performed using antibiotics as a selection agent [12].

Antibiotics as a selection agent work by inhibiting protein synthesis is disrupting the function of ribosomal subunits that reversibly inhibit protein synthesis by inhibiting the translocation of the tRNA-peptide complex from the location of the amino acid to the peptide site, consequently the polypeptide chain cannot be prolonged, but the resistance of the antibiotic will have no effect against cells that have a marker-resistant gene. Antibiotic selection agencies have different lethal doses of each type to be used as concentration selectors, as well as for each plant having different tolerance limits for each type of antibiotic. Pre-transformation selection can facilitate the transformation, by knowing the lethal dose of antibiotic selection agency so that the selection process becomes efficient. For that we need a series of studies to obtain the concentration and type of antibiotic selection agency that is effectively used in selecting callus transformation of wasabi defensin gene with mediation *Agrobaterium tumefaciens*.

2. Research Methods

2.1. Place and Time

The research was conducted at the Laboratory of Plant Biosciences and Reproduction Biotechnology, Department of Agronomy, Faculty of Agriculture, Hasanuddin University, Makassar. The study took place from April to November 2017.

2.2. Materials

Materials needed were plantlets of Chrysanthemum var: Pasopati, inorganic and organic chemical material components for Murashige and Skoog (MS) media, Luria Bertani (LB) media, growth regulators α-Naphthalenacetic Acid (NAA), Benzyl Amino Purin (BAP), and several kind of antibiotics (kanamycin, hygromycin, paromomycin, spectinomycin, choramphenicol, meropenem).

2.3. Implementation of Research

This experiment was carried out by testing the concentration and kind of antibiotic selection agency on chrysanthemum leaf explants to determine lethal doses of various types and concentrations of antibiotics used. The media used were MS medium with incorporation of NAA (2 mg L^{-1}) BAP (2 mg L^{-1}), placed in Erlenmeyer flask then sterilized. Addition of antibiotics is done in the laminar air flow after the medium has a temperature of 40°C then poured into culture bottles. The concentration of antibiotic selection agent that is examined were: with no antibiotics; with Kanamycin (25 mg L^{-1}, 50 mg L^{-1}, 100 mg L^{-1}, 150 mg L^{-1}, and 200 mg L^{-1}); Hygromycin (5 mg L^{-1}, 10 mg L^{-1}, 25 mg L^{-1}, 50 mg L^{-1}, and 75 mg L^{-1}); Paromomycin (10 mg L^{-1}, 25 mg L^{-1}, 50 mg L^{-1}, 75 mg L^{-1}, and 100 mgL^{-1}).

2.4. Methods and Analysis

Testing the concentration and type of antibiotic selection agency on Chrysanthemum leaf explants were to determine lethal doses of various types and concentrations of antibiotics used. Testing was done by planting leaf disc slices on media in previous experiment that is NAA (2 mg L^{-1}) and BAP 2 mg L^{-1} with addition of various antibiotic selection agents. The experiments were prepared in a Completely Randomized Design with Kanamycin treatment (Km): 25, 50, 100, 150 and 200 (mg L^{-1}); Hygromycin (Hyg): 5, 10, 25, 50 and 75 (mg L^{-1}); Paromomycin (Par): 10, 25, 50, 75 and 100 (mg L^{-1}). Each treatment was replicated 5 times so that there were 75 bottles of culture, each bottle consisted of 5 leaves of whole leaves 375 ring pieces of leaves. Data were analysed using statistic to determine the best treatment further tests. Data were performed by bar chart.

3. Result and Discussion

3.1. Result

3.1.1. Percentage (%) of Explants Life on Determining Concentration and Kind of Antibiotic Selection. Figure 1 shows the highest percentage of live explant on control which is not different from K1, K2, K3, H1, H2, P1, P2, P3 and P4 that is 100%.

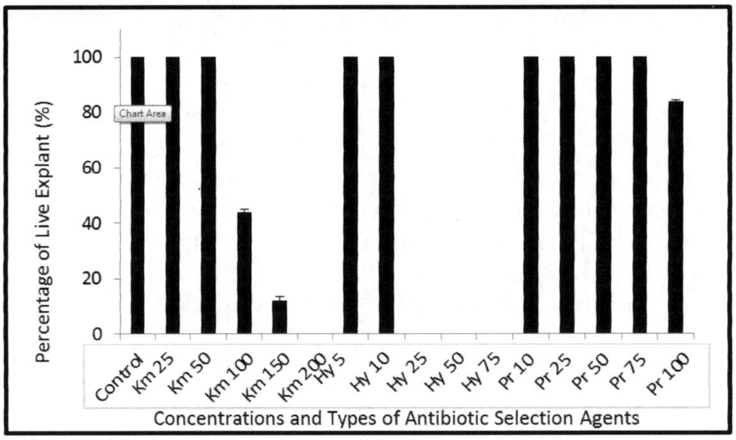

Figure 1. Average live of explant percentage of experimental results of determination of concentrations and types of antibiotics

3.1.2. Percentage (%) Smooth Explants on Determination of Concentrations and Types of Antibiotics Selection Agents. Figure 2 shows the results of the highest percentage of the highest controlled explant explant with K1, H1, H2, P1, P2 and P3 that is 100%.

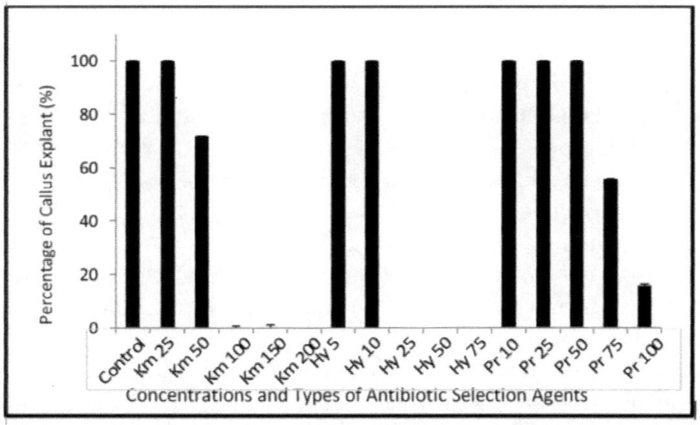

Figure 2. Average percentage of blunt explant results of experiments determining the concentration and type of antibiotics

3.1.3. Weight explants callus on Determining Concentration and type Antibiotic Selection Agent (g). Figure 3 shows the highest yield of callus on control (1.022 g) and in the treatment of paromomycin antibiotic gave the highest callus result of 0.174 g.

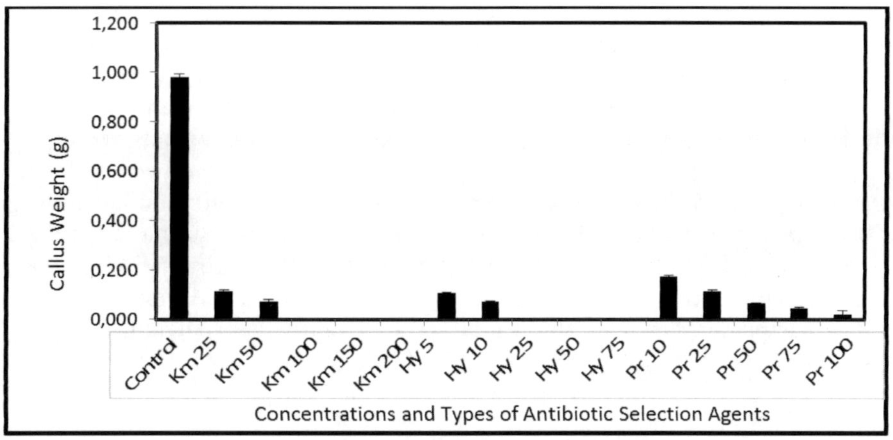

Figure 3. Graph of average percentage of blunt explant results of experiments determining the concentration and type of antibiotics

3.2. Discussion

Lethal testing of doses of various types of antibiotics selection agent was done for subsequent use in the selection medium of callus / plant transformant and nontransforman where the results obtained that the higher concentration of antibiotic in explant without carrier resistance gene will inhibit cell division in leaf explant even will cause death on explant but each plant has different tolerance limits to each type of antibiotic, seen in the results of experiments determining the concentration and type of antibiotic selection agency on the percentage of explants that grew in concentration is still within the limits of chrysanthemum tolerance that control without antibiotics can grow 100% in the administration of kanamycin concentrations of 25 mg L^{-1} and 50 mg L^{-1} can still grow up to 100%, the administration of hygromycin at concentrations of 5 mg L^{-1} and 10 mg L^{-1} still grows 100% and paromomycin administration all given concentrations are able to grow up to 100% except at concentrations of 100 mg L^{-1} growth decreased 16%.

For each concentration given to each type of antibiotic effect on the growth of explant, as well as the percentage of explant smooth concentration that is still within the limits tolerant explant can remain to grow and smooth but when the concentration of the stress on explant will affect the process of growth and development explant , ranging from explant will survive but not smooth to cause death. In general, antibiotics have lethal doses which may inhibit plant growth or delay, according to Anggraito [13], antibiotic poisoning causes explants to respond more to stress than to form buds.

The result of dead explant percentage shows lethal dose, where kanamycin at concentration of 100 mg L^{-1} explant yellowish green and not smooth finally experience nekrotik or die, explant color change is sign of cell death, for hygromycin that start from 25 mg L^{-1} causing death on explant in a fast time while for paromomycin the highest concentration of 100 mg L^{-1} is not yet in lethal doses because explant still able to live up to 84%, explant very little (16%) only explant brown (dead). Explant experiencing stress will affect the explant not able to form callus even death, although the time required each concentration and type of antibiotic is different.

In high concentrations of higromycin, the growth of explant is inhibited to cause death from excessive stress, besides it can also inhibit the transformant cells, so the growth or regeneration takes a long time. Higromycin inhibits metabolic processes by binding to the 80S ribosome resulting in translational error of mRNA [14]. The concentration of high concentrations of higromycin will begin with the explant from color change from green to yellowish green and then brownish browning.

Kanamycin at concentrations of 100 mg L^{-1} causes the plants to turn yellowish green and calluses are not formed, these concentrations have explant signs that they will inhibit the formation of new cells or callus until finally explants die. In contrast to hygromycin, kanamycin takes longer and high concentrations can cause toxicity to chrysanthemum explants. Kanamycin is a selection agency commonly used to select whether the transformed gene is successful.

Giving antibiotics paromomycin in this experiment is not toxic seen from the results the percentage of explants callus or not experience stress and callus produced by the weight of callus was highest after the control is Par 10: 0174 g, while the highest concentration of paromomycin of 100 mg L^{-1} remains formed a little callus and stresses experienced by explant only visible on the outskirts of explants brown this is what caused very rarely using paromomycin as the selection agent because it requires higher concentrations to be used as the selection agent. Among the three antibiotic agents who attempted hygromycin have toxic effects that high compared to kanamycin and paromomycin for hygromycin inhibit the binding 80S ribosomal subunit so that there was an error of translation, to kanamycin 40S subunit and paromomycin 16S subunit, thus seen from the number of ribosomal subunits bound hygromycin may hamper explant mensitesis more protein causes the metabolism process stalled so explant die.

4. Conclusion

Selection of antibiotic agents that have a critical value for Hygromycin was 25 mg L^{-1} and Kanamycin was 100 mg L^{-1}, which can make the Chrysanthemum explant experienced necrosis. Paromomycin at 100 mg L^{-1} was only able to damage the explant periphery so that when used as the selection agent could produce more escapes.

References

[1] Takatsu Y 2005 *Transgenic Chrysanthemum (Dendranthema grandiflorum (Ramat.) Kitamura) Expressing A Rice Chitinase Gene Shows Enhanced Resistance To Fungal Diseases* (Japan: Agris FAO Japan) http://agris.fao.org/agris-search/search.do?recordID=JP2002005344. Accessed 21 April 2014

[2] Kim Y, Lim S, Yoda H, Choi Y and Sano H 2011 Simultaneous activation of salicylate production and fungal resistance in transgenic chrysanthemum producing caffeine *Plant Signaling & Behavior* **6** (3) 409-412

[3] Toguri T, Ogawa T, Kakitani M, Tukuhara M and Yoshioka M M 2003 Agrobacterium-mediated transformation of chrysanthemum (*Dendrathema grandiflora*) plants with a disease resistance gene (pac1) *Plant Biotech.* **20** (2) 121-127

[4] Xu G, Chen S and Chen F 2010 Transgenic chrysanthemum plants expressing a harpinxoo gene

demonstrate induced resistance to alternaria leaf spot and accelerated development *Russ. J. of Plant Physiol.* **57** (4) 548-553.

[5] Shinoyama H, Mochizuki A, Komano M, Nomura Y and Nagai T 2003 Insect resistance in transgenic chrysanthemum [*Dendranthema x grandiflorum* (Ramat.) Kitamura] by the introduction of a modified δ-endotoxin gene of *Bacillus thuringiensis Breed. Sci.* **53** 359-367

[6] Pedras M S C, Sorensen J L, Okanga F I and Zaharia I 1999 Wasalexins A and B, new phytoalexins from wasabi: isolations and synthesis and antifungal activity *Bioorgan Medic CChem Lett.* **9** 3015-3020

[7] Saitoh H, Kiba A, Nishihara M, Yamamura S, Suzuki K K and Terauchi R 2001 Production of antimicrobial defensin in Nicotiana Benthamina with a potato virus X vector *Mol. Plant Microbe Interact.* **14** 111-115

[8] Kiba A, Saitoh H, Nishihara M, Omiya K and Yamamura S 2003 C-terminal domain of a hevein-like protein from Wasabia japonica has potent antimicrobial activity *Plant Cell Physiol.* **44** (3) 296-303

[9] Kanzaki H, Nirasawa S, Saitoh H, Ito M, Nishihara M, Terauchi R and Nakamura I 2002 Overexpression of the wasabi defensin gene confers enhanced resistance to blast fungus (*Magnaporthe grisea*) in transgenic rice *Theor. Appl. Genet.* **105** 809-814

[10] Sjahril R, Chin D P, Yamamura S, Khan R S, Nakamura I, Amemiya Y and Mii M 2006 Transgenic phalaenopsis plants with resistance to erwiniacarotovora produced by introducing wasabi defensin gene using agrobacterium method *Plant Biotech.* **23**191-194

[11] Acquaah G 2007 *Principles of Plant Genetics and Breeding* (Australia: Blackwell Publishing) 569 p

[12] Arencibia A, Molina P, De la Riva G and Selman G 1995 Production of transgenic sugarcane (*Saccharum officinarum* L.) plants by intact cell electroporation *Plant Cell Rep.* **14** 305-9

[13] Anggraito Y U 2012 *Transformasi Genetik Nicotiana benthamiana L. dan Kedelai dengan Gen MaMt2 Penyandi Metallothionein Tipe II dari Melastoma malabathricum L.* Disertasi S3 (Bogor: Program Studi Biologi, Sekolah Pascasarjana Institut Pertanian) 71 p

[14] Bashir K, Rafiq M, Fatima T, Husnain T and Riazuddin S 2004 Hygromycin based selection of transformants in a local inbred line of *Zea mays* (L). *Pak. J. of Biol. Sci.* **7** 318-323

IOP Conf. Series: Earth and Environmental Science **157** (2018) 012045 doi:10.1088/1755-1315/157/1/012045

Interaction between genotypes, environment and season (G x E x S) on anthocyanin corn in lowland zone of Indonesia

M Yasin H G, M Isnaeni, Faesal, and M Azrai

Indonesia Cereal Research Institute (ICERI), Jl. Dr. Ratulangi No. 274 Maros, South Sulawesi, Indonesia.

E-mail: *hg_yasin@yahoo.co.id*

Abstract. Interaction between genotypes (G), environment (E) and season (S) on anthocyanin corn has been studied under lowland zone in Indonesia. The experiment were conducted using Randomized Complete Block Design (RCBD) based on location namely Maros, Bajeng, and Polman with three replications. Ten populations of open pollinated variety (OPV) of anthocyanins purple corn included checks were used in two planting seasons (dry and rainy) in 2015/2016. The objective of the experiment is to determine which of the population is stable and has high yield to be promoted as candidate for new OPV varieties. Genotypes were planted in four rows of 5.0 m length, spacing of 75 cm x 20 cm, one plant per hill, and applied with Urea, Ponska (300 and 200 kg Ha^{-1}). Population was then selected by t test in model $Y_{ij}=\mu+\beta_i I_j+\delta_{ij}$, $i\neq j$ (Y: yield, μ: mean, β: regression coefficient, I: environmental index, δ: deviation from regression), and stability parameter $b_i=\Sigma Y_i I_j/I^2_j$. The results show that there was a significant interaction of GxExS on PMU(S1).Synth.F.C1 and PPU(S1).F.C1 with yield ranging between 6.70-8.48 t ha^{-1} and grain potential found in the locations of Maros and Bajeng in the rainy season. Content of anthocyanins were between 37.15 and 51.92 μg per 100 g sample.

1. Introduction

Corn with purple coloured grain can be defined as a specialty corn which contains high content of nutritious anthocyanins. Originated from Mesoamerica, the corn with a pigment that gives colour of red, purple, and blue in the grain have health beneficial to prevent viral, cholesterol, heart disease, obesity, and cancer [1, 2]. Yasin *et al.* [3] reported that the corn has many excellent characters and higher economic value, that makes it as an importance raw material for food industries, textiles, paper making and feed. Germplasms had been selected to increase cycle of selection by self (S1, S2 to S4/S5). Anthocyanins corn is a recessive characteristic, production must be isolated from normal/dent to prevent losing the periculiar starch and colour properties. The variety evaluation trial under some environments have been conducted and continued to provide guidance for selecting the best genotype and estimated yield. Several studies to test the yield stability of Quality Protein Maize (QPM) crops were conducted [3, 4] in seven environments under lowland tropical zone which resulted in 11 entries that found to be stable and had highly significant effect of GxExS.

The objective of the experiment is to evaluate the yield stability of ten anthocyanins corn under three environments during dry and rainy planting seasons in 2015/2016. The best population will be determined as the potential genotype for new open pollinated variety (OPV) of anthocyanins corn.

2. Methodology

The experiment was conducted in Maros (e1), Bajeng (e2) and Polman (e3) regencies in South Sulawesi Province of Indonesia. The locations are under lowland tropical zone <50 m above the sea level (asl). The genotypes (g) were 10 populations of anthocyanins corn with seed colours purple to black (table 1) based on Pixley *et al.*[5] that genetic population of specialty corn could be increase by initial selection of colour seeds. The experiment was set in Randomized Complete Block Design (RCBD) with three replications in dry and rainy season in 2015-2016. The genotypes were planted in four rows per se, spacing 75 x 20 cm, fertilized with Urea and Ponska with dosage of 300 and 200 kg Ha^{-1}, respectively.

Table 1. Seed texture, colour, origin and cycles population of anthocyanins corn planted in the two planting seasons in 2015/2016 in three locations under lowland of tropical zone condition.

Genotype (g)	Dry season, 2015		Rainy season, 2016	
	Seed texture	Seed colour	Seed origin	Cycles population
g1. PTU(S1)F.C0	flint	purple to black	China	C0
g2. PTU(S1)D.C0	flint	purple to black	China	C0
g3. PMU(S1)SYN.D.C1	semi dent	purple	Manado	Syntetik C1
g4. PMU(S1)SYN.F.C1	flint	purple	Manado	Syntetik C1
g5. PPH.FS.C1	flint	purple to black	Lembah Palu	C1
g6. PPH.(S1).C1	flint	purple to black	Lembah Palu	C1
g7. PVU.FS.C0	flint	purple	Vietnam	C0
Check				
g8. PLP.C0 (check 1)	semi flint	purple	Lembah Palu	C0
g9. PLM.C0 (check 2)	flint	White	Pulut Maros	C0
g10. URI1 (check 3)	flint	White	UPBS	C0

PTU : purple waxy of China; URI : waxy corn 1(check, white); PLM : Local waxy of Maros
PPH : black waxy of Palu; PMU : purple waxy of Manado; flint : pearl, dent:horse teeth
PVU : purple waxy of Vietnam; PLP : purple waxy of Palu

The variable observed were yield, plant and ear height, period of tasseling and silking, weight of 1000 seeds, water content and shelling percentage. The data were analyzed by single effect of genotype (G), two factors interaction between genotype and environment (E) (GxE) by polled analysis ei, i=1,2,3 on dry and rainy [3, 6], three factors interaction between genotype, environment and season (S) (GxExS) by simple regression analysis with xi as environmental index.

Coefficient of simple regression of Gi was calculated using: $Y = \beta_0 + \beta_1$ x. bi (i= 1,2, . . .,10), bi = $(\sum\sum xij.\ \acute{y})/\sum\acute{y}^2$, where ý : grand mean of all population and xij : mean of each population gi [10].

The hypothesis of stability was $H_o{:}\beta1=1$ vs $H_1{:}\beta1\neq1$ by d.f : n-1. t.cal.:(β-1)/sd ; sd : VS^2 ; S^2: Varians : $1/n[\sum xi^2-(\sum xi)^2/n]$.

Different of population in each location was compared to control with three checks by Least Signicance Difference (LSD) test in level of significant of 95%. LSD: $t(18)\sqrt{(2s^2/3)}$, s^2 : error mean square. Yield of grain in 15% water content was predicted by: Y: ((10000/7.5)*(100-**wa**)/85)*(shelling)*ear weight on harvested. Genotype selected if hypotesis of $H_0{:}\beta i\neq1$ was rejected, and average Yi>Ýt [2, 7].

3. Results and Discussion

3.1. GxE Interaction Analysis

The result analysis of variance and the mean square of observed variable and coefficient of variance value under dry and rainy season is shown in table 2. The analysis shows that under dry season there were significant interaction of GxE on variable of yield, tasseling, and weight of 1000 seeds. The single effect of genotype (G) was significant on yield, plant height, silking period, and shelling percentage. In the rainy season, GxE was significant on plant height and water content of seeds only, and single effect of G shows that all variable were significant except of plant and ear height.

Table 2. The mean square and coefficient of variance (CV) of GxE interaction on anthocyanins corn in dry and rainy season in 2015/2016

Variable	E	U/E	G	GxE	Error	CV(%)
Dry season 2015						
Yield	10,544*	0,275	4,997*	1,530*	0,660	12,4
Plant height	4699,547*	135,890	156,044*	75,877	85,908	4,9
Ear height	2196,466*	1,805	28,173	42,730	54,802	8,1
Tasseling	12,878*	1,056	1,328	2,051*	1,228	2,3
Silking period	13,478*	0,256	1,888*	0,910	1,095	2,0
Water moisture	151,800*	26,455	3,084	5,290	5,026	8,3
Shelling percentage	14,042*	1,778	5.404*	3,845	4,418	2,7
Weight 1000 seeds	112,678	488,778	621,685	1330,715*	983,049	10,42
Rainy season 2016						
Yield	3,987	3,029	9,934*	0,584	0,750	12,9
Plant height	547,600*	69,433	89,723	157,205*	89,396	4,9
Ear height	26,678	41,956	42,326	62,011	82,412	9,6
Tasseling	0,900*	0,500	8,099*	1,147	1,290	2,3
Silking period	2,178*	0,900	9,723*	1,042	1,369	2,2
Water moisture	257,387*	19,709	24,792*	12,899*	11,072	10,3
Shelling percentage	0,289	1,349	4,760*	1,528	6,360	3,19
Weight 1000 seeds	578,011	638,344	2303,190*	371,838	919,005	10,02

* significant different on 95% level n : 90

3.2. GxExS Interaction Analysis

Table 3 shows analysis of three factors, GxExS, with significant variable were yield, plant height and water content of seeds. The results indicate that yield of all genotypes were different under three locations in Maros, Bajeng and Polman The CV found for all variable was <20,0% and be assume that there are could be selected of Gi, i =1, 2, 3, . . ,10 for new corn varieties [3]. The experiment of QPM of yellow grain from CIMMYT showed that population could be adapted in Maros after selected by S1 families and genetically improved by one cycles of selection, yield was 8.0-10.0 t Ha^{-1} [11]. The significant Interaction of GxExS on three variables could indicate that genotypes yield, plant height and water content of seeds at harvest were significantly different under three locations.

Table 3. The mean square, coefficient of variance (CV) and interaction GxExS on anthocyanins corn in 2015/2016.

Variable	E	R/E	G	ExG	S	ExS	GxS	GxExS	Error	CV (%)
Yield	12.542	1.901	12.040	0.967	0.591	1.990	2.890	1.147*	0.742	13.0
Plant height	1658.860	69.704	165.426	121.716	789.188	3588.287	79.340	111.366*	90.173	4.9
Ear height	983.382	27.138	27.277	54.614	268.156	1231.761	43.222	50.127	65.871	8.7
Tasseling	3.489	0.567	2.393	1.859	2.689	10.289	7.035	1.338	1.245	2.3
Silking period	2.617	0.428	4.388	0.907	1.606	13.039	7.223	1.045	1.206	2.1
Water moisture	222.913	9.937	10.468	7.739	460.160	186.275	17.409	10.450*	9.911	10.2
Shelling, %	8.434	1.757	4.046	3.058	11.300	5.879	6.119	2.316	5.178	2.8
1000 seeds	232.267	206.961	1578.154	731.730	107.339	458.422	1348.722	970.823	949.441	10.2

* Significant different on 95% level, n : 180

3.3. Yield Stability

The analysis of yield stability shown in table 4 shows significant differences were found in dry season and there are four candidate of genotypes (G1, G3, G4, and G7) that were significant by statistical t test, while three checks were not significant. However, in the rainy season, G1 and G3 were significant. The result indicates that population G1 (PTU(S1)F.C0 could become as a candidate for promising new OPV. The significant result indicates that the genotype yield was stable in the three locations under two seasons in rainy and dry seasons and assuming that the condition in the locations were favorable for plant growth, like well fertilized, no weeds and pest-disease resulted in maximum grain yield obtained. The statistical analysis of G1 shows that b=1, H_1:b≠1.0 was rejected and R^2:99.2%.

Table 4. Parameters of yield stability of anthocyanins corn. 2015-2016

	R	R^2	B	Se.β	MSe	t.cal.
Dry 2015						
G1. PTU(S1)F.C0	0.996	0.992	3.050	0.340	0.001	6.029*
G2. PTU(S1)D.C0	-0.462	0.213	-1.397	2.677	5.037	0.894
G3. PMU(S1)SYN.D.C1	0.906	0.820	2.469	1.156	0.939	1.271*
G4. PMU(S1)SYN.F.C1	0.966	0.934	1.374	0.366	0.094	1.023*
G5. PPH.FS.C1	-0.559	0.313	-0.426	0.631	0.280	2.260
G6. PPH.(S1).C1	-0.474	0.225	-0.774	1.437	1.451	1.235
G7. PVU.FS.C0	0.973	0.946	1.915	0.457	0.147	2.003*
Check						
G8. PLP.C0 (check 1)	0.980	0.950	2.600	0.024	0.001	66.040
G9. PLM.C0 (check 2)	-0.961	0.923	-0.575	0.167	0.020	9.456
G10. URI1 (check 3)	0.813	0.661	0.828	0.597	0.243	0.300
Rainy 2016						
G1. PTU(S1)F.C0	0.961	0.924	2.549	0.731	0.142	2.119*
G2. PTU(S1)D.C0	0.281	0.079	0.815	2.780	2.054	0.066
G3. PMU(S1)SYN.D.C1	0.876	0.768	2.330	1.285	0.439	1.035*
G4. PMU(S1)SYN.F.C1	0.414	0.171	0.069	1.153	0.006	0.807
G5. PPH.FS.C1	-0.991	0.982	-1.347	0.101	0.009	23.237
G6. PPH.(S1).C1	0.546	0.290	0.772	1.184	0.373	0.193
G7. PVU.FS.C0	0.669	0.447	1.834	2.040	1.106	0.409
Check						
G8. PLP.C0 (check 1)	0.997	0.995	1.678	0.122	0.004	5.568
G9. PLM.C0 (check 2)	0.979	0.994	3.747	0.302	0.024	9.083
G10. URI1 (check 3)	0.848	0.719	1.273	0.796	0.160	1.275

*: significant different on 95% level

3.4 . Agronomic and Yield Variables

Grain yield of the genotypes including checks and statistical test by LSD of all variable in the three environments are shown in table 5. Population of G1.PTU(S1)F.C0 and G2.PTU(S1)D.C0 found to be significant in E1, E2 and E3 in rainy season, the highest yield in Maros experimental farm was 8.48 t ha^{-1} shown by G1.PTU(S1)F.C0 which differed by 50.3% than the best yield obtained by the check genotypes in rainy season and 21.3% in dry season. This indicates that G1 was the candidate for new OPV variety of anthocyanin corn. In the dry season the G1 candidate was significant in Maros experimental farm but not significant in Bajeng and Polman.

Observation on agronomic characters of the plant ie. plant and ear height, tasseling and flowering (table 6) shows that plant height was not significant but ear height was significant between genotypes. The position of ear height was found around middle from plant height. This position of ear plant was suitable for selection of corn. Shelling percentage of G1. PTU(S1)F.C0 was higher 8.4% than check. Variable of anthesis silking interval (ASI) shows that the range of flowering and tasseling periods was

2.0 to 4.0 days. The different value of flowering and tasseling defines ASI value. ASI value < 3.0 days was very suitable for maximal grain formation while ASI value > 6.0 days showed no grain formed. Experiment results of Djamaluddin and Yasin [11] found that the ASI values of population MS-2 that ranged 5-6 days were associated with yield 2-3 t ha^{-1}. Gambin *et al.* [12] found that drought tolerant inbred line with ASI of 6.0 days experincing decline in yield by 58-69%. Water content of seeds ranged between 27-34.0% and G1.PTU(S1)F.C0 found to be less than check 36.0%. The result is similar with shelling percentage of 79.3%.

Table 5. The yield potential and statistical test of LSD on anthocyanins corn

Genotype (G)	**Dry 2015**			**Rainy 2016**		
	E1	E2	E3	E1	E2	E3
G1. PTU(S1)F.C0	7.95abc	7.15	6.94	8.48abc	8.17abc	7.24bc
G2. PTU(S1)D.C0	7.69abc	7.21	6.76	7.65abc	7.86abc	6.35bc
G3. PMU(S1)SYN.D.C1	6.69	7.37	7.84bc	7.06abc	7.91abc	5.75bc
G4. PMU(S1)SYN.F.C1	7.96abc	6.28	7.80bc	7.46abc	7.76abc	7.91abc
G5. PPH.FS.C1	6.97	7.70b	6.17	6.95abc	7.57abc	5.60
G6. PPH.(S1).C1	6.86	5.77	6.67	7.40abc	7.92abc	6.66bc
G7. PVU.FS.C0	6.75	7.78b	6.71	6.64bc	6.62ac	6.94bc
Check						
G8. PLP.C0 (check 1)	6.56	6.76	7.13	5.64	5.60	6.57
G9. PLM.C0 (check 2)	5.80	6.17	5.93	4.86	5.97	4.99
G10. URI1 (check 3)	5.88	6.86	6.16	4.26	5.25	4.76
CV (%)	12.62	15.25	17.18	13.08	11.81	12.08
LSD 5%	1.06	1.28	1.42	1.05	1.01	0.92
LSD 1%	1.45	1.75	1.95	1.44	1.39	1.26

a,b,c : significant vs check 1, 2 and 3
E1 : Expm Farm of Maros,
E2 : Expm Farm of Bajeng,
E3 : Farmer's field of Polman-Sulbar

Table 6. The average of variable of anthocyanins corns planted in Maros (E1), Bajeng (E2) and Polman (E3) under two seasons in 2015/2016

E/G	Plant height (cm)	Ear height (cm)	Tasselling (days)	Flowering (days)	Water content (%)	Shelling (%)
E1. Maros						
G1. PTU(S1)F.C0	191.9	99.2	48.8	51.3	30.6	79.3
G2. PTU(S1)D.C0	194.3	94.4	48.5	51.3	29.9	73.6
G3. PMU(S1)SYN.D.C1	201.7	98.5	49.0	51.5	29.7	74.0
G4. PMU(S1)SYN.F.C1	196.4	91.9	48.2	51.0	31.0	76.7
G5. PPH.FS.C1	198.5	96.0	48.3	52.0	30.1	77.7
G6. PPH.(S1).C1	193.3	95.5	48.5	51.8	31.0	77.3
G7. PVU.FS.C0	195.3	95.4	48.5	51.5	31.1	75.7
G8. PLP.C0 (check 1)	193.5	92.9	48.8	52.2	31.1	77.3
G9. PLM.C0 (check 2)	197.6	96.8	48.3	51.5	30.6	75.7
G10. URI1 (check 3)	195.2	96.2	49.7	52.7	31.1	73.1
CV (%)	3.0	3.9	2.0	2.0	7.8	4.8
LSD (5%)	10.1	6.4	1.7	1.6	4.1	6.3

E2. Bajeng						
G1. PTU(S1)F.C0	194.2	97.9	49.3	52.0	29.5	79.7
G2. PTU(S1)D.C0	194.3	101.6	48.8	51.8	30.8	78.5
G3. PMU(S1)SYN.D.C1	193.5	91.0	49.5	52.7	29.5	78.5
G4. PMU(S1)SYN.F.C1	195.6	94.6	48.8	51.8	28.5	79.2
G5. PPH.FS.C1	187.6	91.6	49.8	52.8	29.3	78.0
G6. PPH.(S1).C1	189.3	95.1	48.7	51.7	27.7	78.7
G7. PVU.FS.C0	194.4	94.3	48.5	51.7	27.3	79.3
G8. PLP.C0 (check 1)	201.9	99.7	48.5	51.7	29.0	79.5
G9. PLM.C0 (check 2)	193.4	93.5	49.0	52.0	28.4	79.4
G10. URI1 (check 3)	191.8	91.9	49.7	52.8	28.6	79.7
CV (%)	3.3	6.3	1.7	1.4	7.8	1.8
LSD (5%)	11.0	10.3	1.4	1.2	3.9	2.4
E3. Polman						
G1. PTU(S1)F.C0	181.9	86.0	47.5	50.7	33.1	75.4
G2. PTU(S1)D.C0	179.4	87.6	47.7	50.3	33.0	78.6
G3. PMU(S1)SYN.D.C1	187.7	84.3	48.0	50.7	33.9	76.7
G4. PMU(S1)SYN.F.C1	191.8	84.4	48.8	51.8	32.9	76.7
G5. PPH.FS.C1	176.9	87.7	50.0	52.7	33.9	78.5
G6. PPH.(S1).C1	178.4	88.4	48.8	52.0	32.6	79.9
G7. PVU.FS.C0	181.8	85.1	48.7	51.8	33.5	76.1
G8. PLP.C0 (check 1)	182.4	86.9	49.3	52.2	34.4	77.2
G9. PLM.C0 (check 2)	175.9	86.9	49.5	52.5	34.2	78.0
G10. URI1 (check 3)	180.2	86.1	49.3	52.3	33.0	76.8
CV (%)	5.1	6.9	2.1	1.4	4.6	2.3
LSD (5%)	15.9	10.2	1.8	1.3	2.6	3.0

3.5. Nutrition

The result of ICERI Laboratorium shows that the three genotypes that contain anthocyanins 51.92 μg g^{-1} were PMU(S1).Synth.F.C1, PMU(S1).Synth.D.C1, and PPH.(S1).C1. Compared to content of this pigment in the check genotype (12.10 μg g^{-1}), there are difference of 329.1%. According to Widowati *et al.* [13] and Wong [14], the amylose content are <10.0% as long as that the anthocyanins population is assumed also as waxy corn.

4. Conclusion

Three factors significant interaction of genetic, environment and season (GxExS) was found on grain yield parameter with highest yield of 8.17 t ha^{-1} obtained by genotype G1. PMU(S1)SYN.D.C1 in Maros which was 21.3-50.3% higher than the three check genotypes. Two genotypes, G1.PTU(S1)F.C0 and G2.PMU (S1)SYN.D.C1, were found to be stable (β=1) and showed mean yield of Gi:i:1,2 higher than mean total Gij and can be concluded as potential candidate for new superior variety of open pollinated variety of anthocyanins corn with anthocyanin content of 51.92 μg g^{-1} which is 329.1% higher than the check genotype.

References

[1] Harakotr B, Bhalang S, Ratchada T, Marvin P S and Kamol L 2014 Anthocyanins and antioxidant activity in coloured waxy corn at different maturation stages. *J. Funct. Foods* **9**: 109-118

[2] Mary A L 2004 Anthocyanins and human health. An invitro investigative approach. *J. Biomed. and Biotech* **5** 306-313

IOP Conf. Series: Earth and Environmental Science **157** (2018) 012045 doi:10.1088/1755-1315/157/1/012045

[3] Yasin H G M, Sumarno and Nur A 2015. *Perakitan Varietas Unggul Jagung Fungsional*. (Bogor: Puslitbangtan Tanaman Pangan. Badan Penelitian dan Pengembangan Pertanian) p.60
[4] Crossa J, Cornelius P L, and Yan W 2002 Biplots of linier-bilinier models for studying crossover genotype x environment interaction *Crop Sci.* **42** 619-633
[5] Pixley K, Palacios N, Rocheford T, Bahu R and Yan J 2010 Agriculture for nutrition: maize biofertification strategies and progress *Proc. of 10th Asian Regional Maize Workshop* October 20-23, 2008. Makassar Indonesia
[6] Sharma J R 2008 *Statistical and Biometrical Techniques in Plant Breeding* (New Delhi: New Age Internationl (P) Limited, Publishers) p. 104
[7] Hallauer A R and Miranda Fo J B 1988 *Quantitative Genetics in Maize Breeding. 2nd* (Iowa : State University Press/Ames) p. 358
[8] Meseka S K, Menkir A, Ibrahim A E S and Ajalan S O 2006 Genetic analysis of performance of maize inbred lines selected for tolerance to drought under low nitrogen *MAYDICA* **51** (3)
[9] Yasin H G M and Mejaya M J 2016 *Rancangan Statistik Khusus Pemuliaan Jagung (Kasus Jagung Fungsional QPM, Provit A, dan Pulut)* (Jakarta: IAARD Press. Badan Litbang Pertanian)
[10] Zhao X, Corrales M, Zhang C, Hu X, Ma Y and Tausher B 2008. Composition of thernal stability of anthocyanins fromchinese purple corn (*Zea mays*. L) *J. Agric. and Food Chem.* **22** 8625-6
[11] Djamaluddin and Yasin H G M 2008 Konversi inbred tetua jagung hibrida menggunakan donor jagung QPM gen opaque 2 *J. Pen. Pert. Tan. Pangan* **27** (1)18-23
[12] Gambin B L, Barras L, and Otegui M E 2007 Is maize kernel size limited by its capacity to expand. *MAYDICA* **52** (4) 434
[13] Widowati S, Santosa S, and Suarni 2006 Mutu gizi dan sifat fungsional jagung. *Pros. Seminar dan Lokakarya Nasional Jagung* (Bogor : Pusat Penelitian dan Pengembangan Tanaman Pangan. Badan Litbang Pertanian) p. 343 – 350
[14] Wong C 2016 The benefits of anthocyanins. Holistic Health Dietary Supplements (USA : Cultivates Consistency)

IOP Conf. Series: Earth and Environmental Science **157** (2018) 012046 doi:10.1088/1755-1315/157/1/012046

Effect of heavy ion beam irradiation on germination of local Toraja rice seed (M1-M2) mutant generation

R Sjahril[1], M Riadi[2], Rafiuddin[3], T Sato[4], K Toriyama[4], T Abe[5] and Trisnawaty A R[6]

[1] Laboratory of Plant Bioscience and Reproduction Biotechnology, Hasanuddin University, Jl. Perintis Kemerdekaan KM 10, Makassar, 90245, Indonesia.
[2] Laboratory of Plant Breeding and Seed Sciences, Hasanuddin University, Jl. Perintis Kemerdekaan KM 10, Makassar, 90245, Indonesia.
[3] Laboratory of Plant Physiology and Nutrition, Hasanuddin University, Jl. Perintis Kemerdekaan KM 10, Makassar, 90245, Indonesia.
[4]Graduate School of Agricultural Science, Tohoku University, Sendai, Japan
[5] Radiation Biology Team, Nishina Center for Accelerator-Based Science RIKEN, 2 – 1 Hirosawa, Wako, Saitama, 351-0198, Japan.
[6]Agrotechnology, Graduate School of Faculty of Agriculture, Hasanuddin University, Jl. Perintis Kemerdekaan KM 10, Makassar, 90245, Indonesia.

E-mail: *rinaldi-sjahril@agri.unhas.ac.id*

Abstract. Local rice in general has several weaknesses among others, long life, high plant posture and low yield result. The character is a limiting factor that causes farmers low interest to grow local rice. It is feared this will cause the lack of local rice cultivars as germplasm materials. Therefore, there is an effort to create a diversity of morphological characters, as the character of selection, especially related to the age of harvest and plant posture. One method is through breeding mutation by irradiation using ion beam. The objective of this research is to evaluate seed germination resulted after irradiation using ion beam in two varieties of Toraja local rice. The study was prepared based on a randomized block design pattern consisting of six treatments by testing two local Toraja rice varieties namely Pare Ambok and Pare Lea treated with ion beam irradiation of Argon and Carbon ion and control plant as comparison. Each grain from one panicle was germinated in one line method on a Ø15 cm Petri dish and transplanted into small plastic bags. Each treatment was repeated as much as 20 times which was then considered as a strain. The results showed that irradiation using Argon ion in local rice seed of Pare Ambok variety and of Pare Lea varieties produce better seedlings sprouts than irradiation using Carbon ion. Further M2 seed germination shows uniqueness in some seedlings produced such as lighter leaf color, albinism, wrinkled leaf, etc. which could prove potential mutant lines in tested M2 lines seed.

1. Introduction

Tana Toraja is one of area in South Sulawesi that has local rice germplasm diversity. Based on the results of a survey of local rice in Tana Toraja District conducted by Sahardi *et al.* [1] found around 29 local varieties of Toraja rice each of which has exotic characteristics such as brown rice, red, aromatic and glutinous. Local rice is germplasm that has potential as a source of genes that control important properties in rice plants.

IOP Conf. Series: Earth and Environmental Science **157** (2018) 012046 doi:10.1088/1755-1315/157/1/012046

One of the local rice varieties used for generations as part of the traditions and culture of Tana Toraja people is black rice and red rice. The use of black and red rice in Toraja cultural rituals helps preserve these traditional rice varieties amidst the rapid use of introduced varieties.

Black rice and red rice contain anthocyanin compound, water-soluble pigment, which has antioxidant activity. Black rice and red rice are known to increase the body's resistance to disease, repair damaged liver cells, prevent kidney function, prevent cancer / tumor, slow aging, as an antioxidant, clean cholesterol in the blood, and prevent anaemia and many other benefits.

Despite having richer health benefits compared to other rice varieties, black rice and brown rice is not popular for extensive cultivation by farmers, since it can take up to 5 to 7 months before the rice can be harvested.

This type of rice is only cultivated on a limited scale on a small scale because people are still using the rice in cultural rituals. Using biotechnology, plant breeders can utilize the black rice varieties of Tana Toraja, as potential sources of germplasm to assemble new rice varieties whose rice contains anthocyanin.

Local rice in general has some disadvantages such as long life, high plant posture and low yielding results [2]. The character is a limiting factor that may cause farmers' interest to grow local rice lower so it is feared will cause the lack of availability of local rice cultivars as germplasm materials.

To optimize the potential of local rice as a source of germplasm for plant breeding, the first step that needs to be done is to improve the character of local rice varieties. Improvement of the character can be done in various ways, either conventionally or by mutation induction one of them through the use of heavy ion beam. Mutations allow the production of local rice mutant genotypes that have good quantitative and qualitative characteristics.

Research for the improvement of rice varieties with radiation induction has been widely practiced. Atomita 4 rice is a 110-120 day old rice varieties that have been released by the National Atomic Energy Agency (BATAN) induced mutation with gamma-ray radiation of 200 Gray from Cisadane rice with harvest age of 135-140 days [3].

Mustikarini [4] suggests that 200 Gray gamma ray radiations can accelerate the age of rice harvest of Bangka local red rice. Sobrizal [5] re-examined atoms of Atomita 4 with 200 Gy radiations and produced a shorter mutant. Ismachin [6] states that Pandan Wangi rice varieties in gamma-ray radiation produce mutant strains that have short-lived, broad adaptation properties, do not alter the texture and flavor of the rice.

Kadir's research [7] describes the genotypes of red rice mutants resulting from gamma-ray radiation gripped with drought through PEG administration can respond well to stress.

As the times progressed, breeding techniques are now developed using ion beams. This ionic file is safer, does not damage the endosperm, because the dose is low so that the mutation induction rate is higher.

Plant breeding with heavy ion beam utilization is a unique technology in Japan. Heavy ion beam is generated by speeding up atomic ions using particle accelerators. The existing facilities at RIKEN Japan have high performance to create new cultivars for plants and microba. Now the Radiation Biology Team at RIKEN is closely linked with more than 150 parties in research cooperation, including agricultural experiment stations, universities and private companies.

Compared with gamma rays, ion beam irradiation has several advantages such as higher mutation rates, lower doses with high survival rates, high mutation induction rates and various other variations, ion beams can be focused and regulated through penetrating embryo networks, and does not damage the endosperm [8]. Thus, induction of mutation in local rice with radiation irradiation is expected to produce mutant rice which has better characteristic than its original variety.

This research will be the basis for creating new varieties of rice that can be a clear proof of Unhas's contribution to the national agricultural sector. Based on these matters, it is necessary to determine the effect of ion beam irradiation on the germination of local Toraja rice seed.

2. Materials and method

The research was conducted at the Breeding and Seed Technology Laboratory for the seed testing stage and at the Screen House Department of Agricultural Cultivation, Faculty of Agriculture Unhas from March to June 2016.

The material used in this research is Toraja local rice seed, Ambok and Lea varieties that have been irradiated in RIKEN Nishina Center, Japan, planting media in the form of soil mix and manure (2:1 comparison), water for watering, and paper label. The tools used include, plastic tubs for soaking seeds, plastic pots for nurseries, and stationery.

The study was arranged in a Randomized Block Design consisting of six treatments by testing two local Toraja rice varieties namely Pare Ambok, and Pare Lea treated with ion beam irradiation of the argon ion (10 Gy) and carbon (10 Gy) and control as a comparison with the method of planting panicle into the line. Each treatment was repeated as much as 20 times which then is considered as a strain.

3. Results and Discussion

3.1. Germination Rate

This stage showed the effect of irradiation on seed growth germination. From this stage it is hoped that information can be obtained about the effects of irradiation ion types on seed germination in both M1 and M2 generation.

Germination is the proportion of the germinated seeds of some seeds added in the optimum growth medium at a predetermined time and expressed in percent. The germination percentage of each treatment show in table 1.

Table 1. Percentage of germination rate (%) on M1 and M2 generation.

Cultivar	M0	M1			M2		
	Before Irradiation	Irradiation Ion			Irradiation Ion		
		Control	Carbon	Argon	Control	Carbon	Argon
Pare Ambok	85	38	54	55	83	85	90
Pare Lea	98	96	97	98	84	93	94

The highest percentage of germination is found in the treatment of argon ion irradiation on Pare Lea varieties (98%) on M1 and M2 generation. While the lowest germination rate is in the treatment of Ambok varieties without irradiation. Irradiation treatment with carbon ions causes decrease of germination about 31% on Pare Ambok and 1% in Pare Lea is presented in table 1.

3.2. Number of Mutants Plant

The observation of chlorophyll mutation in M2 plant was done by observing the color of the leaves 10 days after sowing until the plant was transplanted to the rice field. The irradiation treatment with ion beam also gives rise to chlorophyll mutations as seen in population M2 (figure 2). The number of plants undergoing chlorophyll mutations (albino) is presented in table 2.

Table 2. Number of plants undergoing chlorophyll mutations

Treatments	Number of plants	Number of mutant	Mutant frequency (%)
Pare Ambok-Control	0	0	0
Pare Ambok-Carbon	414	8	2
Pare Ambok-Argon	90	1	1
Pare Lea-Control	0	0	0
Pare Lea-Carbon	1619	46	3
Pare Lea-Argon	1034	22	2
Number of mutan M2	**3157**	**77**	**2**

Figure 1. Condition of rice seedlings growth

Overall, the number and frequency of chlorophyll mutation emerging in the M2 population are found in the irradiation treatment of carbon ions in both Pare Lea and Pare Ambok cultivars (table 1). The high frequency of mutations induced by carbon ion irradiation is due to the energy fired by larger Carbon ions that is 135 MeV / u compared to the energy at Argon ions which is only 95 MeV / u [9]. This allows the magnitude of the damage caused by the release of energy from the fired ion so that the variety (mutant) generated also more and more. The chlorophyll mutation occurring is a mutagenic effect of irradiated beam ion beam indicating that the treatment is quite effective in creating genetic diversity in the M2 population at the nursery level.

Figure 2. Rice seeds that have mutated chlorophyll (albino) in the nursery

4. Conclusion

The results showed that irradiation using argon ion in local rice seed of Pare Ambok and Pare Lea cultivars had better germination rate compared to irradiation using carbon ion. While the number of mutants were produced more in carbon ion irradiation treatment.

References

[1] Sahardi, Herniwati and Djufry F 2013 Diversity of morphological characteristics of local rice germplasm of Tana Toraja highlands, South Sulawesi *Proc. of the National Seminar on Genetic Resources of Agriculture* p134–43.

[2] Wahdah R and Langai B F 2009 Observation of local rice varieties in tidal lands of South Kalimantan *J. Agroscientiae* **16** (3) 177–84

[3] Wahyudi R, Agus and Shiddiq M 2012 Description of improved varieties of mutation breeding results: Rice, Soybean, Green Bean, Cotton (Jakarta: Nuclear Science and Dissemination Center. National Nuclear Power Agency) 28 pages.

[4] Mustikarini E D, Zasari M and Kartika 2012 *First Mutant Selection (M1) of Red Rice Accession Local Bangka with Treatment Gamma Radiation Dose 200 Gray Bachelor thesis* (Pangkal Pinang: Department of Agrotechnology, Bangka Belitung University)

[5] Sobrizal, Sutisna Sanjaya, Carkum, Ismachin 2004 Dwarf-rice muted irradiated gamma rays of 0.2 kGy on varieties of atomita 4 *Minutes of the Scientific Meeting on Research and Development of Isotope and Radiation Applications National Nuclear Energy Agency* (Jakarta: Puslitbang Isotopes and Radiation Technology) p 35.

[6] Ismachin M and Sobrizal 2006 A significant contribution of mutation techniques to rice breeding in Indonesia *Plant Mutation Rep.* **1** (1) 18-21

[7] Kadir 2011 Respons of mutant genotypes rice to results for irradiation of gamma ray to drought stress *J. Agrivigor* **10** (3) 235–246 ISSN 14122286

[8] Yazid M and Muryono H 2000 Study on the characterization of accelerators for Applications in the field of biotechnology *Proc. of National Seminar on Accelerator Technology and Its Application* vol 2 No 1 pp 139–47

[9] Hayashiah Y, Takehisa, Kazama Y, Ichida H, Ryuto H, Fukunishi N, Abe T, Kamba C and Sato T 2007 Effects of ion beam irradiation on mutation induction in rice. Cyclotrons and Their Applications 2007 *Eighteenth International Conference.*

The Phenotype performance of M3 red rice mutant (*Oryza sativa* L.)

N Kasim[1], R Sjahril[1], M Riadi[1] and F Arbie[2].

[1]Department of Agronomy, Faculty of Agriculture, Hasanuddin University, Campus of UNHAS Tamalanrea, Perintis Kemerdekaan KM 10 Makassar, 90245, South Sulawesi, Indonesia.
[2]Agrotechnology Studies Program, Faculty of Agriculture, Hasanuddin University, Campus of UNHAS Tamalanrea, Perintis Kemerdekaan KM 10 Makassar, 90245, South Sulawesi, Indonesia.

E-mail: *nina_nurlina@yahoo.com*

Abstract. Local rice genotype generally has colour, flavour and scent more preferred by consumers, yet unfortunately it has long-lived planting period and low production. Therefore, the plant breeding practices in rice needs to be implemented for better rice varieties which are superior in terms of both quality and quantity. Our findings describe the growth character performance and the production of red rice mutant from M3 generation. This study was conducted in the Agriculture Faculty wetlands, Hasanuddin University, Makassar, by using ANOVA test with some red rice mutant genotypes *i.e.* 7 genotypes mutants (G1, G2, G3, G4, G5, G6 and G7) and controls/parent-plants (not the mutant). Results show that there were difference in growth performance and production of red rice mutant. Each parameter observed on each genotype had different results. Mutants produced best response in tillers production were G4 mutant with the tillers grain weight at 99.2 g, whereas by the results of the analysis of rank, mutants showed the best overall response were found in G6 mutants.

1. Introduction

Local red rice is more favored by the community because it has advantages such as having a delicious taste of rice, containing vitamins and minerals higher than white rice, but the existence of local rice has been rivaled by new varieties that have potential high yield [1]. Plant breeding is used to improve and increase the genetic diversity of existing varieties to make it a superior crop in some traits both in quality and quantity so that the breeder can do the selection according to the desired crop characteristics. Diversity can be enlarged by several methods, one method that can be used is a mutation technique commonly known by the name of plant breeding mutations [2].

Mutation is one of the methods of plant breeding that can change the genetic crops randomly and is inherited. There are two mutation methods : natural mutation and induced mutation, both mutations have no differences, this is in accordance with Harten's statement [3] which suggests that there is no difference between the spontaneous mutation and the induced mutation because both can cause genetic variation to be the basis of plant selection, both natural selection (evolution) and artificial selection (breeding).

Mutation techniques in plant breeding programs will more quickly obtain the desired results when compared with the conventional techniques. Using a mutation technique, one of the properties of a variety can be improved without changing the other's properties. If a trait to be repaired is controlled by a linkage gene or closely bound to other genes then the problem can only be solved by using mutation techniques.

The induction mutation technique is the most effective breeding method for the repair of one or more undesirable properties and capable of producing mutants that have different properties from the parent as well as having wide genetic diversity. The mutation technique is complementary with other techniques so that the technique can be used in conjunction with other techniques such as hybridization and biotechnology [4].

Local rice is essentially long-lived and low-yielding but at present there is research that is able to change the age of rice to be shorter. Indonesian breeders successfully transformed 180-day-old (6-month) rice with production of 2-3 tons ha^{-1} to be 105 days old with 6-8 tons ha^{-1} productivity such as red rice varieties Aek Sibundong, a local varieties of North Sumatra released by Indonesian Center for Rice Research in 2006 [5].

The study of Shaleh in 2011 [6] produced the first generation of Suta seed red rice seedlings (M1) from the parent seeds that were irradiated in PATIR-BATAN (Center for Applications of Radioactive Isotopes and Technologies - National Nuclear Energy Agency), from the M1 planting conducted in the Village Karattuang, Bantaeng sub-district, Bantaeng district obtained second generation red rice mutant (M2), then after second generation mutant plantation (M2) conducted in Kera-kera village, Tamalanrea sub-district, Makassar, third generation of the red rice was obtained (M3). Based on the description above, a further research was conducted to study the growth and production performance of the third generation of Sinjai originated red rice genotypes [6].

2. Research Methods

The research was conducted in wetlands, Faculty of Agriculture, Hasanuddin University, Makassar from May to October 2015. The materials used were local red rice seed and red rice mutant seed (M3) from Sinjai, sand, compost, urea fertilizer, SP-36, KCl and pesticide. The tool used in this research is hoe, hand tractor, water pump machine, sickle, machetes, scales, bucket, container for sowing, wire counter, digital camera, hand sprayer, raffia strap, fishing net, fishing net, moisture tester, Skates and stationery.

This study used ANOVA test with several types of red rice mutants. The mathematical model used as follows:

1. Hypothesis H0: $\mu 1 = \mu 2 = \mu 3$ dan H1 : $\mu 1 \neq \mu 2 \neq \mu 3$
2. Statistical F Test (*analysis of variances*)
3. Level $\alpha = 0{,}05$
4. Areas of critism : count of F $\leq$ table of F or count of F> table of F

$$F = \frac{MS\ between\ Groups}{MS\ Within\ Groups}$$

In this study, local red rice from Sinjai was used, with eight genotypes: control / parent (not mutant), mutant G1, mutant G2, mutant G3, mutant G4, mutant G5, mutant G6, and mutant G7.

3. Result And Discussion

3.1. Plant Height

The results of the average measurements of the plant height of red rice mutant from Sinjai third generation are shown in table 1. Mutant G5 is a genotype that has the highest number of plant height of 142.9 cm and G2 mutant is a genotype that has the average number of the lowest (108.4 cm).

Table 1. Results of analysis of Plant Height (cm) of red rice mutant from Sinjai third generation.

genotype	genotype 1 (141.2)	genotype 2 (108.4)	genotype 3 (137.8)	genotype 4 (138.9)	genotype 5 (142.9)	genotype 6 (136.9)	genotype 7 (138.7)
genotype 2 (108.4)	32.8*						
genotype 3 (137.8)	3.4 ^{ns}	-29.4*					
genotype 4 (138.9)	2.3 ^{ns}	-30.5*	-1.0 ^{ns}				
genotype 5 (142.9)	-1.7 ^{ns}	-34.5*	-5.1*	-4.0 ^{ns}			
genotype 6 (136.9)	4.3 ^{ns}	-28.5*	0.9 ^{ns}	2.0 ^{ns}	6.0*		
genotype 7 (138.7)	2.5 ^{ns}	-30.3*	-0.9 ^{ns}	0.2 ^{ns}	4.2 ^{ns}	-1.8 ^{ns}	
Control (108.4)	32.8*	0.0 ^{ns}	29.4*	30.5*	34.5*	28.5*	30.3*

ns = Not significant

3.2. Productive Tillers

Results of observation of productive tillers of red rice mutants from the third generation are as follows.

Table 2. Results of analysis of productive tillers of red rice mutant from the third generation

genotype	genotype 1 (37.6)	genotype 2 (21.8)	genotype 3 (16.9)	genotype 4 (25.5)	genotype 5 (21.2)	genotype 6 (15.5)	genotype 7 (20.1)
genotype 2 (21.8)	15.8**						
genotype 3 (16.9)	20.7**	4.9**					
genotype 4 (25.5)	12.1**	-3.7 ^{ns}	-8.6**				
genotype 5 (21.2)	16.4**	0.6 ^{ns}	-4.3*	4.3*			
genotype 6 (15.5)	22.1**	6.3**	1.4 ^{ns}	10.0**	5.7**		
genotype 7 (20.1)	17.5**	1.7 ^{ns}	-3.2 ^{ns}	5.4**	1.1 ^{ns}	-4.6*	
Control (21.8)	15.8**	0.0 ^{ns}	-4.9*	3.7 ^{ns}	-0.6 ^{ns}	-6.3**	-1.7 ^{ns}

ns= Not significant

Table 2 shows that, mutant genotypes 2-7 (G2, G3, G4, G5, G6 and G7) and Control have different productive tillers with G1 mutants. Mutant G1 is a genotype that has the highest average productive tillers value of 37.6.

3.3. Length of Panicle

The result of observation length of panicle of red rice mutant of third generation that is as follows.

Table 3. The result of analysis length of panicle (cm) of red rice mutants from the third generation.

Genotype	genotype 1 (27.5)	genotype 2 (28.0)	genotype 3 (28.0)	genotype 4 (27.2)	genotype 5 (28.6)	genotype 6 (28.9)	genotype 7 (28.9)
genotype 2 (28.0)	-0.4 ^{ns}						
genotype 3 (28.0)	-0.5 ^{ns}	0.0 ^{ns}					
genotype 4 (27.2)	0.3 ^{ns}	0.7 ^{ns}	0.7 ^{ns}				
genotype 5 (28.6)	-1.1 ^{ns}	-0.7 ^{ns}	-0.6 ^{ns}	-1.4*			
genotype 6 (28.9)	-1.4 ^{ns}	-1.0 ^{ns}	-1.0 ^{ns}	-1.7**	-0.3 ^{ns}		
genotype 7 (28.9)	-1.4 ^{ns}	-1.0 ^{ns}	-1.0 ^{ns}	-1.7**	-0.3 ^{ns}	0.0 ^{ns}	
control (26.9)	0.6 ^{ns}	1.0 ^{ns}	1.0 ^{ns}	0.3 ^{ns}	1.7**	2.0**	2.0**

ns= Not significant

Table 3 shows that, mutant G4 is very different from mutant G6 and G7 mutant. The controls differred significantly ($p<0.05$) from the mutants G5, G6, and G7. Mutants G6 and G7 are genotype that had the average value of the highest panicle length of 28.9 cm.

3.4. Number of grain per panicle

The result of observation of grain of rice paddy mutant rice product of the third generation that is as follows.

Table 4. Results of the analysis of the number of grains (g) of red rice mutant from the third generation

Genotype		**genotype 1 (235.6)**	**genotype 2 (207.4)**	**genotype 3 (211.3)**	**genotype 4 (245.7)**	**genotype 5 (242.3)**	**genotype 6 (256.0)**	**genotype 7 (243.9)**
genotype 2	**(207.4)**	28.3 ns						
genotype 3	**(211.3)**	24.4 ns	-3.9 ns					
genotype 4	**(245.7)**	-10.1 ns	-38.3**	-34.4**				
genotype 5	**(242.3)**	-6.7 ns	-34.9**	-31.0**	3.4 ns			
genotype 6	**(256.0)**	-20.4 ns	-48.7**	-44.8**	-10.4 ns	-13.8 ns		
genotype 7	**(243.9)**	-8.3 ns	-36.5**	-32.6*	1.8 ns	-1.6 ns	12.2 ns	
Control	**(158.9)**	76.7**	48.5**	52.4**	86.8**	83.4**	97.1**	85.0**

ns= not significant

Table 4 shows that, all mutants (G1, G2, G3, G4, G5, G6 and G7) have an average number of different diamond grains very significant with control. Mutant G6 is a genotype that has an average value of the highest grain that is 256.0 grains.

3.5. Weight of grain per hill

The result of observation of grain weight of red rice mutant of sinjai origin of third generation is as follows.

Table 5. Results of grain weight (g) analysis of red rice mutant from the third generation.

genotype		**genotype 1 (71.4)**	**genotype 2 (87.4)**	**genotype 3 (70.3)**	**genotype 4 (99.3)**	**genotype 5 (53.6)**	**genotype 6 (64.2)**	**genotype 7 (87.4)**
genotype 2	**(87.4)**	-15.9 ns						
genotype 3	**(70.3)**	1.1 ns	17.0 ns					
genotype 4	**(99.3)**	-27.8*	-11.9 ns	-28.9**				
genotype 5	**(53.6)**	17.8 ns	33.7**	16.7 ns	45.6**			
genotype 6	**(64.2)**	7.2 ns	23.2**	6.1 ns	35.0	-10.6 ns		
genotype 7	**(87.4)**	-15.9 ns	0.0 ns	-17.1 ns	11.9 ns	-33.8**	-23.2**	
control	**(85.3)**	-13.9 ns	2.0 ns	-15.0 ns	13.9 ns	-31.7**	-21.1*	2.1 ns

ns=not significant

Table 5 shows that, in grain weight of the G4 mutant axis, it is the genotype with highest weights of grain of 99.3 g. Mutant G4 is very different from mutant G3, G5, and G6 but not significantly different from mutant G2, G7, and Control.

3.6. Discussion

The results show that irradiation treatment had an effect on the growth of red rice mutant from Sinjai in the third generation (M3). Based on the results of statistical analysis using two middle values test, the anova table shows that all genotypes have a very significant effect on all observation parameters. Therefore, Least Significant Difference (LSD) advanced test was carried out to see the relation between one genotype with other genotype.

Changes that occur in plants can be caused as a result of the treatment, but can also be caused by external factors or planting environment, so that the involvement of external influences should be minimized as possible, so that the changes in the characters that appear are really changes that result from the radiation treatment given to plants.

The treatment of gamma ray irradiation on rice plants can affect the genotype of rice crops. Changes in plant genotypes can be seen from the measurement differences on each observed character observed. The irradiation carried out leads to an increase or even decrease in the observed character measured quantitatively.

The result of plant height analysis in table 1 is Control and mutant G2 is different genotype very real with other genotipe. Mutant G2 is a mutant with the shortest plant height indicating that the mutant's height is better than that of other mutants. This is in accordance with the statement of Peng *et al* [7]

IOP Conf. Series: Earth and Environmental Science **157** (2018) 012047 doi:10.1088/1755-1315/157/1/012047

which states that the ideal rice plant height is about 90 cm to 100 cm, with the potential height of the loading will decrease compared to the high crop. Although the expected height of the plant has not been achieved yet with an increase in the height of the treated plants shows that there is a change in the high phenotype of plants in rice plants.

Yoshida [8] divided the number of tillers into three categories (slightly (<17), medium (17-24) and many (> 24), Irawan and Purbayanti also suggested that the number of tillers in each clump is generally small, but in some cultivars Medium or large number of tillers. The statement is seen in the research that has been done that the seeds of rice plants decreased and increased compared to control. Table 2 shows that, mutant G1 has more productive tillers than the other genotype of 37.6.

Productive tillers is one component of the results that directly affect the high low grain yield [1]. Increased productivity of rice crops is associated with the number of productive tillers, because the tillers directly produce rice strands producing grain or paddy, this is seen in G4 mutants, G2 mutants, controls, G7 mutants, G3 mutants, and G6 mutants that have number of seedlings and but the statement is not seen in the mutant G1 and mutant G5 which has the most productive tiller but has the lowest production, it can be seen in tables 2 and 5.

The length of panicle increased after irradiation. In table 3, the highest panicle length was found in mutant G6 and mutant G7 that was 28.9 cm, while the lowest panicle length was in Control which was 26.9 cm. According to Australian Government [8] which states that the length of panicle can be categorized based on the size of short panicles (<20 cm), medium panicles (20 cm - 30 cm) and long panicles (> 30 cm). All genotypes are categorized in moderate panicles because of their length ranging from 20 cm - 30 cm. In Table 3 it can be seen that the length of panicles that undergo a change that is mutant G5, mutant G6 and Mutant G7. Changes that occur in the three mutants is the occurrence of panicle length increase and very different with the real control.

Based on table 4 it can be seen that the genotype that the highest average number of grain of probes is the G6 mutant that is 256 and the genotype which has the lowest number of seeds of the panicle is 158.9. This indicates that the average number of grain probes increases in each mutant so that the irradiation treatment has an effect on the number of seeds because it is not the same as the parent.

The weights of the grains of the grains show how much grain is produced in one clump. Based on the mean of grain weight of cultivars, there is an increase and decrease of grain weight in gamma ray irradiation treatment. Based on LSD test in table 5 it can be seen that G4 mutant has the highest mean of grain weight of 99.2 g with an average of 25.5 panicles, whereas mutant G5 has the lowest mean of grain weights of 53.6 g with an average of 21.2 panicles. The occurrence of diversity of weeds of rice grains shows that irradiation causes the effect of mutation on the plant so that the mutant of rice plant of Sinjai originated in the third generation.

The irradiation treatment causes some mutants to increase in diversity and it is expected that the grain weight character of this clump can be inherited to the fourth generation mutant so that further research is needed to know the stability of the character.

Differences occur both in growth and production due to the effects of irradiation so that in the third generation there has been mutants that grow uniformly in each genotype, although there are still some mutants that have not uniform growth such as mutant G1, mutant G3 and mutant G5. The uniformity of each mutant is evident from the height of the plant, the number of tillers, and the time of harvest.

4. Conclusion

Based on the results of research that has been done, it can be concluded that :

- Mutants G1, G2, G3, G4, G5, G6, G7 and control are significantly different based on observation parameters except on the number of book parameters indicating that there is a difference in the growth and production performance of the sinjai red rice mutants in the third generation.
- The mutant that produces the best response based on the cultivar production is the G4 mutant with the weight of the axle of grain 99.2 g while in the rank analysis, the mutant showing the best response is the G6 mutant.

IOP Conf. Series: Earth and Environmental Science **157** (2018) 012047 doi:10.1088/1755-1315/157/1/012047

Reference

[1] Yoshida S 1981 *Fundamentals of Rice Crop Science* (Los Banos: International Rice research Institute)

[2] Department Of Health and Ageing Office of the Gene Technology Legulator. 2005. *The Biology and Ecology Of Rice (Oryza sativa L.)* (Australia: Australian Government. Department of Health and Ageing_Office of the Gene Technology Legulator) pp 1-26

[3] Harten Van 1998 *Mutation Breeding: Theory and Practical Application* (Cambridge: Cambridge University Press.)

[4] International Atomic Energy Agency 1971 *Rice Breeding With Induced Mutations III* (Vienna: International Atomic Energy Agency)

[5] Ro P V and At D H 2001 *Improvement Of Traditional Local Rice Varieties Through Induced Mutations Using Nuclear Techniques* Vietnam.

[6] Shaleh R M 2013 *Aplikasi Sinar Gamma Terhadap Keragaan Karakter, Tiga Genotipe Padi Lokal Sulawesi Selatan Generasi M1* [in Indonesian] *Skripsi* (Makassar: Jurusan Budidaya Pertanian Universitas Hasanuddin)

[7] Peng S, Khush G S, Virk P, Tang Q and Zou Y 2008 Progress in ideotype breeding to increase rice yield potential *Field Crop Res.* **108** 32-38

[8] Australian Government 2005 *The Biology and Ecology of Rice (Oriza sativa L.) in Australia* (Australia: Office of the Gene Technology Regulator)

IOP Conf. Series: Earth and Environmental Science **157** (2018) 012048 doi:10.1088/1755-1315/157/1/012048

Evaluation of some new plant type of upland rice (*Oryza sativa* L.) lines derived from cross breeding for the growth and yield characteristics

G R Sadimantara, W Nuraida, N W S Suliartini and Muhidin

Department of Agrotechnology, Faculty of Agriculture, Halu Oleo University Kendari, Southeast Sulawesi, 93232 Indonesia

E-mail: *gusti5@yahoo.com*

Abstract. Improving rice production through cross breeding is one of the efforts to overcome a major challenge in future need of rice consumers. An experiment was conducted to evaluate different upland rice lines for their growth and yield characteristics. Four rice lines including GS44-2 (V1), GS16-2 (V2), GS44-1 (V3), and GS16-1 (V4) were evaluated in a randomized complete block design (RCBD). The variables observed were vegetative character and generative character including grain yield. Data on various growth and yield characteristics revealed there were growth and yield characteristics different among breeding lines of local upland rice. Plant height at age 84 DAP, GS16-2 line was the tallest (92.71 cm) and GS44-1 line the shortest one (70.07 cm). Breeding line of GS16-2 produced higher leaf area (61.0 cm2), panicle length (30.94 cm), grain number per panicle (230.60), and filled grain number (190.25). The lowest unfilled grain percentage (26.67 %) was also recorded from GS16-2 lines. The highest number of tillers, number of productive tillers, grain dry weight, and grain yield were obtained from GS44-2 lines. Further, grain yield component such as grain dry weight and 1000 grain weight recorded positive and significant correlation with number of productive tillers per hill.

1. Introduction

Rice (*Oryza sativa* L.) is second most widely grown cereal crop and the staple food for more than half of world's population, providing two thirds of calorie intake for more than three billion people in Asia, Africa and Latin America [1]. Rice grain contains 75 to 80% starch, 12% water, and 7% protein [2, 3]. Rice is the most important source of staple food in Indonesia, followed by corn and tubers. Rice production in Indonesia is almost entirely dependent on lowland rice, so that the production sustainability is difficult to maintain. This is due to the large conversion of land functions, the degradation of soil fertility, and the disruption of water availability. Therefore, the development of rice cultivation in addition to paddy rice is very necessary especially upland rice [4]. Indonesia is one of the main rice producer but the main primary consumer also [5]. The Indonesian government has two main police to maintain stability of rice supply [6], through the increasing rice production from existing farm while decreasing rice consumption level and promoting local food such as sago [7].

Upland rice in Indonesia generally cultivated in upland areas with low yield (2 ton ha^{-1}), low quality and aged (5-6 months), so that new potential gene sources for the development of high yielding and early maturity varieties necessary to be considered. Other factors that also lead to low production of upland rice are the lack of adaptive promising lines, which have good rice quality, medium yield potential, pest and disease resistance, and tolerance to mineral pressure.

The development of high potential yield of upland rice is also aimed to preserve the genetic potential of local upland rice [6] that has undergone genetic erosion, especially the local aromatic and flavor upland rice. In the previous study, obtained the best progeny derived from 33/Wagamba and its reciprocal cross due to several advantages [8] such as performance of upright growth, plant height, number of tillers, and dark green leaves. Furthermore, reported eight promising progeny of upland rice that are tolerant to drought and high potential yield [9]. Genetic diversity information is needed to obtain new varieties that are expected. In order to achieve the objective of selection, it must be known between the agronomic characters, the yield and yield components, so that the selection of one or more characters can be done [10]. The objectives of this study were to evaluate different new plant type of upland rice lines for their growth and yield characteristics.

2. Materials and Methods

Four new plant types of upland rice lines including GS44-2 (V1), GS16-2 (V2), GS44-1 (V3), GS16-1 (V4) were grown at the experiment station of Faculty of Agriculture, Halu Oleo University in a randomized complete block design (RCBD) with four replications. The variables observed were plant height, leaf area, number of tillers, number of productive tillers, flag leaf width, panicle length, number of grains per panicle, number of filled grain per panicle, percentage of unfilled grain, 1000 grain weight, grain dry weight , yield (ton ha^{-1}), and plant biomass. The variances of the data were analyzed by analysis of variance (ANOVA) with SPSS software 13.0 the means were compared using Duncan's Multiple Range Test (DMRT) at the 5% level. The correlation of yield components and yield was analysed using Pearson's correlation coefficient analysis.

3. Results and Discussion

The average of plant growth characteristics of upland rice lines, *i.e.* plant height, panicle length, leaf area, number of tillers, and flag leaf width were varied and significantly different among four new type upland rice lines (table 1). The same result with yield components, *i.e.* number of productive tiller, number of grain per panicle, filled grain per panicle, percentage of unfilled grain, 1.000 grain weight, and grain dry weight (table 2).

3.1. Plant Height

Plant height showed significantly difference among the new plant type upland rice lines (table 1 and figure1). The tallest plant was GS16-2 (92.71 cm) and the shortest was GS44-1 line with 70.01 cm height. According to IRRI [11], plant height is divided into three categories, namely short (<110 cm), medium (110-130 cm), and tall (>130 cm).

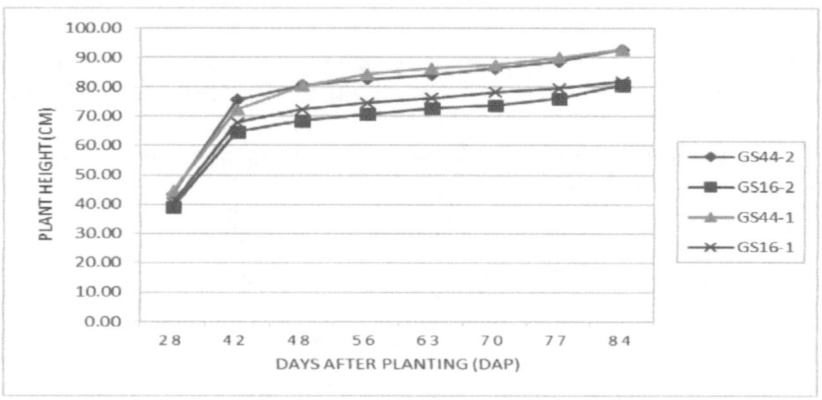

Figure 1. Growth of plant height of four new type upland rice lines

Based on these categories, four upland rice lines evaluated in the present study were short (70.01 cm-92.71 cm). Tall plant is not desired because it is sensitive to lodging, and lodging will reduce grain yield [12]. Furthermore, Bhadru *et al* [13] stated that the plant height is highly correlated with the level of lodging and ease of harvest, so it is one of the important characters in influencing the levels of farmer's acceptance on new cultivar

3.2. Leaf Area

Leaf area measured at 84 day after planting and recorded different among upland rice lines. The lines with the highest leaf area was GS16-2 (61.00 cm^2), significantly different with GS44-2 and GS44-1 lines but was not significantly different with GS16-1 line. The narrowest was GS44-1 line (36.32 cm^2).

3.3. Number of Tillers

The tested lines have average tillers number significant difference each other. GS44-2 was the line with the highest tillers number and significantly different with GS16-1 and GS16-2 lines. The lines with the lowest number of tiller was GS16-2. Wang *et al.* [14] reported that the unequal distribution of photo-synthetically active radiation (PAR) was the source of heterogenity in individual tiller yields, in that early emerging superior tillers pre-empted the uppermost light source, and shaded the late emerging tillers under limited light conditions.

Table 1. Average of plant growth characteristic of upland rice lines

Rice lines	Plant height	Leaf area	Number of tillers	Flag leaf width	Panicle length
GS44-2	80.73 b	45.39 b	25.20 a	1.33 b	29.62 b
GS16-2	92.71 a	61.00 a	12.60 b	1.62 a	30.94 a
GS44-1	70.01 c	36.32 c	24.95 a	1.20 c	28.07 c
GS16-1	89.44 a	57.53 a	14.45 b	1.53 a	28.08 c

Means followed by the same letter are not significantly different at 5 % level by DMRT

3.4. Flag Leaf Width

Flag leaf width at maturity obtained significant difference among upland rice lines which ranging from 1.20-1.62 cm. The lines with the widest flag leaf was GS16-2, while the narrowest was GS44-1 line. The GS16-2 line was significantly different with GS44-2 and GS44-1 lines, respectively.

3.5. Panicle Length

The tested lines have panicle length ranging from 28.07 to 30.94 cm. The lines with the longest panicle was GS16-2; while the shortest was GS44-1 line. Rice plants with long panicles potentially have high number of grain total and high yield because there is a positive correlation between panicle lengths with the number of grains per panicle [15] and weight of 1,000 grain [16].

3.6. Number of Productive Tillers

Number of productive tillers varied significantly among upland rice lines (table 2). The average numbers of productive tillers ranging from 10.45-18.90 tillers with average of 14.33 tillers.

Table 2. Average of yield components of upland rice lines

Rice lines	Productive tillers	Number of grain/ panicle	Number of filled grain	Percentage of unfilled grain	Weight of 1000 grain	Grain dry weight
GS44-2	18.90 a	182.30 bc	149.80 bc	21.70 bc	28.63 b	61.77 a
GS16-2	10.45 b	230.60 a	190.25 a	21.21 cd	26.67 d	45.74 c
GS44-1	16.35 a	167.75 c	138.60 c	21.03 d	30.80 a	55.57 b
GS16-1	11.65 b	200.80 b	162.90 b	23.26 a	27.20 c	41.19 d

Means followed by the same letter are not significantly different at 5 % level by DMRT

The highest number of productive tiller was GS44-2 line, while GS16-2 line recorded the lowest number among lines. Low number of tiller follows the low productivity [17].

3.7. Number of grains per panicle

Range of total grain per panicle was from 167.75 (GS44-1) to 230.60 grains (GS16-2). Average numbers of total grain were more than 195 grains per panicle. The GS16-2 line was significantly different from the GS16-1 line, GS44-2 line, and GS44-1 line. Haryanto *et al* [15] reported that in addition to the number of tillers, panicle length and 1.000 grain weight, the high number of grains total per panicle would also increase rice yield.

3.8. Number of filled grain per panicle

Numbers of filled grains per panicle range from 138.60 to 190.25 grains. GS16-2 had the highest filled grains that was significantly different with GS16-1, GS44-1, and GS44-2 lines. The lowest was GS44-1 with the number of grains of 138.60 grains. In addition to panicle length, the number of grains per panicle is one of yield components that affect the productivity of rice [17]. The difference between the number of grains per panicle and the amount of unfilled grain per panicle is thought to be caused by the genetic influence of each different line. But apart from genetic influences, environmental factors also affect the number of grains per panicle and the amount of unfilled grain per panicle.

3.9. Percentage of unfilled grain

The percentage of unfilled grain ranged from 21.03% to 23.26% and different among the lines. The highest and the lowest percentage of unfilled grain were GS44-1 and GS16-2 lines, respectively. The grain yield is determined by the number of panicles per unit area, the number of grains per panicle, the fertility of the grains and the weight of the grain. Each growth stage contributes to the rice yield.

3.10. Weight of 1000 grain

Weight of 1.000 grain ranged from 26.67 g to 30.80 g. The highest weight was recorded in line GS44-1 (30.80 g) followed by GS44-2 (28.63 g), GS16-1 (27.20) and GS16-2 (26.67 g). The GS44-1 line was found significantly different from the GS44-2, GS16-1, and GS16-2 lines.

3.11. Grain dry weight

Grain dry weight showed significant difference among the upland rice lines tested. Average of grain dry weight of GS44-2 line was the highest compared with GS44-1, GS16-2, and GS16-1 lines. The lowest weight of dried grains recorded on GS16-1 line.

3.12. Grain Yield (t/ha)

Average of grain yield was 5.11 t/ha, while it ranged from 4.12 t/ha to 6.18 t/ha and all of the lines significantly different each other. The highest grain yield (6.18 t/ha) was observed in GS44-2, while the lowest grain yield (4.12 t/ha) was observed in GS16-1 (Fig. 2). Factors that affect the rice yield are not only genetic, but environmental factors also affect the yield of rice plants.

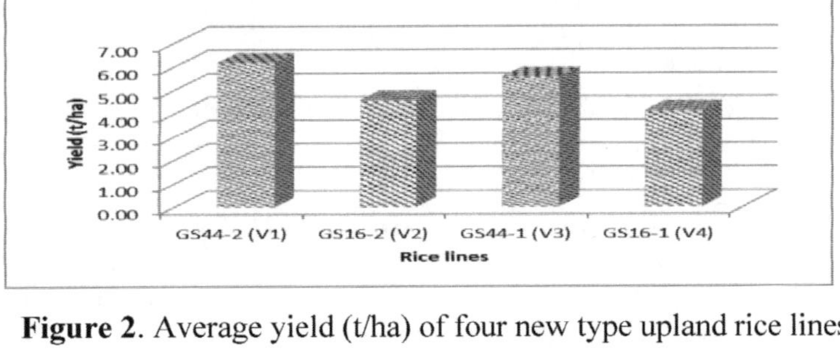

Figure 2. Average yield (t/ha) of four new type upland rice lines

3.13. Characters Correlation
Correlations are measures of the intensity of association between characters. Correlation between characters can be used to determine correlated responses in selection activities. By using correlation analysis, it can be seen whether the agronomic characters and yield component have a positive correlation or negative correlation to the yield of four upland rice lines. As well as on the number of productive tillers and the grain dry weight which has a correlation value of 0.817 (Table 3).

Table 3. Pearson correlation analysis of yield components and yield

	PH	LA	TN	PTN	FLW	PL	GDW	GN	FGN	PUG	TGW
PH	1										
LA	0.947**	1									
TN	-0.840**	-0.806**	1								
PTN	-0.664**	-0.608**	0.781**	1							
FLW	0.95**	0.912**	-0.872**	-0.717**	1						
PL	0.539*	0.444*	-0.412	-0.231	0.5*	1					
DGW	-0.636**	-0.677**	0.835**	0.817**	-0.695**	0.029	1				
GN	0.851**	0.765**	-0.809**	-0.664**	0.792**	0.675**	-0.564*	1			
FGN	0.807**	0.713**	-0.763**	-0.622**	0.739**	0.676**	-0.509*	0.994**	1		
PUG	0.916**	0.889**	-0.892**	-0.754**	0.913**	0.545*	-0.737**	0.846**	0.784**	1	
TGW	-0.925**	-0.873**	0.793**	0.622**	-0.885**	-0.519*	0.662**	-0.777**	-0.719**	-0.926**	1

*PH=plant height, LA=leaf area, TN=tillers number, PTN=productive tillers number, FLW=flag leaf width, PL=panicle length, GDW=grain dry weight, GN=grain number/panicle, FGN=filled grain number, PUG=percentage unfilled grains, TGW=1000-grain weight. *= significant different; **=highly significant different.*

The more number of productive tillers the more dried grain will be produced. Whereas if there are two properties indicating the negative correlation that means a large increase or increase in number of properties will be followed by a decrease in the size or number of other properties. Furthermore, the yield component is a quantitative trait that affects the grain yield, so the high yield is highly depends on the yield components [18]. The correlation coefficient can be either expressed as the total effect of a yield component on the grain yield, directly or indirectly generated by genetic factors, environmental factors, and interactions between the two. Each character that is correlated with the grain yield is broken down into two components: direct and indirect influences. Grain yield was positively associated and not significantly with days to maturity [19].

4. Conclusions
The new plant type of upland rice lines derived from crosses have different growth and yield characteristics. Plant height at age 84 DAP, GS16-2 line was the tallest (92.71 cm) and GS44-1 line the shortest one (70.07 cm). Breeding line of GS16-2 produced higher leaf area (61.0 cm^2), panicle length (30.94 cm), grain number per panicle (230.60), and filled grain number (190.25). The lowest unfilled grain percentage (26.67 %) was also recorded from GS16-2 lines. The highest number of tillers, number of productive tillers, grain dry weight, and grain yield were obtained from GS44-2 lines. Further, grain yield component such as grain dry weight and 1000 grain weight recorded positive and significant correlation with number of productive tillers per hill.

Acknowledgments
The authors extend their gratitude to the Directorate General of Research, Ministry of Research, Technology and Higher Education of the Republic of Indonesia for providing research grant under the MP3EI Scheme in the budget year of 2017 to support this research.

References

[1] Lestari AP, Abdullah B, Junaedi A and Aswidinnoor H 2015 Agronomics characteristics and its correlation of new plant type promising rice lines *Buletin Plasma Nutfah* **17** (2) 96-103

[2] Oko AO, Ubi BE, Efisue AA and Dambaba N 2012 Comparative analysis of the chemical nutrient composition of selected local and newly introduced rice varieties grown in Ebonyi State of Nigeria *Int. J. Agric.For.* **2** (2) 16-23

[3] Hossain S, Maksudu HMD and Jamilur RJ 2015 Genetic variability, correlation and path coefficient analysis of morphological traits in some extinct local Aman rice (*Oryza sativa* L.) *J. Rice Res* **3** 158

[4] Syaiful SA, Syam'un E, Dachlan A, Jusoff K and Haerani N 2013 The effect of inoculating nitrogen fixing bacteria on production of rice *World Applied Sciences Journal* **26** (26) 94-99

[5] Muhidin, Jusoff K, Elkawakib S, Musa Y, Kaimuddin, Meisanti, Sadimantara GR and Baka LR 2013 The development of upland red rice under shade trees *World Applied Sciences Journal* **24** (1) 23-30

[6] Kadidaa B, Sadimantara GR, Suaib, Safuan LO and Muhidin 2017 Genetic diversity of local upland rice (*Oryza sativa* L) genotypes based on agronomic traits and yield potential in marginal land of North Buton Indonesia *Asian Journal of Crop Science* **9** (4) 109-117

[7] Muhidin, Leomo S, Alam S and Wijayanto T 2016 Comparative studies on different agroecosystem base on soil physicochemical properties to development of sago palm on dryland *Int. J. of ChemTech Res.* **9** (8) 511-18

[8] Sadimantara GR, Muhidin and Cahyono E 2014 Genetic analysis on some agro-morphological characters of hybrid progenies from cultivated paddy rice and local upland rice *Advanced Studies in Biology* **6** (1) 7–18

[9] Sadimantara GR, Muhidin, Ginting S and Suliartini NWS 2016 The potential yield of some superior breeding lines of upland rice of Southeast Sulawesi Indonesia *Biosci., Biotech. Res. Asia* **13** (4) 1867-1870

[10] Zen S 2002 Genetic parameters of agronomic character of paddy rice promising lines. *Stigma* **10** (4) 325-330 (*in Indonesian, with English abstract*)

[11] International Rice Research Institute (IRRI) 2002 *Standard Evaluation System for Rice* (International Rice Research Institute. Philippines) 56 p

[12] Shahidullah SM, Hanafi MM, Ashrafuzzaman M, Uddin MK and Meon S 2009 Analysis of lodging parameters in aromatic rice *Achieves on Agronomy and Soil Sci* **55** 525-533

[13] Bhadru D, Reddy DL and Ramesha MS 2011 Correlation and path coefficient analysis of yield and yield contributing traits in rice hybrids and their parental lines *Electronic Journal of Plant Breeding* **2** (1) 112-116

[14] Wang Y, Ren T, Lu JW, Ming R, Li PF, Saddam H, Cong RH and Li XK 2016 Heterogeneity in rice tillers yield associated with tillers formation and nitrogen fertilizer *Agronomy Journal* **108** 1717–1725

[15] Haryanto TAD, Suwarto and Yoshida T 2008 Yield stability of aromatic upland rice with high yielding ability in Indonesia *Plant Prod. Sci.* **11** 96-103

[16] Akinwale MG, Gregorio G, Nwilene F, Akinyele BO, Ogunbayo SA and Odiyi AC 2011 Heritability and correlation coefficient analysis for yield and its components in rice (*Oryza sativa* L.) *Afr. J. Plant. Sci* **5** (3) 207-212

[17] Hossain MF, Bhuiya MSU and Ahmed M 2005 Morphological and agronomic attributes of some local and modern aromatic rice varieties of Bangladesh *Asian J. Plant Sci* **4** 664-666

[18] Ranawake AL and Amarasinghe UGS 2014 Relationship of yield and yield related traits of some traditional rice cultivars in Sri Langka as described by correlation analysis *Journal of Scientific Research & Reports* **3** (18) 2395-2403

[19] Khan AS, Imran M and Ashfaq M 2009 Estimation of genetic variability and correlation for grain yield components in rice (*Oryza sativa* L.) *Am-Euras. J. Agric. Environ. Sci* **6** (5) 585-590

IOP Conf. Series: Earth and Environmental Science **157** (2018) 012049 doi:10.1088/1755-1315/157/1/012049

Genetic by environment interactions and stability of tropical wheat lines in Indonesian medium-plains

A Nur[1], K Syahruddin[1], M Azrai[1] and M Farid[2]

[1]Indonesia Cereal Research Institute (ICERI), Jl. Dr.Ratulangi No. 274 Maros, South Sulawesi, Indonesia.

[2]Department of Agronomy, Faculty of Agriculture, Hasanuddin University, Jl. Perintis kemerdekaan KM. 10 90245 Makassar, South Sulawesi, Indonesia.

E-mail: iceriamin76@gmail.com

Abstract. Wheat can adapt well in Indonesia in altitude of 1000-3000 m above sea level (asl) with production ranging 3-5 t/ha. Recently, Indonesia has released several varieties of wheat, but only a few varieties have an adaptation zone under 1000 m asl, which competes to horticultural crops. Therefore, it is important to study the genetic x environment (G x E) interactions and the yield stability of tropical wheat genotypes in the low-medium plains. The study was arranged using Randomized Completed Block Design with 3 replications and 9 genotypes at six locations. The data was analyzed for variant, variant combined and yield stability using STAR programs. The results showed that the environment, genotypes, and interaction of G X E had a highly significant effect on yield and yield components, except for seed weight/spike trait. Further analysis result showed the G4 genotype has significantly greater than the control for trait of days to flower, days to harvest, plant height and a number of panicles/meter. The G1, G4 and G5 genotypes have significantly higher yields than controls. The G4 and G5 genotypes have wide adaptability and G1 genotype has specific adaptability.

1. Introduction

Wheat in Indonesia has high economic value. This is reflected in the pattern of consumption and import of wheat in the form of seed and flour continue raises along with the increasing number of population, especially in urban areas. Currently, Indonesia is the second largest importer of wheat after Egypt. Based on data of USDA (2015 – 2017) for wheat import estimated wheat imports reached 10.7 million tons, increased approximately 9.8% from the year 2013 as much as 7.17 million tons [1].

The policy of wheat import to meet the increasing of public need is deemed less precise, considering the magnitude of foreign exchange depleted and the nation's economic dependence on the wheat exporting countries. Therefore, the effort to grow wheat in Indonesia is very important in order to reduce the dependence of wheat imports. However, the problem of unavailability of local genetic resources and limited introduced genetic resources made the development of wheat crops in Indonesia hampered. Wheat crops can basically adapt well in Indonesia at the altitude of 1000-3000 m above sea level (asl) with production reaching 3 ton ha^{-1} in Malino (1350 m asl) and 4 ton ha^{-1} in Tosari [2]. However, the planting of wheat on that altitude competes with horticultural products which have higher economic value, so it is important to produce wheat genotypes that can adapt in medium plains. The genotypes arose from the decline of wheat growing altitude was decreased wheat production due to stunted plant growth by the increasing temperature and humidity in medium plains.

The multilocation test aims to determine the yield and adaptability of the genotype in different locations and to examine the interaction of the environment x genotypes. According to [3], the G × E interaction information is very important for the countries with wide bio-geophysical variability such as Indonesia. The genotype adaptability can be observed by studying the interactions between the environment x genotypes or the genotype x season [4]. The interaction information of the genotype x environment is useful to know the stability of the phenotype yields in order to ease the selection of varieties in the plant breeding program [5]. Genotype x Environment interaction decrease the genetic progression in plant breeding program by minimizing the association between the phenotype and genotype values, therefore multi-environment is essential to identify the superior genotypes in the final selection cycles [6].

The approach method to assessing the ability of G x E interactions is to use the methods proposed by Finlay and Wilkinson [4] and Eberhart and Russel [7]. This study aims to determine the stability of yield components and the yield of nine genotypes on eight test locations by looking at the G x E interactions using the models introduced by Finlay and Wilkinson [4] and Eberhart and Russel [7].

2. Material and Methods

The research was conducted at six locations in Indonesia: Garut and Cisarua-West Java (2014), 600 m asl, Malino and Jeneponto-South Sulawesi (2014) Malino and Jeneponto-South Sulawesi (2015) 800 m asl, Malang-East Java (2015) altitude 600 m asl and Salatiga-Central Java (2015) 600 m asl. The experiment was conducted using Randomized Completed Block Design (RCBD) with three replications. Size of the plot was 1.5 m x 5 m. Genetic material used six genotypes and three national varieties as control (Guri3 Agritan, Guri5 Agritan and Dewata). Each genotype planted in 6 rows along 5 m with an intercropping distance 25 cm, the seeds are drawn in rows. The plants were fertilized by 150 $kg.ha^{-1}$ Urea, 200 $kg.ha^{-1}$ SP-36 and 100 $kg.ha^{-1}$ KCl in the 10 Days-After-Planting (DAP) and the second fertilization using 150 $kg.ha^{-1}$ Urea at 30 DAP. Before planting the seeds were given carbaryl insecticide 85% and at the time of planting the runway was given 17 $kg.ha^{-1}$ of carbofuran.

The observation of agronomical trait includes days to flower (day), days to harvest (day), plant height (cm), panicle length (cm), number of spikelet, number of empty floret, void floret percentage, seed weight (g)/panicle, number of $spike.m^{-2}$, number of seeds/panicles, weight of 1000 seeds (g), and yield ($ton.ha^{-1}$). The data were analyzed using variance analysis, combined variance analysis following methods proposed by Singh and Chaudhary [8] with STAR program. If F-test shows significance of G x E interactions, continued with yield stability following methods in Finlay and Wilkinson [4] as well as Eberhart and Russel [7].

3. Result and Discussion

Combined variance analysis (table 1) showed that the environmental influence is highly independent to all observed trait. The genotypes have highly significant effect on the trait of days to flower, days to harvest, plant height, panicle length, number of panicles per meter, number of spikelet per panicle, number of seeds per panicle, weight of 1000 seeds and yield and have a significant effect on weight of seed/panicle. The interaction of G x E also had a highly significant effect on all observed trait. The same result is also shown by Nur [9] which showed that environmental influences are highly significant in almost all agronomic and yield trait excepted in the trait of days to flower. The study also showed that the interaction of G x E did not significantly affected the plant height, days to harvest, panicle length and number of panicles/m^2, contrasted to the recent study results which were highly different in those traits.

The results of this study indicated the entire agronomic and yield traits were strongly influenced by environment factor and G x E interactions. G x E interaction can assist in identifying trait that contribute to better plant performance and the environment that helps to evaluate the genotypes [10]. Wheat is a very sensitive to the environmental changes [11]. The results of Mohamed [12] showed that environment, genotype and the G x E interaction effected to yield trait. The Declining elevation of wheat cultivation caused heat stress [13] and high humidity. Heat stress was a major factor in decreasing yield in Egypt [14]. Heat stress also affected wheat productivity in tropical regions around the world [15]. Heat stress caused some agronomic trait of wheat change [16].

Highly significant of G x E interaction whether on the environment, the genotypes, and the interactions showed differences in response between genotypes and the response of each genotypes to the environments (table 2). On the agronomical trait, only the number of spikelet/panicle was not significantly different to the controls, but remained significantly different between the genotypes (table 2).

Table 1. Mean square of variance combined analysis on the agronomic trait and yield.

Traits	(E)	(G)	(G x E)	CV
Days to flower (HST)	1,177.13**	787.62**	167.08**	6,64
Days to harvest (HST)	1,218.16**	363.02**	64.62**	4,63
Plant height (cm)	2,316.67**	481.69**	123.08**	5,66
A number of spike/m^2	61,000.44**	43,269.50**	4,691.94**	18,4
A number of spike	57.62**	27.82**	8.22**	7,9
Spike length (cm)	7.28**	5.57**	1.43**	6,8
A number of seed/m^2	3,237.23**	174.24**	279.27**	13,3
Seed weight/spike (g)	6,24**	0,41*	0,46**	16,9
Weight of 1 L seeds (g)	330631.80**	20,645.91**	12954.28**	9.4
Weight of 1000 seeds (g)	826,62**	51,62**	32,39**	9,5
Yield (t/ha^{-1})	2,90**	1,46**	0,79**	15,5

Description:** = F : 1%; * = F : 5%; CV= variance coefficient

Table 2. Average of tropical wheat genotype trait on the Indonesian medium plains, 2014-2015.

Genotypes	PH	DF	DH	SPM	SL	SPk
	--cm--	---days---			--cm--	
G1	80.85^{b}	55.42^{abc}	99.94^{abc}	214.63^{c}	16.96	9.16
G2	77.33^{abc}	54.27^{abc}	99.05^{abc}	216.19^{c}	18.59^{ab}	8.77
G3	73.37^{abc}	53.58^{abc}	97.25^{abc}	209.12^{c}	16.05	8.19
G4	75.70^{abc}	57.17^{abc}	100.26^{abc}	284.59^{abc}	15.92	8.28
G5	84.12	62.11^{abc}	103.73^{bc}	320.50^{abc}	17.07^{a}	8.55
G6	74.61^{abc}	55.46^{abc}	98.94^{abc}	248.3 ^{c}	15.64	7.80
Guri3 (a)	82.42	59.78	105.74	251.37	16.65	8.99
Guri5 (b)	85.55	69.38	107.89	226.92	17.99	8.79
Dewata (c)	82.95	67.08	106,65	181.45	18.30	9.25
Mean	79.60	59.40	102.2	239.23	17.02	8.64
SE	0.92	0,80	0.96	8.96	0.13	0.32
5%LSD	2.57	2.25	2.70	25.08	0.38	0.91
CV (%)	5.6	6.6	4,6	18.36	7.86	9.43

Description: PH: Plant Height (cm), DF: Days to flower (Days), DH: Days to Harvest (Days), SPM: Number of spike/m^2, SL: Spike Length (cm), SPk: Number of Spikelet/panicle
* The numbers in the same column followed by the letter are not significantly different to controls at the 5% level by Least Significant Different (LSD)

In agronomic trait, only G5 genotype showed non-significant different to the controls for plant height trait. For Indonesia medium-plains area, wind is a limitation factor causing many plants in turn decreasing crop production, because the strong wind will lay down the plant and causing unstable production. Therefore, short wheat crop is preferred because it is more wind resistant, although the weakness short plants is ease of attacked by rat [17]. In long panicle trait, the G2 genotypes showed significant differences with the controls of Guri3 and Guri5, while the G5 genotypes showed only significant differences to the Guri3.

The G4 and G5 genotypes indicated for higher number of panicles compared to controls.. The number of panicles indicates the potential of the genotype may form a seed, which is highly correlated with the ability of each panicle to produce fertile pollen, where fertile pollen is affected by ambient temperature. According to Rawson [18] high yielding in wheat crops is determined by the number of panicles per unit with productive tillers. Table 3 showed the trait of yield and yield components, 1000 seeds weight and weight of 1 litre of seed showing significant differences compared to control. This indicates that the influence of the environment in this location was considerable influenced the wheat yield in medium-plains area. These results are supported by Mohamed [12], that the environment and G x E interactions are positively correlated to wheat yields.

Table 3. The average of tropical wheat yield component traits (2014-2015).

Genotypes	NSS	SwS	W1000	W1L
			---g----	
G1	43.00	1.57	35.85 ^{abc}	717.093 ^{abc}
G2	43.39	1.49	33.88	677.537
G3	41.02	1.42	33.96	679.241
G4	41.59	1.26	30.27	605.389
G5	42.29	1.41	33.56	671.167
G6	37.21	1.26	33.85	677.000
Guri3 (a)	43.08	1.45	33.28	665.611
Guri5 (b)	47.45	1.66	32.77	655.370
Dewata (c)	43.53	1.48	33.90	678.074
Mean	42.51	1.44	33.48	669.610
SE	1.36	0.05	0.65	12.91
5% LSD	1.15	1.38	1.80	36.13
CV(%)	13.31	16.84	9.50	9,40

Description: value followed by different letters in the same column significantly different at LSD 5%. NSS: Number of seed/spike, SwS: Seed Weight/Spike., W1000: weight of 1000 seeds, W1L: Weight of 1litre of seed.

Tabel 4. Yield average at 12% (tons.ha^{-1}) moisture content of wheat (2014 – 2015).

Genotypes	Yield (ton.ha^{-1})								Average
	L1	L2	L3	L4	L5	L6	L7	L8	
G1	2.48^{a}	3.01^{a}	3.26^{ab}	2.33^{a}	1.96^{b}	2.72^{a}	1.77^{d}	1.69^{bc}	2.40^{a}
G2	2.01^{bc}	2.76^{ab}	2.61^{ad}	1.44^{d}	1.58^{bc}	1.86^{bd}	1.65^{d}	1.64^{bc}	1.94^{cd}
G3	1.55^{d}	2.28^{bc}	2.15^{cd}	1.51^{cd}	1.43^{c}	1.56^{cd}	2.32^{c}	0.88^{e}	1.71^{d}
G4	2.25^{ab}	2.13^{c}	3.49^{a}	1.77^{bc}	0.74^{d}	2.18^{ac}	2.93^{a}	2.72^{a}	2.28^{ab}
G5	2.23^{ab}	2.07^{c}	2.55^{ad}	1.88^{b}	2.43^{a}	1.56^{cd}	2.70^{b}	2.55^{a}	2.24^{ab}
G6	1.49^{d}	1.35^{d}	0.93^{e}	1.96^{b}	1.89^{b}	2.40^{ab}	2.30^{c}	2.83^{a}	1.89^{cd}
Guri3	2.05^{ac}	2.11^{c}	1.83^{de}	1.67^{bd}	1.77^{bc}	1.49^{de}	3.08^{a}	1.22^{de}	1.90^{cd}
Guri5	2.20^{ab}	2.42^{ac}	2.94^{ac}	1.49^{cd}	0.87^{d}	1.83^{bd}	2.62^{b}	1.84^{b}	2.03^{bc}
Dewata	1.67^{cd}	1.91^{cd}	2.32^{bd}	1.38^{d}	1.55^{bc}	0.93^{e}	2.57^{b}	1.40^{cd}	1.72^{d}
Genotype	**	**	**	**	**	**	**	**	
CV	13.2	15,9	20	10,7	15,2	19,6	4,1	11,3	
LSD	0,45	0,61	0,9	0,32	0,42	0,63	0,20	0,36	
Average	1.99^{bc}	2.23^{ab}	2.45^{a}	1.71^{de}	1.58^{e}	1.84^{ce}	2.44^{a}	1.86^{cd}	

Description: L1 = Garut-WJ 2014; L2 = Jeneponto-SS 2014; L3 = Malino-SS 2014, L4 = Cisarua-WJ 2014, L5 = Malang-EJ 2015; L6 = Salatiga-CJ 2015; L7 = Jeneponto-SS 2015, L8 = Malino-SS 2015.

In order to obtain varieties that have high-yield potential, direct selection by choosing a long panicle, short stalk and more number of spikelet/panicle traits. This is consistent with the results of Walton (1971) study in [19], where wheat selection should be directed to the trait and also increase the flag leaf and panicle length traits. Longer flag leaves and shorter stem stalk lead to increased yield.

To determine the stability and adaptability of genotypes in various environments, a stability analysis of yields is necessary (table 4). It is important to identify which genotypes are widely adapted, so that they can be developed in larger environments or specific genotypes developed at specific locations. The G5 genotype was significantly superior to all controls in many locations. The breeding approach for selecting high-yielding genotype is determined by the purpose of assembling varieties, for instance environmentally-specific varieties [20] or stable varieties and adapting to large environments [21]. Based on this, the yield stability analysis by Finlay and Wilkinson [4] method involved all environments and seasons to determine the adaptability level of genotypes as presented in table 5.

According to Finlay and Wilkinson [4] a genotypes having a regression coefficient (bi) = 1 and the yield average higher than the total mean, it is expressed as a stable genotypes with high adaptability to all environments. From the results of the analysis in some environments it is shown that almost all genotypes have bi values that were not significantly different to 1 except in the G5 genotypes with bi = 0.53, so it can be said that almost all test genotypes are stable in all locations. This suggests that the G5 genotype has static stability where plant performance will be the same in all environments and shows an increase in resistance to environmental changes or in other words insensitive to the environment, so can be adapted in suboptimal environments. While the G2, G3, G4 and G6 genotypes have dynamic stability because the performance of plants is not same in all environments and have increased sensitivity to environmental changes, which causes plants adapted to optimal environment. In table 5 showed that the G2 genotypes as a stable genotypes with high adaptability to all environments. G1 and G5 genotypes have higher yield than the average total yield, but quite unstable in all environments.

Table 5. Average and stability of wheat yield based on environment combined analysis

Genotypes	Range	Av	bi	SE	Sdi	GCI	GCR	SD	R^{2}(%)
G1	1,8 – 3,3	2.4	0.67	0.66	0.57	0.29	0.08	0.33	4.00
G2	1,4 – 2,8	1.9	0.94	0.47	0.41	0.14	0.00	0.17	0.00
G3	1,4 – 2,3	1.7	1.21	0.38	0.33	0.10	0.03	0.11	5.00
G4	0,7 – 3,5	2.3	2.07	0.58	0.51	0.34	0.85	0.26	36.00
G5	1,6 – 2,7	2.2	0.53	0.44	0.38	0.15	0.17	0.14	16.00
G6	0,9 – 2,8	1.9	-0.8*	0.70	0.61	0.66	2.44	0.37	52.00
Guri3	1,2 – 3,1	1.9	1.08	0.53	0.46	0.18	0.00	0.21	0.00
Guri5	0,9 – 2,9	2.0	1.9*	0.23	0.2	0.13	0.65	0.04	73.00
Dewata	0,9 – 2,6	1.7	1.37	0.36	0.31	0.10	0.10	0.10	15 .00
Average		2.0	1.12	0.48	0.42	0.23	0.48	0.19	22.33

Description: bi = genotype average regression on environment index, * = significantly different to 1; Av = Average; SE = Standard Error, GCI = Genotypes contribution to mean squared interaction, GCR = Genotypes contribution to regression components on G x E interaction; SD = mean squared deviation, R^2 = *Squared Correlation*

4. Conclusion

- There were a highly significant interaction in the environment, genotype and GxE interaction to all traits.
- G1 genotype has a potential yield and average yield of 3.3 $tons.ha^{-1}$ and 2.4 $tons.ha^{-1}$, higher than all controls. This genotype has a shorter days to flower and days to harvest and higher weight of 1000 seeds compared to control. This genotype has adaptive yield in the suboptimal environment based on yield stability analysis.

- G4 genotype has the yield potential and an average yield around 3.5 tons.ha^{-1} and 2.3 tons.ha^{-1}, significantly superior to controls. This genotype has adaptive yield in an optimal environment based on yield stability analysis.
- The G5 genotype has a yield potential and an average yield around 2.7 tons.ha^{-1}of 2.2 tons.ha^{-1}, the mean yield was superior to the Dewata variety and relatively equal to Guri5 variety. The genotype has adaptive yield in the sub-optimal environment based on yield stability analysis.

References

[1] USDA 2018 *Grain: world market and trade* (United State: Foreign Agricultural Service/USDA) 61p.
[2] Hamdani M, Widodo S, Ismail and Dahlan M M 2002 *Evaluasi galur gandum introduksi dan CIMMYT* Pros. Kongres IV dan Simposium Nasional PERIPI (Yogyakarta : UGM)
[3] Baihaki A and Wicaksana N 2005 Interaksi genotip x lingkungan, Adaptabilitas dan stabilitas hasil dalam pengembangan tanaman varietas unggul di Indonesia *Zuriat* **16** (1) 1 – 8
[4] Finlay K W and Wilkinson G N 1963 The analysis of adaptation in plant breeding program *Aust. J. Agric. Res.* **13** 742-754
[5] Admassu S, Nigussie M and Zelleke H 2008 Genotype Enviroment Interaction and Stabililty Analysis for Grain Yield of Maize (*Zea mays*. L) in Ethiopia *Asian J of Plant Sci.***7** (2) 163-169
[6] Hassan M S, Mohamed G I, El-Said R 2013 Stability analysis of grain weight and its component of some durum wheat genotypes (*Triticum durum* L) under different environments *Asian J crop sci.* **2** 179-189
[7] Eberhart S A and Russel W A 1966 Stability parameters for comparing varieties *Crop Sci.* **6** 36-40
[8] Singh R K and Chaudhary B D 1985 *Biometrical Methods in Quantitative Genetics Analysis* (New Delhi : Kalyani Publisher) p 318.
[9] Nur A 2014 Interaksi Genetik x Lingkungan dan Variabilitas Genetik Galur Gandum Introduksi (*Triticum aestivum* L) di Agroekosistem Tropika *J Agrobiogen* **10** (3) 93 – 100
[10] Yan W, Cornelius P L, Crossa J and Hunt L A 2001 Two types of GGE biplots for analyzing multi-environment trial data *Crop. Sci.* **41** 656 – 663
[11] Kang M S 2004 Breeding: Genotype-by-environment interaction for crop cultivar development. *Adv. Agron.* **62** 199-252
[12] Mohamed N E M 2013 Genotype by environment interactions for grain yield in bread wheat (*Triticum aestivum* L.). *J. Plant Breed. Crop Sci* **5** (7) 150-157
[13] Altuhaish K A, Miftahuddin, Trikoesoemaningtyas and Sudirman Y 2014 Field Adaptation of some introduced Wheat (*Triticum aestivum* L) Genotypes in Two Altitudes of Tropical Agro Ecosystem Environment of Indonesia *HAYATI J Biosci* **21** (1) 31 -38
[14] Kheiralla K A, El-Morshidy M A and Zakeria M M 2001 Inheritance of earliness and yield in bread wheat under favorable and late sowing dates 2nd *Plant Breed. Cont.*, October, Assiut Univ. 2001 (1) p 219-239
[15] Ashraf M and Harris P J C 2005 *Abiotic stress: Plant Resistance through breeding and molecular approaches* (New York : Haworth Press Inc) pp 3-18.
[16] Yamin M 2014 *Pendugaan komponen ragam karakter agronomi gandum (Triticum aestivum L,) dan identifikasi marka Simple Sequence Repeat (SSR) terpaut suhu tinggi menggunakan Bulk Segregant analysis (BSA)* (Bogor :IPB)
[17] Nur A 2015 Perbaikan genetik gandum tropis toleran suhu tinggi dan permasalahan pengembangannya pada dataran rendah *J. Pen. Pert. Tan. Pangan* **34** (1) 19 -30
[18] Rawson H M 1987 *Effect of high temperature on the development and yield of wheat practices to reduce deleterions effects* (*Wheat production in tropical environments ed.* Klatt A R *A Proc.of Int. Conf.*) January 19-23 1987 CIMMYT pp 44-62
[19] Bahar H, Sumartono and Nasrullah 1988 Sidik lintas beberapa karakter tanaman terigu

(*Triticum aestivum* L.) *Berkala Penelitian Pasca Sarjana UGM* **2** 145 – 153
[20] Kang M S and Miller J D 1984 Genotype x environment interactions for cane and sugar yield and their implications in sugarcane breeding *Crop Sci.* **24** 435-440
[21] Subandi 1981 Genotype x environment interactions in corn varieties test *Food Crop* **65** 1-9

IOP Conf. Series: Earth and Environmental Science **157** (2018) 012050 doi:10.1088/1755-1315/157/1/012050

Heritability and path coefficient analysis for important characters of yield component related to grain yield in M4 red rice mutant

M Riadi[1], R Sjahril[2], N Kasim[3] and R H Diarjo[4]

[1] Laboratory of Plant Breeding and Seed Sciences, Faculty of Agriculture, Hasanuddin University, Makassar, 90245, Indonesia.
[2] Laboratory of Plant Bio-science and Reproduction Biotechnology, Faculty of Agriculture, Hasanuddin University, Makassar, 90245, Indonesia.
[3] Laboratory of Plant Physiology and Nutrition, Faculty of Agriculture, Hasanuddin University, Makassar, 90245, Indonesia.
[4] Undergraduate course, Agro-technology Program Study, Hasanuddin University, Makassar, 90245, Indonesia.

E-mail: *riadimuh@yahoo.co.id*

Abstract. Formation of new red rice cultivar through mutation breeding is reported in this study. Mutant lines of fourth generation have been planted and analysed. The purpose of the experiment were to determine the value of heritability and the direct and indirect effects of the characters of yield components on the grain. The experiments were prepared based on a randomized block design. Seven red rice mutant lines of fourth generation and one non-mutant origin as control were used as treatments. Data from yield components used were the number of productive tillers per hill, the number of grains per panicle, the panicle density,and grain weight per hill. Results showed that the characters of the productive tillers number per hill, the number of grains per panicle, the panicle density, and the grain weight per hill gave all high heritability values, *i.e.* 0.83, 0.83, 0.62, and 0.52, respectively. Based on the path coefficient analysis against these characters, that there were only two very important character and effective for screening in increasing grain yield per hectare of red rice cultivars, *i.e.*: the character of productive tillers number per hill and grain weight per hill.

1. Introduction

Rice (*Oryza sativa* L.) is one of the most important staple food crops, as a source of food for more than half of the world's population. It is estimated that the world should produce 60% more rice by 2030 than it produced in 1995 [1]. In Indonesia, rice is the main food for most Indonesians. The average rice consumption in rural households were higher than the average consumption of rice in urban households. For rural households, there were a tendency that higher incomes, spending on rice consumption were greater. The opposite occured in urban households [2].

South Sulawesi Province is one of the provinces in Indonesia known as the national rice barn. Rice cultivation has become part of the life and cultural tradition of the community for generations. In some districts in the province of South Sulawesi is still cultivated several types of local rice. One of the local rice species found and still cultivated in some districts is red rice. According to Saxena [3], the red rice has a red pigment, ranging from light to dark red in the pericarp and testa or husk. The portion is recognized to contain polyphenols and anthocyanins and has antioxidant. The results of Muntana and Prasong [4] found that the antioxidant activity of all rice bran extracts from white rice, brown rice and black rice Thailand showed high antioxidant efficiency in the following order: brown rice> black rice> white rice.

Now, red rice becoming more popular and demand for that rice is expected to increase in the future along with the increasing level of people's welfare and awareness of the importance of healthy food. According to Rohman *et al.* [5], public awareness to consume brown rice is generally still associated with diabetes (diabetes mellitus) because of its high glycemic index. Based on the function of red rice which is very important for health, efforts are needed to promote and expand its cultivation area. However, on the other hand, there is still a weakness of the red rice, which is long-lived and the production is still low compared to the national superior rice production, resulting in the interest of farmers to cultivate it is still low [6]. The establishment of a new red rice cultivars, which is short-live and high-yielding through mutation breeding activity has been done and in this research has been done planting of mutant lines of fourth generation (M4).

Grain yield is the main criterion always used by rice plant breeders in producing new varieties with high yield potential. However, when selecting for very varied characters such as grain yields, the direct selection is not always efficient because the grain yield is a polygenic character that is strongly influenced by the environment. In the grain yield improvement program, information about the interrelationships between characters, the direct and indirect effects contributed by each character to the grain yield will be an additional advantage in assisting the selection process. Path coefficient analysis can be used to determine the direct and indirect effects of the those characters to the grain yield, thus providing a clear understanding of the relationship of the characters of grain yield components with the grain yield. Ultimately, such an analysis can help breeders to design a selection strategy to improve grain yield. The aim of this research is to determine heritability and path coefficient values of the important characters of yield componentsto the grain yield per hectare and to formulate selection criteria of red rice cultivar with high grain yield per hectare.

2. Materials and method

The experiment was conducted in irrigated rice field located in Alletengae Village, Bantimurung District, Maros Regency, South Sulawesi Province at altitude ± 25 m above sea level with Alluvial soil type. The study took place from May to September 2016. The experiment was prepared based on a Randomized Block Design with three replications. The treatments were seven mutant red rice mutant lines at fourth generation and one origin (not mutant) line as a control. Seeds of all lines used are first seeded.

Soil preparation is done by hand tractor until sludge structure is formed. Layout of experimental plots measuring 3 m x 2 m was done using a hoe and completed before the seedlings ready for transplanted. After the seedlings are 15 days old, the seedlings are transplanted on the prepared fields. Planting is done manually. Spacing used is 25 cm x 25 cm with one seedling per planting hole. The condition of the land at the time of planting is muddy. Weeding done twice. The first weeding is done after the plants are 15 days after transplanting and the second weeding is done after the plants are 30 days old. Fertilization is done three times. The first fertilization with 1/3 dose Urea (100 kg/ha), SP-36 (200 kg/ha), and KCl (100 kg/ha) is done at age 7 days after planting. The second fertilization using 1/3 dose of Urea (100 kg/ha) was done at age 45 days after planting and third fertilization using 1/3 dose of Urea (100 kg/ha) done at flowering age 50%. Pest control using Furadan 3 G according to recommended dosage, done before planting by way of sprinkling on the surface of the soil. Harvest is done after 2/3 part of the panicle has yellowed. Harvesting is done manually.

Observations were made on each of 12 plant samples from all treatments and replications. The characters of yield components observed were number of productive tillers per hill, number of grains per panicle, panicle density (grains.cm^{-1}), grain weight per hill (g), and grain yield per hectare (t). The panicle density data was obtained from the division between the number of grains per panicle with panicle length, while the grain yield data per hectare was obtained from converting grain yield (g) per plot at 12% moisture grain of all treatments and replications. Data were tabulated using MS Excel 2007. The analysis of variance, correlation coefficient and path coefficient analysis were performed using SPSS version 22 program. The analysis of variance was used to calculate heritability value.

3. Results and discussion

The results from analysis of variance on the characters of yield components and grain yield per hectare are presented in table 1. In table 1, it shows that there are significantly different between red rice mutant lines in terms of the characters of yield components, *i.e.* the number of productive tillers per hill, the number of grains per panicle, the panicle density, and grain weight per hill. The existence of differences between mutant lines, illustrates the great diversity among the mutant lines, so that selection activities can be performed to determine the superior mutant lines based on those characters.

Based on the analysis of variance can be determined the genotypic variance (Vg), phenotypic variance (Vp), genotypic coefficient variation (GCV), phenotypic coefficient variation (PCV) and heritability value (h^2), are presented in table 2. Phenotypic variance (Vp) separated into two distinct components, namely the genotypic variance (Vg) and the variance of environments (Ve) used to clarify the understanding of the variation patterns of the observed characters. The values of the phenotypic coefficient variation (PCV) wereslightly higher than the genotypic coefficient variation (GCV) for all observed characters, indicating an environmental influence to some degree of phenotypic expression of the characters. The coefficient variation of phenotypic values (PCV) obtained for all characters ranged from 21.02% to 54.86%, whereas the coefficient variation of genotypic values (GCV) for all characters ranged from 16.50% to 51.50%. According to Akinwale *et al.* [7], Konate *et al.* [8] and Sumanth *et al.* [9], the values of the phenotypic coefficient variation (PCV) and the genotypic coefficient variation (GCV) greater than 20% are categorized as high and those characters can be used as selection criteria in the improvement of mutant lines.

Table 1. Analysis of variance of the yield components characters and grain yield per hectare from 8 red rice mutant lines at fourth generation (M4).

Source of variation	Degree of freedom	Characters				
		Number of productive tillers per hill	Number of grains per panicle	Panicle density	Grain weight per hill	Grain yield per hectare
Replications	2	3,617ns	227,391ns	4,059*	1,602ns	0,681ns
Mutant lines	7	82,815**	3212,531**	3,119*	4,006*	3,131**
Error	14	5,401	205,573	0,536	0,956	0,135
CV (%)		14,02	9,91	13,02	19,50	18,90

ns= not significant; *= significant at 5% level; **= significant at 1% level

The values of heritability are the ratio between the genotypic variance value to the phenotypic variance value can help to determine whether the variation of all observed characters are more controlled by genetic factors or more controlled by environmental factors. The heritability values obtained from the observed characters, i.e. the number of productive tillers per hill, the number of grains per panicle, the panicle density, and the grain weight per hill are all 0.83, 0.83, 0.62, and 0.52. Those means indicate that all of the observed charactersare greater controlled by genetic factors than environmental factors and the selection made of these characters can be achieved through the phenotypic of those characters. This is in line with the opinion of Akinwale *et al.* [7], Naseem *et al.* [10] and Konate *et al.* [8] that characters with high heritability values are more controlled by genetic factors and those characters can be used as selection criteria for selecting superior lines based on the phenotypic characters.

Table 2. Genotypic variance (Vg), phenotypic variance (Vp), genotypic coefficient variation (GCV), phenotypic coefficient variation (PCV) and heritability values (h^2) of the yield components characters and grain yield per hectares from 8 red rice mutant lines at fourth generation (M4)

Characters	Minimum	Average	Maximum	Vg	Vp	GCV (%)	PCV (%)	h^2
X1	8,08	16,57	23,06	25,80	31,21	30,65	33,71	0,83
X2	115,22	144,61	214,14	1002,32	1207,89	21,89	24,03	0,83
X3	4,53	5,62	7,98	0,86	1,40	16,50	21,02	0,62
X4	14,54	28,17	41,46	1,02	1,97	28,94	48,96	0,52
Y	0,44	1,94	3,41	1,00	1,13	51,50	54,86	0,88

X1= Number of productive tillers per hill; X2= Number of grains per panicle; X3= Panicle density (grain.cm^{-1}); X4= Grain weight per hill (g); and Y= Grain yield per hectare (t).

The result of path analysis is presented in Table 3, showed that the number of productive tillers per hill, the number of grains per panicle, and the grain weight per hill werethe characters that havesignificant positive direct effects on the grain yield per hectare with the values of direct effects (C) were 0.729*, 0.916*, and 0.637* respectively, whereas the panicle density was a character that has a non signicant negative direct effect on the grain yield per hectare with the value of direct effect -0.497^{tn} on the grain yield per hectare. These means that, increasing of the characters that have significant positive direct effects will increase the grain yield per hectare, whereas for the character that have a negative direct effect will decrease the grain yield per hectare. Similar result were associated with a significant positive direct effect of the number of productive tillers per hill on the grain yield per hectare reported by Badri *et al.*[11], whereas the results of other studies related to the significant positive direct effects of the number of productive tillers per plant on the grain yield per plant are reported by Vanisree *et al.* [12], Ashok *et al.* [13] and Kalyan *et al.* [14]. The results of the research showed that there weretwo important characters of yield components were considered to be effective as selection criteria in increasing grain yield per hectare at fouth generation (M4) of red rice mutant lines, i.e. the number of productive tillers per hill and the grain weight per hill.

Table 3. Direct, indirect and total effects of the characters of yield components to the grain yield per hectare

Characters of yield components	Direct effects	Indirect effects through				Total effects
		X1	X2	X3	X4	
X1	0,729*	-	-0,594	0,214	0,438	0,786**
X2	0,916*	-0,473	-	-0,375	-0,261	-0,193ns
X3	-0,497ns	-0,314	0,691	-	-0,032	-0,152ns
X4	0,637*	0,501	-0,376	0,025	-	0,787**

X1 = Number of productive tillers per hill; X2 = Number of grains per panicle; X3 = Panicle density; X4 = Grain weight per hill; ns = not significant; * = significant at level 5%.

In this study, the number of grains per panicle has a significant positive direct effect on the grain yield per hectare, but because the indirect effects were all negative and cause the total effect becomes negative, causing the character cannot be used as selection criteria to choose the superior red rice mutant lines at fourth generation (M4) with high grain yield per hectare. The panicle density is a representation of the ratio between the number of grains per panicle to the length of panicle has a non significant negative direct effect on the grain yield per hectare, so it also cannot be used as selection criteria to choose the superior red rice mutant lines with high grain yield per hectare. The magnitude of the panicle density does not necessarily represent a large number of grains and the long panicle length, because the same value of the panicle density can be obtained from less number of grains with shorter panicles.

4. Conclusion

The heritability values of the yield components characters observed, *i.e.* the number of productive tillers per hill, the number of grains per panicle, the panicle density, and the grain weight per hill were all high with the heritability values are 0.83, 0.83, 0.62 and 0.52. Based onthe path coefficient analysis toward characters of yield components on the grains yield per hectare obtained three characters which showed direct positive effects to grainsyield per hectare, that is the number of productive tillers per hill, the number of grains per panicle and the grain weight per hill with path coefficient valueswere 0.729, 0.916 and 0.637, respectively, but the number of grains per panicle showed negative effect (-0.497) with grain yield per hectare. This is indicates that there are only two very important and effective characters for selection in increasing the grain yield per hectare in red rice cultvar of fourth generations (M4), *i.e.* the number of productive tillers per hill and the grain weight per hill.

References

[1] Babu V R, Shreya K, Dangi K S, Usharani G and Shankar A S 2012 Correlation and path analysis studies in popular rice hybrids of India *Int. J. Sci. Res. Publ.* **2** 1–5

[2] Yunita and Riswani 2013 Behavior of household rice consumption in different income level *Int. J. Humanit. Manag. Sci.* **1** 220–4

[3] Saxena A 2014 Save the red rice: a unique gift of nature *Int. J. Curr. Res. Biosci. Plant Biol.* **1** 32–4

[4] Muntana N and Prasong S 2010 Study on total phenolic contents and their antioxidant activities of Thai white, red and black rice bran extracts *Pakistan J. Biol. Sci.* **13** 170

[5] Rohman A, Helmiyati S, Hapsari M and Setyaningrum L D 2014 Rice in health and nutrition *Int. food Res. J.* **21**

[6] Mau Y S, Markus J E R, Shirly S, Oematan S, Ndiwa A S S, Handoko D D, Nasution A And Makbul K 2017 Genetic diversity of red and black upland rice accessions from East Nusa Tenggara, Indonesia as revealed by agro-morphological characters *Biodiversitas J. Biol. Divers.* **18**

[7] Akinwale M G, Gregorio G, Nwilene F, Akinyele B O, Ogunbayo S A and Odiyi A C 2011 Heritability and correlation coefficient analysis for yield and its components in rice (*Oryza sativa* L.) *African J. plant Sci.* **5** 207–12

[8] Konate A K, Zongo A, Kam H, Sanni A and Audebert A 2016 Genetic variability and correlation analysis of rice (*Oryza sativa* L.) inbred lines based on agro-morphological traits *African J. Agric. Res.* **11** 3340–6

[9] Sumanth V, Suresh B G, Ram B J and Srujana G 2017 Estimation of genetic variability, heritability and genetic advance for grain yield components in rice (Oryza sativa L.) J. Pharmacogn *Phytochem.* **6** 1437–9

[10] Naseem I, Khan A S and Akhter M 2014 Correlation and path coefficient studies of some yield related traits in rice (*Oryza sativa* L.) *Int. J. Sci. Res. Publ.* **4** 1–5

[11] Badri A, Rassam G, Dadkhah A and Mohaddesi A 2016 Path Coefficient Analysis for The Yield-Related Traits of Rice Lines in North Iran *Ann. West Univ. Timisoara Ser. Biol.* **19** 119

[12] Vanisree S, Swapna K, Raju C D, Raju C S and Sreedhar M 2013 Genetic variability and selection criteria in rice *J. Biol. Sci. Opin.* **1** 341–6

[13] Ashok S, Jyothula D P B and Rantnababu D 2016 Character association and path analysis for yield components and grain quality parameters of rice (*Oryza sativa* L.) *Int. J. Agric. Sci. Res.* **6** 253–8

[14] Kalyan B, Krishna K V R and Rao L V S 2017 Path Coefficient Analysis for Yield and Yield Contributing Traits in Rice (*Oryza sativa* L.) Genotypes *Int. J. Curr. Microbiol. App. Sci.* **6** 2680–7

IOP Conf. Series: Earth and Environmental Science **157** (2018) 012051 doi:10.1088/1755-1315/157/1/012051

Assessment and selection of M3 generation of wheat mutants adaptive in lowland

Nasaruddin, M Farid, Y Musa and H Iswoyo

Department of Agronomy, Faculty of Agriculture, University of Hasanuddin, Tamalanrea Campus, Jl. Perintis Kemerdekaan KM. 10 Makassar, 90245, Indonesia.

E-mail: *nnasaruddin@gmail.com*

Abstract. Wheat is a sub-tropical plant which adaptability in the tropics in areas with altitudes over 1000 m above sea level (asl). Cultivation on the highlands is in competition with vegetables and fruits in terms of land use, therefore, it is essential to have wheat genotypes adaptive to the lowlands. This study aimed to test and select offspring of three adaptive mutant on lowland with high temperature ranged from 28 to 33°C. The benefit of the research is to obtain a potential strain of adaptive wheat mutants in the lowlands. The experiment was conducted at Experimental farm of Agriculture Faculty, Hasanuddin University Makassar at an altitude below 50 m from sea level. The study was conducted from March to June 2017. The experimental study applied a randomized block design with three replications. The treatments consisted of 20 mutants of wheat with 4 varieties used for comparison. They were Nias, Selayar, Dewata, and Munal. Results showed that the wheat mutant N.300 4.3.6 (1.92 t per hectare) gave highest growth and production and was significantly different from all other mutants and comparing varieties. The mutant genotypes that gave higher growth and production than all comparing varieties were N.350 3.1.4 (1.86 t per hectare), N.200 2.3.3 (1.67 t per hectare), N. 250 4.2.1 (1.48 t per hectare), N.200 2.4.B.6 (1.36 t per hectare), N.250 4.5.2 (1.26 t per hectare), N.350 3.2.2 (1.26 t per hectare), and N.350 3.1.3 (1.22 t per hectare). Mutants that produce higher yields than the three comparing varieties (Dewata, Selayar and Munal) with yields > 1 t per hectare, yet lower than Nias varieties (1.15 t per hectare) are N.200 2.5.2 (1.11 t per hectare), N.250 4.6.2 (1.14 t per hectare), and N.350 3.6.2 (1.07 t per hectare. In conclusion, there are 11 potential mutant genotypes for further experiment in order to develop adaptive wheat to lowlands.

1. Introduction

Food diversity program is an alternative effort for national food self-sufficiency. Nowadays, all demands for domestic flour are fulfilled by importing wheat seeds which will then be processed into flour. Import of wheat to Indonesia in 2010 was 5.2 million ton, in 2011 was increasing to 6.3 million ton, in 2012 was 7.4 million ton, and in 2016 the import was already reaching 8.1 million ton [1]. Increase in wheat import burdened state finance and therefore require food diversification through development of wheat in Indonesia.

Various challenges in the up to downstream development of local wheat have not yet been fully managed by farmers and field officers. These challenges being: wheat is not a priority commodity, ecology of wheat growing field is the altitude of at least 800 m from sea level [2], thus in competition with highland vegetables and fruits. In the last few years, Indonesian government has set development program of wheat in the areas considered to be appropriate, such as in the highland and middle land where the temperature is relatively lower. For this purpose, various domestic wheat development activities are carried out through the application of cultivation technology in accordance with agro-climate conditions in Indonesia [3], and seeking alternative development in lowland to middle land areas on unproductive lands. Therefore this study was conducted to obtain lowland wheat varieties from local Indonesian wheat crops that have been irradiated with gamma rays through mutation breeding and mutant heredity tests that can adapt to the lowlands.

The study aimed to perform population heredity test of the Mutant three (M3), population formation for the subsequent generation up to Mutant five (M5). Also, to test the formation of family through adaptive selection of genotypes (mutants) adaptive to the lowlands high temperature with range between 28 and 33°C during dry season and high rainfall during rainy season; and tests on adaptation, stability and multilocation.

2. Research Methodology

The research was conducted in the experimental farming of Agriculture Faculty, University of Hasanuddin, on a location of 50 m asl. The experiment was performed from March to June 2017. The experiment was designed with randomized block design consisting of 3 repetitions. Treatments were 21 different wheat mutants: g1= Munal 1.7.1, g2 = N.350 3.6.2 , g3 =N.300 3.6.1, g4 = N.250 4.2.1, g5 = N.350 3.1.3, g6 = N.350 3.2.2, g7 = N.250 4.5.2, g8 = N.350 3.1.4, g9 = Munal 200 1.7.1, g10 = N.250 4.6.2, g11 = N.350 3.8.9, g,12 = S. 6.4.1,g13 = S.300 8.3.1, g14 = N.250 4.4.2, g15 = N.350 3.8.10, g16 = N.350 3.5.10, g17 = S. 8.4.2, g18 = N.200 2.4.B.6, g19 = N.200 2.5.2,g20 = N.200 2.3.3, g21 = N.300 4.3.64. Comparing varieties were g22 = Dewata, g23 = Selayar, g24 = Nias, and g25= Munal. Experiment was conducted on 1 m x 3 m plot with 5 cm x 25 cm of plants spacing. Fertilizers applied to the plants at 10 days after planting were 150 kg per hectare of Urea, 200 kg per hectare of SP36 and 100 kg per hectare of KCl. More fertilizers were applied at 30 days after planting with 150 kg per hectare of urea. Observation variables in this research were number of productive tillers, length of the panicle, number of the spikelet per panicle, number of seeds per panicle, weight of the seeds per panicle and the production per hectare. In addition, heritability value of each parameter was calculated.

3. Results and Discussion

3.1. Results

Parameters values of each genotype observed in the experiment including number of productive tillers, panicle length, number of spikelet in each panicle, number of seeds per panicle, weight of 100 Seeds and yields per hectare are shown in table 1.

Statistical analysis showed that genetic material had a very significant effect on the number of productive tillers, the length of panicle, the number of spikelet per panicle, the number seeds per panicle, the weight of the seeds per panicle and the production per hectare. The Least Significant Difference test of 0.05 in table 1 shows that the genetic material N.300 4.3.6 (g21) has an average number of productive tillers, panicle length, number of spikelet probes, number of seeds per panicle, the seeds weight per panicle and the highest yield per hectare when compared to other genetic materials.

Table 2 shows that all observed characters, whether vegetative, generative to production, have moderate to high heritability values based on the index values of each character. Heritability analysis in table 2 shows that there are six parameters having high heritability value and one having medium heritability value. The highest heritability value is shown in production parameters *i.e.* 93.48%.

3.2. Discussion

Treatment N.300 4.3.6 (g21) shows a good potential for development in high temperature regions. This is because the given irradiation dose is not too high. Excessively high doses of irradiation can cause the plant to become abnormal and damage the genetic makeup of plants. This is in accordance with [4] which states that the higher the dose, the more mutations occur and the more damage may occur.

IOP Conf. Series: Earth and Environmental Science **157** (2018) 012051 doi:10.1088/1755-1315/157/1/012051

Table 1. Number of productive tillers, panicle length, number of spikelet in each panicle, number of seeds per panicle, weight of 100 Seeds and yields per hectare.

Treatment (Genotype materials)	Number of productive tillers	Panicle length (cm)	Number of Spikelet in each panicle	Number of seeds per panicle	Weight of seed per panicle (g)	Weight of 100 Seeds (g)	Yield productio n (kg)
g1 (M 1.7.1)	4,77^{bc}	6,63^{ef}	11,93^{c}	20,07^{e}	1,94^{d}	3,13^{d}	0,53^{g}
g2 (N.350 3.6.2)	4,47^{bc}	7,74^{c}	12,03^{c}	27,07^{cd}	2,30^{bc}	4,54^{bc}	1,07^{e}
g3 (N.300 3.6.1)	4,67^{bc}	7,13^{de}	12,03^{c}	20,27^{e}	1,75^{ed}	4,34^{bc}	0,53^{g}
g4 (N.250 4.2.1)	5,47^{a}	8,47^{ab}	14,23^{ab}	30,67^{bc}	2,45^{ab}	5,59^{a}	1,48^{c}
g5 (N.350 3.1.3)	5,27^{ab}	8,36^{ab}	13,83^{ab}	37,27^{ab}	2,46^{ab}	4,45^{bc}	1,22^{de}
g6 (N.350 3.2.2)	5,27^{ab}	8,18^{ab}	13,53^{b}	29,37^{c}	2,40^{b}	5,48^{ab}	1,26^{d}
g7 (N.250 4.5.2)	5,17^{ab}	8,11^{bc}	13,13^{bc}	37,27^{ab}	2,51^{ab}	5,31^{ab}	1,26^{d}
g8 (N.350 3.1.4)	5,17^{ab}	8,20^{ab}	13,23^{bc}	34,27^{b}	2,51^{ab}	5,59^{a}	1,86^{a}
g9 (M. 200 1.7.1)	3,57^{d}	6,24^{f}	8,83^{d}	18,77^{e}	2,00^{cd}	3,13^{d}	0,51^{g}
g10 (N.250 4.6.2)	5,27^{ab}	8,47^{ab}	13,43^{b}	31,47^{bc}	2,41^{b}	5,39^{ab}	1,14^{de}
g11 (N.350 3.8.9)	4,57^{bc}	7,83^{bc}	12,53^{bc}	27,47^{cd}	2,38^{b}	4,94^{b}	0,87^{f}
g12 (S. 6.4.1)	4,47^{bc}	6,79^{e}	13,13^{bc}	23,77^{de}	2,08^{cd}	3,13^{d}	0,51^{g}
g13 (S.300 8.3.1)	4,87^{b}	7,28^{d}	12,83^{bc}	20,27^{e}	1,66^{e}	3,13^{d}	0,51^{g}
g14 (N.250 4.4.2)	4,27^{c}	7,84^{bc}	13,03^{bc}	28,87^{cd}	2,35^{bc}	4,73^{bc}	0,77^{fg}
g15(N.350 3.8.10)	3,77^{cd}	7,57^{cd}	11,63^{c}	19,57^{e}	1,53^{e}	3,13^{d}	0,51^{g}
g16 (N.350 3.5.10)	3,77^{cd}	7,45^{cd}	11,73^{c}	23,17^{de}	2,07^{cd}	3,13^{d}	0,51^{g}
g17 (S. 8.4.2)	4,97^{ab}	7,43^{cd}	12,93^{bc}	19,87^{e}	2,06^{cd}	4,25^{c}	0,51^{g}
g18(N.200 2.4.B.6)	5,27^{ab}	8,25^{ab}	13,73^{ab}	37,67^{ab}	2,53^{ab}	5,54^{ab}	1,36^{cd}
g19 (N.200 2.5.2)	5,07^{ab}	8,14^{b}	13,33^{b}	32,47^{bc}	2,47^{ab}	5,45^{ab}	1,11^{e}
g20 (N.200 2.3.3)	5,37^{ab}	8,38^{ab}	14,63^{ab}	38,17^{ab}	2,61^{ab}	5,62^{a}	1,67^{b}
g21 (N.300 4.3.6)	5,47^{a}	8,55^{a}	14,83^{a}	40,47^{a}	2,64^{a}	5,70^{a}	1,92^{a}
Average	4,80	7,76	12,89	28,49	2,24	4,56	1,00
g22. DEWATA	4,87^{b}	8,04^{bc}	13,13^{bc}	20,77^{e}	1,93^{d}	3,13^{d}	0,51^{g}
g23. SELAYAR	4,97^{ab}	7,57^{cd}	11,93^{c}	22,27^{de}	2,27^{bc}	4,92^{b}	0,64^{g}
g24. NIAS	5,07^{ab}	8,29^{ab}	13,93^{ab}	30,37^{bc}	2,39^{b}	5,53^{ab}	1,15^{de}
g25. MUNAL	4,47^{bc}	7,65^{cd}	12,83^{bc}	24,97^{d}	2,16^{c}	4,25^{c}	0,51^{g}
Average	4,84	7,88	12,96	24,59	2,19	4,46	0,7
Genotype	**	**	**	**	**	**	**
CoV (%)	9,61	4,34	8,31	12,29	7,56	11,49	12,69
LSD	0,54	0,39	1,24	3,97	0,20	0,61	0,14

*Numbers followed by same letter in the same column (a, b, c, d, e, f, g) denote the insignificant difference at test level of BNT=0,05

** CoV = Coefficient of Variance

Tabel 2. Heritability value of wheat M3 mutant strains

No	Observation variables	h2 value (%)	Remarks
1	Productive yields	50,33	High
2	Length of panicle	74,87	High
3	Number of Spikelet in each panicle	48,71	Medium
4	Number of seeds per panicle	79,00	High
5	Weight of seed per panicle	74,33	High
6	Weight of 100 Seeds	77,04	High
7	Yield production	93,48	High

$0 < h^2 \leq 20$ (low); $21 < h^2 \leq 50$ (medium); $50 < h^2 \leq 100$ (high)

The number of productive tillers that formed during the vegetative period generally decreased due to the stresses of the growing environment due to the decrease of elevation from high to lowlands. According to Nur *et al.* [5] the growing environment of wheat flour from highland to lowland in wet tropical environments leads to a decrease in seed germination and a decrease in the number of productive tillers of each genotype of wheat. The results showed that genetic material N.300 4.3.6 (g21) had the most productive tillers.

The results of the variance indicated that for panicle length parameter, the mutant genotypes N.300 4.3.6 (g21) had the average best panicle length (8.55 cm) and were significantly different from all genotypes except g4, g5, g6, g8, g10, g18, g20 and g24. For the parameter of spikelet number per panicle, N.300 4.3.6 (g21) has average spikelet number more than any other genotypes (14.83) and significantly different from all genotypes except g4, g5, g18, g20 and g24. The formation of panicle, panicle length and the amount of spikelet in wheat is affected by the assimilates content when the wheat is in its vegetative phase. It is in line with Nur [6] that the panicle length character and the number of spikelet is determined by the supply of assimilates in the vegetative phase as the source to form the panicles in the generative phase. If the source is insufficient in the formation of a sink (generative phase), then the spikelet formation becomes low.

For the parameter of seeds number per panicle, the weight of 100 seeds, seeds production per hectare shows that N.300 4.3.6 (g21), has a higher mean compared to other genetic materials. This is due to the high temperature stress causing the seeds to fail to form which will affect the number of seeds produced in panicles and also affect the weight of the seeds of the barn. This is in accordance with [7] which states that wheat productivity has a close relation with air temperature[8]. At the Celsius scale a degree increase in average air temperature, the yield of wheat will drop 504 kg per hectare.

Heritability analysis is a quantitative benchmark to determine the phenotype difference of a character caused by genetic or environmental factors, therefore, it can give an idea whether the observed character is more influenced by genetic or environmental factors. Table 2 shows that almost all morphological characters have high heritability values. In the parameters of the number of productive tillers (50.33%), the length of panicle (74.87%), the number of the seeds per panicle (79.00%) seeds weight per panicle (74.33%), weight of 100 seeds (77.04%) and yield production (93.48%) belongs to high heritability. While the number of spikelet per panicle (48.71%) is classified as moderate heritability. This is stated in Basir [9] and Ferris *et al.* [10] that characters with high heritability will increase the effectiveness of selection in endurance testing because the observed character is a reflection of the influence of genetic factors rather than environmental influences.

4. Conclusions

Based on the results obtained, this result drew the following conclusions:

- Wheat mutant genotype N.300 4.3.6 gave best growth and highest yield (1.92 tons per hectare) and is significantly different from all other mutants and comparing varieties. The mutant genotypes that gave higher growth and production compared to all comparing varieties were N.350 3.1.4 (1.86 tons per hectare), N.200 2.3.3 (1.67 tons per hectare), N. 250 4.2.1 (1.48 tons per hectare), N.200 2.4.B.6 (1.36 tons per hectare), N.250 4.5.2 (1.26 tons per hectare), N.350 3.2.2 (1.26 tons per hectare), and N.350 3.1.3 (1.22 tons per hectare). Mutants that produce higher yields than three of comparing varieties (Dewata, Selayar and Munal) with yields > 1 tons per hectare, but lower than Nias varieties (1.15 tons per hectare) are N.200 2.5.2 (1.11 tons per hectare), N.250 4.6.2 (1.14 tons per hectare), and N.350 3.6.2 (1.07 tons per hectare). Thus, there are 11 potential mutant genotypes to be selected further in an effort for adaptive wheat development in the lowlands
- High heritability prediction value, followed by heritability more than 50% possessed by the characters in: productive tiller, long panicle, number of seeds of per panicle, seeds weight per panicle, weight of 1000 seeds and production

References

[1] Asosiasi Produsen Terigu Indonesia 2016 http://bataviase.co.id/node/436332.Html Accessed on Februari 17 2017

[2] Dubin H J and Rajaram S 1996 Breeding disease-resistant wheats for tropical highlands and lowlands *Annu. Rev. Phytopathol.* **34** 503–26

[3] Sovan M 2002 *Postharvest Handling of Wheat Coordination Meeting of the Development of Wheat* (Pasuruan: Directorate general of Cereals for Production of Food Crops)

[4] Mugiono 2001 *Plant Breeding With Mutation Technique* (Jakarta)

[5] Nur A, Trikoesoemaningtyas, Khumaida N and Sujiprihati S 2010 *Phenology of Growth and Production of Wheat in a Wet Tropical Environment* (National Cereals Week)

[6] Nur A 2013 Adaptation of high temperature tolerant of wheat crops (*Triticum aestivum* L.) and increased genetic variability through mutation induction by using gamma ray irradiation (Bogor: Bogor Agricultural Institute)

[7] Aggarwal P K 1991 *Simulating Growth, Development, and Yield of Wheat in Warmer Areas Wheat for Non-Tradisional Warm Area* (Brazil)

[8] Saunders D A 1988 *Characterization of tropical wheat environments: Identification of production constraints and progress achieved: South and Southeast Asia. Wheat Production Constraints in Tropical Environments Proc. of the International Conference* Chiang Mai (Thailand) 19-23 Jan 1987

[9] Basir M 2001 Utilization of Heritability Value and Correlation Coefficient to Determine Selection Indicators *J. Agrivigor* **1** 1–6

[10] Ferris R, Ellis R H, Wheeler T R and Hadley P 1998 Effect of high temperature stress at anthesis on grain yield and biomass of field grown crops of wheat *Ann. of Bot.* **82** 631-639

IOP Conf. Series: Earth and Environmental Science **157** (2018) 012052 doi:10.1088/1755-1315/157/1/012052

Characterization of Siarang Dairi local corn, Regency of Dairi, North Sumatera Province

L Z Nasution[1] and N Buri[2]

[1]Assessment Institute of Agricultural Technology (AIAT) North Sumatera, Jend. A.H. Nasution No.1B Street, Medan, 20143 Indonesia.
[2]AIAT Gorontalo, Kopi No. 270 Street, Lioheluma Kabila, Bone Bolango, Gorontalo, 96183, Indonesia.

E-mail: *lellyzulhaida@yahoo.co.id*; *nanangburi@gmail.com*

Abstract. Corn is one of the carbohydrate-producing plants in addition to rice and wheat which contains nutrients that can be used for humans and animals. Local plants have an important role for genetic resources and can create new varieties. So it needs to be preserved and utilized. The purpose of this research is to characterize the local maize of Siarang Dairi. This research uses qualitative method is descriptive analysis. The data collected include the quantitative and qualitative character of agronomic and morphological characters using 5 plant samples. The results showed that the local Siarang Dairi maize had a root falling percentage of 0% and the stem fall percentage of 3.07%. The high of the plant ranged from 236 - 277 cm. The length of the cob were from 16 - 19 cm with an almost homogen diameter of 4 – 4.5 cm. The uppermost form of cylindrical and cylindrical conical. Grain lengths range from 6 - 8 mm, with grain widths ranging from 5 - 9 mm, and grain thicknesses ranging from 3 - 5 mm. The top grain surface shape is jagged, round, and flat. Specific color grains are unique and varied from brown, orange and red color.

1. Introduction

Corn is one of the 7 (seven) main commodities targeted by the Ministry of Agriculture in achieving self-sufficiency program and increasing production in 2017. In 2015, North Sumatra has corn harvest area of 243,772 ha with production of 1,519,407 tons. One of the districts with hybrid and local corn potential is Dairi with harvested area of 39,261 ha and production of 259,033 tons [1].

In the 90's until now, corn commodities began to grow rapidly and interest farmers in Indonesia. Government agencies, especially the Directorate General of Food Crops through a Special Effort program of corn commodities will expand Indonesian Agency for Agriculture Research and Development products throughout Indonesia, whether integrated with palm or soybean. Private companies are also starting to compete for corn development across the archipelago with the latest technology. So that more and more technological findings to create a commercial corn and create concern in disappearance of local varieties quality that are adaptive in certain areas.

The local corn of Siarang Dairi grown in Dairi district, especially in Sidikalang, Parbuluan, and Sumbul subdistricts has different characteristics from maize which is more commonly known to the common farmers. The Siarang Dairi local corn has a harvest time of about 5-6 months of the year and has different color characteristics in 1 corncob because it is free sided maize. The varieties of free-sided maize have advantages such as adaptability in every type of land, low seed prices and can be used for generations [2].

The local production of Siarang Dairi corn ranges from 1.25 tons / ha. Local corn yields that still use previous harvested seeds indicate low seed quality. The low production of local maize caused by unavailability of quality seed when needed and marginal development of the corn [3]. In the 2000s, the production of superior maize varieties could reach 10-12 tons / ha and hybrid maize reached 7 - 8 tons / ha compared to local corn production ranging from 2.5 to 4 tons / ha [4]. Use of local varieties in particular has a tendency for farmers to grow more high-yielding corn varieties, thereby erode the availability of locally useful genes.

Local plants have an important role for genetic resources and can create new varieties. So it needs to be preserved and utilized as well as possible by corn plant breeders for the assembling of superior varieties. IPGRI defines genetic resources as plant genetic material that has value as a source for present and future generations [5].

AIAT of North Sumatra has been doing activities of local plant characterization in the last ± 5 years. This characterization program is an activity supported by both government and local government to maintain and preserve local adaptive genes that have certain advantages. The main objectives of breeding maize are to increase the potential of genetic outcomes, early age, improve pest and disease resistance, improve tolerance to abiotic stress, and nutritional quality analysis [6]. One of the preferred properties of maize varieties is aged to avoid drought at the end of the growth period [7].

Characterization is one of the important stages in the formation of superior varieties that aim to know the important characters of economic value and as a characteristic of the relevant varieties [8]. The purpose of this study was to identify the characters of the local Siarang Dairi maize, Dairi District, North Sumatra whose genes can be used for the assembling of new improved varieties.

2. Methodology

The material used was local corn Siarang Dairi with 5 sample plants. The tools used are meter, ruler, sample plastics, instrument scales, and stationery.

The research used qualitative method with descriptive analysis, including exploration, collection, and characterization. Exploration was done by obtaining the Siarang Dairi local corn and observed the morphological characteristics of the plant. Characterization was conducted by referring to Food Crop Characterization Guidelines especially corn morphology [9].

Siarang Dairi corn samples were taken from Sidikalang Subdistrict plants that had been harvested. Planting was done back in the Garden Collection of AIAT North Sumatra with sandy loam. The data collected included the quantitative and qualitative characteristics of agronomic and plant morphological characteristics. Agronomic characters observed were: plant height. While morphological characters observed included leaf, totally of leaves on top of cob, totally of leaves / plant stem, leaf length, leaf width, leaf orientation after flowering, leaf tip shape, angle between leaf and stem, direction of leaf and stem strands, ligula condition, root volume, stem, panicle length, length of footstalk, branch range of panicle, totally of primary branches on panicle, totally of secondary branches on panicle, totally of tertiary branches on panicle, panicle size after maternity, lateral branch direction, and the angle between the main axis and lateral branch is below from 1/3 panicle, stem colour, percentage of root falling plant, percentage of stuck plant, leaf midrib, panicle type, cob covering length of corncob, length of cob stalk, cob diameter, diameter of corncob, rachis diameter, corncob/rachis index, cornhusk/seed index, rachis/seed index, seed closing index, totally of branches, totally of seeds/row, color of corncob colour, the uppermost form of cob, and the plant grains include grain length, grain width, grain thickness, grain surface shape, pericarp colour, aleuron colour, and endosperm colour. The collected data is tabulated and analyzed descriptively.

3. Result and discussion

3.1. The Agronomic Character and Plant Morphology

The agronomic character of the plant can be seen from plant height in the field taken randomly from 5 plant samples. As well as plant morphological characters characterized on vegetative and generative phases include characterization of leaves, stems, panicles, corncob, and grains of Siarang Dairi corn (table 1).

IOP Conf. Series: Earth and Environmental Science **157** (2018) 012052 doi:10.1088/1755-1315/157/1/012052

Table 1. Vegetative character plant

Character	Plants				
	1	2	3	4	5
Plant height (cm)	263	277	257	236	263
Leaf	Wide	Wide	Wide	Wide	Wide
Totally of leaves on the top (leaves)	6	6	6	6	9
Totally of leaves/plant	20	26	18	25	28
Leaf length (cm)	88	96	67	83	105
Leaf width (cm)	6.5	8	8	7.5	6
Leaf orientation after flowering	straight	straight	straight	straight	straight
Leaf tip shape	Pointed	Pointed	Pointed	Pointed	Pointed
Angle between leaf and stem	Small ± 25°	Medium ± 50°	Medium ± 50°	Medium ± 50°	Medium ± 50°
Direction of leaf and stem strands	straight	slightly curved	arched strongly	arched strongly	curved
Ligula condition	Exist	No exist	exist	exist	exist
Root volume	Medium	Medium	Medium	Medium	Medium
Panicle length (cm)	21	25	33	42	28
Length of footstalk (cm)	18	13	11	9	9
Branch range of panicle (cm)	5	5	6	5	5
Totally of primary branches on panicle (leaves)	5	4	0	6	5
Totally of secondary branches on panicle (leaves)	2	2	0	2	3
Totally of tertiary branches on panicle (leaves)	2	2	0	3	1
Panicle size after maturity	Medium	Medium	Medium	Small	Small
Lateral branch direction	slightly curved	slightly curved	Curved	slightly curved	Curved
The angle between the main axis and the lateral branch is below from 1/3 panicle	Medium ± 50°	Medium ± 50°	Small ± 25°	Medium ± 50°	Medium ± 50°
Stem colour	Green	Green	Green	Green	Green
Percentage of root falling plant	0 %	0 %	0 %	0 %	0 %
Percentage of stuck plant	3.07 %	0 %	3.07 %	3.07 %	3.07 %
Leaf midrib	Medium	Medium	Medium	Medium	Medium
Panicle type	Primery-Sekunder	Primery-Sekunder	Primery-Sekunder	Primery-Sekunder	Primery-Sekunder

Figure 1. Plant growth	**Figure 2.** Corncob condition

The characteristic morphology characterized on the local corn plant of Siarang Dairi were characters in the vegetative and generative phases. This is the result of the characterization of Siarang Dairi local corn conducted.

3.1.1. Plant height. Observation of plant height were carried out in vegetative phases aged 15 week after planting. The observed morphological results of the local maize of Siarang Dairi maize shown in table 1, indicate that the height of the 5 plant samples ranged from 236 to 277 cm. Uniformity of the height of the plants (figure 1) is possible because of environmental factors one of which is soil fertility. Priyanto *et al* [7], stated that crop yield is determined by the interaction between genotype and environmental factors such as soil fertility, weather, soil moisture, and technical culture.

3.1.2. Leaf morphology. Observation of plant leaves was carried out in the generative phases, where the plants had already yielded fruit. Leaf morphology observed included leaf length, leaf width, number of leaves on top of cob, total number of leaves, leaf orientation after flowering, leaf tip shape, angle between leaf blade and stem, leaf and stem direction, and leaf and feather tongue leaf midrib. The results of the observed morphological characterization of local maize of Siarang Dairi in table 1, show that the plants have leaf lengths ranging from 67 - 105 cm with observation data on leaf width, leaf orientation after flowering, leaf tip shape, angle between leaf and stem, and stem, the presence of ligula condition, and the leaf midrib were almost homogen.

3.1.3. Panicle. Observation of a panicle was conducted in generative phases where the panicle was perfectly formed. The results of the observed morphological characterization of the local Siarang Dairi corn in table 1, show that the plant has a long panicle ranging from 21 - 42 cm with a stalk length ranging from 9 - 18 cm. The panicle type, the spacing of the panicle, the totally of primary, secondary, and tertiary branches on the panicle, the size of the panicle after maturity, the lateral branch direction, and the angle between the main axis and the lateral branch below of 1/3 panicles show almost homogen.

3.1.4. Stem. Observation of stem morphology observed included the colour of the stem and the percentage of stuck plant. Table 1 shows that the overall stem colour was green, with the percentage of stuck plants was 3.07%. Based on the agronomic character values of this plant, there is a tendency of the plant to fall in the occurrence of strong winds.

3.1.5. Root. Observation of root morphology observed included root volume and percentage of root fall plant. The results of characterization of observed local root morphology of Siarang Dairi maize in table 1 show that root volume hds medium quantity, with root falling percentage of 0%. This shows the ability or compactness of the roots to withstand plants from extreme weather factors.

3.1.6. Concorb Morphology. The appearance morphology representing Siarang Dairi corncob is presented in the figure 2, 3 and 4.

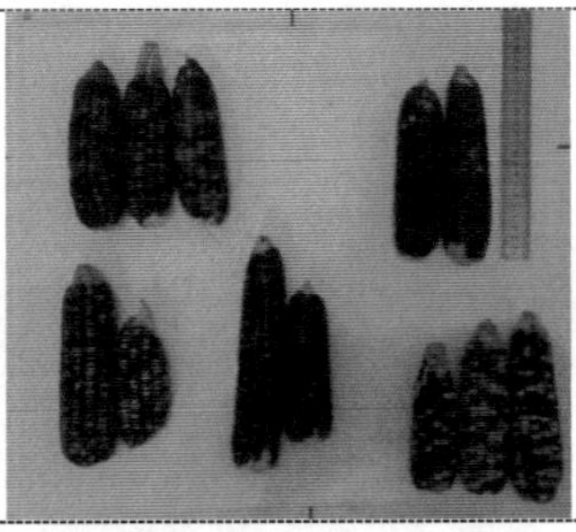

Figure 3. Height of cob

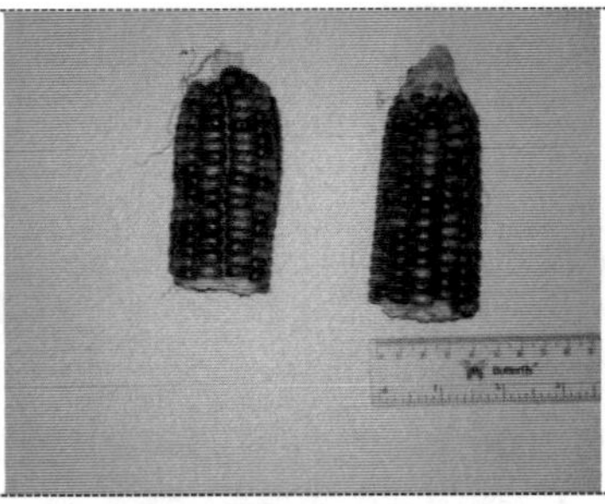

Figure 4. Diameter of cob

Table 2. Character Of Morphology Corncob

Character	Plant				
	1	2	3	4	5
Length of cob (cm)	19	17	19	16	16
Length of cob stalk (cm)	1.5	8	1.5	3	1
Cob diameter	4	4	4.5	4	4
Corncob diameter	3	3	3	2.5	3
Rachis diameter	1.5	2.5	2	1.5	2
Corncob/rachis index	2	1.2	1.5	1.67	1.5
Cornhusk/seed index	0.45	1.6	0.95	0.5	0.8
Rachis/seed index	0.525	0.75	0.7	0.6	0.8
Seed closing Index	2.475	2.25	2.3	1.9	
Totally of branches	0	0	0	0	0
Totally of seed/row	34	34	30	30	33
Corncob colour	White	White	Brown	Brown	White
The uppermost form of cob	Cylindrical	Cylindrical	Cylinder cone	Cylinder cone	Cylinder cone
Angular angle of existence	Medium ± 50°	Medium ± 50°	Medium ± 50°	Medium ± 50°	Medium ± 50°

Observation of corncob morphology observed included length of cob, length of cob stalk, diameter of cob, corncob diameter, rachis diameter, corncob/rachis index, cornhusk/seed index, rachis/seed index, seed closing index, totally of branches, totally of seeds/row, corncob colour, the uppermost form of cob, the angle of existence of the cob and the results are shown in Table 2. The length of cob ranged from 16 - 19 cm with a nearly homogen diameter of 4 - 4.5 cm. The cob does not have branches with totally of seeds/rows ranging from 30 - 34 pieces. The homogen angular angle of cob existence is ± 50°.

3.1.7. Morphology of grain. The appearance morphology of the Siarang Dairi corn grains is presented in the figure 5 below:

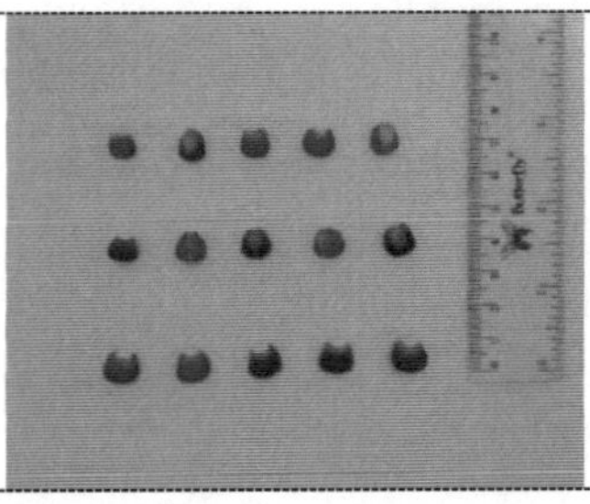

Figure 5. The length of grains

Table 3. Character Of Morphology Grains

Character	Plant				
	1	2	3	4	5
Grain length (mm)	7	6	7	8	8
Grain width (mm)	7	5	6	9	9
Grain thickness (mm)	4,5	3	3	4	5
Grain surface shape	Jagged	Flat	Flat	Circular	Flat
Pericarp colour	Brown	Red	Red	Orange	Red
Aleuron colour	Silvery	Silvery	Silvery	Silvery	Silvery
Endosperm colour	White	Slow yellow	Orange	White	Krem

Observation of grain morphology observed include grain length, grain width, grain thickness, grain surface shape, pericarp colour, aleuron colour, and endosperm colour, and the results are shown in table 3. The grain lengths ranging from 6 - 8 mm with varying widths ranging from 5 - 9 mm. The thickness of the grains ranges from 3 - 5 mm. Variations of grain surface shape include flat, round and serrated. Pericarp colors include brown, red and orange with varied endosperm colours that are white, light yellow, orange, and beige. While the aleuron colour has a homogen of silvery.

3.1.8. Grain Colour

Siarang Dairi corn was analyzed based on the colour classification contained in 1 corncob. The colour classification is shown in figure 6 (a-f).

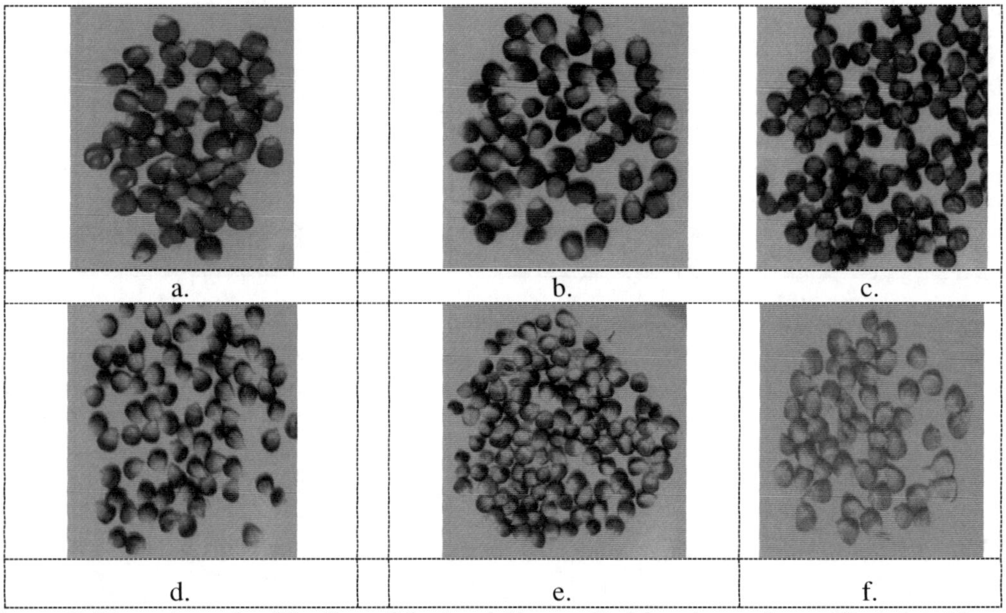

Figure 6. The colour classification of the local Siarang Dairi maize grain: a. Chocolate Young chromo; b. Old chocolate (orange in the middle); c. Dark brown; d. Dark Brown (in the middle is light brown whitish); e. Blackish and f. Yellow

3.2. Protein and Lipid analysis of Local Siarang Dairi Grain

The protein and lipid content of the local Siarang Dairi corn are shown in table 4. The results show that the content of these compound in the grain of the plant vary between colour categorization of the grain. The highest protein content showed by chocolate young chromo grain (8.64%) while dark brown coloured grain with light brown whitish in the middle had the highest lipid content.

IOP Conf. Series: Earth and Environmental Science **157** (2018) 012052 doi:10.1088/1755-1315/157/1/012052

Table 4. Protein and lipid content of the local Siarang Dairi maize grain

Colour	Protein (% w/w)	Lipid level (% w/w)
Chocolate young chromo	8.64	4.04
Old chocolate (orange in the middle)	8.30	3.48
Dark brown	8.28	3.76
Dark Brown (in the middle is light brown whitish)	8.17	4.10
Blackish	8.32	3.56
Yellow	8.58	3.79

5. Conclusion

- Characterization of morphology of local corn Siarang Dairi include plant height ranging from 236 - 277 cm and a leaf length ranging from 67 - 105 cm. Leaf width, leaf orientation after flowering, leaf tip shape, angle between leaf blade and stem, leaf and stem leaf direction, leaf tongue, and almost uniform leaf midrib. The length of panicle ranges from 21 - 42 cm with a long stalk ranging from 9 - 18 cm.
- The length of the cob ranged from 16 - 19 cm with almost homogen diameter of 4 - 4.5 cm. Corncob does not have branches with the totally of seeds/rows ranging from 30 - 34 pieces. The homogen angular angle of corncob existence is ± 50 °.
- The length of grains ranging from 6 - 8 mm with a wide variation ranging from 5 - 9 mm. The thickness of the grains ranges from 3 - 5 mm. Variations of grain surface shape include flat, round and serrated. Pericarp colours include brown, red and orange with varied endosperm colours that are white, light yellow, orange, and beige. While the aleuron colour has a uniformity of silvery.

Reference

[1] BPS North Sumatera Province 2015 *Dairi In Number* (Medan: BPS Sumatera Utara)
[2] Makarim A K, Sjaifullah, Partohardjono S, Hasanah M, and Setyono A 1999 *Methodology of Research and Assessment of Food Crops, Horticulture and Estate Crops* (Jakarta: Indonesian Agency for Agriculture Research and Development Social Economic Department) 160 p
[3] Maemunah and Lapanjang I 2002 *J. Agroland* **9** (1)
[4] http://pangan.litbang.pertanian.go.id/files/PerakitanVarietas.pdf Assembling of Superior Variety of Functional Corn accsessed date 13th July 2017
[5] IPGRI 1993 *Diversity for Development* (Rome: International Plant Genetic Resources Institute)
[6] Budiarti S G 2007 Germplasm of corn as source of genes in breeding program *Bull. of Germplasm* **13** (1) 1 – 9
[7] Priyanto S B, Iriani R N, and Andi T M 2016. *J. Food Crop Research* **35** (2) 125 – 132
[8] Siswati A, Basuki N, and Arifin N S 2015 *J. Plant Production* **3** (1) 19 – 26
[9] Ministry of Agriculture 2004 *Food Crop Characterization Guidelines : Corn dan Sorghum* (Bogor: Indonesian Agency for Agriculture Research and Development Germplasm National Commission)

Social capital on poultry farms in South Sulawesi, Indonesia

V Sri Lestari[1], Natsir[2], I W Patrick[3], H M Ali[4], M Asya[5] and S N Sirajuddin[1]

[1] Socio-Economics Department, Hasanuddin University, Jl. Perintis Kemerdekaan Km. 10, Makassar, 90245, South Sulawesi, Indonesia.
[2] Animal Nutrition Department, Hasanuddin University, Jl. Perintis Kemerdekaan Km. 10, Makassar, 90245, South Sulawesi, Indonesia.
[3]. Institute for Rural Futures, University of New England, Armidale, NSW 2351, Australia.
[4] Food Technology Department, Hasanuddin University, Jl. Perintis Kemerdekaan Km 10, Makassar, 90245, South Sulawesi, Indonesia.
[5] Faculty of Animal Science , West Sulawesi University, Jalan Prof. Dr. Baharuddin Lopa, S.H., Majene, 91412, West Sulawesi, Indonesia.

E-mail: *veronicasrilestari@unhas.ac.id*

Abstract. Social capital plays an important role in the development of poultry farms in South Sulawesi. Poultry farms consisted of laying hen and broiler farms. Most of laying hen farms were located in Sidrap Regency, while broiler farms were located in Maros regency. The aim of this research was to know social capital on beef cattle farms in South Sulawesi. Population of this research was 120 farmers which consisted of 6o were laying hen farmers and 60 were broiler farmers. Variable of social capital was mutual trust, reciprocity, shared norms and linkage. The data were collected from observation and depth interview by using questionnaire. There were 10 questions. The answer was scored by using Likert scale ranging from 1 refer to strongly agree; 2 refer to agree; 3 refer to not sure; 4 refer to disagree and 5 refer to strongly disagree. The data were analyzed descriptively using frequency distribution. The research revealed that mutual trust and shared norms members of the group in broiler farms have higher level than that on laying hen farms, on the other hand, linkage or net working members of the group among laying hen farmers has higher level than that on broiler farms.

1. Introduction

Demand for poultry product such as chicken meat and eggs increases steadily year by year. However, the domestic production of chicken meat and eggs cannot fulfil the local demand. Therefore, Indonesian government should import chicken meat and eggs from other countries. To overcome the loss of the foreign exchange, the development of poultry farms in Indonesia should be supported by a good regulation, poultry farmers participation, private and government intervention [1].

Poultry has the best conversion rate of feed to human food in terms of per kg meat or eggs produced [2]. The success of poultry farms were influenced not only by technical factors, but also non-technical factors such as social capital. According to Putnam [3] and Subejo [4], social capital is a characteristic of social organization such as trust, norm, and networking that facilitate the occurrence of coordination and cooperation for mutual benefit.

Social capital is formed and developed through the existence and life experiences of human beings. Each person in a different social environment such as a different family or location will have different views. Further, each community has its own concepts and values that are formed through the accumulation of beliefs, moral standards and other individual contacts inside that community [5]. Hasbullah [6] added that social capital is anything related to cooperation in society or nation. To achieve a better purpose of life, sustained by the values and norms that become the main elements such as trust, community participation, reciprocity, collective rule in a society or nation. Dinda [7] offered a more general definition, calling social capital "a broad term containing the social networks and norms that generate shared understandings, trust and reciprocity, which underpin cooperation and collective action for mutual benefits, and creates the basis for economic prosperity".

According to Salman [8], social capital can be divided into some form. It is in the form of cognitive social capital: something inherent to the individual in the form of cognition for mutual trust and principled reciprocity; it is in the form of social capital structural: something inherent to the structure of values and norms that emphasize togetherness, associations and organizations and governance; and it is in the form of relational social capital: something born of social relationships in the form of social networks.

Syahyuti [9] found that social capital in a society can be strengthened but requires the support of certain resources. In order to create good social and institutional relationships so community members must support it. Social capital acts as a glue that binds all the people in the society. In order for social capital to grow well there is a need for shared values, beliefs, and organizing roles expressed in personal relationships. According to Flora [10] and Flora *et al.* [11], an important aspect of community quality of life is social capital, which includes mutual trust, reciprocity, and shared norms and identity. In general, communities with greater social capital provide greater quality of life.

Based on previous studies [12], social capital was a significant factor determining the level of agricultural income. Munasib and Jordan [13] found that community involvement had a positive effect on the decision to adopt sustainable agricultural practices, and, secondly, it also had a positive effect on the extent to which farmers adopt these practices.

The most populous laying hen in South Sulawesi was in Sidrap Regency. In 2013, there were 4,041,027 tail. Unfortunately, the population decreased to be 3.48 million tail in 2016. On the other hand, the most progress poultry farms in South Sulawesi was broiler farms which was located in Maros regency. Since 2011, the population increased by 185%. Total population was 18,497,399 tail in 2011, then it increased steadily to be 52,651,682 tail in 2016 [14].

According to central bureau of statistics [14], the role of broiler chicken (broilers) is very important to meet the needs of society for meat as material nutritious food, this is remembering the population of chickens is quite large and its maintenance is almost on all corners of the country. The development of poultry farms in South Sulawesi was supported by local government, animal husbandry services, private sectors and poultry farmers. Most of poultry farmers became a member of farmers group.

The purpose of this research was to know social capital on poultry farms in South Sulawesi.

2. Research Method

This research was conducted on poultry farms which consisted of laying hen farms and broiler farms. Laying hen farms were mostly located in Sidrap regency, while broiler farms were located in Maros regency. Population of this research was based on FAO data. Sample were consisted of 120 poultry farmers which consisted of 60 laying hen farmers and 60 broiler farmers. The data were collected through observation and depth interview by using questionnaire.

The element of social capital was consisted of trust, norms and networking (linkage). There were 10 questions. The answer was scored by using Likert scale which started from 1 refer to strongly agree, 2 refer to agree, 3 refer to doubt or not sure, 4 refer to disagree and 5 refer to strongly disagree. The data were analysed descriptively using frequency distribution [16].

3. Results and Discussion

Table 1. showed that on average respondents were in productive age both to laying hen farmers and broiler farmers. Managing poultry farms need a lot of work such as cleaning the cage, feeding, picking

IOP Conf. Series: Earth and Environmental Science **157** (2018) 012053 doi:10.1088/1755-1315/157/1/012053

the eggs and marketing. According to their education, on average, poultry farmers spent more than 9 years at school. This indicated that both laying hen farmers and broiler farms finished their elementary school. In other words, they were literate. The education level of respondents was better than that of Riduwan and Akdon [16] which majority (68.85%) of native chicken farmers just finished from elementary school. With regard to gender, male was dominant than female for both laying hen farmers and broiler farmers. Most of female in rural area were busy with their domestic job in their family, such as take care of their children, cooking, washing, and cleaning the house. Looking at farms experience, both laying hen farmers and broiler farmers spent more than 5 years in managing their farms, this indicate that they have enough experience. Usually farmers go their experience from their parents. According to the number of chicken they have, on average for both laying hen farmers and broiler farmers have chicken less than 10,000 tail, this indicate that they were smallholder farmers.

Table 1. Characteristic of respondents

No	Characteristics of respondents	Laying Hen Farmers	Broiler Farmers
1	Age (year)		
	• Average	44.67	45.23
	• Minimum	23	26
	• Maximum	80	80
2	Education (year)		
	• Average	10.2	11.07
	• Minimum	0	0
	• Maximum	17	26
3	Gender (person)		
	• Male	48	53
	• Female	2	7
4	Number of chicken (tail)		
	• Average	5,875	4,338
	• Minimum	1,000	300
	• Maximum	40,000	50,000
5	Farm experience (year)		
	• Average	8.2	6.43
	• Minimum	1.0	1.0
	• Maximum	20.0	23.0

On table 2, it can be seen that mutual trust and norm variables on broiler farms were higher than that of laying hen farms. This was because currently, the majority of broiler farmers is not working independently but united with integrated partner companies. According to [18], the partnership scheme is a partnership between the partner farmer and partner companies. The partner farmer groups act as plasma, while the partner companies act as the core. In the broiler partnership that is currently running, the partner or core companies provide farming production facilities such as DOC (Day Old Chick), feed, medicines, vitamins, vaccines, technical supervisor, and products marketing, while the plasma provide the chicken house and worker. The aim of this partnership scheme is to help the broiler farmers who have limited funds. However, essentially partnership is business cooperation for certain objectives and each party should have equal interest and position.

Table 2. Category of Social Capital on Poultry Farms

No	Variable	Laying Hen Farmers		Category	Broiler Farmers		Category
		Score	(%)		Score	%	
1	Trust	3.26	63.0	High	3.63	69.8	High
2	Norm	2.98	57.13	Enough	3.60	60.5	High
3	Linkage	4.00	60.33	High	2.98	58.67	Enough
	Social Capital		60.8	High		64.20	High

During its journey, the broiler partnership experience an up and down. It because of the system of broiler raising management should be well conducted to reach optimum results as broiler age is relatively short between 32 to 36 days of raising period [19]. Yunus [20] argued that broiler harvesting carried out at 32-36 days, when the chicken have average weight of 1.75 kg/head. Samarokoon and Samarasinghe [21] added that as the harvesting time increased, it would cause the increasing of FCR. Therefore, the result of this research indicated that trust and norms of broiler farmers were categorized as "high".

On the other hand, linkage of member group on laying hen farms was higher than that on broiler farms. This was because the breeders of laying hen farms were independent breeders. Everything from the procurement of DOC, feed, chicken house, vitamins, vaccines, medicine, worker, technical supervisor and marketing were provided by themselves. So the network with other laying hen breeders must be strong. Total social capital for both poultry farms were categorized as "high". Although social capital was high, it is also important to understand whether or not this social capital can be turned into action (i.e. whether or not there is a high level of "agency" within the group). This group agency variable was determined from an understanding of how groups would respond to issues or difficulties facing the group and individuals.

Table 3. Farmers response to HPAI (bird flu) within the farms

Farmer response	Laying hen farmers	Broiler farmers
Group members would help each other	43 (71.67%)	40 (66.67%)
Group would ask for help from government	13 (21.67%)	10 (16.67%)
Everyone would deal with the situation by themselves	4 (6.67)	7 (11.67%)
Group would seek help from wider community	0	3 (5.0%)

Based on table 3, if there was an issue facing the group (for example Avian influenza), 71.67% of laying hen farmers and 66.67% of broiler farmers would like to help each other rather than asking for help from government. Helping each other was the characteristics of poultry farmers. The result of the recent research was supported by Patrick *et al.* [22] who found that beef cattle farmers in Bali and Lombok do the same respond.

Overall, social capital on poultry farms both for laying hen farmers and broiler farmers in South Sulawesi were categorized as high.

Acknowledgements
Authors would like to appreciate to ACIAR (Australia Centre for International Agriculture Research) for the financial support for doing this research.

References
[1] Fahmid I M 2004 *Gagalnya politik pangan di bawah rezim orde baru: kajian ekonomi politik pangan di Indonesia* (Jakarta: Yayasan Sandi Kota-ISPEI).
[2] Vaarst M, Steenfeldt S and Horsted K 2015 Sustainable development perspectives of poultry production. *World's Poul. Sci.* **71** 609-18.

[3] Putnam R D 1993 *Making Demoracy Work : Civic Traditions in Modern Italy* (USA: Priceton University Press)
[4] Subejo 2004 Peranan social kapital dalam pembangunan ekonomi: Suatu pengantar untuk studi social capital di pedesaan Indonesia *Majalah Agro Ekonomi* **11** (1) 79
[5] Anh N T N 2014 Social capital, leadership reputation and labour force motivation in enterprise *The J. of Comparative Econ. Studies.* **9** 89-97
[6] Hasbullah J 2006 *Social capital (Menuju keunggulan budaya manusia Indonesia.)* (Jakarta: MR-United Press)
[7] Dinda S 2008. Social capital in the creation of human capital and economic growth: A productive consumption approach *J. of Socio-Econ.* **37** 2020-33
[8] Salman D 2016 *Sosiologi desa: Revolusi senyap dan tarian kompleksita.* (Makassar: Inninawa)
[9] Syahyuti 2008 The role of social capital in agricultural trade *Agro Economic Forum* **26** 1 Pusat Sosial Ekonomi dan Kebijakan Pertanian Badan Penelitian dan Pengembangan Pertanian. Kementerian Pertanian Republik Indonesia
[10] Flora J L 1998 Social capital and communities of place *Rural Sociology* **63** 481-506
[11] Flora J L, Sharp J, Flora C and Newlon B 1997 Entrepreneurial social infrastructure and locally-initiated economic development *Socio. Quarterly* **38** 623-45
[12] Wolz A, Fritzch J and Reinsberg K 2005 *The Impact of Social Capital on Farm and Household Income: Results of A Survey among Individual Farmers in Poland* The 94th EAAE Seminar: From household to firms with independent legal status: the spectrum of institutional units in the development of Europe agriculture Ashford (UK) 9-10 April 2005
[13] Munasib A B A and Jordan J L 2011 The efffect of social capital on the choice to use sustainable agricultural practices *J. of Agric. and App.Econ.* **43** 2 213–27
[14] Central Bureau Statistics 2016 South Sulawesi in Figure Makassar Indonesia
[15] Ratnasari R, Sarengat W and Setiadi A 2015 Analisis pendapatan peternak ayam broiler pada sistem kemitraan di kecamatan Gunung Jati kota Semarang *Animal Agric J.* **4** 1 47-53
[16] Riduwan and Akdon 2009 *Rumus dan Data dalam Analisis Statistika.* (Bandung: Alfabeta)
[17] Putra RAR, Ariyadi SB, Kurnia N and Haryadi FT 2017 Pengaruh modal sosial terhadap tingkat kesejahteraan rumah tangga peternak: Studi kasus pada kelompok peternak ayam kampong Ngudi Mulyo Gunung Kidul *Buletin Peternakan.* **41** (3) 349-354.
[18] Salam T, Muis M and Dan AENR 2006 Financial analysis on broiler farming in partnership system *J. of Agric. System.* **2** (1) 32-9
[19] Nurtini S, Muzayyanah M A U, Haryadi F T and Hakim A 2017 Performance of broiler farmer in partnerships system at Surakarta Indonesia *J. of Adv Agric Tech.* **4** (2)
[20] Yunus 2009 *Efficiency production analysis on broiler farming in partnership and independent system in Palu, Central Sulawesi Province,* Postgraduate Program, (Semarang: Universitas Diponegoro)
[21] Samarokoon S M R and Samarasinghe K 2012 Strategies to improve the cost effectiveness of broiler production *Tropical Agric.l Research.* **23** 4 338-46
[22] Patrick I W, Marshall G R, Ambarawati I G A A and Abdurachman M 2010 Social capital and cattle marketing chains in Bali and Lombok Indonesia *ACIAR Technical Reports* No. 74 Australian Centre for International Agricultural Research Canberra 78 p

IOP Conf. Series: Earth and Environmental Science **157** (2018) 012054 doi:10.1088/1755-1315/157/1/012054

The innovative characteristics and obstruction of technology adoption for management of integrated plants (PTT) of corn in Gowa Regency Indonesia

M H Jamil[1], Y Musa[2], A N Tenriawaru[1] and N E Rahayu[3]

[1]Department Socio-Economics, Study Program Agribusiness, Faculty of Agriculture, Hasanuddin University. Jl. Perintis Kemerdekaan Km 10, 90245, Makassar, South Sulawesi, Indonesia.
[2]Department Agronomy, Faculty of Agriculture, Hasanuddin University. Jl. Perintis Kemerdekaan Km 10, 90245, Makassar, South Sulawesi, Indonesia.
[3] Indonesia Cereal Research Institute (ICERI), Jl. Dr.Ratulangi No. 274 Maros, South Sulawesi, Indonesia

E-mail: *hattaj@yahoo.com*

Abstract. The research aimed to analyze the effects of the farmer's characteristic, innovation characteristics, and the obstruction faced in the technology adoption for the management of integrated plants corn in Gowa Regency. The method used was explanative in character. Respondents comprised 80 corn farmers chosen randomly. Data were collected using the interviews method which were then quantified using likers scale. The data was analyzed by logistic binary regression. The research results indicated that the farmer's characteristics which consisted of the age, education, experience, and the land area had no significant effect on the technology adoption of maize integrated crops management (PTT). The obstruction of the adoption, which consisted of the limited capital, availability of inputs, and intensity of counseling had a significant effect on the adoption of maize integrated crops management. While the farmer's knowledge had no significant effect on the adoption of maize integrated crops management. The variable of the limited capital had a positive coefficient, the more available the farmer's capital the higher was the chance of farmers to adopt technology integrated crops management. The higher of the extension intensity, the higher of farmer's chance to adopt the technology of the maize integrated corps management.

1. Introduction

Integrated crop management technology (PTT) in corn commodities has increased corn productivity in the range of 4 to 5 tons per hectare [1]. Compared to the productivity that combines high-quality varieties of free-sized corn or hybrids and innovative application of innovative technologies through the PTT approach, productivity is achieved between 7 to 9 tons per hectare [2]. Productivity gaps at farmers' level with the results of the study indicate low adoption of PTT at the farm level. Referring to the theory of Inovation-Difusion; the speed of adoption of an innovation is influenced by innovation characteristics consisting of relative advantages, compatibility, complexity, trialability, and observability. Research conducted in 2015 by the Agency for Agricultural Research and Development, Ministry of Agriculture of the Republic of Indonesia, showed that efforts to increase corn production

IOP Conf. Series: Earth and Environmental Science **157** (2018) 012054 doi:10.1088/1755-1315/157/1/012054

still encountered problems namely the adoption of PTT components have not been intact and constraints of capital limitations and availability of production facilities.

The above facts are also shown by Dahlan's research results [3] where farmers in Bantaeng District 73.5% only adopt 1 to 3 basic components, while farmers who adopt 4 to 5 basic components as much as 26.5% of total respondents 83 people. These results indicate that the adoption of farmers on the component of PTT maize technology is still low. This fact reinforced the discovery of [4], which states that of all the basic components of PTT corn technology, balanced fertilizer technology components and the lowest planting distance adopted by farmers, respectively 21.7% and 37.3% of the total respondents.

The low adoption of PTT components has been identified by Ginting [5] where technological obstacles adopted by farmers are: (a) technical factors i.e. technological effect on yield improvement, income, and farming profits not yet believed by farmers. Limited information and infrastructure networks, causing outside information and knowledge to be unacceptable to farmers; (b) factors, namely the lack of knowledge of agricultural technology dissemination systems (extension, demonstration plots and performances technology) and a low level of education and knowledge of farmers making it difficult to translate the benefits of new technologies; (c) social factors where generally small farmers are afraid of low selling price risk, inefficient marketing system, and high transportation cost [6].

According to Rogers [7], from the communication side, the low adoption of PTT is caused by the inhibiting factor of technology adoption that is the inhibiting factor of communication in the form of Source, Message, Channel and Recipient / Client. Adoption will not take place if there are obstacles to the communication process between extension workers and farmers [8]. Based on the above description, it is very interesting to analyze the effect of innovation characteristics and barriers to adoption of maize PTT technology in Gowa regency, South Sulawesi, Indonesia.

2. Research Design

The design of this study is explanatory, *i.e.* research that examines the relationship between variables and test the hypotheses that have been formulated and explain the influence between independent variables (characteristics of innovation and adoption barriers) to the dependent variable (adoption of basic components of PTT corn technology) using logistic binary regression analysis.

The study was conducted in three sub-districts with two typologies, namely lowland (Bontomarannu and Bontonompo sub-districts) and highlands (Parangloe sub-district) Gowa District, South Sulawesi, Indonesia. The research population is farmers who are members of farmer group and have conducted PTT corn activity with certain quality and characteristic according to criteria to be achieved in this research [9]. The samples were drawn using a simple random sampling method preceded by a proportional sampling method using the Slovin formula, resulting in 80 samples.

Data collection techniques using questionnaires, formulated with questions containing all indicators of the dimensions of research variables. Prior to the first data collection conducted testing instrument (questionnaire) to test the validity and reliability of the instrument. Next is a structured interview and record the answers given by respondents.

Data processing is done through stages; editing of the data that has been collected by sorting to maintain the validity, reliability and accuracy; then coding and scoring in accordance with the classification of data that has been determined using Likert scale; followed by data entry using SPSS 21.00 program. Meanwhile, to determine the effect of innovation characteristics and constraints faced in adopting the basic components of integrated crop management technology (PTT) maize used logistic regression analysis Binary.

3. Results and Discussion

3.1. Test Validity and Reliability

Validity test is done by correlating the score of each item with the total score of each variable.

IOP Conf. Series: Earth and Environmental Science **157** (2018) 012054 doi:10.1088/1755-1315/157/1/012054

Decision-making is based on value r_{count} (*Corrected Item-Total Correlation*) > r_{table} of 0.2199, for df = 80 - 2 = 78; α = 0.05 then the question item is valid and vice versa. Data processing shows all research questions declared valid, because the value of r_{count} (Corrected Item-Total Correlation) > r $_{table}$ of 0.2199 (table 1).

Table 1. Validity Test Results of Innovation Characteristics and Adoption Barriers

Consept	Dimension	Value of r count	Value of r table	Explanation
Innovation Characteristics	Relative Advantage	0.440 – 0.742	0.29	Valid
	Compatibility	0.539 – 0.699		
	Complexity	0.433 – 0.691		
	Trial	0.357 – 0.448		
	Observability	0.330 – 0.515		
Adoption Barriers	Limited of Capital	0.486 – 0.595	0.29	valid
	Limited of production facilities	0.578 – 0.640		
	Knowledge	0.651 – 0.671		
	Intensity of Counseling	0.468 – 0.588		

Reliability test was carried out using Alpha Cronbach method. Alpha Cronbach value> r table of 0.2199, for df = 80 - 2 = 78; α = 0.05 then the question item is reliable and vice versa. The test results show that each questionnaire question in this study is stated reliable, because the value of Alpha Cronbach> r table of 0.2199 (table 2).

Table 2. Reliability Test Results of Innovation Characteristics and Adoption Barriers

Consept	Alpha Cronbach	Value of r table	Explanation
Innovation Characteristics	0.741	0.2199	Reliable
Adoption Barriers	0.644	0.2199	Reliable

3.2. Logistic Binary Regression Test

The results show; variables of innovation characteristics with relative profit sub-variables, compatibility, complexity, trialability and observability have a significance value of 0.564 each; 0.201; 0.263; 0.214 and 0.161 is greater than α = 0.05. This shows that the variable characteristic of innovation does not significantly influence the adoption of integrated corn plantation integrated management (PTT) component of maize.

While the variable of adoption constraint with sub variable of capital limit, and availability of production and intensity of extension have significance value equal to 0.000 and 0.006 and 0.012 <α = 0.05. This indicates that the adoption barrier variable consisting of capital limitations, availability of production facilities and intensity of extension significantly influence the farmer's decision to adopt the basic components of PTT corn completely, while the knowledge of farmers has a significance value of 0.663> α = 0.05 which has a meaning that the knowledge of farmers does not have a significant influence on the farmers' decision to adopt the basic components of PTT maize completely.

The variable of capital limitation has positive coefficient value, which means that the more available the farmer's capital, the higher the chance of the farmer to adopt the basic component of integrated corn plant management technology with probability equal to 423,525 times compared with the farmer with limited capital. While variable of production facility have positive coefficient value. The more available and timely means of production, means the higher the chance of farmers adopting the complete PTT maize technology with a probability value of 24,460 times. Likewise, the extension intensity variable has a positive coefficient value. The more intensively held counseling the higher the

IOP Conf. Series: Earth and Environmental Science **157** (2018) 012054 doi:10.1088/1755-1315/157/1/012054

chance of farmers to adopt the basic components of PTT maize technology complete with probability value of 25,183 times.

This study shows that innovation characteristics consisting of relative advantage, compatibility, complexity, trialability, and observability have no significant effect on the adoption of basic components of PTT maize completely. Nevertheless, the farmers of the respondents have a positive perception on the characteristics of innovation, because the application of basic components of PTT has been in accordance with the needs of farmers who want increased production of corn crops and income increase. The basic components of PTT are very easy to understand by farmers, but for the application of the plant population fulfillment of 60,000-75,000 plants / ha through planting spacing of 70 x 40 cm for 2 seeds of plants / holes and 70 x 20 cm for 1 seed of plant / difficult to apply in Parangloe and Bontolempangan sub-districts, due to the sloping and rocky soil topography which makes it difficult to adjust the spacing according to recommendation. Farmers' decision to adopt the basic components of maize PTT is related to their owned capital, availability of seed and fertilizer and intensity of counseling. In addition, adoption barriers consisting of capital limitations, availability of inputs and intensity of extension significantly influence the adoption of basic components of PTT corn completely. While the knowledge of farmers has no significant influence on the adoption of basic components of PTT corn completely. This shows that in the corn farming, farmers are very concerned about the capital, the availability of production facilities in the form of seeds and fertilizers and intensively counseling held.

The use of high quality and superior varieties of corn seeds is the key to success in the application of maize PTT. Of course this is accompanied by plant spacing arrangements to meet the plant population of 66,000-75,000 plants / ha and the application of balanced fertilizer as recommended. The acquisition of high yielding corn seeds and fertilizers is obtained by farmers with high prices, which is not suitable with the limited capital of farmers. While at harvest time, the price received does not match the capital they spend. The higher the price of seeds and fertilizers, the more difficult the farmers to resume the planting of superior corn seed varieties and difficult to use fertilizer as recommended without any subsidy from the government.

According to Schiffman and Kanuk [9] and Joni *et al.* [10], price is one of several factors that influence consumers in buying seed decisions. Likewise, with Sumiati in [11], which states that one of the factors affecting the farmer's decision to use seed is the price of seed. Purwantini and Sejati [12] argued that the availability of production facilities is not a constraint, precisely the problem of limited capital becomes the dominant factor. Capital is very important role in technology adoption especially related to the cost required to use new technology [13]. Research by Irawan *et al.* [14], shows that farmers do not plant superior seeds because they lack the capital to buy superior seeds, and other production costs such as fertilizer in Malaysia. While the case study in Sumatra also shows that the lack of capital is a major constraining factor for farmers to grow superior rubber seeds [8].

The availability of production facilities is one of the obstacles in adopting complete maize PTT. Seed production facilities and fertilizers are always available on time and in exact quantities, but can only be obtained by ordering advance to the chairman of gapoktan in Parangloe, Bontonompo and Bontolempangan areas. However, in Bontomarannu area, this production suggestion can be obtained more easily through retailers around their area.

Manure (organic fertilizer) that is needed in large numbers in the highlands is difficult because of the limited availability of transportation facilities to distribute it. In addition, some farmers think the condition of land in the highlands is still quite fertile and does not require manure. Irawan *et al.* [14] showed that farmer inputs needed by farmers in general are always available in the market, especially for the main types of inputs such as urea and pesticides. Meanwhile, according Supriadi and Chamala [8], the availability of fertilizers and pesticides is sufficient, but with a relatively high price. This causes farmers who have limited capital choose not to use fertilizer in accordance with the recommended standard.

The intensity of counseling at the research sites is routinely conducted, both at the agricultural extension office (BPP), at the home of one farmer and at the farm site. Field guides (extension workers) play an active role in disseminating information and solving problems related to farmer cultivation. This indicates that stepping up of agricultural extension (based on social learning process), which becomes a push factor towards independency group in finding new innovation [15]. In addition, stepping up agricultural and non-agricultural extension services, could also significantly reduce poverty, though indirectly [16]. They establish intimacy with farmers, so even though outside the schedule of meetings, farmers can still communicate intensively with extension workers. However, access to information on agricultural technology through mass media (television, newspapers, agricultural and radio magazines) has not been effectively implemented by farmers. Most farmers think that the information is more easily obtained through extension workers.

3. Conclusion

The characteristics of innovation has no significant effect on the adoption of PTT corn technology. Adoption barriers consisting of limited capital, availability of means of production, and intensity of counseling have significant influence with positive coefficient value on adoption of PTT maize technology, while farmers' knowledge has no significant effect. Strengthening of farming capital through the provision of low-interest credit assistance needs to be provided by the government. In addition, agricultural insurance for farmers will minimize losses to the failure of production and basic pricing for Besides, the guarantee of availability of production facilities and extension are regularly important to be maintained.

References

[1] BPS 2015 www.bps.go.id. Accessed January 4, 2015

[2] Saenong S and Subandi 2002 *The concept of PTT on Corn Crops Paper presented on Technical Development and Management of PTT Palawija in Balitkabi* (Malang: Balitkabi) 21 – 22

[3] Dahlan S S 2014 *Effect of Human Capital and Entrepreneurship on Adoption Component of Integrated Crop Management Technology (PTT) on Corn Farmers In Bantaeng Regency. (Thesis)* (Makassar: Hasanuddin University)

[4] Kasimin S 2013 Product linkages and performers in the development of leading horticultural agribusiness in Aceh Province *J. of Management & Agribusiness* **10** (2)

[5] Ginting B 2006 *Technology Adoption by Farmers* (Indonesia: Faculty of Agriculture University of North Sumatra)

[6] Fahmid I M 2013 Social innovation among ethnics in cocoa farming at Sulawesi, Indonesia *J. of Biology, Agriculture and Healthcare* **3** (15)

[7] Rogers E M 2003 *Diffusion of Innovations fifth ed.* (New York: New York Press)

[8] Supriadi M and Chamala S 1998 Technology attributes and adoption: improving the effectivennes of indonesian smallholders's rubber research *J. of Int. Agric. and Ext. Edu.* **5** (3) 45-54

[9] Schiffman L G and Kanuk L L 2008 *Consumer behavior* ed. 7 (Jakarta: PT. Index)

[10] Joni M M A, Rubzen M F R and Batt P J 2001 *Factors Influencing a Farmer's Decision to Purchase Seed Potatoes in East Java Paper presented at the 45th Annual Conference of the Australian Agricultural and Resource Economics Society* (Adelaide, South Australia)

[11] Ministry of Agriculture, Directorate General of Food Crops 2015 *Technical Guidelines of GPT PTT 2015 Implementation* (Jakarta: Ministry of Agriculture)

[12] Purwantini T B and True W K 2013 *Role of Agribusiness Supporting Institution in Rice Farming Proc. of the 33rd World Food Day National Seminar "Optimizing Local Resources Through Food Diversification Toward Food Self-Reliance and Nutrition Improvement of the Society Toward the 2015 Asian Economic Community"*

[13] Feder G, Just R E and Zilberman D 1985 *Adoption of Agricultural Innovations in Developing Countries*; a Survey-Economy. Dev, Cult, Change

[14] Irawan B P et al. 2007 *National Farmers' Panel (PATANAS) Analysis of Agricultural and Rural Development Indicators Research Report* (Bogor: Agricultural Policy and Analysis Center)
[15] Ekasari Z K, Saleh S A M, Jusoff K, Salman D, Akhsan, Kasirang A, Arsyad M, Amrawaty A A and Fudjaja L 2013 communication pattern and conflict in agricultural extension *Asian Soc. Sci.* **9** (5) 27-33
[16] Thalyta E Y, Mustajab M, Arsyad M 2010 Abstracts of doctoral theses on the Indonesian economy *Bull. of Indonesian Eco. Studies* **46** (2) 251-254 DOI: 10.1080/00074918.2010.486113

IOP Conf. Series: Earth and Environmental Science **157** (2018) 012055 doi:10.1088/1755-1315/157/1/012055

Attitude and perception of farmers to the implementation of conservation farming in the mountainous area of South Sulawesi

N Busthanul[1], Y Lumoindong[1], M Syafiuddin[2], Heliawaty[1], N Lanuhu[1], T Ibrahim[1] and R R Ambrosius[1]

[1] Dept of Social Economy, Faculty of Agriculture, Hasanuddin University, Makassar, Indonesia 90245
[2] Dept of Soil Science, Faculty of Agriculture, Hasanuddin University, Makassar, Indonesia 90245

E-mail: *nurbaya_busthanul@agri.unhas.ac.id*

Abstract. Farmers' attitudes and perceptions may be the cause of ineffective implementation of conservation farming for agriculture sustainability due to vary of implementing of conservation techniques. The purpose of this research is to know the attitude and perception of farmer toward the application of conservation technique and to know correlation between farmer attitude and perception toward the application of conservation technique. The research was carried out in Kanreapia Village, Tombolo Pao District, Gowa Regency, South Sulawesi Province, Indonesia. Sampling was done by randomly with 30 farmers; using non-parametric statistics with quantitative and qualitative descriptive data analysis approach, using Likert scale. The result showed that farmer attitude and perception toward conservation technique implementation which having the highest category (appropriate) is seasonal crop rotation, while the lowest with less appropriate category is the processing of land according to the contour and the cultivation of the plants accordingly. There is a very strong relationship between farmer attitude and perception. The implications of the findings are that improvements the implementation of conservation farming techniques should be made through improved perceptions.

1. Introduction

Implementation of sustainable agriculture is needed in the framework of food security and resilience of the world population. However, whether or not the implementation is determined by the attitude of farmers. Being attitude is determined by perception [1] and [2]. Therefore, a correct perception of conservation awareness is essential for the sustainability of agriculture. This is especially urgent in the central areas of agricultural food production especially on agricultural land with intensive processing in upland areas and slopes [3]. One area that has the characteristics of such areas is the production center of Kanreapia highland vegetables.

The area located in Gowa regency, South Sulawesi, Indonesia, in addition to being processed intensively, also has many critical lands that require conservation measures. According to the Provincial Forestry Office of South Sulawesi, currently the area of critical land is about 28,000 ha. Critical land is located in several areas including Tombolo Pao District about 6,830 ha, Tinggimoncong District 5,590 ha, Parangloe District 5,403 ha, and the rest spread in other districts.

The critical land is part of the total critical land in South Sulawesi Province which recorded 920,452 ha in 2011.

Measures or treatment of farmers on conservation will differ across regions, in accordance with land type, land area, availability of on farm and off farm, and policies of local government programs [4]. The attitude or tendency of farmers to do or not to take conservation measures on their agricultural activities is driven by many factors. In this case the opportunity to increase profit or welfare may be the most important thing [5]. In the same word, appropriate conservation activities can increase farmers' production and income [6]. The purpose of this research is to know the attitude and perception of farmer toward the application of conservation technique and to know correlation between farmer attitude and perception toward the application of conservation technique. The research was carried out in Kanreapia Village, Tombolo Pao District, Gowa Regency, South Sulawesi Province, Indonesia.

2. Material and Method

2.1. Data collection

The research was carried out in Kanreapian Village, Sub-district Tombolo Pao, District Gowa, South Sulawesi, Indonesia in October to November 20016. Collecting data was done through observation and interviews directly with the respondents who were conservation farmers while secondary data were supporting data obtained through literature review.

This study uses non-parametric statistics, with data analysis approach descriptively quantitative and qualitative, using Likert scale to measure low, medium, or high attitudes and perceptions of farmers on the implementation of conservation farming. Data types are interval scales [7, 8, 9] and [10]. For "attitudes" are measured three variables: (i) whether to know about, (ii) whether to do, and (iii) whether to do (accordingly theory); while "perceptions" are measured three variables: (i) the level of importance to do, (ii) the level of effect/benefit to their farm business, and (iii) whether there is or not influence of other people's knowledge on the implementation of the nine factors of conservation action. These nine factors include: (1) aging of terraces or beds on sloped land, (2) Planting plants or turf for terrace reinforcement, (3) processing of land according to contour lines, (4) Cultivation of plants according to contour lines, (5) seasonal crop rotation , (6) planting annual crops on high slopes, (7) use of mulch (or manure), (8) maintenance of terraces (or beds), and (9) making drainage canals.

2.2. Data analysis

To measure per indicator (question) of each variable can be done by determining the interval and category setting as follows [11]: The maximum value is 3 x 3 = 9 while the minimum is 1 x 3 = 3. Thus the data interval = (9-3) / 3 = 2; then the value categories are obtained: 3-5: Low; 6-7: Medium; and 8-9: Height.

To determine each individual category of farmer against 9 conservation implementation variables, the maximum value is 9 x 9 = 81 while the minimum value is 9 x 3 = 27. Thus the interval is (81-27) / 3 = 18, so the category is 27-45; low, 46-63: medium, and 64-81: high.

In the same way, to measure the total variables of farmer attitudes and perceptions on the application of conservation farming with 30 respondents, conducted by determining the interval as follows: The maximum value is 9 x 30 = 270 while the minimum value is 3 x 30 = 90. Thus the data interval = (270-90) / 3 = 60; then the value categories are obtained: 90-150: Low; 151-210: Medium; and 211-270: Height. Furthermore, with the same event, with 27 questions then to determine the level of cumulative attitude and perception of farmers (recapitulation) obtained categories: 810-1350: Low; 1351-1890: Medium and 1891-2430: High.

3. Results and Discussion

3.1. Farmer attitude

3.1.1. Attitudes of farmers individually. Attitudes of farmer individually partially and cumulatively of nine variables of implementation of conservation farming techniques are shown in table 1. Table 1 showed, the implementation of conservation farming techniques has been addressed as appropriate or high level. There are 28 of 30 farmers (93.3%) of farmers whose attitudes are at a high level. This clearly indicates the awareness of farmers on the importance of implementing conservation agriculture. There is no farmer in the low level and only 2 farmers (6.7%) in medium level.

Table 1. Partial and total score of conservation technique factors of farmers individually

No.	Name	Gender	Age (yr)	Conservation technique factors*) (1)	(2)	(3)	(4)	(5)	(6)	(7)	(8)	(9)	Total
1	Karaeng Bonto	M	36	9	9	9	9	9	6	8	8	7	74
2	Muh. Jufri Basari	M	39	9	9	6	6	9	8	9	8	9	73
3	Haji. Jurhan	M	54	9	9	9	9	9	8	8	8	8	77
4	Khaerul Rizal	M	35	9	9	6	6	8	6	8	9	9	70
5	Jamaluddin	M	28	9	9	6	6	9	8	9	8	8	72
6	Usman	M	50	9	9	9	9	9	7	8	8	8	76
7	Harifin	M	46	6	7	6	6	9	8	8	8	8	66
8	Bahri	M	32	7	6	6	6	9	8	7	8	7	64
9	Mappi	M	39	7	5	9	9	9	8	8	8	8	71
10	Hawani	F	28	6	7	6	6	9	8	8	8	8	66
11	Jumarang	M	34	9	9	9	9	8	6	7	9	8	74
12	Baharuddin	M	47	9	9	6	6	9	8	7	8	7	69
13	Harding	M	29	7	5	9	9	9	8	7	9	8	71
14	Labo	M	33	9	9	6	6	8	8	9	8	9	72
15	Taufia	M	25	9	9	5	5	9	8	8	7	9	69
16	Dg. Rambu	M	28	9	9	6	6	9	8	9	8	8	72
17	Amir Cabaru	M	37	9	9	6	6	9	8	9	8	8	72
18	Sirajuddin Dg	M	31	9	9	6	6	9	6	8	9	9	71
19	Rate	M	52	9	9	9	9	8	8	7	9	7	75
20	Haji Cacing	M	46	6	7	6	6	9	8	7	9	8	66
21	Sumaryanto	M	43	9	9	6	6	9	6	9	9	9	72
22	Ujung Rahmat	F	40	9	9	6	6	9	8	8	8	7	70
23	Rahimin	F	25	7	5	5	5	9	8	9	7	8	63
24	Rusnawati	M	31	9	9	6	6	9	8	8	8	7	70
25	Sudirman	M	45	9	9	6	6	9	8	8	8	8	71
26	Sanusi	M	38	9	9	5	5	9	8	8	8	8	69
27	Muh. Syakir	M	42	9	9	6	6	9	8	8	9	9	73
28	Abd. Hakim	F	37	7	6	6	6	9	8	8	8	9	67
29	Dg. Samma	M	44	9	9	6	6	9	8	8	9	8	72
30	Mustafa Latif	M	27	6	6	5	5	9	8	8	8	8	63
	Total			248	243	197	197	266	229	241	247	242	2110

Category: High level (64-81) : 28 (93,3%) farmers
Medium level (26-63) : 6 (6,7 %) farmers
Low level (27-45) : 0 (0 %) farmer
Total : 30 (100%) farmers

(1) Aging of terraces or beds on sloped land
(2) Planting plants or turf for terrace reinforcement
(3) Processing of land according to contour line
(4) Cultivation of plants according to contour lines
(5) Seasonal crop rotation
(6) Planting annual crops on high slopes
(7) Use of mulch
(8) Maintenance of terraces/beds
(9) Making drainage channels

There is no information that explains the motivation of farmers in the research location, but is suspected due to economic motive factor. Usually the main factors that motivate farmers to improve their cultivation are income and welfare [5]. In accordance with [6], implementation of conservation farming techniques can increase farmer production and income. Furthermore, it shown that there is variation of attitude values both individual and cumulative variables. This may be due to the diversity of conservation issues facing farms, as reported by Hayley [5] and Lightenberg [6].

3.1.2 Attitudes of farmers in groups. The data of farmer attitudes group in the nine variables (factors) of implementation are shown in Table 2. From Table 2 it can be shown that, among the nine variables in the implementation of conservation farming techniques, seven (77.8%) are categorized appropriate match and only two (22.2%) are in the less appropriate category. The highest is the rotation of seasonal crops, while the lowest and less appropriate is the cultivation of land according to contour lines and cultivation of plants according to the contour line. There is no factor categorized not appropriate (not match). The corresponding attitude on the crop rotation variables is motivated by the growing season (climates) that corresponds to the needs of plant growth and commodity price developments. Likewise, with the low attitude in processing of land according to contour and cultivation of plants according to contour, also due to the adjustment to the climatic conditions with the demands of cultivated crops. At certain times they deliberately do so against the principle of soil conservation by cultivating the soil or making a bed and planting it in the direction of the slope (cutting the contour lines) so that the beds do not hold water in the rainy season [12] and [13]. Generally they plant vegetables that are not resistant to soil saturated water conditions (high humidity) such as potatoes, tomatoes, carrots, onions and kola or mustard greens. The position of the transverse bed can also result in the collapse of these beds during heavy rains.

Table 2. Values and categories of farmer attitudes toward conservation variables

No.	Variable of Conservation Technique	Value	Category
1	Aging of terraces or beds on sloped land	248	Match
2	Planting plants or turf for terrace reinforcement	243	Match
3	Processing of land according to contour lines	197	Less match
4	Cultivation of plants according to contour lines	197	Less match
5	Seasonal crop rotation	266	Match
6	Planting annual crops on high slopes	214	Match
7	Use of mulch	241	Match
8	Maintenance of terraces/beds	247	Match
9	Making drainage channels	233	Match

3.2. Farmers Perception

3.2.1. Perceptions of Farmers in Group. The values of perception and category are shown in table 3. As in "attitude", the highest value in perception is achieved in the seasonal crop rotation factor, while

Table 3. Values and categories of farmer perception toward conservation variables

No.	Variable of Conservation Technique	Value	Category
1	Aging of terraces or beds on sloped land	244	Match
2	Planting plants or turf reinforcement	240	Match
3	Processing of land according to contour line	193	Less match
4	Cultivation of plants according to contour lines	189	Less match
5	Seasonal crop rotation	257	Match
6	Planting annual crops on high slopes	217	Match
7	Use of mulch	244	Match
8	Maintenance of terraces/bunds	255	Match
9	Making drainage channels	252	Match

the lowest value with the "less appropriate" category is the processing of land according to contour lines and cultivation of plants according to contour lines. There is an indication of a strong relationship between attitudes and perceptions.

3.3. Correlation of Attitude and Perception

Perception is the most important thing in attitude formation. Simply can be described that the action starts from the attitude while attitude may be based on perception. Therefore, to understand more about the formation of attitudes toward one action, need to be understood more in the relationship attitude with perception. The correlation test results showed that the correlation coefficient of r = 0.97 shows a very strong relationship between farmer attitude and perception in Kanreapia. A correlation greater than 0.8 is generally described as strong, whereas a correlation less than 0.5 is weak [14]. These values vary from the type of data being examined. A study utilizing 'cores' of scientific data may require a study of social science data [15]. Harmony of attitudes and high perceptions in this case may be a very good tool for the neglect of attitude (further action) in the future because perception can be changed or motivated by filling individuals with knowledge, although there is no absolute guarantee. On the other hand, good perceptions of the KPHP Model of the Poigar Model do not have a significant effect on community behavior; the influential is education [1]. In general, however, the farmer's decision to adopt a new agricultural technology, depends on many factors; and one of them is perception [16] and [17], including less pro-poor policy interventions [18] agricultural extension sustainability [19] and providing transfer-source income [20].

4. Conclusion

The result showed that farmer attitude and perception toward the implementation of conservation farming techniques which having the highest category (appropriate) is seasonal crop rotation, while the lowest with less appropriate category is the processing of land according to the contour and the cultivation of the plants accordingly. There is a very strong relationship between farmer attitude and perception. The implications of the findings are that improvements the implementation of conservation farming techniques should be made through improved perceptions.

References

[1] Irawan A and Halawane J E 2017 Analisis persepsi dan perilaku masyarakat terhadap keberadaan kawasan kphp model *Poigar* **14** 71–82
[2] Asafu-Adjaye J 2008 Factors affecting the adoption of soil conservation measures: A case study of Fijian cane farmers *J. Agric. Resour. Econ.* **33** 99-117
[3] Fahmid I M 2004 Gagalnya politik pangan di bawah rezim orde baru: kajian ekonomi politik pangan di Indonesia (Jakarta: Yayasan Sandi Kota-ISPEI)
[4] Boisvert C H and R N 2009 Distinguishing between whole-farm vs. partial-farm participation in the conservation reserve program *Land Econ.* **85** 144–61
[5] Chouinard H H, Paterson T, Wanderschneider P R and A M O 2008 Will farmers trade profits for stewardship? heterogeneous motivation for farm practice selection *Land Econ.* **84** 66–82
[6] Lichtenberg B G S and E 2001 Priorities in cost sharing and water conservation: A revealed preference study
[7] Joshi A, Kale S, Chandel S and Pal D 2015 Likert scale: Explored and explained *Br. J. Appl. Sci. Technol.* **7** 396–403
[8] Subedi B P 2016 Using Likert type data in social science research: Confusion, issues and challenges *Int. J. Contemp. Appl. Sci.* **3** 2308–1365
[9] Boone H N J and Boone D A 2012 Analyzing Likert data *J. Ext.* **50** 30
[10] Conner D, Miller J, Zia A, Wang Q and Darby H 2016 Conjoint analysis of farmers' response to conservation incentives *Sustain.* **8** 1–15
[11] Taufik M, Putra Y S and Hayati N 2015 The utilization of global digital elevation model for watershed management a case study: Bungbuntu sub watershed, Pamekasan *Procedia Environ. Sci.* **24** 297–302

[12] Lastro-Bravo X B , Hubbard M C , Garrod G D and Tolon-Becerra 2015. What drives farmers' participation in EU agri-environmental schemes? Results from a qualitative meta-analysis *Environmental Science & Policy*, **54** 1-9

[13] Faharani S S, Fard F S and Asoodar A 2016 Effects of contour farming on runoff and soil erosion reduction: A review study *Elixir Agr.* **101** 44089-93

[14] Sugiyono S 2009 *Statistik untuk penelitian* (Bandung: Alfabeta)

[15] Samuel M and Okey L E 2015 the Relevance and significance of correlation in social science research *Int. J. of Socio. and Anthro. Res.* **1** 3

[16] Manoppo C N, Amanah S, Asngari P S and Tjitropranoto P 2017Persepsi Perempuan terhadap pemanfaatan pekarangan mendukung diversifikasi pangan di Sulawesi Utara *Jurnal Penyuluhan* **13** 1

[17] Bagheri A, Shabanali Fami H, Rezvanfar A, Asadi A and Yazdani S 2008 Perceptions of paddy farmers towards sustainable agricultural technologies: Case of haraz catchments area in Mazandaran province of Iran *Am. J. Appl. Sci.* **5** 1384–91

[18] Thalyta E Y, Mustajab M, Arsyad M 2010 Abstracts of doctoral theses on the Indonesian economy *Bull. of Indonesian Eco. Studies* **46** (2) 251-254 DOI:10.1080/00074918.2010.486113

[19] Ekasari Z K, Saleh S A M, Jusoff K, Salman D, Akhsan, Kasirang A, Arsyad M, Amrawaty A A and Fudjaja L 2013 Communication pattern and conflict in agricultural extension *Asian Soc. Sci.* **9** (5) 27-33

[20] Arsyad M, Nuddin A, Zamhuri M Y, Yusuf S 2014 The Poverty Reality of Coastal and Agriculture: How Severe the Seaweed Farmers and Cocoa Smallholders Are? *Int. J. Agr. Syst.* **2** (2) 119-131

IOP Conf. Series: Earth and Environmental Science **157** (2018) 012056 doi:10.1088/1755-1315/157/1/012056

Indonesian jellyfish as potential for raw materials of food and drug

S Yusuf[1], I M Fahmid[1], N Abdullah[1] and Zulhaeriah[2]

[1] Hasanuddin University, Jalan Perintis Kemerdekaan KM 10, Makassar, 90245, Indonesia.
[2] Institute for Social and Political Economic Issue, Jl. Ance Dg. Ngoyo No.88 Makassar, 90231, Indonesia.

E-mail: *s.yusuf69@gmail.com*

Abstract. Jellyfish used to be considered as a pest of fish and a nuisance to fishing operations. Yet, forty years ago this jellyfish was found to be materials of food, medicine and cosmetics and the utilization of jellyfish is now familiar in Indonesia after being imported by China and Japan industry. This study aims to determine the potential development of jellyfish commodities as food and drugs from Indonesia with the target to improve the welfare of fishermen. This research used methods of rapid observation, limited interview, processing with immersion experiment and desiccation. In addition, various literatures were also used to enrich the knowledge about jellyfish business. Observation showed that the appearance of jellyfish in Indonesian waters varies based on the fertility of the waters affected by oceanographic conditions. Jellyfish contains low calorie and fat content, high protein and minerals as well as total collagen. Thus, jellyfish is a nutritious food source to be developed into food supplements, nutricosmetics and functional foods. Due to its large size, the jellyfish from Bunyu Island is more viable than jellyfish from Suppa Pinrang to be exported as raw material. Therefore, the manufacture of food and medicines from jellyfish materials is possible to be done in Indonesia.

1. Introduction

Jellyfish contains many essential substances beneficial as raw material for food, cosmetics and drugs. For that feature, countries such as China and Japan are until now still importing semidried jellyfish from other country, where Indonesia served as one of jellyfish exporters in South East Asia. A high demand of annually tens or even hundreds of tones from various regions in Indonesia is recognized. It is noticed that until several years ago Japan imported 5,400-10,000 tones of jellyfish products annually [1].

Due to its higher protein quality and low calorie, jellyfish is a nutritional food source to be developed as supplementary diet, nutricosmetics and functional foods [2]. The potential development of jellyfish is quite high, however, the knowledge on its distribution and economical value is still limited due to research on jellyfish in Indonesia is yet to be developed. The distribution of jellyfish potential is only acknowledged when there is the utilization (catching and processing) by the community in specific location. On the other hand, ecologically the availability of jellyfish is relevant with wet season [1]. This in turn restricted the utilization and the annual availability of jellyfish raw material.

Jellyfish is either on the third and or fourth level of marine ecosystem food chain. As the third level, jellyfish feed on zooplankton or microorganism, whereas as the fourth level, a big size jellyfish diet beside zooplankton include larvae, juvenile or even pelagic breed. Provided this ecological process is sustainable, a high number of big fish population will disappear, which will reduce the fisheries productivity *in situ*, thus the fishermen income will be reduced as the population of adult fish decline. It can be predicted that when millions of jellyfish in the waters become pest to any fish larvae and juveniles [2] the fishermen will suffer much loss as the fish catch declined. Fishermen could not even go fishing as jellyfish get into their fishing nets.

Finally, since the end of the 1980ies jellyfish became the alternative source of income other than fish. Therefore, the utilization of abundant jellyfish became the main target to meet the international market demand. The utilization of jellyfish becomes a new model in increasing fisheries economy. However, the constraint of semidried processing with still higher cost that takes quite a long time (30 days) and lack of hygiene - bacteria contaminated resulted in low product quality. This research aimed to create potential in the utilization of jellyfish from various regions in Indonesia by modeling an efficient and cheap jellyfish processing to be an internationally marketable product. This paper is part of our proposal submit with the same theme of '*Transforming a marine nuisance into an asset: using jellyfish to improve health and the economy in Indonesia*'.

2. The Utilization of Jellyfish Indonesia

The abundance period of jellyfish is varied on oceanographic conditions [3] and the waters fertility. Jellyfish is distributed in various locations especially at the bays and in the waters which often with higher eutrophication as well as higher pelagic fish stock [4]. The list of location in the table 1 below presents a very high abundance of jellyfish. Several locations were recorded [1] *i.e.* in Kalimantan Island of South and East Kalimantan, in Halmahera of Bacan Island, and generally in West Java coast and Sumatera Island of Medan and Bangka Island.

Presently there are 13 points of jellyfish utilization in Indonesia which spread out from the Island of Sumatera, Java, Kalimantan, Sulawesi, Nusa Tenggara and North Moluccas (figure 1 and table 1). There has not been any information on the distribution of jellyfish utilization in the larger area other than those mentioned above. For example in the whole coasts of Papua, East Nusa Tenggara, Central and West Kalimantan, Central and North Sulawesi are estimated to possess the water characteristics suitable for jellyfish that served as a potential new development location of jellyfish utilization.

The jellyfish is generally available in the dry season toward and during the wet season. This is as at that time the waters condition is undergoing an eutrophication during the monsoon except in areas where upwelling occurs on the dry season. Several species of jellyfish simultaneously occurred in the time range of August-November and some occurred almost the whole year from May to December depends on the season, type, and location as well as eutrophication condition [1]. In around Kalimantan jellyfish occurred from August, which also occurred in part of West Java and in Central Java, while in Nusa Tenggara it occurred in this October. In July jellyfish occurred in Halmahera, Muncar of West Java and Medan of Sumatera (table 1).

Figure 1. Location of jellyfish utilization in Indonesia

Tabel 1. Distribution of utilization location, month and season of jellyfish abundance in Indonesia

Specific Location	Region	Time of Abundance	Seasons
Bay of Cempi,	Sumbawa Island	October-December	Wet
Bay of Saleh,	Sumbawa Island	October-February	Wet
Bunyu Island Tarakan	North Borneo	August-January	Dry-Wet
Sebatik Island	North Borneo	August -January	Dry-Wet
Coastal of Balikpapan	East Borneo*	August -November	Dry-Wet
Kota Baru	South Borneo*	August -November	Dry-Wet
Bacan Island	Halmahera*	July-August	Wet
Coastal of Tuban	East Java*	March-May Sept-November	Dry-Wet
Coastal of Cirebon	West Java*	August -November	Dry-Wet
Coastal of Muncar	West Java *	July-November	Dry-Wet
Coastal of Cilacap	Central Java*	August-November	Dry-Wet
Coastal of Medan	Sumatra*	June-November	Dry-Wet
Coastal of Bangka	Sumatra	No data	No data

Note: * locations were as recorded by Omori and Nakano (2001)

3. Physical Characteristics

The jellyfish were sampled from two locations, i.e. Tanjung Suppa waters of Pinrang Regency in South Sulawesi Province and Bunyu Island of Tarakan Regency in North Kalimantan Province. All samples show linear relation between weight and jellyfish umbrella diameter with linear equation as $Y = 0.011x - 13.39$, with correlation strength (R^2) of 90 percent. However, those two jellyfish samples were different in size, both in diameter and weight. Samples from Pinrang waters were smaller with 9-12.5 cm diameters and weight of 1800-3200 gram. While samples from Bunyu Island waters were bigger with umbrella diameters of 31-53 cm and weighted between 3400 and 6200 gram (figure 2).

Further, jellyfish samples from Bunyu Island were more suitable to be utilized as raw material according to the jellyfish business required criteria. The jellyfish being traded are those with minimum size of 30 cm in diameter. Thus, the jellyfish from Suppa Pinrang can not be utilized as export commodity due to their small size. It should be noted though that those two jellyfish samples from the waters of Bunyu Island and Suppa Pinrang were of different species.

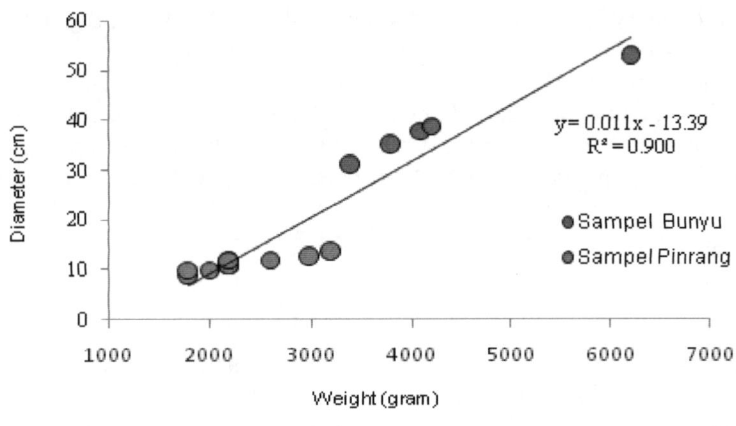

Figure 2. Correlation between umbrella diameter and weight of jellyfish

IOP Conf. Series: Earth and Environmental Science **157** (2018) 012056 doi:10.1088/1755-1315/157/1/012056

4. Processing

The processing of raw jellyfish to be a half-ready product should be carefully done so as not reducing its chemical content [5]. The jellyfish umbrella should be simply processed to get a semidried raw material. The jellyfish are soaked in a mixture of salt and alum for 30 days which show positive result of elastic and firm jellyfish texture. The remained mixture of alum and salt are cleansed and dried using controlled media of modified wood oven box with a controlled temperature. Kogov *et al.* [6] compared several methods used for the study of jellyfish biochemical composition and found that during drying in the oven at 60°C, the content of C, N and the total content of amino acids in the jellyfish tissue decreases 10 times as compared with cool dehumidification [5, 6].

5. Proximate and Micromineral Content

In three jellyfish species, *Acromitus hardenbergi*, *Rhopilema hispidum* and *Rhopilema esculentum,* it was found that especially for parts of umbrella (cup) and arm also mouth contain the proximate composition of calorific value, collagen content, amino acid profile, chemical score and elemental constituent [3]. Other research [7] measured the chemical content of the moon jelly *Aurelia aurita*. Generally all jellyfish contain low calorie and fat, high protein and mineral, as well as the total collagen (table 2).

Tabel 2. Chemical composition of the moon jelly*Aurelia aurita* [7].

Proximate parameters	Content (%) in dry weight
Water Content	68,67
Ash Content	12,81
Protein Content	11,09
Fat Content	0,30
Carbohydrate	7,13
Micro Mineral	Content (ppm) in dry weight
Calcium (Ca)	11,1
Magnesium (Mg)	3652,6
Natrium (na)	111209,4
Kalium (K)	2959

The proximate content of the river jellyfish *Acromitus hardenbergi*, the sand jellyfish *Rhopilema hispidum* and the red jellyfish *Rhopilema esculentum* [3] are 1.0-4.9 kcal/gram; fat 0.4-1.8%; protein 20-53.9%; mineral 15.9-57.2 %; collagen 122.64 – 693.92 mg/g in dry weight. Moreover the dominant amino acids in the jellyfish umbrella and tentacle are glycine, glutamate, threonine, proline, aspartate and arginine, while the major elements are sodium, potassium, chlorine, magnesium, sulphur, zinc and silicon. As it contains higher protein quality, low calorie, jellyfish is a nutritious food source to be developed into supplementary food, nutricosmetics and functional foods [3]. High collagen content that is the protein contained in the body which play important role in the stamina and in building the tissue are bone, tooth, joint, muscle and skin. Acid amino content give various important use to the body.

6. Jellyfish Market Development

Jellyfish fishing in various countries is millions of dollar business. Its annual trade value of 25.5 US million dollars is accounted from the Philippines, Vietnam, Thailand, Malaysia, Indonesia, Singapore and Myanmar. This effort really helps the fishermen and businessmen as well as the state in increasing their fisheries industry income. The biggest importers are China and Japan, but in 2001 Japan is more dominant by 50% of the world jellyfish [1]. The market price varies on the population size and condition. The larger the jellyfish, the better is the price. The jellyfish with higher stinger force is cheaper while the regular umbrella shape is more expensive. The wholesale price of A premium grade jellyfish umbrella in Asia is USD $ 10.00-12.00 per pound, with minimum diameter of 18 inches and its color should be white to creamy, and at the same time it is of crispy texture yet tender [8].

Potential of jellyfish business in Indonesia is more developed as the eutrophication potential of waters in the coastal and bay regions are high. This is supported by the labor intensive fishermen working system and skilled labor is needed to increase processing quality. Thus the semidried jellyfish product can be accepted by the international market. However, processing study is required to make new processed product of raw jellyfish to satisfy Indonesia's food and cosmetics industries.

Acknowledgments
We wishing to acknowledge to Maulida Suaeb, Urfan, Winwien, Ferdinan, Syahrullah, who were active in this jellyfish research and sampling in Bunyu and Pinrang.

References

[1] Omori M and Nakano M 2001 Jellyfish fisheries in Southeast Asia *Hydrobiologia* **451** 19-26
[2] Purcell J E 2005 Climate effects on formation of jellyfish and ctenophore blooms: A review *J. Mar. Biol. Assoc. U. K.* **85** 461–476
[3] Nicholas M, Fatimah H K, Yusoffa M D, Jamilah C B, Basria M, Maznaha I, Chana K W and Nishikawad J 2015 Nutritional composition and total collagen content of three commercially important edible jellyfish *J. of Food Chem.* https://doi.org/ 10.1016/j.foodchem Accessed on 2015.09.09
[4] Lynam C P, Gibbons M J, Bjørn E. Axelsen B E, Sparks C A J, Coetzee J, Heywood B G and Brierley A S 2006 Jellyfish overtake fish in a heavily fished ecosystem *Curr. Bio.* **16** (13) 492-493
[5] Yuferova A A 2017 The impact of different drying modes of scyphozoan jellyfish *Rhopilema esculentum* and *Aurelia aurita* on the protein and carbohydrate components in their composition and the possibility of their use as dried prepared food *J. of Food Proc. Eng.* doi: 10.1111/jfpe.12326
[6] Kogovšek T, Tinta T, Klun K and Malej A 2014 Jellyfish biochemical composition: importance of standardised sample processing *Mar. Eco. Prog. Series* doi.org/10.3354/ meps10959
[7] Nurokhmatunnisa, Pujiyanti D and Arifah A N 2013 *Utilization of Jellyfish (Aurelia aurita) as Functional Food to Increase Human Body Stamina Report of Student Creativity* (Indonesia: Bogor Agriculture University Indonesia)
[8] Hsieh Y H P, Leong F M and Rudloe J 2001 Jellyfish as food *Hydrobiologia* **451** 11-17

IOP Conf. Series: Earth and Environmental Science **157** (2018) 012057 doi:10.1088/1755-1315/157/1/012057

Improving agricultural commodity supply-chain to promote economic activities in rural area

R Padjung

Department of Agronomy Hasanuddin University, Jl. Perintis Kemerdekaan Km 10, Makassar, 90245, Indonesia.

E-mal: *rusnadi@unhas.ac.id, rusnadi2015@gmail.com*

Abstract. Long supply chain of agricultural commodities has become concern to governments particularly in large countries such as Indonesia as it causes high price disparity between farm-gate and retailer. Policies to overcome such problem are usually by shortening the chain, by which farmers sell the products directly to retailers. Using an action research in AEDEF (Aceh Economic Development Financing Facilities) Program, conducted in the province of Nangro Aceh Darussalam (NAD) Indonesia, the paper shows that shortening the commodity supply chain is not the best solution to such problem, as it causes loss of jobs in the villages. High price disparity between farm-gate and retailer is not necessary brought about by long supply-chain but by the efficiency of the chain instead. Efficiency of the chain can be improved by creating enabling business environment such that every actors and players work in a fair manner. This can be achieved by transparency in price and quality grade. With development achieved in Information and Communication Technology (ICT), having a good and reliable flow of such information is not difficult. In addition to information flow, the availability and quality of infrastructure to support flow of goods from farm-gate to end-user is of reasonably important.

1. Introduction

Most agricultural commodities in Indonesia have to pass long supply chain before they arrive to consumers. From farmer to consumers it may pass through collectors, village and district scale traders, wholesalers, distributors, retailers. Along the flow, goods or products may be cleaned, sorted, dried and undergone other simple processing, transported, and manufactured. As in the case of cocoa for example, before it get into consumer, cocoa beans and cocoa products have to pass through local collectors, local traders, in-country processors, in-country manufacturers, and exporters before they get into importers, regional and multinational processors, manufacturers, and chocolate – confectionary market. In Sulawesi alone there are more than 9,000 local collectors, more than 1,000 local traders, 10 – 20 small scale exporters, 6 – 8 medium and big scale exporters, 5 multinational affiliates, 13 local processors, and some local manufacturers serving about 400,000 small-holders farmers [1].

Typical supply-chain of vegetable in Indonesia involve collectors, trade wholesaler, supermarket – wholesaler, interisland trader, and food industries before it gets into supermarket and or traditional retailers. In South Sulawesi, 85% of chili produces go to collectors and only 10% go straight to local retailer and 5% to wholesaler. Half of chili at collectors are sold to wholesaler and another half to interisland traders. Those at wholesaler are distributed to interisland traders, dry chili wholesalers, food industries, traditional retailer and supermarket [2]. In East Java, 90% of chili produced by farmers goes to collectors, only 10% goes straight to traditional market, and none to wholesaler. The produces then go to retailers, supermarkets, food industries, and interisland traders [2].

As of rice, the main staple food in Indonesia, the supply chain consist of collectors, rural cooperatives, millers, BULOG, wholesalers, provincial market, interisland traders, wholesaler market traders, and retail stores. Un-hulled rice are bought, dried, stored, milled and finally sold by tier 1 actors which are millers and rural cooperatives to wholesaler, BULOG and provincial markets before distributed to the inter-island traders, wholesaler market traders, retail stores and end consumers [3].

In general, supply chain of agricultural commodities in Indonesia is very long and complex involving many actors. Such a complex and long chain lead to a concern by government as it cause high price disparity between farm-gate and retailer. Policy to overcome such problem is by shortening the chain, by which farmers or farmer group sell the products directly to retailer or to consumers. Shortening the commodity supply chain, however, may not be the best solution to such problem as it causes loss of jobs or employment in the villages. Those live in the rural area or villages are not only farmers, but also traders, middle-man, collectors, and those who work on home-scale industries and other off-farm activities. As un-employment and poverty increasingly become a major problem in rural area, shortening the chain may even worsen the problem.

Poverty and unemployment are among main problems in rural area or villages. As per September 2016, there are 17.28 million people in rural area are poor or 13.96%, which is much higher than 10.49 million those in the cities or 7.73% [4]. Although the figure has significantly decreased by 4.53 million in the last ten years which was 24.81 million in 2006 [5], poverty still become a major problem in the village as the rate of decrease in poverty in rural area is much slower than that in urban. The decrease of poverty in rural area in the last ten years was from 21.81% in 2006 to 13.96% in 2016 (decrease by 36%), while in urban area was from 13.47% in 2006 to 7.73% in 2016 (decrease by 43%).

In 2016, un-employment rate in rural area is 4% [6]. For the last 20 years, un-employment in Indonesia has increased from only 2.5% in 1990 to 7.41 in 2010 [7]. In some cases, the increase of unemployment in rural area is brought about by failure of harvest due to climate [8].

Facing high price disparity between farm-gate and consumer and complex agricultural commodity supply chain, government of Indonesia proposing to shorten the chain. For rice, farmers or farmer groups will sell the product directly to BULOG and BULOG will then distribute it to retailer or wholesaler, and so there is no middle man or collectors needed [9]. Such approach will certainly cause the loss of many jobs in rural area which in turns will increase poverty, hence shortening the supply chain may not be a good solution

High price disparity between farm-gate and retailer is not necessary brought about by long supply-chain but by the efficiency of the chain instead. Efficiency of the chain can be improved by creating enabling business environment such that every actors and players work in a fair manner. This can be achieved by transparency in price and quality of goods or services. Therefore, flow of information of price and quality of goods from end-user all the way along the chain to farmers occurs without any distortion. With development achieved in Information and Communication Technology (ICT), having a good and reliable flow of such information is not difficult. In addition to information flow, the availability and quality of infrastructure to support flow of goods from farm-gate to end-user such as transportation, telecommunication, electric power, is of reasonably important.

The purpose of this action research is to show that job absorption in rural area can be increased and so poverty is decreased by promoting economic activities in rural area trough improvement of agricultural supply chain, instead of shortening it

2. Methods

The action research is part of AEDEF (Aceh Economic Development Financing Facilities) Program, conducted in the province of Nangro Aceh Darussalam (NAD) Indonesia in 2009 – 2012 financed by Aceh Post Tsunami Multi-Donor Trust Fund for Aceh and Nias (MDTF), where I become a program leader. The program was implemented by eight Sub Implementing Entities (SIE), i.e. International NGO, each with one sub program with specific scope and objective [10]. In this paper, I choose two SIE, i.e Swisscontact that worked on *Peningkatan Ekonomi Kakao Aceh* (PEKA, Aceh Cocoa Economy Development) and International Organization for Migration (IOM) that worked on Sustainable

Economic Growth for Aceh (SEGA). The nature of these sub programs are basically improvement of agricultural commodities supply chain for cocoa by Swisscontact and for coffee by IOM.

The scope area of PEKA subprogram covers 5 districts in NAD province, i.e. Pidie Jaya, Bireun, Aceh Tamiang, Aceh Barat Daya and Aceh Tenggara district. These districts are central producing area of cocoa in the province. SEGA sub-program by IOM covers two districts, i.e Aceh Tengah and Bener Meriah district. These two districts are also central producing area of coffee in the province.

3. Improving Cocoa Supply Chain

Cocoa supply chain in Aceh involves farmers, farmer group, cooperatives, collectors, wholesaler, district wholesaler, processor, and exporters. Therefore, increasing economy in cocoa producing area cannot be achieved merely by working with farmers but instead working with all actors in cocoa supply chain.

As was discussed in the introduction, high price disparity between farm-gate and consumer is not brought merely by long supply chain but by fairness and equity of each actor in the chain. Hence, creating enabling business environment is vital to improve the quality of the supply chain. To create enabling business environment, two important activities were introduced in Aceh *i.e.* Establishment of Aceh Cocoa Forum, and Development of 5 District Based Cocoa Master Plans. With a cocoa forum, any unfairness arise among actors can be resolved. The forum also established internet website (*http://forumkakaoaceh.or.id/*) and the SMS price information system. The SMS price information system will further promote a transparent and fair cocoa trade. District Based Cocoa Master Plans also cover policy, system and procedure, and regulation required for enabling business environment in cocoa. The needs for infrastructure and utilities development to guaranty a good flow of cocoa product from farm-gate to end-user are also identified.

Collectors, middleman, and other smallholders cocoa enterprise are also villagers, and hence shortening the supply chain will mean loss of jobs for them. What need to be done is to increase their efficiency with proper knowledge, information, skill, management, and access to finance. Therefore, 112 SCEs were organized and supported with 112 nurseries, compost choppers, and equipment as well as trained in management, marketing, and compost and seedling business, In addition, 58 traders (collectors and middleman) were trained and provided with trading equipment (moisture meters and digital scales) and office equipment. Noted that moisture meters and scales are two important trading equipment to guaranty a fair trade. Along with those interventions, 105 cocoa farmers, 16 traders, 13 government representatives and 7 bank staff participated in a discussion on marketing and access to finances/credit.

By nature, farmers are not entrepreneur. Instead of taking middleman roles to sell their product directly to retailers, farmers can increase their income by increasing productivity and quality of harvest. For this purpose, the sub-program trained 12,540 farmers in Farmer Field Schools (FFS). Along with this, 1.2 million cocoa trees in 1,516 hectares were rehabilitated through side grafting, and cocoa farmers were trained in farm evaluation and farm rehabilitation. In addition, 5 District Cocoa Clinics (DCC) along with Cocoa Clonal Garden were established.

4. Improved Cocoa Supply Chain

As a result of intervention made in improving cocoa supply chain, 116 small scale cocoa enterprises (SCEs) have been linked to regional and national cocoa buyers. In addition, 58 traders actively involved in cocoa value chain business. The engagement of SCEs to regional and national cocoa buyer has increase the quality of supply-chain, hence the flow of cocoa product from farm to national buyer become effective. Active involvement of traders in cocoa value chain business has stimulated economic activities in rural and increased community income that will in turn decreased poverty. About 76% of these 58 traders stated that their profit has increased, and 69% of their employees were benefitting from increased income.

IOP Conf. Series: Earth and Environmental Science **157** (2018) 012057 doi:10.1088/1755-1315/157/1/012057

As for farmers, although they are not directly involved in trading as a middleman or they do not sell the product directly to retailer or consumer, their income has increased due to increase in cocoa yield and quality. Cocoa yield has increased 150%, from 400 kg $ha^{-1}year^{-1}$ to 1,000 kg ha^{-1} $year^{-1}$.

5. Improving Coffee Supply Chain

Typical supply chains and or activities of coffee in Indonesia, including in Aceh are pulping, collectors, hulling, drying, warehouse, cupping test, roaster, and export. If the products are exported as beans, the supply chain include cherry parchment, cherry / parchment drying for dry process or washing / fermentation for wet process, dry or wet process green bean, and export, that involve farmers, collectors and middleman, processors, and exporters. Action to improve supply chain of coffee in Aceh in this sub-program, therefore, covers intervention on all actors and processes in the chain.

In relation to such complex supply chain in coffee, the intervention made in the sub-program include improving coffee processing business, establishing coffee Arabica network, strengthening linkages with buyers, stimulating coffee trading, and increasing coffee productivity and quality. Intervention made in improving coffee processing business were provision of matching grant to 78 SME's, 37 grantees Aceh Tengah district and 38 in Bener Meriah as well as trainings and technical assistance to 200 SMEs and cooperatives. The matching grants were mainly provided for construction such as drying floor, warehouse, fence, etc, as well as for transportation and machinery such as pulper and huller, roasting machine, coffee packaging/sachet, etc. The training and technical assistance covered business plan development, accounting, business ethics, and environmental aspects.

Establishment of coffee Arabica network were made through participation in socialization of Fair Trade USA Standards, participation in local, national, international exhibitions and conferences, strengthening MPKG (*Masyarakat Perlindungan Kopi Gayo* / Gayo Coffee Protection Community), one of the local organizations to protect coffee gayo brand name. In addition, a partnership with Specialty Coffee Association of Indonesia (SCAI) and the Aceh Coffee Forum was established to build a stronger network for Arabica Gayo on the national level and on the international level. Strengthening linkages with buyers covered establishment of auction market, and establishment of research nursery. A private managed auction platform fully registered with Bappepti and KBI, where Arabica Gayo was included in I-Pasar as an independent platform.

In stimulating coffee trading, the sub-program trained 36 cooperative agriculture workers in the area of administration, finance, coffee production, cooperative management, quality testing and fair tradestandards, and established a Warehouse Receipt System (WRS). The WRS is certified by SUCOFINDO and is the first one in Indonesia for Arabica Coffee. The warehouse provides storage capacity up to 800 MT. Within the premises a certified quality control laboratory was set up.

The activities of actors in the supply chain are very much affected by bean production and quality. Therefore, intervention made in increasing coffee productivity and quality, were intended not only to increase income of farmers but also to increase the size of coffee business and industry such that activities of all actors in the chain increase. To increase coffee productivity and quality, 50 Agricultural Extension Workers (AEWs) are trained to give Direct Support to Farmers, 50 Farmer Field Schools (FFS) were performed to trained 1,821 coffee farmers in land preparation, coffee production, pest & disease management, post-harvest management, quality testing, and the sale of products, 50 coffee nurseries were established and supported, and 36 Cooperative Agriculture Workers (CAWs) were trained in the area of administration, finance, coffee production, cooperative management, quality testing and Fair Trade standards.

6. Improved Coffee Supply Chain

As a result of intervention made in improving coffee supply chain, there have been additional of 193 job opportunities related to SMEs activities, a number that will continuously grow along with increase in coffee bean yield. In addition, Warehouse Receipt System (WRS) is on place preventing price fall, a private managed auction platform fully registered with Bappepti and KBI and has been working, and the partnership agreement with and certification by Fair Trade US has increase coffee bean market globally.

Coffee seedling business has generated income in the amount of IDR 1,400 million along with significant increase in bean production. In addition, coffee bean quality has also increased as indicated by increase in the cup quality score from 77 to 83. Such increase in cup quality score mean significant increase in price. Increase in bean production and quality caused increase not only in farmers income but also the size of coffee industry. As the supply chain occurs in the village, and all actors in the business / chain are villagers economic activities in the village and income of the villagers increase. Although survey on poverty has not been undertaken, but increase in job opportunity and income will certainly decrease poverty.

7. Conclusion

Solution to long and complex agricultural supply chain is not by shortening the chain because it will cause loss of jobs of villagers that will in turn decreasing villagers income and increasing poverty in the village. An alternative solution to such problem is by improving quality of the supply chain by mean of working with all actors in the chain, creating enabling business environment, providing actors with training on technical skill and management, and facilitating actors in accessing credit and finance. By improving quality of supply chain, the size of business increase, the supply chain become effective more actors involve in the supply chain business, and economic activity in the village increase. These will in turn increase employment, increase village income, and reduce poverty.

Acknowledgments

I would like to acknowledge Euroconsult Mott MacDonald who has served as Project Management Consultant, Swiss contact who has been implementing PEKA subprogram, International Organization for Migration (IOM) who has been implementing SEGA Program, and to Bappeda Aceh who has been served as Project Management'

References

[1] Panlibuton H and Meyer M 2004 *Value chain assessment: Indonesia Cocoa* (Washington: United Satate Agency for International Development) 54 p.

[2] White B, Morey P, Natawidjaya R, and Morgan W 2007 *Vegetable value chains in Eastern Indonesia–a focus on chilli* SADI-ACIAR Research Report (Canberra: Australian Center for International Agricultural Research) 73 p.

[3] Spire 2013 *Indonesia market assesment for agricultural value-chain payments via mobile technology* 98 p

[4] SPI (Serikat Petani Indonesia) 2017. Profil Kemiskinan di Indonesia 2016: Dalam Angka Berkurang, Namun di Desa Makin Dalam dan Parah. http://www.spi.or.id/profil-kemiskinan-di-indonesia-2016-dalam-angka-berkurang-namun-di-desa-makin-dalam-dan-parah/. Accessed on September 10, 2017

[5] BPS 2017 *Poverty Rate and Gini Ratio* 1970-2017 https://www.bps.go.id/link Tabel-Statis/view/id/1494. Accessed on September 10, 2017

[6] BPS 2017 *Keadaan ketenagakerjaan Agustus 2016 dibandingkan degan Agustus 2017* https://www.bps.go.id/brs/view/1230. Accessed on September 10 2017

[7] BPS 2017 Jumlah Angkatan Kerja, Penduduk Bekerja, Pengangguran, TPAK dan TPT, 1986–2013. https://www.bps.go.id/linkTabelStatis/view/id/973. Accessed on September 1, 2017

[8] Andrias D 2017 Faktor Alam Pengaruhi Angka Pengangguran di Pedesaan. https://tirto.id/faktor-alam-pengaruhi-angka-pengangguran-di-pedesaan-coaw. Accessed on October 2017

[9] Republika 2016 *Mentan akan Perpendek Rantai Distribusi Beras untuk Basmi Mafia* http://www.republika.co.id/berita/ekonomi/makro/16/03/02/o3e2fu382-mentan-akan-perpendek-rantai-distribusi-beras-untuk-basmi-mafia. Accessed October 2017.

[10] Euroconsult Mott MacDonald 2012 *Aceh Economic Development Financing Facilities*. Final Report 96 p

IOP Conf. Series: Earth and Environmental Science **157** (2018) 012058 doi:10.1088/1755-1315/157/1/012058

Impact of life expectancy, literacy rate, opened unemployment rate and gross domestic regional income per capita on poverty in the districts/city in Central Sulawesi Province

A D Tombolotutu[1], M A Djirimu[2], M Lutfi[2] and F Anggadini[3]

[1]Muhammadiyah University of Palu, Jl. Jabal Nur No.1, Talise, Mantikulore, Kota Palu, 94118, Central Sulawesi, Indonesia.
[2]Department of Economics, Tadulako University, Jl. Soekarno Hatta KM.9, Tondo, Mantikulore, Kota Palu, 94148, Central Sulawesi, Indonesia.
[3]Statistic Central Agency of Palu, Jalan Baruga No.19, Palu Selatan, Tanamodindi, Mantikulore, Kota Palu, 94111, Central Sulawesi, Indonesia.

E-mail: *ahlis.djirimu66@gmail.com, itaahlis97@gmail.com*

Abstract. Research was conducted in several districts/city in Central Sulawesi Province in order to determine the effect of life expectancy, literacy rate, opened unemployment rate, and gross domestic regional income per capita on poverty at the districts/city in the province. The analysis used is Panel Data Regression. The results show that first, life expectancy and gross domestic regional income have a negative and significant impact on the poverty level in the districts/city in the Province. Second, the opened unemployment rate has a positive and significant effect on the poverty level in the districts/city in the province. Third, literacy rates show a positive effect and insignificant effect on the poverty level in the districts/city in the Province of Central Sulawesi. Fourth, these four variables simultaneously affect the poverty in the districts/city in Central Sulawesi

1. Background

Poverty is a complex issue that is influenced by a variety of interrelated factors, such as income level, unemployment, health, education, access to goods and services, location, geography, gender, location and environment [1, 2]. The issue of poverty is the focus of development in each country.

Astrini *et al.* [3] research focus on the effect of GDRP, education, and unemployment on poverty in Bali. In order to achieve these objectives, multiple regression analysis is done. The following result are obtained firstly, GDRP has no significant and negative effect on poverty. Secondly, education has significant and negative effect on poverty. Thirdly, unemployment has a positive and significant impact on poverty [4].

The poverty line according to the Central Agency of Statistics [5] which is used in this measurement equivalent to the amount of Rupiah per capita per month required to meet the daily needs of 2,100 kilo calories per capita plus some non-food commodities such as housing, clothing, education, and health.Bhattacharyya and Resosudarmo [6] state that per capita income has a positive

IOP Conf. Series: Earth and Environmental Science **157** (2018) 012058 doi:10.1088/1755-1315/157/1/012058

and insignificant effect because with the high per capita income owned by each province, it will increase the poverty rate but not significant change. This is because per capita income is usually used only for the improvement of infrastructure of the region as concerned [4].

The first target in the MDGs is to eradicate poverty and hunger, the poverty rate of the population living below the national poverty line is between 8% and 10%. The MDGs target in accordance with the Public Budget Policy of Central Sulawesi Province in 2011-2016 period outlined in the Medium Term Development Planning (RPJMD) 2011-2016 period. Development of education and health are the two pillars to form human capital in economic development is an investment in the long term. The achievement of development goals in education and health, can improve the quality and productivity of the population, with the population of productivity growth is the engine of economic growth and welfare of the population.

Central Sulawesi RPJMD targets for life expectancy (AHH) in 2013 were 71.90. But the reality in 2013 AHH of Central Sulawesi is 67.21. This shows that the life expectancy of the population in Central Sulawesi, about 67.21 years. One of several characteristics of poverty is education. Education as a determinant of poverty significantly and positively affects the per capita consumption of families and poverty. One of several measurements for the level of education is the literacy rate (AMH). In line with the MDGs by 2015 the target of education for all, Central Sulawesi province RPJMD targets for achieving the literacy rate in 2013 amounted to 97.69. The target cannot be realized, because in 2013 the literacy rate amounted of Central Sulawesi to 96.22. This figure shows that approximately 96.22% of the population in Central Sulawesi has the ability to read and write (literacy).

The level of income is the element that determines the wealth of a society. Public income reaches its maximum when the conditions of full employment level of usage (full employment) can be realized. Unemployment will cause the effect of reducing public revenue, and it will reduce the level of prosperity that has been achieved. Economic growth is crucial to reducing poverty in the region [7]. The high economic growth needed to accelerate change towards economic structure of the regional economy is increasing and dynamic industry, characterized by powerful and advanced, and has a strong agricultural base sectoral growth is potentially huge [8, 9]. Economic growth is also necessary to mobilize and spur development in other fields as well as the main force of development in order to increase people's income and address socio-economic imbalances.The study was conducted to determine what factors affecting poverty in the districts / city in Central Sulawesi province. The formulation of the problem statement, namely: does Life Expectancy (AHH), literacy rate (AMH), Unemployment Rate (TPT), and the Gross Domestic Income (PnDRB per capita) affect the level of poverty in the districts / city in Central Sulawesi province.

2. Research Method

This study examined the level of poverty throughout the Districts / City in the province of Central Sulawesi, namely: Banggai, Banggai Islands, Morowali, Poso, Tojo Una-Una, ParigiMoutong, Donggala, Toli-Toli, Buol, Sigi and the Municipality of Palu in 2010-2013 periods.Researchers provide understanding and explanation of each term used as follows:

Poverty (Y). According to the Central Statistics Agency (BPS) is the percentage of population below the poverty line in each district /city in Central Sulawesi. In this study examined poverty is the percentage of people living below the poverty line (P_0). **Life Expectancy (X_1)**. Estimates of the number of years of life or the life of residents in an area measured by life expectancy expressed in years. Because of the AHH the data we can see the performance of the government in improving the health of residents in an area, whether an increase or decrease from year to year. **Literacy Rate (X_2).** Literacy Rate (AMH) represents the percentage of population aged 15 years and over who can read and write. AMH is an important indicator to look at the extent to which the population of a region open to knowledge, skills, eases in communication. **Unemployment Rate (X_3).** The working age population that is included in the group of unemployed measured by the Unemployment Rate (UR) expressed as a percent. Because of the data we can see the UR successful performance of government in employment programs in an area, whether an increase or decrease from year to year. **Per Capita**

IOP Conf. Series: Earth and Environmental Science **157** (2018) 012058 doi:10.1088/1755-1315/157/1/012058

Gross Regional Domestic Product (X_4). Gross Regional Domestic Product at constant prices is used because it can be a means of comparison are considered more accurate to measure the level of prosperity of the people of interstate / inter-regional / inter-sectoral.

3. Data Analysis

The data used in the econometric analysis consists of three types of time series data, cross section and panel data. At the time series data, some variables will be observed within a certain time, while the cross-section data, some variables were collected from multiple sample units within a certain time point. Data panel is a combination of time series data (time series) and data cross-section (cross section).Panel data regression model in this study is to use the dependent variable poverty rate (PR), while the independent variable is the life expectancy, literacy rate, Unemployment Rate, Gross Domestic Regional Income. These variables could be formulated as follow:

$$Y=\alpha + \beta_1X_{1it}+\beta_2X_{2it}+\beta_3X_{3it}+\beta_4X_{4it}+\varepsilon_{it}(1)$$

Where:
Y = the poverty rate (%);
X_1=variable life expectancy (years);
X_2 = variable literacy rate (%);
X_3 = variable Unemployment (%);
X_4 = variable Per Capita Income (in million rupiah);
i = unit of cross section;
t= unit of time series;
α= constant
β= coefficient
ε= residual

The model can be expressed in the form of a log linear model through the transformation of variables. Transformation is done by logarithm equation, so that the model is transformed into shape linear.

$$Y_{it} = \alpha+\beta_1 \text{ Log } X_{1it} + \beta_2 \text{ Log } X_{2it} + \beta_3 \text{ Log } X_{3it} + \beta_4 \text{ Log } X_{4it} + \varepsilon_{it}(2)$$

4. Results and Discussion

The relationship between Y and X_2 is likely to have a very weak correlation or no correlation. Best Regression Model Selection Panel Data Regression Results Double log regression analysis using the program E-views are as follows:

4.1. Fixed Effect Regression Model Results

$$\text{Log}(Y_{it})=14{,}2-7{,}3\text{Log}(AHH_{it})+1{,}9\text{Log}(AMH_{it})+0{,}03\text{Log}(TPT_{it})-0{,}52\text{Log}(PnDRBPerKAP_{it})+v_{it}$$
(2,0980)** (-3,0264)*** (0,5161) (2,1656)**(-5,2315)***

Adjusted R^2=0.9886
*** Significant at α = 1% level
** Significant at α = 5% level

4.2. Results Regression Model Common Effect

$$\text{Log}(Y_{it})=16{,}24-8{,}4\text{Log}(AHH_{it})+0{,}55\text{Log}(AMH_{it})+0{,}04\text{Log}(TPT_{it})-0{,}13\text{Log}(PnDRBPerKAP_{it})+v_{it}$$
(6,4965)***(-6,5317)*** (0,4004) (0,6770)(-1,1737)

IOP Conf. Series: Earth and Environmental Science **157** (2018) 012058 doi:10.1088/1755-1315/157/1/012058

Adjusted $R^2 = 0.7668$
*** Significant at $\alpha = 1\%$ level

4.3. Random Effect Model Regression Results

Log(Y$_{it}$)=16,76–6,79Log(AHH$_{it}$)+0,12Log(AMH$_{it}$)+0,03Log(TPT$_{it}$)-0,52Log(PnDRBPerKAP$_{it}$)+v$_{it}$
(3,8479)*** (-3,9478)*** (0,0504) (2,2876)** (-6,2445)***

Adjusted $R^2 = 0.8735$
*** Significant at $\alpha = 1\%$ level
** Significant at $\alpha = 5\%$ level

Based on a series of model significance testing that has been done, it can be determined that the model used to estimate the regression model of poverty in the districts / City in Central Sulawesi province is a model Random Effect.

Log(Y$_{it}$)=16,766,79-Log(AHH$_{it}$)+0,12Log(AMH$_{it}$)+0,03Log(TPT$_{it}$)-0,52Log(PnDRBperKAP$_{it}$)+v$_{it}$
(3,8479)*** (-3,9478)*** (0,0504) (2,2876)**(-6,2445)***

Adjusted $R^2 = 0.8735$
*** Significant at $\alpha = 1\%$ level
** Significant at $\alpha = 5\%$ level

Equation regression results the effect of life expectancy, literacy rate, opened unemployment rate, and Gross Domestic Regional Income per capita on poverty in the districts / city in Central Sulawesi by using Random Effect Model, the value of the regression coefficients for each variable in the study by the equation as follows:

Log(Y$_{it}$)=16,76–6,79Log(AHH$_{it}$)+0,12Log(AMH$_{it}$)+0,03Log(TPT$_{it}$)–,52Log(PnDRBPerKAP$_{it}$)+v$_{it}$
(3,8479)*** (-3,9478)*** (0,0504) (2,2876)** (-6,2445)***

Adjusted $R^2 = 0.8735$
*** Significant at $\alpha = 1\%$ level
** Significant at $\alpha = 5\%$ level respectively

The regression coefficient of -6.79, life expectancy is partially elasticity of poverty to life expectancy. This figure shows that the condition of ceteris paribus, when life expectancy increased by 1%, then the average poverty will decrease by 6.79%.Variable Life Expectancy (AHH) showed a negative and significant effect on poverty in Central Sulawesi. The higher the life expectancy, the level of increased quality of public health. This is consistent with the theory of the cycle of poverty which states that the more qualified public health shown by the increasing life expectancy (AHH). The productivity level increased community can encourage economic growth which in turn will lower the poverty level, meaning that the higher the life expectancy, the poverty rate would decrease. Health development effort is to meet one of the people's basic rights, namely the right to medical care. Health development is seen as an investment to improve the quality of human resources and support economic development. Life expectancy increased reflecting an improvement in the quality of health districts / city in Central Sulawesi, both of facilities and health services.

Education has an important role in shaping the ability of developing countries to absorb modern technology and develop the capacity to create growth and sustainable development [10]. Education is closely related to poverty. People who are better educated tend to have better income, because highly

IOP Conf. Series: Earth and Environmental Science **157** (2018) 012058 doi:10.1088/1755-1315/157/1/012058

educated people have a better chance to get a job with a wage rate that is higher than the less educated. People who have a good level of education have fewer opportunities to be poor than those who educated. One of the few characteristics of literacy education is shown by indicators of literacy rate.The regression coefficient of 0.12 Literacy Rate is partially elasticity of poverty to literacy rate. This figure shows that the condition of ceteris paribus, when the literacy rate rose by 1%, then the average poverty will rise by 0.12%.Based on the survey results revealed that literacy correlated positively and significantly to poverty in Central Sulawesi. This show is not enough to have the ability to read and write one can escape poverty. Someone who can read and write (literacy) if not followed with adequate skills and abilities, not necessarily increase productivity. Someone who has a high productivity will gain a better welfare, so they can get out of the shackles of poverty. Just like [11] identified that increasing family remittance income ($\beta = 0.258$) are the next most important potential means that could be used to address poverty directly, to continue a better social change [12].This leads us to argue that transfer income can be still considered in maintaining daily life, means helping them move out poverty [13].

The regression coefficient of 0.03 Opened Unemployment Rate is partially elasticity of poverty on Opened Unemployment Rate. This figure shows that the condition of ceteris paribus, when the Opened Unemployment Rate rose by 1%, then the average poverty will rise by 0.03%.The regression results show that the unemployment rate have a positive influence and significant impact on poverty levels 11 Districts / City in the province of Central Sulawesi. Residents included in the group there are several kinds of opened unemployment rate, those who are looking for work, they are preparing for a business, they are not looking for work because they feel it is impossible to get a job and the last one they already have a job but have not started working.

One of the factors that lead to low levels of the population lives in the province of Central Sulawesi is the lack of efficient use of labor. Residents who have jobs sometimes do not correspond with the level of expertise, so the results are not optimal. This labor is categorized as unemployed apparent. Residents who have high levels of education and skills are low, generally worked odd jobs; it is characterized by the lowest income levels anyway. High population growth rate to encourage the greater number of workers who are unemployed. The government experienced problems of lack of funds or lack of investment levels to absorb the labor force is unemployed.

GDRIP per capita in the region reflect the average income of the community's ability to meet his needs, especially basic necessities. Fulfillment of basic needs is an indicator of the welfare of the community aspect of income distribution in the region. Poverty levels are not only dealing with aspects of revenue capacity, but also linked to the equal distribution of public revenue in an area [10].The regression coefficient of Gross Domestic Regional Income Per Capita (PnDRB per capita) of -0.52 is partially elasticity of poverty to PnDRB per capita. This figure shows that the condition of ceteris paribus, if PnDRB per capita rose by 1%, then the average poverty will decrease by 0.52%.The regression results indicate that the Gross Domestic Regional Income per capita (PnDRB) gives a negative and significant effect, according to the hypothesis, and the results. Variable of GDRIP a significant negative effect on poverty. Poverty in Indonesia would be lower if economic growth occurs. The higher the GDRIP growth, the decline in poverty rapidly. Reduction of poverty is almost always followed by an increase in average per capita income or standard of living, and vice versa poverty increases when GDRIP declined.

5. Conclusion

Poverty in the districts / city in Central Sulawesi province in the 2010-2013 periods has decreased every year, but the pace of poverty reduction is quite slow. Life Expectancy and Gross Domestic Regional Income Per Capita have significant negative effect on poverty in the districts / city in Central Sulawesi Province in 2010-2013 periods; The literacy rate has no effect on poverty at the districts /city in Central Sulawesi province in 2010-2013; Opened Unemployment Rate has positive and significant impact on poverty in the districts / city in Central Sulawesi province in 2010-2013 periods; Life Expectancy, Literacy Rate, Unemployment Rate, and Gross Domestic Regional Income Per Capita have overall effect on poverty. The research suggest that the government may promote health by increasing health facilities evenly, not just concentrated in one area alone, and assign qualified health personnel in areas that are still difficult to reach by rural communities and islands; districts / city in Central Sulawesi province should be able to create quality economic growth. Attempts to do the government is creating jobs in labor-intensive and upgrading the informal sector to reduce poverty in the districts / city in Central Sulawesi, that economic growth is achieved and the problem of unemployment can be reduced.

References

[1] Dollar D and Kraay A 2002 Growth Is Good For The Poor *J. of Econ* **7** 195-225
[2] Alesina A and Rodrik D 1994 Distributive Politics and Economic Growth Quart. *J. of Econ.* **109** 465-490
[3] Astrini, Myanti NM, Purbadharmaja and Putu I B 2013 The Effect of GDRP, Education, and Unemployment on Poverty in Bali Province. *E-Econ of Dev. J.* **2**(8) 384-392
[4] Fahmid I M 2013 Relation Between Economic Growth and Distribution of Income Per Capita And Human Development Index Among Five Major Islands and Provinces In Sulawesi Island, Indonesia *World Applied Sciences Journal* **28** (28)
[5] Statistic Central Agency 2000 *Map of The Poor (Poverty Map)* (Indonesia: BPS)
[6] Bhattacharyya S and Resosudarmo B P 2013 *Growth, Growth Acceleration and the Poor: Lesson from Indonesia* (Oxford: Centre for the Study of African Economies, Department of Economics, Oxford University)
[7] Barro R 2000 Inequality and Growth In A Panel of Countries *J. of Econ. Growth* **5**(2) 5-32
[8] Christian L and Demery L 2007 *Down to Earth: Agriculture and Poverty Reduction in Africa* (Washington DC: The World Bank)
[9] Datt G and Ravaillon M 1999 *When is Growth Pro-Poor? Manuscript* (Washington DC: The World Bank)
[10] Todaro M P and Smith SC 2015 *Economic Development, 09 Edition* (UK: Pearson Education Limited)
[11] Thalyta E Y, Mustajab M, Arsyad M 2010 Abstracts of doctoral theses on the Indonesian economy *Bull. of Indonesian Eco. Studies* **46** (2) 251-254 DOI: 10.1080/00074918.2010.486113
[12] Ekasari Z K, Saleh S A M, Jusoff K, Salman D, Akhsan, Kasirang A, Arsyad M, Amrawaty A A and Fudjaja L 2013 communication pattern and conflict in agricultural extension *Asian Soc. Sci.* **9** (5) 27-33
[13] Arsyad M, Kawamura Y, Yusuf S, Jamil M H, Nuddin A and Laapo A 2017 Can Access to Social Facility Help Poverty Reduction in Agricultural Sector? Evidence from Indonesia *Int. J. Agr. Syst.* **1** (2) 154-165

IOP Conf. Series: Earth and Environmental Science **157** (2018) 012059 doi:10.1088/1755-1315/157/1/012059

Application of capital social of Bali cattle farmers that participate in the partnership system in Barru Regency, South Sulawesi Province

S N Sirajuddin, A R Siregar and P Mappigau

Dept. Socio Economics, Fac. Animal Husbandry, Hasanuddin University, Jalan Perintis Kemerdekaan KM 10, Makassar 90245, South Sulawesi Indonesia.

E-mail: *sitti.nurani@unhas.ac.id*

Abstract. There are four models of partnership that is centralized models, multipartite models, intermediary models and informal model application in all livestock commodities, including beef cattle. Partnership in the beef cattle business has been done in Barruie the program showroom cattle (SRS).This study aimed to known application the social capital of beef cattle breeders who followed the partnership system (program showroom cattle) in Barru. This research was conducted in April 2017 in the district Tanete Riaja. The population is all the farmers in Barru Regency who joined the partnership system (showroom program) and the sample is beef cattle breeders who followed the partnership system in Tanete Riaja district, Barru regency. This type of research is quantitative descriptive. This type of data is quantitative and qualitative. The resource data are primary data and secondary data. Data analysis uses descriptive statistical analysis with Likert scale. The results research show that social capital (trust, linkage, norm) of beef cattle breeders who joined the partnership system (cattle showroom program) at high scale

1. Introduction

Rural development is a strategy to state development in order to reduce the disparities of urban and rural development. Between agricultural development and rural development to very close relationship as well as agricultural development is very closely related to the development of farms so that farming needs with agribusiness system is very consistent applied in rural communities because it can touch the community with little capital and can develop the area of rural areas [1]

Beef cattle agribusiness is defined as an activity that integrates simultaneous agricultural development with the development of industrial sector and related services in a cluster of beef cattle industry [2]. According to Siregar and Ilham [3] for the development of agribusiness system can accommodate the objective to improve product competitiveness and also involve the middle to bottom farmers there are three alternative activities that can be done that is 1) vertically managed integration professionally by a private company, 2) Vertical integration carried out by farmers jointly incorporated in the container of cooperatives or other organizations and 3) a combination of the two or known partnership business system

The livestock partnership system undertaken provides several advantages, in line with Strohm and Hoefller [4] opinion that partnerships are very popular in developing countries because they provide some advantages, in accordance with research conducted by Majid and Hasan [5] and Wang *et al.* [6] who obtained the result that the broiler chicken partnership system provides an economic advantage that is contributing positively to the production and supply chain efficiency and the effect is significant on the welfare of farmer farmers. In addition, according to Covey and Stennis [7], the partnership

IOP Conf. Series: Earth and Environmental Science **157** (2018) 012059 doi:10.1088/1755-1315/157/1/012059

system reduces risks for producers, price risks and reduces risk which is a basic incentive for producers.

From the partnership system, the cattle showroom program since 2011 in Barru District between the beef cattle farmers who are members of livestock groups as plasma and local government as the core, this indicates the need for cooperation and communication that runs between the two sides.

Social capital has been believed to have a major impact on the community and its members. As the World Bank argues that social capital refers to the institutional dimension, the relationships created, and the norms that shape the quality and quantity of social relationships in society [8]. Adler and Kwon [9] mentioned that social capital is a description of the internal attachment that color the collective structure and provide cohesiveness and mutual benefits of social dynamics processes that occur in society. With this matter, it is necessary to study about the application of elements of social capital in partnership of beef cattle in Barru Regency

2. Material and Method

The research was conducted in May-June 2017 in Tenete Riaja Sub-district, Barru Regency. The population is all cattle ranchers following the partnership system in Barru Regency. The sample is a cattle rancher participating in a partnership program in Tanete Riaja Sub-district, Barru Regency. Data retrieval with in depth interview and using questionnaire. Data analysis is descriptive statistically by using Likert Scale

3. Results and Discussion

3.1. Characteristics of Respondents

Based on the results of research conducted on 28 respondents from farmers then obtained some characteristics such as gender, age, education level, the number of family dependents and breeding experience that can be seen in table 1

Table 1. Classification of cattle ranchers who follow the cattle partnership system in TaneteRiaja Sub-district, Barru Regency

No	Parameter	Number(people)	Percentage
1	Education		
	a. Finished primary school	4	14.28
	b. Finished junior high school	6	21.42
	c. Graduated from high school	16	57.14
	d. College	2	7.14
2	Dependents of the family (people)		
	0	2	7.14
	2-5	24	85.71
	6-10	2	7.14
3	Long breeding (year)		
	2-10	14	50.00
	11-22	8	28.57
	23-33	4	14.28
	34-50	2	7.14
4	Length Following Partnership Program		
	1-2	10	35.71
	3-4	14	50.00
	5-6	4	14.28

Source: Primary data, 2017

Table 1 show the classification of respondents by age level indicates that 100% of respondents are of productive age in Tanete Riaja Sub-district of Barru Regency with an age range of 15-64 years. These conditions support the livestock business that is run because it requires good physical skills so as to maximize in developing the farm business. This is in accordance with the opinion [10] which states that the ability of a breeder's work is greatly influenced by the age level. The more productive the age of the breeder the more enthusiastic the new things that have not been known. In addition, age also affects the physical condition and motivation of farmers. The age of the population is grouped into 3 (1) 0-14 years old named young age / unproductive age, (2) age 15-64 years old named age / working age / productive age, and (3) age 65 years and above is called old age / unproductive age / old age. Age is one indicator that shows a person's physical ability. People with an older age are physically weaker than those who are younger. The age of a farmer can affect their work productivity in livestock business activities, Age will influence the breeders in studying, understanding and adopting innovation in animal husbandry business, Age is also closely related to the mind set of farmers in determining the management system that will be applied in business activities farms.

Table 1 shows the number of respondents who follow the partnership system (cow showroom program) in Tanete Riaja Sub-district by gender, male 100.00%. This is because in the business of beef cattle ranches require a larger manpower and generally men stronger work than women, but did not rule out the possibility for women to be able to do it. That almost all men who have reached working age are engaged in economic activities because men are the main breadwinners in the family.The women can work or assist in farming crops activities.

Based on table 1. Most of respondent's education level in Tanete Riaja Sub-district is at SMA level (57. 14%) with 16 people. This will affect the mind set in making financing decisions on business. The education will increase knowledge, develop attitudes and grow the interests of farmers, especially in the face of change.

Table 1. It also shows that the number of family dependents ranges from 2 to 5 people. The number of respondents who do not have dependents is 1 respondent with percentage (7.14%), and the respondent has family dependent ranged between 2 until 5 people 12 respondents with percentage (85.71%) and number of dependent 6 to 10 people 2 respondents with percentage (7.14%). Given these facts it can be seen that the availability of manpower or human resources in the application of partnership system (cow showroom program) is quite available, this is in the opinion of [11, 12], which states that most small businesses households use household members as labour work or human resources.

Table 1 show that the duration of beef cattle rising in respondents in Tanete Riaja Sub-district of Barru Regency is 2 to 10 years. The most respondents are respondents who have experience of breeding between 2 years to 10 years that is as many as 14 people (60%), while respondents who have the lowest experience is between 34 years to 50 years as many as 2 people (7.14%). In general, respondents have had sufficient experience in processing their business so that with the experience, respondents are able to overcome the problems that occur. This is in the opinion of [13] which states that experience is a factor that affects a person's ability in running his business. The experience of farming / rising is an important capital for the success of a farming activity. Differences in the level of experience of each farmer will also differ their mind set in applying innovation in business activities 30 in the farm. Application of technology and good management will influence the behaviour of farmers trying to do their farm business. The longer the experience of raising a person's skills will be higher and quality. It is seen that the participation of farmers who follow the partnership system (program showroom of beef cattle) to the respondents in Tanete Riaja Sub-district of Barru Regency. The most respondents who followed the old system of partnership is 3 to 4 years as many as 14 people (50.00%). Participation is the mental and emotional involvement of people in group situations and encourages them to make a contribution for group goals

3.2. Implementation Social Capital of Cattle Breeders Following Partnership System

Social capital in farmer-breeder communities is the product of human relationships with each other, especially intimate and consistent relationships. Social capital refers at the same time elements of the parameters are trust, network and norms that have the potential to productivity society so that social capital is cumulative and increase by it [14]. According to [15], social capital in a community, organization or group is an accumulation of individual capital which depends on collective capital which can be utilized by all community members. The relational / networking network is the central point in social capital, which with the network will be able to provide collective ownership to the members of the organization.

IOP Conf. Series: Earth and Environmental Science **157** (2018) 012059 doi:10.1088/1755-1315/157/1/012059

The social capital component studied in cattle ranchers following the partnership system is trust, network and norm.

3.2.1. Trust. A desire to take risks in social relationships based on the belief that others will do what is expected and always act in a pattern of mutually supportive actions is called trust. The trust among members within the group and the trust of the members with the group management assessed in this study can be seen in table 2

Table 2. Trust between members in groups

No	Category	Score	Frequency	Percentage	Bobot
1	High	3	28	100	84
2	Medium	2	0	0	0
3	Low	1	0	0	0

Source: Primary Data, 2017

Table 2 shows that in group activities the mutual trust between members and beliefs towards group leaders is high; this means that each member in a group trusts and gives full confidence to the other members and also the group leader. The form of trust that occurs is in terms of help, group meetings and group activities. This is in line with [16] opinion that trust is a trusting attitude in society that allows the community to unite with others and contribute to the improvement of social capital, that empowerment management Which based on community participation by prioritizing participation is believed to foster a sense of belonging

3.2.2. Network. The network in social capital theory means as follows:

- There is a bond between nodes (people or groups linked to the media). This social relationship is bound by mutual trust in the form of a strategy in a moralistic form. Trust is bound by a norm that binds both parties
- There is work between nodes (people or groups) through the media of social relations into one cooperation. Bilateral symbolic beliefs and interpersonal beliefs fall into this category
- Just as with a net tied to each other a work is done between the nodes and holds the load together.

The description of social capital regarding the network in the group of farmers who follow the partnership system can be seen in Table 3

Table 3. Networks in groups of farmers who participate in partnership system

No	Categori	Score	Frequensi	Percentage	Weight
1	High	3	12	42,8	36
2	medium	2	15	53,6	30
3	Low	1	1	3,6	1

Source: primary Data, 2017

Table 3 shows that the network of farmer groups that follow the partnership system (showroom cattle program) is in the medium category (53.6%), meaning that the breeders in the group mostly create social networks and only one person does not network in groups. Social networks formed between fellow members in the group of information networks and networks help. Members of the group help each other and exchange information and knowledge that they know to other members as well as the group leader exchange information with members of his group, this is in accordance with the opinion [17] that the social network occurs due to the relationship of individuals and communities. A strong social network between members of the group is absolutely necessary in maintaining synergy and cohesiveness and creating close relationships among fellow group members as well as with other parties. Participations and cooperation are forms of social capital that can be developed in rural development

3.2.3. Norms. Norms in groups of farmers who follow the partnership system has a role to smooth the business activities and create a good community. In accordance with the opinion of [18] that the norm as a rule that has become habitual and institutional based on mutual agreement. Understanding the values of norms will create a community that harmonious and peaceful so that activities or business that runs can run well. The norms performed can be seen in table 4

Table 4. Norms on farmers who follow the partnership system

No	Category	Score	Frequensi	Percentage	Weight
1	High	3	28	100	84
2	Medium	2	0	0	0
3	Low	1	0	0	0

Source : Primary Data,2017

Table 4 shows that members of breeder groups following the partnership system have high norms in compliance with group rules. This means breeders run the rules agreed upon and apply sanctions. This is in accordance with the opinion of [19] which states that there are some functions of the rules in society, such as (1) guidelines in society behave. Norma contains rules of social behaviour in social interaction, (2) maintaining community members harmony. Norma regulates that differences in society do not create chaos and disorder, (3) social control systems. The behaviour of community members is controlled and controlled by the prevailing rules. This is also in line with [20] opinion that the adaptive capacity building process is determined through the use of resources / potential and modification of the institutional / regulatory or norm system

4. Conclusions and recommendations

The application of social capital (trust, network, and norm) to beef cattle ranchers following the partnership system in Barru District is in the high level category. With this matter, the local government of Barru regency needs to strengthen social capital so that the partnership system will be run according to the objective

Acknowledgment

This research was supported by a grant from Hasanuddin University through the research scheme of The World Class University – Acceleration of International Publication or WCU-PPI – 2017.

References

[1] Sirajuddin S N, Lestari V S and Siregar A R 2016 The limitations and benefits of partnership sharing system of corporate cattle market (CCM). *American-Eurasian J. of Sust. Agr.* **11** (1) 11-14

[2] Suryana 2009 Pengembangan usaha ternak sapi potong berorientasi agribisnis dengan pola kemitraan *J. Litbang Pertanian* **28** 1 29-37

[3] Siregar and Ilham 2003 Upaya peningkatan efisiensi usaha ternak ditinjau dari aspek agribisnis yang berdaya saing *FAE.* **21** 1 57-66

[4] Strohm K and Hoefller N 2006 *Contract Farming in Kenya: Theory, Evidence from Selected Value Chains & Implications for Development Cooperation* Research Report (Kenya: Ministry of Agriculture)

[5] Majid R B and Hasan S 2014 Performance at broiler contract farmers: A case study in Perak, Malaysia *UMK Procedia International Agr. Marketing Conf.* **1** 16-25

[6] Wang H H, Wang Y and Delgado M S 2014 The transition to modern agriculture: Contract farming in developing economies *Am. J. Agr. Econ.* **96** 5 1257-1271

[7] Covey and Stennis E 1985 Analysis of the rough rice futures contracts. *Agricultural Economics Research Report 156* (USA: Mississipi State University)

[8] Cahyono B 2015 Peran modal sosial dalam peningkatan kesejahteraan masyarakat petani tembakau di Kabupaten Wonosobo *Ekobis.* **15** 1 1-16

[9] Adler PS and Kwon SW 2000 *Social capital: The good, the bad, and the ugly*. In E. Lesser (Ed.), Knowledge and social capital (Boston: Butterworth).

[10] Swastha and Sukotjo1997 *Pengantar bisnis modern (Pengantar ekonomi perusahaan modern)*. (Yogyakarta: Liberty Offset)

[11] Suratiyah 2009 *Ilmu Usaha Tani* (Jakarta: Penebar Swadaya).

[12] Daniel M 2002 *Pengantar ekonomi pertanian untuk perencanaan*. (Jakarta: Universitas Indonesia Press)

[13] Syafaat, Simatupang N P, Mardiantodan S and Pranaji T 2003 Konsep pengembangan wilayah berbasis Agribisnis dalam rangka pemberdayaan petani *Forum penelitian Agroekonomi* **21** (1) 130-137

[14] Handayani, Gayatri S and Mulyatno B 2005 Pendapatan tenaga kerja keluarga pada usaha ternak sapi potong di kecamatan Toroh kabupaten Grobogan *Mediagro* **1** (2) 38-44

[15] Suharto 2013 *Islam, modal social dan pengentasan kemiskinan Indonesia social economic outlook*. (Jakarta: Dompet Duafa).

[16] Fukuyama 1995 *Trust: The Social Virtues and the Creation of Prosperity* (NewYork: The Free Press)

[17] Kusumastuti 2015 Modal sosial dan mekanisme adaptasi masyarakat pedesaan dalam pengelolaan dan pembangunan infrastruktur *Jurnal Sosiologi* **20** 1 81-97

[18] Priyono and Utami 2012 Penguatan modal sosial dalam upaya meningkatkan pendapatan peternak sapi potong di kabupaten Banjarnegara *Suryaagritama* **1** 1 1-10

[19] Peeling and High C 2005 Understanding adaptation: What can social capital offer assessment of adaptive capacity? *Global Environmental Change.* **15** (4) 308-319 doi:10.1016/j.gloenvcha.2005.02.001

[20] Smith B and Wandel J 2006 Adaptation, adaptive capacity and vulnerability *Glob. Env. Change* **16** 3 282-292. doi:10.1016/j/gloenvcha.2206.03.008

Role of joined farmer groups in enhancing production and farmers income

M Arsyad[1], Rahmadanih[1], S Bulkis[1], Hasnah[1], A Sulili[1], Darwis[1], A Bustan[2] and M Aswad[3]

[1] Department of Socio-economics of Agriculture/Agribusiness, Faculty of Agriculture, Hasanuddin University, Jl. Perintis Kemerdekaan KM 10, Makassar, 90245, Indonesia.
[2]Faculty of Education, Palangkaraya University, Jalan Yos Sudarso, Jekan Raya, Palangka Raya, 74874, Central Kalimantan, Indonesia.
[3]Institute for Research, Community Services and Quality Assurance, Universitas Sulawesi Barat, Jalan Prof. Dr. Baharuddin Lopa, S.H., Majene, West Sulawesi, 91412, Indonesia.

E-mail: *arsyad@unhas.ac.id*

Abstract. Production and farmers income still becomes a global issue in economic development. The study aims to (1) describe the implementation of the role of Joined farmer groups (called *Gapoktan*) in accordance its function and (2) to analyze the role of Gapoktan in increasing production and farming income. The study was conducted in Camba Sub District, Maros District, South Sulawesi, Indonesia in 2017 and choosing *Aspana* Gapoktan as Case Unit. Data collection is done by a combination of qualitative and quantitative methods. Qualitative data were collected by Focus Group Discussions method, indepth interview and document study while quantitative data was collected through survey method on 60 farmers respondents. The results showed that, (1) Aspana Gapoktan has implemented a role related to its function as a business unit in the provision of production facilities and farming as well as marketing but has not implemented roles related to its function as a processing business unit, and saving and loan (2) Gapoktan role in increasing production and income of farming is facilitating procurement of farm inputs and agricultural production tools for farmers and developing various commodities in farming activities, especially horticultural crops. More than 44.00% of farmers perceived that their production increased about 10.00% - 25.00% and more than 68.00% of farmers perceived that their income increased by about 10.00% - 25.00% for the last three years. It is necessary to increase the role of Gapoktan through (1) the procurement of horticultural product processing industry and (2) doing savings and loan activities by utilizing 40.00% of funds managed by Gapoktan or through the formation of cooperatives under the management of Gapoktan.

1. Introduction

One of livelihood rural communities source is agriculture-based activities [1]. However, farm production and income are still a problem in economic development [2]. One effort to improve farm income is strengthening farmer institutions [3-6]. Combined Farmer Group (Gapoktan) is one of farmer institution that can give contribution to increase production and income of farmer family. This institution is developed to fulfill the feasibility of economy scale and business efficiency, so that it functions as a unit of the provider of production facilities and infrastructure, farm / production unit, processing business unit, marketing business unit and micro finance / savings and loan business unit [7].

At the sametime, the role of Gapoktan in increasing production and income of smallholders are still becomes serious issue to solve. In other words, it is really hard to explain whether or not Gapoktan has truly function as business unit, marketing unit, and microfinance unit for the farmers in rural area, including its implementation in improving farmer's welfare. Therefore, specific objective of the research is to (1) describe the implementation of the role of Gapoktan in accordance with its function, and (2) to analyze the role of Gapoktan in increasing production and income of farming.

2. Method

The research was conducted in Camba Subdistrict, Maros District, with consideration of Camba Subdistrict area in highland ecosystem which has potential in developing food crop (especially horticultural crops). The result design is a case study, by selecting Aspana Gapoktan as the case unit. The research employed qualitative approach. To clarify the analysis on the second objective, the qualitative data is supported with quantitative data so that it is used the mixed of combining qualitative and quantitative methods [8, 9]. Qualitative Data Collection is conducted through: (1) Indepth-interview to informants who determine purposively. (2) Focus Group Discussion (FGD) and (3) document studies. While Quantitative Data Collection is done through Survey Methods on farm households on two selected Poktans. Informants consisted of farmers, farmers group (Poktan) and Gapoktan officials and Agricultural Extension Workers (PPL). The number of informants as a whole consists of 9 people while the total number of FGD participants consists of 12 people. The farmers' sample (respondent) as a whole is 60 people, consisting of 25 members of Poktan-Duppa Siruntu and 35 members of Poktan-Pammase.

3. Results and Discussion

3.1. Description of Aspana Gapoktan

Gapoktan is a merging of several farmer groups (Poktans), where the farmer group is mostly formed from several years before the formation of Gapoktan. The establishment of Poktan in Camba Sub-district of Maros Regency, officially confirmed in 2017 [10]. While the formation Gapoktan started in 2008 and one of Gapoktan formed in that year is Aspana's Gapoktan. The resources owned / controlled by Aspana's Gapoktan until 2016 [11] are consists of *Human resources* covering: (a) the number of members / farmers involved (439 persons) and (b) the education level of the gapoktan's Administrator (elementary to senior high school); *Natural resources* in the form of: (a) semi-technical irrigated rice field (115 ha), non-technical irrigated rice fields (165 ha) and 68 ha of rain-fed rice fields; *Physical resources* consist of: (a) facilities / equipment and technologies owned or accessible to group members; (b) Hand tractor 35 units, (c) threshing machine (280 units), sickle (587 units) and farmer group (11 groups); *Financial resources* in the form of receivables. However, due to incomplete group administrative records (including finances), the amount of receivables is not known with certainty; Organizational *structure i.e.* rules / norms that exist in Gapoktan.

3.2. Role of Gapoktan in Increasing Production and Income Farming

The role of Gapoktan referred to in this case is related to the five functions contained in the Regulation of the Ministry of Agriculture, Republic of Indonesia (Permentan RI) No. 67 of 2016 namely as (1) Provider Facilities and Production Infrastructure Business Unit, (2) Unit Farming / Production, (3) Processing Business Unit, (4) Marketing Business Unit and (5) Savings and Loan Business Unit [7]. To increase farming production and household income of farmers should be Aspana's Gapoktan performs minimal role in accordance with the five functions.

3.2.1. Role of Gapoktan as Provider of Facilities and Production Infrastructure. The role of the Aspana Gapoktan that is routinely carried out in relation to this function is the way farmers work in the provision of production facilities depending on subsidized and certified seeds. In addition, they facilitate procurement of agricultural tools and machinery for farmer groups to be used by the farmers who are members. Especially in the provision of subsidized fertilizers, the first step is to facilitate Poktan in preparing the Group's Definitive Needs Plan (RDKK) of production inputs (Saprodi) together with its members, in accordance with the area of managed land. RDKK which has been prepared will be proposed to Gapoktan to be forwarded to the competent party in providing subsidized fertilizer. Farmer-level fertilizer receipts are sometimes late (not timely). In this condition, the poktan management strives in such a way that the saprodi can be accepted by its members with a delay that is not too long. The trick is to confirm with Gapoktan. Furthermore, Gapoktan make direct confirmation to relevant stakeholders, for example Agricultural Extension Field (PPL) who served in each Village.

Saprodi needs proposed by Poktan to Gapoktan Aspana, among others, in the form of subsidized fertilizer consisting of Urea, SP3, SP-36, ZA, NPK and Organic Fertilizer. The amount of fertilizer proposed is always the same as the amount received by farmers. Based on data on RDKK Poktan [11], the realization of subsidized fertilizer demand in 2016 for planting area of 267.05 hectares consists of urea fertilizer 133,525 kg, SP-36, ZA and NPK 53,410 kg each and organic fertilizer as much as 267,050 kg. This number is spread across 11 farmer groups. The amount of fertilizer proposed depends on the area of land managed by the farmers of each Poktan member. For time efficiency, fertilizer demand is proposed once for two growing seasons. Efforts Gapoktan to increase production and income of farmers through the facilitation of the procurement of agricultural production facilities (fertilizer), causing farmers tend to more easily access production facilities, especially subsidized fertilizer. Even more than 56.00% of farmers perceiving more easily access current production facilities when compared to the previous three years (table 1).

Table 1. Farmers' Perception on Increasing Access to Production Input

No	Perception	Aspana Gapoktan			
		Poktan Duppa Siruntu		Poktan Pammase	
		n	%	n	%
1	Easier	14	56.00	31	88.57
2	The same	11	44.00	4	11.43
3	More difficult	0	0.00	0	0.00
	Total	25	100.00	35	100.00

The ease to access production input (saprodi) tends to differ between farmers from one poktan and other poktan farmers. Farmers from Poktan Pammase, for example, tend to be more accessible to fertilizers when compared to farmers who are members of Poktan Duppa Siruntu. This condition is related to the distribution of fertilizer, which sometimes comes late. Although subsidized fertilizers sometimes have not arrived at the time of need, however, farmers have benefited from subsidized fertilizers that have been facilitated by Poktan and Gapoktan. The benefits are not only for rice farming but in the last three years, especially for horticultural crops

3.2.2. The Role of Gapoktan as Production Business Unit. Referring to Permentan RI [6], each gapoktan ideally has a business unit that produces commodities to meet the needs of its members and market needs so as to guarantee the quantity, quality, and continuity of the results. The type of commodities cultivated by farmers within the range of Gapoktan Aspana consists of rice, corn, beans and horticulture. The overview of rice and corn and peanut production is shown in table 2.

IOP Conf. Series: Earth and Environmental Science **157** (2018) 012060 doi:10.1088/1755-1315/157/1/012060

Table 2. Productivity of Rice, Corn and Peanut Crops cultivated by Farmers in Aspana Gapoktan

No	Description	Rice	Corn	Peanuts
1	Extent of planting / harvesting (ha)	463.00	87.00	128.00
2	Production (ton)	3,357.00	324,00	232,00
3	Productivity (ton/hectare)	7.25	3.72	1.82

Source: [12]

In addition to rice crops and crops, the most developed crops of Gapoktan farmer group members are horticultural crops. In cultivating horticulture, farmers are also assisted by the means of production in the form of fertilizer subsidies by the government. Submission of fertilizer demand is done every rice growing season through RDKK at each Poktan and then forwarded to Gapoktan for follow up to company. That is, fertilizer subsidy is not only intended for rice plants but also for other types of crops (crops and horticulture) gradually. The dominant crops cultivated in early 2017 are red chili, tomato and watermelon (table 3)

Table 3. The Production of Several Horticultural commodities in Aspana's Gapoktan, 2016

No	Description	Red Chili	Tomato	Watermelon
1	Extent of planting / harvesting (hectare)	40.13	8.30	
2	Number of stems			2,000,000
3	Production (ton)	313.00	65.00	7,00
4	Productivity (ton/hectare)	5.47	7.80	

Source: [12]

Development of horticultural crops facilitated by Gapoktan through Special Effort (UPSUS) Horticulture program has an impact on increasing production of Farming mainly on commodities of large red pepper and fruit tomatoes (apple tomatoes). All farmers perceived that the production of farming in general increased compared with the previous three years. Similarly, the production of horticultural crops is shown in table 4.

Table 4. Farmers' Perceptions of Increasing Horticultural Crop Production

No	Farmers' Perceptions	Aspana Gapoktan			
		Poktan Duppa Siruntu		Poktan Pammase	
		n	%	n	%
1	Increase < 10%	14	56.00	11	31.43
2	Increase 10%-25%	11	44.00	24	68.57
3	Increase >25%	0	0.00	0	0.00
	Total	25	100.00	35	100.00

Especially on horticultural commodities, more than 44.00% of farmers perceived that the increase in production ranges from 10.00% - 25.00%. Differences in production increase as shown in Table 4, tend to be related to environmental conditions, especially irrigation. Both rice and palawija crops, as well as horticulture, can contribute to farmers' income if farmers are in the desired condition with adequate land and air conditions. The description of farm income in several commodities shown in table 5.

IOP Conf. Series: Earth and Environmental Science **157** (2018) 012060 doi:10.1088/1755-1315/157/1/012060

Table 5. Farming Revenues in Some Commodities Cultivated by Poktan Members in Aspana Gapoktan

No	Commodities cultivated	Farm costs (IDR/hectare)	
		Highest	Lowest
1	Rice	2,050,000	1,500,000
2	Corn	1,600,000	1,450,000
3	Peanuts	3,850,000	3,400,000
4	Watermelon	6,300,000	5,600,000
5	Red Chili	7,500,000	7,250,000
6	Tomato	2,750,000	2,250,000

Table 5 clearly shows that, there is a difference in farm income for the same commodity. This tends to be related to differences in crop maintenance (including use of inputs) and irrigation conditions. The highest revenue, contributed by chillies ranged from IDR 7,250,000 – IDR 7,500,000 per hectare. Nevertheless, in general these commodities contribute a higher income when compared to the previous three years, especially to horticultural commodities (table 6). Based on the perception of farmers, it is indicating that all farmers experience increasing income of farming. More than 68.00 percent of farmers perceived that their income increased 10% - 25% when compared with the previous three years.

Table 6. Farmers' Perceptions of Increasing Farm Income

No	Farmers' Perceptions	Aspana Gapoktan			
		Poktan Duppa Siruntu		Poktan Pammase	
		n	%	n	%
1	Increase < 10%	8	32.00	4	11.43
2	Increase 10%-25%	17	68.00	31	88.57
3	Increase >25%	0	0.00	0	0.00
	Total	25	100.00	35	100.00

3.2.3. Role of Aspana Gapoktan as Processing Business Unit. Based on Permentan RI [6], Aspana Gapoktan ideally plays a role in providing services, either in the form of agricultural machinery and technological tools in processing of commodity production, including processing, sorting / grading and packing to increase the added value of the product. In fact, until early 2017, Aspana's Gapoktan has not more given to the role of agricultural processing. Processing of farming results done by poktan members so far in order to increase the added value of farm production is still done simply based on their respective experience. For paddy (grain) for example, farmers have sufficient experience for this, both in terms of drainage and in terms of milling.

Especially on horticulture, postharvest handlers are only cleaning and grading (there is no industry that process from primary product to secondary product). Thus, the household income of farmers is directly related to the profit of farming. If the profit of farming is low or not increases then household income is also low and does not increase. Household income can also be affected by new employment opportunities for household members. Thus, there should be a new job opening with poktan and Gapoktan related to its function as management of production facilities, in the upstream sector and downstream sector handling through outsourcing of results. Based on the results of FGD and indepth interview, the need perceived by Gapoktan based on the potential of horticulture production is chili and tomato processing industry to be sauce accompanied by training and mentoring for the workforce that will be involved in the processing industry.

3.2.4. Role of Gapoktan as Marketing and Microfinance Business Unit. Referring to Permentan RI [7], Gapoktan Aspana ideally serves to provide services/facilitation of agricultural marketing of its members, whether in the form of network development and business partnership with other parties, as well as direct marketing. The current condition, Gapoktan Aspana does not provide many roles related to this because farmers in general have experienced marketing the results of their farming based on their choice. Gapoktan in this case only plays a role if there is a price jump (up or down drastically) even though in general farmers already know the prevailing price phenomenon.

The problem that farmers often encounter with the marketing of chilies and tomatoes is that sometimes the price decreases and is considered cheap but still has to be sold because the product cannot stand it. Therefore, [14, 15] and based on available resources, it is necessary to strengthen the institutional role by: Improvement of physical resources in the form of procurement Horticulture crop processing industry (chili sauce and tomato sauce), accompanied by a training activity of making sauce. Thus, horticultural production can be marketed in Gapoktan for further processed into ready-to-consume products.

Referring to Permentan RI [7], Gapoktan Aspana ideally plays facilitate the capitalization of Farming to members through credit / capital of Farming or from self finance Farmers / residual business results. In fact, Gapoktan Aspana currently has no productive economic activities either in the form of savings and loan (SPP) members or in other economic enterprises (UMKM). Productive activities in the form of capital lending for farm input payments take place in the first year until the third year after Gapoktan receives funding in 2008. However, because the cooperation (mutual understanding) of poktan and gapoktan members in the management & utilization of aid funds are still lacking, then the activity is not going well.

4. Conclusion

Aspana Gapoktan has not run its role optimally in accordance with the general function of gapoktan according to Permentan no. 67 years 2016. The role of Gapoktan associated with its function as a business unit provider of production facilities and farming unit (production) is quite good. These roles make farmers more accessible to production facilities and all farmers have increased production compared to the previous three years. On the other hand, the role of Aspana Gapoktan Aspan related to its function as a business unit of processing and marketing and savings and loans, is still very less. Therefore, it is necessary to increase the management of financial resources through savings and loan activities by utilizing 40.00% of funds managed by Gapoktan or through the establishment of cooperatives under the management of Gapoktan; increase of physical resources in the form of procurement Horticulture crop processing industry (chili sauce and tomato sauce) should be accompanied by training activities of making sauce. It is clear that Gapoktan has a function as facilities and infrastructure provider, production, marketing and microfinance business units.

References

[1] Chambers R and Conway G R 1991 *Sustainable rural livelihoods: Practical concepts for the 21st century* (London: Institute of Development Studies DP).

[2] Fahmid I M 2004 *Gagalnya politik pangan di bawah rezim orde baru: kajian ekonomi politik pangan di Indonesia* (Jakarta: Yayasan Sandi Kota-ISPEI).

[3] Anantanyu S 2011 Kelembagaan petani: Peran dan strategi pengembangan kapasitasnya *Jurnal Sosial Ekonomi Pertanian dan Agribisnis* **7** 2

[4] Ratna DP, Wuradji N and Djazifah E R 2012. Pemberdayaan petani melalui gabungan kelompok tani (Gapoktan) *Jurnal Diklus* **XVI** 02

[5] Rahmadanih, Bulkis S, Amrullah A, Rukka R M. and Arsyad M 2015 Strengthening institutional model of women-farmers group in developing household food diversification *Int. J. of Agr. Syst.* **3** 29-40

[6] Bulkis S, Summase I, Rahmadanih and Fahmid IM 2016 *Model Pengembangan Klaster Industri Kakao di Sulawesi Selatan (Tahun III)* Research Report (Makassar: LP2M UNHAS)

[7] Kementerian Pertanian RI 2016 Peraturan Menteri Pertanian Republik Indonesia Nomor 67/Permentan/SM.050/12/2016 Tentang Pembinaan Kelembagaan Petani (Jakarta: Kementerian Pertanian RI)

[8] Tashakkori A and Teddlie C (ed.) 2003 *Handbook of Mixed Methods in Social & Behavioral Research* (USA: SAGE Publications).

[9] Creswell JW and Clark VLP 2007 *Designing and Conducting Mixed Methods Research.* (USA: SAGE Publications).

[10] Bupati Maros 2007 Keputusan bupati Maros tentang pengukuhan kelompok tani-nelayan kabupaten Maros tahun 2007 Maros Sulawesi Selatan

[11] BPP-KP 2017 *Program penyuluhan UPT-BPP dan ketahanan pangan kecamatan Camba tahun 2017.* (Indonesia: Badan Pelaksana Penyuluhan dan Ketahanan Pangan (BPP-KP) Kabupaten Maros)

[12] Gapoktan-Aspana 2016 Rencana defenitif kebutuhan kelompok terhadap pupuk bersubsidi. Kecamatan Camba. Kabupaten Maros

[13] UPT-BPPKP Kecamatan Camba 2017 Program penyuluhan desa Cendrana kecamatan Camba tahun 2017 (Maros:UPT-BPPKP)

[14] Uphoff N 1986 *Local institution development: An analytical source book with cases.* (West Hartford, Connecticut: Kumarian Press).

[15] Uphoff N 1992 *Local institution & participation for sustainable development* (London: IIED).

IOP Conf. Series: Earth and Environmental Science **157** (2018) 012061 doi:10.1088/1755-1315/157/1/012061

Sustainable integrated farming system: A solution for national food security and sovereignty

M Ansar and Fathurrahman

Agriculture Faculty, Tadulako University, Kampus Bumi Tadulako Tondo, Jl. Soekarno Hatta KM. 9, Palu, 94118, Indonesia.

E-mail: *apasigai@yahoo.com*

Abstract. This paper provides a comprehensive review of literature related to food security. The world food crisis is a threat to all countries, including Indonesia. The problem of food security in Indonesia is still happening, particularly, aspects of production and increasingly unbalanced food availability. Due to the increasing rate of population growth, land functional shift, degradation of land resources and water, as well as environmental pollution and climate change. Food production has not been able to meet the needs of the population continuously. Therefore, the food policy paradigm applied in Indonesia must change from food security to food independence. Thus, Indonesia is not dependent on other countries. Food diversification is one of the best policies to be implemented in achieving food independence and anticipating the food crisis. Food diversification utilizes land optimally by developing an integrated farming system. The integrated farming system is an efficient and environmentally agricultural system. It is able to utilize sustainable agriculture development, followed by the development of participatory technology (Participatory Technology Development) which refers to the local wisdom of the community.

1. Introduction

Food has become a primary need of human beings to ensure their life. Indonesia is one of the developing countries known for its agriculture. Most of the people do farming for a living. Indonesia has a massive land being potential for farming, although food security still becomes a solemn problem requiring a serious attention. Laws on food issues are made to amend that food is not only a strategic and crucial matter, but it has also become a part of human right. As such, on condition that the state does not fulfill people's need of food, it is said to violate human right.

Food security in Indonesia has recently been worse in as much as vigorous change of agriculture land function for use beyond agricultural sectors, a situation under which the government need to be more sensitive. Managing food security and its sustainable development is one of the biggest challenges in Indonesia. Countries are not able to provide sufficient quantities of nutritious food to these people so that they can live healthily. It is strongly believed that there is enough food to feed everyone adequately but the problem is distribution and management [1]

This has drawn the attention of many international bodies, not to mention Food and Agriculture Organization (FAO), which argue that Indonesia sits at an insecure levelin the Global Hunger Index. This is predicted to be worsen in line with the increase in the number of its population and, in the long run, food scarcity may occur because of several reasons such as environmental damage, land conversion, high price of fossil fuel, and climate change. Climate changes have started showing its impact on water resources and agricultural yield worldwide. Water and food security are the key challenges under climate change as both are highly vulnerable to continuously changing climatic patterns. Climate changes have started showing its impact on water resources and agricultural yield worldwide. Agricultural sector and food securities are threatened and if the basic adaptive measures such as changes in crop pattern, crop breeding and types and innovative technologies, which use less water are not used global food production [2]

There is a new urgency to improve the accuracy of predicting climate change impact on crop yields because the balance between food supply and demand is shifting abruptly from surplus to deficit [3]. This reversal is being driven by a rapid rise in petroleum prices and, in response, a massive global expansion of biofuel production from maize, oilseed, and sugar crops. Soon the price of these commodities will be determined by their value as feedstock for biofuel rather than their importance as human food or livestock feed.

Food autonomy is a prerequisite of food security. Such security becomes intricate to achieve if a state and its people do not have an authority on the consumption and production process of the food. Food security means every individual is able physically and economically to fulfill sufficient, secure, and nutritious food for an active and healthy life. National food security is reached when the food need of every individual household is fulfilled, which is portrayed in the safety, equality, accessibility, and availability of sufficient food both in quantity and quality. Of these, production and availability of food are still problematic in ensuring food security. Food production has not been able to continually fulfillthe people's need because of a relatively slower food production growth than the demand.

In recent years, Indonesia faces food security problem. For this, it is important to find out ways to solve it, which includes increase of food availability, stability and accessibility of food to promote competitiveness in the state economy. It is undeniable that to extend the quantity of land for agriculture and to slow down population growth in Indonesia are difficult, but what is important in this matter is to find out the strategy to improve food security by effectively cultivating and making the best ofthe available land. Space management of the land must also be carefully done for further reservation of "permanent farming land", which is regulated through central government regulation or local government regulation.

The other challengesare global warming affecting climate change and food competition for consumption and bioenergy. Consequently, climate is unpredictable, causing production uncertainty or failure in harvest, which severely declines food production. Scarcity and continuous resource usage competition (land and water) results in more difficult food production and ultimately causes food shortage. In fact, Law No.18/2012 on food has entrusted that food provision is intended for fulfilling basic needs of human beings equitably, equally, and continually on the basis of food sovereignty, food autonomy, and food security [4] (Kemenkumham).

This paper provides a comprehensive review of literature related to objective: to explain or to describe the role of integrated agriculture in the handling of national food security.

2. Developing National Food Security

The success of a state development is indicated by the availability of quality human resource those who are physically and mentally strong, healthy, and intelligent. Empirical data indicate that this type of resource is built through good nutrition, and good nutrition is determined by the quantity of intake received. Thus, fulfilling food and nutrition is a necessity to promote health of the society and becomes an investment to boost the quality of human resource.

Establishing food security and nutrition is a part of the national perspective. Indonesia, a nation with a huge number of population and a large coverage of territory, food security and nutrition, has an important agenda in its economic development. The government must continually push development of food security and nutrition through programs that are able to strengthen food security and at the same time improves prosperity of the society. Food shortage is another part of the Indonesia National perspective, which is considered sensitive in the part of the dynamic socio-political life. Hence, it is important for Indonesia to actualize regional food security and nutrition, and food-security-based

IOP Conf. Series: Earth and Environmental Science **157** (2018) 012061 doi:10.1088/1755-1315/157/1/012061

household and individual. Programs intended for nutrition and food security development are to be integrated, measurable in its successand sustainable.

The world food crisis threatens all countries, including Indonesia. The national policy on food implemented in Indonesia should be shifted from food security to food autonomy to avoid food dependency on other countries. One of the policies considered appropriate to be implemented in order to achieve food autonomy as well as in anticipation of food crisis is diversification. It is a process of diversifying food, or efforts to increase consumption of varied foods based on the principle of balanced nutrition. One of the obstacles in developing local foods is there has not been any local food easily accepted and afforded by the people. Apart from this, the people consider rice is the main food for them. One of the ways to develop local food product is by providing incentive for enterpreneurs or industries that use local food as raw materials.

3. Food Security Policy and System

Food security is divided into three: *(i) food availability; (ii) access to food; and (iii) food usage*. Food availability is physical availability of the food in the area, gained either from domestic product, import/trade or food aid. Food availability is indicated by domestic product, which comes from market, food stock owned by trader and government, and food aid from either government or food aid agencies. Food access is the ability of household to get sufficient food, either from own production, purchasing, trade, gift, loan and food aid, or combination of the six. Food availability in a certain area may be sufficient, but not all households have equal access to it either in terms of quantitative or food diversity. Food usage refers to food use by households and individual ability to absorb and metabolize nutritious essences (conversion of nutritious essence by the body efficiently). Food usage also includes strategy of storing, processing and serving the food, including use of water and fuel during the process and hygienity. Another important aspect of food usage is the custom for serving the food, particularly for the individual needing special food, distribution of the food in the household which suits each individual need, and health record of every individual member of the household.

One of the government policies in empowering food security is Acceleration of Food Consumption Diversification (*AFCD*). It is an implementation of strategic plan of the Ministry of Agriculture, one of which is increasing food diversification. Acceleration of diversification on consuming local food is a national move supporting food security of the society on the basis of local wisdom and cooperation of central and local governments as well the people. Activities of *AFCD* include: using house yardas affectively as possible following the concept of Sustainable Food Center (*SFC*). Itis done by empowering women to make the best of the available house yard as a family food source. This concept requires cultivation of varied kinds of plantation that suit the family needs, such as various roots, vegetables, fruits, and livestockand fish as supplement for availability of source of carbohydrates, vitamins, minerals,and protein for the family in an area of housing /community living nearby. This leads to an establishment of an area rich with own produced food as a result of making the best of house yard. The approach to this development is through sustainable farming, such as opening seed plantation by prioritizing local resources and making use of local wisdom, to ensure the nature remains protected. Home gardens are an integral part of local food systems and the agricultural landscape of developing countries all over the world and have endured the test of time [5].

4. Sustainable Integrated Farming System

Integrated farming system is an agricultural system which combines the activities of crops and horticulture, animal husbandary, fishery, forestry and other sciences related to agriculture within an area simultaneously at a time or so. Developing an integrated farming system needs an appropriate technology, because such technology is partly important in sustainable farming and food security. As a tool, it is necessary to think about how to help farmers improve the quality of their crops. Technology also needs to be considered as a solution to overcome and to balance decrease in food production due to decrease of agricultural land. Some countries with small land are still able to overcome their food problem because of technology. Crop production can increase if irrigated areas are expanded or irrigation is intensified, but these may increase the rate of environmental degradation. Since climate change impacts on soil water balance will lead to changes of soil evaporation and plant transpiration, consequently, the crop growth period may shorten in the future impacting on water productivity. Crop yields affected by climate change are projected to be different in various areas, in some areas crop yields will increase, and for other areas it will decrease depending on the latitude of the area and irrigation application [6]. This has encouraged Indonesia to integratedly cultivate its land adopting the so-called Integrated Farming System.

Implementation of integrated farming system promotes higher income, minimizes risk of crops failure of one type of crop, and creation ecological farming system which is able to make the best of local resources as efficiently as possible for the puprpose of sustainable environmentally-friendly farming system. Maximizing use of land, which currently has decreased, encourages integrated farming system of crops, forestry and livestock within the same land space simultaneously. The adoption of integrated poultry-fish-duck farming in different forms such as intensive broiler production, backyard poultry rearing, etc. leads to conservation of natural resources and provision of additional income/employment opportunities, especially to the youths and women workers in the villages. That the integrated poultry-fish system could address issues of sustainability, women empowerment and livelihood security effectively [7].

4.1. Polyculture system

Cultivating land under yearly crop stands sustainably can be done through polyculture system. Polyculture means planting more than one type of plants in an area at a time or almost simultaneously. Some planting patterns of polyculture system include Intercropping and Interplanting, Multiple Cropping, Companion Planting, Mix Cropping, and Alley Cropping. Implementing polyculture, in principle, intends to make land use efficient, minimize land damage, overcome risk of crop failure, increase income of farmers, and guarantee food availability.

Planting intercropping plants of varied kinds of food among coconut trees is profitable. In Lampung is one of the central areas of coconut plantations in Indonesia, covering an area of about 126,129 ha, only about 30% of the total area under coconut trees was used for intercrops. Types of plants for intercrops were maize, bananas, and upland rice. This is because 80 percent of the land under the coconut trees can be used for other plants and livestock [8]. The kinds of intercropping plants that can be planted under this area include plantation crops, food crops, and horticulture depending on the land and the climate of the area. One of the obstacles in using land under coconut stands, besides the percentage of sun light, is fertility of the land. Planting intercrops and livestock in such a land can increase efficiency of land use, and the income of the farmers from intercrops and livestock increases at least 30 percent, and another 30 percent from crops of coconut.

Some research also reported that it is economically profitable when corn plantation is managed well. Planting corn as intercrop plants among coconut stands yealds 80 percent more than planting corn monoculturally [9]. To increase income and efficiency of corn planting needs modification of corn planting system for a more optimal use [10]. Moreover, [11] found that such a modification of corn planting for the purpose of double earnings—that is seeds and stover for animal feed—is economically appropriate to be implemented with a B/C value and an anaverage of more than 2.0.

Such efforts need an integrated and sustainable management with technological innovation suitable with a specific location. One of the breaksthroughs in coconut plantation technology is to move from a traditional farming system of monoculture to a polycultural farming of coconut, including developing food crops among coconut plants, to support community food security.

4.2. Livestock plant system

Farmers have long been employing the concept of integrated farming involving plants and livestock although they still did it traditionally in a small land. Such farming is really potential to develop commercially to support national food security. Integration of livestock in farming system is of importance in: 1) ensuring farming to be sustainable through diversification of activities in order to produce food for the family of the farmers; 2) removing the nutrients and energy between livestock and plants by way of using manure and agricultural wastes as food source to create sustainable farming,

IOP Conf. Series: Earth and Environmental Science **157** (2018) 012061 doi:10.1088/1755-1315/157/1/012061

which is environmentally-friendly [12]. The presence of livestock and food plants in the area of coconut plantation within a farming system will open crop diversification opportunity by including *palawija* (crops planted as 2nd crop in dry season), horticulture, and morage in a single unit of farming. Such a link between plant farming and livestock enables the two activities to provide food in support of food security and autonomy of the community.

The food plants planed producing biomass can be used to feed livestock, while the livestock produce manure which can be used to improve soil quality and assure sustainable soil fertility.This will create efficient, productive, profitable and environmentally-friendly farming system. Animal husbandary needs to be developed integratedly to become part of an"organic farming".

4.3. Forest crop system and livestock (agrosilvopasture)

The growing intensification of farming has been accompanied by degradation of wild lands, including tropical forests and wetlands, at an alarming rate. Further pressure on fragile land has come from associated urbanization. Integrating crops and livestock into existing farming systems is being recognized as a means to address these problems and the smallholder farmers have the possibility of realizing immense benefits from the integrated systems [12]. In Indonesia, agroforestry is often regarded as one of the sustainable integrated farming system. It is a collective term of varied kinds of integrated land use (forestry, agriculture, and/or animal husbandary) existing everywhere in the world, not to mention those found in tropical developing countries, i.e. Indonesia. Such traditional land usage has been developed/practiced by the local community or introduced by many parties within the last three decades [13]. Main constituent component of agroforestry is forest, farming, and/or livestock. By looking at the components, agroforestry can be classified into Agrisilviculture, an agroforestry system that combines between forest components (woody plants) and agricultural components (non-woody plants), and Silvopastural, an agroforestry system that covers forest components (woody plants) and livestock components (pasture).

4.4. Intercropping system

Intercropping is one of the traditional plant breeding methods on dry land. In this model, two or more plant species are planted at the same time on the same field. This model can reduce the risk of crop failure, improve soil fertility, decrease soil erosion and increase farmers' income. It can also improve the use of environmental factors, labor and water as well as surpressing weed attacks and plant diseases. In most cases, intercropping system is more lucrative than monoculture system due to high land productivity and varied commodity type. Economical use of production means and lower risk of failure so that the sustainability of food availability can be guaranteed.

Intercropping system can improve the effectiveness of land use. Intercropping soybean and maize is advantageous. This is associated with a greater total intercrop yield, higher land equivalent ratios greater than 1.0, indicating greater productivity per unit area and greater percentage of land saved, which could be used for other agricultural purposes. In addition, land equivalent coefficient values exceeded 0.25, indicating yield advantage of the intercropping system. Competitive pressures were low, signifying that the crops were found to be complementary and suitable in mixture [14].

As one of the alternative ways in agricultural intensification program, intercropping system is suitable for multiplying agricultural products in less productive areas. Not only can the system increase the harvest amount per year, it can also maintain the fertility of soil by restoring many organic matters and ground covered by canopy plants. Intercropping system gives more benefit than monocropping does. It can reduce production cost due to the efficiency of land use.

Intercropped between leguminosae and cereal crops has been a common combination done by the farmers. The use of legumes like levials as intercropping plants can be profit-making for the staple plants because it produces a lot of nitrogen. The root nodule of the legume crops can bind the free nitrogen from air and this is quite profitable for increasing the nitrogen in the soil and meeting nitrogen needs of other plants. The planting soybeans using right intercropping system can avoid plants from pest attacks, resulting in high crop productivity. Intercropped system between corn and soybean from various soybean cultivars at various planting items as a whole is more advantageous than the monoculture system [15].

4.5. Dry farming system

The development of food crops in addition to rice is often done on dry land, even on critical land. With narrow land ownership, annual water availability remains an obstacle to cropping patterns. As a result, the rate of soil fertility and the tillage is low and is relatively susceptible to erosion in rainy season. For that reason, it is important to seek for alternative technology so that corn farming on dry land can give an extra income for farmers increasing their family food security. The crop production should focus on increasing crop productivity through Integrated Nutrient Management (INM). INM systems more efficient using compost, farm wastes, green manures, sewage, sludge etc. Bio-fertilizer technology needs to be to be made more acceptable to farmers. Microbial strains which can compete with indigenous ones and work over a range of soil climatic conditions need to be isolated and multiplied [16].

Indeed, dry farming system has not been understood in depth while its ecosystem diversity is quite complex. The constraints pertaining to environment, socio-economic condition of society and less touch of adaptive technology cause the quality, productivity and stability of existing farming to be limited. Plant that can be cultivated on dry land is *palawija* crops, and one of the crops is corn. The corn demand both for food and non food tends to be high from year to year. Any efforts to increase corn productivity in the country have been done, including the use of improved varieties, fertilization and spacing on farming areas [17]. According to [15], double cropping pattern is one of the agricultural technology managements that can be applied to minimize the risk of dry land use in food crops development. Pest anticipation, on the other hand, is aimed at reducing risk of pest attacks and diseases in intercropping pattern. In order to avoid a plant from becoming the source of pest and disease for other intercropped plants, it is suggested to plant crops with different pest and disease.

5. Conclusion

The problem of food security in Indonesia is still happening, particularly, aspects of production and increasingly unbalanced food availability. Due to the increasing rate of population growth, land functional shift, degradation of land resources and water, as well as environmental pollution and climate change. Integrated Farming System which combines activities of food-crop farming with horticulture, animal husbandry, fisheries, forestry and other science related to farming on the same field at the same or almost the same time needs to be developed as a solution to food security problem resulting from decreasing food productivity area out of land conversion and climate change. It plays a role in suppressing land degradation and water resources and environmental pollution as well. It is an efficient and environmentally-friendly farming system utilizing the potential of local resources in optimal way for a sustainable development of agriculture. However, the development of integrated farming system needs to be followed by participatory technology development, which is based on local wisdom in society.

References

[1] Fahmid I M 2004 *Gagalnya Politik Pangan di Regim Orde Baru (Kajian Ekonomi Politik Pangan di Indonesia)* (Jakarta, Yayasan Studi Perkotaan-ISPEI)

[2] Misra A K 2014 Climate change and challenges of water and food security *Int. J. Sustain. Built Environ.* **3** (1) 153–165

[3] Cassman K G 2007 Climate change, biofuels, and global food security *Environ. Res. Lett.* **2** (1) p 1100

[4] Kemenkumham 2012 *Undang-undang Nomor 18 Tahun 2012 Tentang Pangan* (Jakarta: Kementerian Hukum dan Hak Azasi Manusia Republik Indonesia)

[5] Galhena D H, Freed R and Maredia K M 2014 Home gardens: a promising approach to enhance household food security and wellbeing *Agriculture & Food Security* **2** (8) 1–12

[6] Kang Y, Khan S and Ma X 2009 Climate change impacts on crop yield, crop water productivity and food security – A review *Progr. in Nat. Sci.* **19** 1665–1674

[7] Gangwar L S, Saran S and Kumar S 2013 Integrated Poultry-Fish Farming Systems for Sustainable Rural Livelihood Security in Kumaon Hills of Uttarakhand Agr. Econ. Res. Rev.

26 (Conference Number) 181-188

[8] Barus J 2013 Intercropping under coconuts in Lampung *J. Lahan Suboptimal* **2** (1) 68-74 (In Indonesia)

[9] Ruskandi 2005 Teknik pemupukan buatan dan kompos pada tanaman sela jagung di antara kelapa *Bul. Tek. Pert.* **10** (266) 73–77

[10] Syafruddin dan Saidah. 2006. Produktivitas jagung dengan pengaturan jarak tanam dan penjarangan tanaman pada lahan kering di Lembah Palu *J. Pen. Pert.* **25** (2) 129– 134

[11] Syafruddin 2011 Modifikasi system pertanaman jagung dan pengolahan brangkasan untuk meningkatkan pendapatan petani di lahan kering *J. Litbang Pert.* **30** (1) 16-22

[12] Obi A 2013 Integration of crops and livestock in the smallholder farming system of the former homelands of South Africa *J. Agric. Sci.* **5** (10) 183–198

[13] Sardjono M A, Djogo T and Arifin H S 2003 *Klasifikasi dan Pola Kombinasi Komponen Agroforestri* (*Bahan Ajaran Agroforestri* 2) (Bogor: World Agroforestry Centre (ICRAF) Southeast Asia Regional Office) http://www.worldagroforestry.org/sea/

[14] Ijoyah M O, Ogar A O and Ojo G O S 2013 Soybean-maize intercropping on yield and system productivity in Makurdi, Central Nigeria *Sci. J. Crop Sci.* **2** 49–55

[15] Turmudi E 2002 Study of crops growth and yield in intercropping system between maize and four soybean varieties on different time planting *J. Ilmu-Ilmu Pertanian Indonesia* **4** (2) 89–96

[16] Singh F, Kumar R and PAL S 2008 Integrated nutrient management in rice-wheat cropping system for sustainable productivity *J. Indian Soc. Soil Sci.* **56** (2) 205–208

[17] Patola E 2008 Analisis pengaruh dosis pupuk urea dan jarak tanam terhadap produktivitas jagung Hibrida P-21 (*Zea mays* L.) *J. Inov. Pertan.* **7** (1) 51–65

IOP Conf. Series: Earth and Environmental Science **157** (2018) 012062 doi:10.1088/1755-1315/157/1/012062

Role of farmer group institutions in increasing farm production and household food security

Rahmadanih, S Bulkis, M Arsyad, A Amrullah and N M Viantika

Department of Agricultural Socio-economics /Agribusiness, Faculty of Agriculture, Hasanuddin University, Jalan Perintis kemerdekaan KM 10, Makassar, 90245, Indonesia.

E-mail: *rahmadani.sosek@yahoo.com*

Abstract. Food productions and food security are still becomes serious issue in agriculture development. The research objectives are (1) Examining the role of Farmer Group in increasing farm production and (2) Analyzing the role of Farmer Group in relationship with the household food security. The research was conducted in Soppeng District, South Sulawesi, Indonesia by taking unit of analysis at household level of Farmer Group members. Sample were 6 farmer groups, consisting of 365 households. The Results showed that the Farmer Group has a good role as a production, collaboration and market unit to increase farming production (score 2.3-2.6), but haven't much impact to increase household food security. Rice production are 4.2 – 6.2 tons per hectare and household food security level are "low and marginal food secure" (mean score 4 and 6). Nevertheless, the role of Farmer Group showed a positive relationship with the level of household food security. So, it is necessarry to increase the role of Farmer Group in order to increase income and household food security based on Farmer Group's needs and potencies through facilitation of (a) seed and water/irrigation availability, (b) preparing organization rules (AD/ART) and (c) appropriate training and education of food and nutrition.

1. Introduction

Food security is still become a global issue in developing country and Indonessia is no exception. There were 47.02 million (19.04 percent) of Indonesian populations in very low food security category and 83.65 million (33.87 percent) falls into low food-security and then 116.31 million (47.09 percent) in food security category [1]. Food security is determinate of food system including food production, while the food production is resultant of various factors. One of that is the role of agricultural institutions in management of agricultural resources, including farmer institutions [2] and [3]. Therefore, it is really necessary to study on food securty in various aspects, including role of stakeholders. One of the important stakeholders is farmers group (called *Poktan*) and/or joined farmers group (*Gapoktan*). The research objectives are (1) examining the role of Farmers Group (Poktan) in increasing farm production and (2) analyzing the role of Poktan in relationship with the household food security.

2. Research Method

Design of the research was cross sectional study conducted in Manorang Salo and Panincong villages of Marioriawa subdisctrict, Soppeng District, South Sulawesi Province. The criteria of village samples were: (1) One village near from capital subdistrict and one village far from capital subdistrict. Data collection was carried out during May-August, 2016. The research was designed with qualitative and quantitative (mixed) method [4]. Qualitative data collected through Indepth-Interview and Focus Group Discussion (FGD); while quantitative data collected through household survey in Poktan level. Number of sample Poktan are 6 Groups consist of 365 households, as clearly shown in table 1.

Table 1. Sample of Regions, Poktans and Households

No	Village		Farmers Group Selected		Households
	Name	Criteria	Poktans Name	Group Gategory	
1	Manorang Salo	Near from capital subdistrict	Mappadeceng I	Poktan-1	70
			Mappadeceng	Poktan-2	65
			Mattarima	Poktan-3	65
2	Panincong	Far from capital subdistrict	Akkajangnge I &II	Poktan-1	63
			Cempae	Poktan-2	60
			Akkajangnge	Poktan-3	42

Data of household food consumption was collected through food list method for 7 days and food recall for 24 hour ago [5, 6]. The household food security level is determined through food consumption score (FCS) approaches [7]. If FCS ≥ 6 is food-secure category, which can be classified in high food security (FCS > 6) and Marginal food security (FCS=6). If FCS < 6 is food-insecure category, which can be classified in low food security (FCS 4-5) and very low food security (FCS < 4). The category of Poktan role consisting is low (score <1.0-1.6), moderate (score of 1.7-2.4) and high (score 2.5 - 3.0). The relationship between the Poktan role and household food security level were analyzed with Spearman's rho Correlations

3. Results and Discussion

3.1. The Role of Farmer Groups in Efforts to Increase Farm Production

One of the important things that can be done by the farmers group as an effort to improve the consumption of food of its members is by increasing the production of farming. Increased production can be pursued through increasing the role of Poktan. Quantitatively, the role of Poktan in carrying out its function (as a production, collaboration and market unit) in Manorang Salo Village is high (score 2.60) while poktan in Panincong Village is moderate (score 2.3-2.4) as shown in table 2.

Table 2. The average poktan's roles, production rates and farm income

No	Description	Poktan of Manorang Salo Village			Poktan of Panincong Village		
		Poktan-1	Poktan-2	Poktan-3	Poktan-1	Poktan-2	Poktan-3
1	The Role of Poktan						
	Mean Score	2.6	2.6	2.6	2.3	2.3	2.4
	Category	High	High	High	Moderate	Moderate	Moderate
2	Rice Productions (ton/hectare)	5.7	5.8	6.2	4.2	4.5	4.7
3	Total cost (IDR/hectare)	5,124.8	4,894.6	4,796.2	4,670.0	4,904.2	4,900.0
4	Income (IDR,000/ hectare)	21,335.2	23,525.4	25,583.8	15,910.00	17,047.8	18,130.0

Table 2 shows that the role of poktan in performing its function is moderate to high. Nevertheless, it has not significantly contributed to the increase of farmers' production and income. On average, rice production of Poktan in Manorang Salo village reached 5.9 tons per hectare while in Panincong Village reached 4.5 tons per hectare, lower than the district-level average production of 6.68 tons per hectare [8]. The low level of production is related to irrigation conditions that are sometimes not sufficient for farming needs. Although rice production is still low but there has increasing 10 to 25 percent compared the productions three years ago (table 3).

Table. 3 Farmers Perception to Increase Production of Farms in Marioriawa Sub District Soppeng District

No	Perception on Increased Production (compared three years ago)	Manorang Salo Village (percent)			Panincong Village (percent)		
		Poktan-1	Poktan-2	Poktan-3	Poktan-1	Poktan-2	Poktan-3
1	Not increased	0.0	0.0	0.0	0.0	0.0	0.0
2	Increased < 10%	21.4	20.0	20.0	68.3	66.7	52.4
3	Increased 10-20%	65.7	66.2	60.0	17.5	18.3	33.3
4	Increased > 20%	12.9	13.8	20.0	14.3	15.0	14.3
	Total	100.0	100.0	100.0	100.0	100.0	100.0

Implementation of the role of farmer groups can improve their activities in managing farming, both rice and crops. As an implication of that role, it can indirectly contribute to the increase of farming production managed over the last three years. The production increase achieved by Poktan members in Manorang Salo dominant village is 10-20 percent while in Panincong Village the dominant is still less than 10 percent

Differences in production levels also cause income differences. The average farm income in Manorang Salo village (IDR. 21,335,200 per hectare – IDR. 25,583,800 per hectare) is higher compared to Panincong Village (IDR. 15,910,000 – IDR. 18,130,000) per hectare. There are 73.8 -80.0 percent poktan members in Manorang Salo with an income increase of more than 20 percent and 69.8 - 73.8 percent in Panincong Village with an income increase of 10.0 - 20.0 percent

Differences in production rates and farm income are related to the different roles performed by poktan, where the average role of poktan in Manorang Salo village is also higher when compared with the role of Poktan in Panincong Village. The role of poktan referred to in this case is related to its function as (1) learning unit, (2) cooperation unit and (3) production unit according to [9].

Based on the results of indepth interviews and FGDs at the sub-district level, the different roles are related to the existence of the Joint Group of Farmers (Gapoktan) is an institution in which the farmers group becomes a larger group. Poktan in Manorang Salo village joined in a Gapoktan called Gapoktan Anugrah and Poktan in Panincong Village joined in Gapoktan Sejahtera. Gapoktan Anugerah more developed when compared with Gapoktan Sejahtera, the management is more active in accommodating the implementation of farmer group functions.

Compared to the previous three years, the role of farmer groups in Soppeng District has been growing, reflected by the development of Gapoktan over the last three years, with the final score of 586 for Gapoktan Anugerah in Manorang Salo Village and score 444 for Gapoktan Sejahtera in Panincong Village [10] . Thus, the high productivity and income of Poktan in Manorang Salo is related to the role of Gapoktan Anugerah especially in facilitating production facilities (fertilizer) and financial capital (for the procurement of pesticides and water pumps) to its members. While Gapoktan Sejahtera does not have financial capital so farmers have to make their own effort in the procurement of pesticides and water pumps, while not all farmers have capital for it. In addition, one of the most prominent forms of institutional strengthening in all Poktan belonging to Gapoktan Anugerah and not

IOP Conf. Series: Earth and Environmental Science **157** (2018) 012062 doi:10.1088/1755-1315/157/1/012062

found in Poktan in Gapoktan Sejahtera is the existence of AD / ART which has been ratified at the BP3K level of the Regency.

Based on the existing problems and potentials, institutional strengthening with [6] strategy is needed: Institutional development strategy can be done through approach (1) Models of the institutional support; (2) Models of the human capacity building in rural: and (3) Models of the institutional capacity strengthening.

3.2. Household Food Security and Poktan Role Related

Generally, food security is defined as: the condition of food fulfillment in the state up to the individual, who is reflected in the availability of quantity and quality of adequate food, safe, diverse, nutritious, equitable, and affordable and not contrary to the religion, beliefs and culture of the community, to be able to live, healthy, active, and productive in a sustainable [11]. Food security is the out put of the food and nutrition system [12]. Food diversification and food security in the household level tend associated with the local food systems [13] and [14]. Therefore, institutional role related to food and nutrition system is very important and one of them is Institutional group of farmer or group of woman farmer [15] and [16].

The role of poktan in Manorang Salo and Panincong Villages are quite good and has contributed to increasing production and income of farmer group members. Given that production and income are the variables associated with household food consumption, indirectly, the role of Poktan can also contribute to the level of household food security of their members. Household food security in this case is analyzed through Food Consumption Scores (FCS) approach consisting of five food groups namely staple food, side dishes, vegetables, fruits and milk [16]. On average, household food security levels are classified as low till marginal food-security level". The mean FCS achieved ranges 4 to 6 (tablc 4).

Table 4. Average Food Consumption Score (FCS) reached by household member of poktan

No	Food Group	FCS score of Manorang Salo			FCS score of Panincong		
		Poktan-1	Poktan-2	Poktan-3	Poktan-1	Poktan-2	Poktan-3
1	Stapple food	2.0	2.0	2.0	2.0	2.0	2.0
2	Dishes	1.0	1.0	2.0	1.0	1.0	1.0
3	Vegetable	2.0	2.0	1.0	1.0	1.0	1.0
4	Fruits	1.0	1.0	1.0	0.0	1.0	1.0
5	Milk	0.0	0.0	0.0	0.0	0.0	0.0
	Total FCS	6.0	6.0	6.0	4.0	5.0	5.0
	Food Security level	Marginal	Marginal	Marginal	Low	Low	Low

Table. 5 Farmers distribution by poktan and household food security level in Marioriawa Sub District, Soppeng District

No	FCS	Food Security Level	Manorang Salo Village (percent)			Panincong Village (percent)		
			Poktan-1	Poktan-2	Poktan-3	Poktan-1	Poktan-2	Poktan-3
1	4	Low food security	12.9	18.5	18.5	57.1	58.3	35.7
2	5	Low food security	28.6	23.1	21.5	31.7	26.7	45.2
3	6	Marginal Food security	51.4	50.8	41.5	11.1	11.7	11.9
4	7	High food security	7.1	7.7	18.5	0.0	3.3	7.1
	Total		100.0	100.0	100.0	100.0	100.0	100.0

On average, household food security level in Manorang Salo is marginal category while in Panincong Village is low category. Nevertheless, there were 7.1 - 18.5 percent of the households in Manorang Salo is classified as high food security (table 5).

Table 4 and 5 shows that in terms of institutional poktan, on average, the level of household food security among poktan members in each village is relatively the same. But in terms of location, the level of household food security of Poktan in Manorang Salo village is better compared to poktan households in Panincong Village. The condition is related where Poktan's located. Manorang Salo is a village adjacent to the capital sub-district. This means that the village is relatively more accessible to the empowerment programs on the one hand (such as counseling/ informal educations of agriculture and food and nutrition) and on the other hand the villagers are more easily accessible to the food available in the sub-district market when compared to the people in Panincong Village. In addition, the production rate and farm income at the three poktan in Manorang Salo village is higher compared to the three poktan in Panincong Village. Because of differences in production and income of Poktan between two villages related to differences in the role of poktan, indirectly the role of poktan is also related to the level of food security of poktan households (table 6).

Table 6. Relationship between poktan role and household food security level in Marioriawa Sub District, Soppeng District

No	Village	Group	Spearman's rho Correlations: Role of Poktan and Food Security		
			Correlation Coefficient	Sig. (2-tailed)	N
1	Manorang Salo	Poktan-1	0.267*	0.026	70
		Poktan-2	0.314*	0.011	65
		Poktan-3	0.335**	0.006	65
2	Panincong	Poktan-1	0.172	0.177	63
		Poktan-2	0.304*	0.018	60
		Poktan-3	0.426**	0.005	42

In table 6, it is seen that the role of poktan is positively related to the level of household food security both in the three Poktans in Manorang Salo Village and the three Poktans in Panincong Village. However, the role of poktan-1 in Panincong Village shows a weak relationship with the household food security levels ($r = 0.172$ and Sig. (2-tailed) = 0.177). The weakness of this relationship is also related to the level of production and income is relatively low when compared with the other five Poktans. The low production is mainly due to the inadequate conditions of irrigation water and the limited use of production facilities. No funds (group cash) can be borrowed for procurement of production facilities. The group does not yet have an organization rules (AD/ART). In addition, in the last two years, there has been no empowerment activities particularly related to the increase of food and nutrition knowledge.

4. Conclusion

In general, poktan institutes have performed their good role in an effort to increase farm production and income. Nevertheless, the role of poktan in Manorang Salo Village is higher when compared with Poktan role in Panincong Village. The low role of poktan in Panincong Village is related to the low role of Gapoktan Institution in facilitating poktan, especially in the provision of financial capital for the procurement of production facilities (seeds and pesticides) and agricultural equipment (water pumps), as well as its AD/ART. The differences in the role of the institute also resulted in differences in farming production and household income of farmer group members and ultimately impacted the differences in food security at the household level of farmer group members. Household food security level of farmer group member in Manorang Salo Village is better compared to Panincong Village. More than 58.5 percent of households in Manorang Salo village and just over 11.1 percent of households in Panincong Village are classified as food security. Nevertheless, on average, the level of food security of farmer group members is at "low till marginal food secure" level. There is no difference in the level of household food security between farmer groups in one village, either in Manorang Salo Village or Panincong Village. In general, the level of household food security is positively related to the role of poktan institutional in an effort to increase the production of farming; both poktan in Manorang Salo Village and Poktan in Panincong Village. This means that the higher the role of poktan in increasing production, the better farming will be and in turn making better level of food security. Based on the problems and potentials in the Poktan environment in the two villages, the alternative strategic to increase the role of Poktan based on potency can be recommended to develop farming production and food security include (a) institutional support model (facilitation of seed and water/irrigation availability), (b) human resources capacity building model (appropriate training and education of food and nutrition) and (c) institutional capacity building model (facilitation of preparing organization rules.

References

[1] Kementerian Pertanian RI 2015 *Rencana Strategis Kementerian Pertanian Tahun 2015*-2019 [in Indonesian] (Jakarta: Kementerian Pertanian RI)
[2] Dimyati A 2007 *Pembinaan Petani dan Kelembagaan Petani* [in Indonesian] (Tlekung-Batu: Balai Penelitian Tanaman Jeruk dan Buah Subtropika)
[3] Darma R, Zain A M M and Amandaria R 2012 Zakat, local social organization, and social capital in rural economic development *Socio. Study* **2** 189-192
[4] Anantanyu S 2011 Kelembagaan petani: peran dan strategi pengembangan kapasitasnya [in Indonesian] *J. SEPA* **7** (2) 102-109
[5] Jelliffe D B and Jelliffe E F P 1989 *Community Nutritional Assesment* (New York: Oxford University Press)
[6] Suhardjo 1989 *Survey Konsumsi Pangan* [in indonesian] (Bogor: Departemen P dan K DIKTI. PAU Pangan dan Gizi IPB)
[7] Hardinsyah, Briawan D, Madanijah S, Dwiriani C M, Atmojo A M and Heryatno Y 1998 *Kajian Kelembagaan untuk Pemantauan Ketahanan Pangan* [in Indonesian] (Jakarta: Kerjasama Pusat Studi Kebijakan Pangan dan Gizi dengan Unicef dan Biro Perencanaan Deptan)
[8] BPS Kabupaten Soppeng, 2016 *Kabupaten Soppeng dalam Angka Tahun 2015* [in Indonesian] (Soppeng: BPS Kabupaten Soppeng Provinsi Sulawesi Selatan)
[9] Kementerian Pertanian RI 2015 *Rencana Strategis Kementerian Pertanian Tahun 2015-2019* [in Indonesian] (Jakarta: Kementerian Pertanian RI)
[10] Bulkis S, Nadja R A, Rahmadanih and Fahmid M 2015 *Development of Poor Household Food Security through Institutional Strengthening and Social Capital at Jeneponto District* (*Proc. International Conference on Social Science and Management)* Kyoto, Japan
[11] Undang-Undang Republik Indonesia Nomor 18 Tahun 2012 tentang Pangan www.hukumonline.com Accessed August 18 2017
[12] Maxwell S and Frankenberger T R 1992 *Husehold Food Security*: *Concepts, Indicators, Measurements* (Rome: UNICEF and IFAD)
[13] Donkers H 2013 Governance for local and regional food systems *J. of Rur. and Comm. Dev.* **8** 178-208
[14] Darma R, Rahmadanih, Tenriawaru A N and Amandari R 2014 Climate change adaptation in the perspectives of food energy crisis and environmental degradation for food sovereignty in indonesia *Int. J. of Agr. Sys.* **2** 77-89
[15] Rahmadanih, Bulkis S, Amrullah A, Rukka R M and M Arsyad 2015 Strengthening institutional model of women-farmers group in developing household food diversification *Int. J. of Agr.*

IOP Conf. Series: Earth and Environmental Science **157** (2018) 012062 doi:10.1088/1755-1315/157/1/012062

Sys. **3** 29-40
[16] Bulkis S, Rahmadanih and Summasse I 2015 *Penguatan Kelembagaan Petani dalam Upaya Meningkatkan Produksi dan Pendapatan Rumahtangga (Research Report)* (Makassar: Fakultas Pertanian Universitas Hasanudidin)

IOP Conf. Series: Earth and Environmental Science **157** (2018) 012063 doi:10.1088/1755-1315/157/1/012063

Factors affecting sustainable dairy production: A case study from Uva Province of Sri Lanka

D Wijethilaka[1], S De Silva[2], R M C Deshapriya[3] and L H P Gunaratne[4]

[1] Postgraduate Institute of Agriculture, University of Peradeniya, Galaha Road, Peradeniya, 20400, Sri Lanka.
[2] Faculty of Economics, Saga University, Honjo 1, Saga-City, Saga-Prefecture, 840-8502 Japan.
[3] Dept. of Animal Science, Faculty of Agriculture, University of Peradeniya, Galaha Road, Peradeniya, 20400, Sri Lanka
[4] Dept. of Agricultural Economics & Business Management, University of Peradeniya, Galaha Road, Peradeniya, 20400, Sri Lanka.

E-mail:. *dedunuwije@gmail.com*

Abstract. Dairy farming has been playing a key role by improving household incomes and food security for rural communities in Sri Lanka. Nevertheless, it has failed to meet the expected self-sufficiency. In 2015, Sri Lanka imported 51percent of the national milk requirement spending US$ 251 million from its debt-ridden economy. This paper aims to analyse socio-economic characteristics of dairy farmers and factors affecting dairy production efficiency in the Uva Province of Sri Lanka, a highly potential area comprising all the dairy value chain actors. Surveyed was conducted to farmers, key informants from input suppliers, collectors, transporters, processors, sellers and support service providers. Result revealed that intensive farmer's milk yields per cow was only 7.97 L/day, which was 35% and 60% higher than the yields of semi-intensive and extensive farmers respectively. The highest profit of Rs. 53.30 per litre was earned by extensive farmers, whereas it was Rs. 47.63 for semi-intensive and Rs. 44.76 for intensive farmers respectively if family labour cost was not taken into the account. The Technical Efficiency Analysis revealed that 37.1% and 20% milk production of intensive farmers and semi-intensive is being loss due to inefficiency and could be increased without any additional inputs. The main factors affecting efficiency in milk production included farmers' socio-economic characteristics and farm characteristics. Based on the results it can be concluded that sustainability dairy production depends on farmer training, collectivizing farmers into farmer societies, culling unproductive male animals, increasing the availability and access to AI/ other breading programs and low-cost quality concentrate feed and other supplements, and, thus appropriate measures should be taken to provide these conditions if Sri Lanka aims to achieve self-sufficiency in milk production.

1. Introduction

Over the years dairy production has been playing a key role by providing livelihoods for rural communities and supplying nutrition to Sri Lankan consumers. Even though dairy farming has been playing a key role by improving household incomes and food security for rural communities it has failed to meet the expected self-sufficiency. In 2015, Sri Lanka imported 51 per cent of the national milk requirement spending US$ 251 million from its debt-ridden economy.

Moreover, Sri Lanka has a great potential to increase its milk production, while securing substantial income for rural milk producers. As stated in the report of Department of Census and Statistics in Sri Lanka, in 2015 the national cattle population was about 1.4 million while buffalo population was about 0.5 million with an estimated 540,100 milking cows and 147,190 milking buffaloes. In this background the study focused on determining factors affecting dairy production efficiency in the Uva Province, one of the nine provinces in Sri Lanka, in order to understand the prevailing causes for current situation in dairy sector and to formulate strategies and recommendations for the future development in the diary sector

2. Methodology

For the purpose of this research the Uva province consisting two districts viz. Badulla and Monaragala was selected purposively considering the relative importance of this province. The two districts presently contribute about 14percent to the total milk production and holds 7.3percent of the cattle populations in Sri Lanka. This province can potentially be developed as successful cases of dairy sector in Sri Lanka as it encompasses all the value chain actors including a large scale commercial dairy plant with a processing capacity of 200,000 litres of milk per day, and many other small and medium dairy processors, collectors, and transporters.

The data for the study were gathered from three forms: 1). field questionnaire survey, 2) key informant interviews and focus group discussions and 3) review of the published and unpublished documents. For the field survey, data were collected primarily through interviews using a semi-structured questionnaire. A total of 300 questionnaires were filled and three questionnaires had to be discarded due to incomplete and non-reliable data. Interviews were conducted with three large milk collectors and one large scale milk processor and one medium scale processor. The field survey was conducted during the period of October 2014 to May 2015.

A total sample of 297 farmers, key informants from input suppliers, collectors, transporters, processors, sellers and support service providers were surveyed, while both quantitative and qualitative analyses were used to draw conclusions. The sample comprised of 278 (93.6%) small scale (up to 10 animals), and 19 (6.4 %) medium to large scale (range from 11 to 60 animals) farmers. Of them, 69 (23.23%), 216 (72.72%) and 12 (4.04%) were intensive, semi-intensive and extensive farmers. Majority of the farmers were men (69.4%), having a mean age of 47 years (range from 21 to 78 years) and with low education standards as most of them (60.9%) have completed only primary level. Most of the farmers (91.24%) owned European crossbred animals, while the herd comprised of 13 percent heifers, 40 percent lactating cows and 11 percent dry cows.

2.1 Technical efficiency

Stochastic frontier approach is widely used in agricultural economics studies [1] because of its ability to estimate technical efficiency while accounting for technical inefficiency factors

Stochastic frontier function is expressed as;

$$\gamma_i = f(x_i\beta) + \varepsilon_i$$
$$\varepsilon_i = v_i - u_i \qquad I = 1,2,3 \ldots\ldots N \qquad (1)$$

Where, γ_i is production of the ith farm, x_i is vector of inputs of production of the ith farm, β is a vector of unknown parameter to be estimated. ε_i is the composed error term consist two independent error terms u and v. v_i is independently and identically distributed random errors of ith producer for all possible random variation in output due to factors outside farmer's control such as weather, animal diseases, and price increases and u_i is a non-negative error term denoting inefficiency of the ith producer [2]. Technical efficiency is estimated using inferential statistical approach. Frontier 4.1 is used to estimate stochastic frontier and technical inefficiency simultaneously.

The explicit Cobb-Douglas stochastic frontier production function is defined in logarithmic form as follows,

IOP Conf. Series: Earth and Environmental Science **157** (2018) 012063 doi:10.1088/1755-1315/157/1/012063

$$In\,Y = \beta_0 + \beta_1 InX_1 + \beta_2 InX_2 + \beta_3 InX_3 + \beta_4 InX_4 + V_i - U_i \quad (2)$$

Where;
Y = milk production in each farm (litres)
X_1 = total number of milking cows (nos)
X_2 = average monthly cost for concentrates (Rs)
X_3 = average monthly cost for supplements (Rs)
X_4 = family labour hours

The model for the technical inefficiency effects in the stochastic frontier of equation (2) is defined as

$$Ui = \delta_0 + \delta_1 + \delta_2 + \delta_3 + \delta_4 + \delta_5 + \delta_6 + \delta_7 + \delta_8 + W_i \quad (3)$$

Where, δ_1 = District, δ_2 = Gender, δ_3 =Age, $\delta 4$ = Education, δ_5 = Member of farmer organization, δ_6 = Main income dairy farming, δ_7 = Main income other than dairy farming, δ_8 = Availability of European cows

Farmer's monthly income, expenditure and profit are analysed to determine the financial status of the farmers. Total income of each farm is calculated based on their income components from dairy activities as well as other sources. However, in this study income from the sale of animals is not taken into consideration as it is repugnant to religious norms [3, 4] and farmer's profit is calculated by excluding family labour as well as exclusive of opportunity cost of family labour.

3. Results and Discussion

3.1. General Characteristics of the Farmers

Out of the total sample of 297 farmers 278 were small-scale farmers (93.6%) and the rest consists of 19 medium/large scale farmers (6.4%). About 70 per cent of the milk producers were men and out of the large scale farmers, the majority of the farmers (84.2%) were men. Age of respondent was range from 21 years to 78 years with mean age of 47.26 ±0.61 years. The large scale farmers were significantly younger than the small scale farmers and this indicates that expansion of scale of the herd will pose difficulty when farmers are getting old. Only three farmers of the sample were illiterate and others have received some school education. However 61.3percent of the farmer's education was limited to primary level of education. It is noteworthy that the majority of small scale farmers have only primary education, while medium and large scale farmers' education were significantly better (χ^2=12.036, P=0.002).

The 297 farmer households in the sample have 1,729 heads of cattle and on the average there were five animals per small scale farms and 21 animals per medium/ large scale farms. Only about 7 per cent of the farmers had more than 10 cattle. There are three types of dairy management systems exist in the area namely; 1) intensive (zero-grazing), 2) semi-intensive (semi-grazing) and 3) extensive (open range grazing). Intensive management system uses zero grazing that is characterized by cut and fed in a shed, farm practices of keeping high yielding cows (usually European crosses) and they are fed with concentrated feed. Extensive management system put to use the open range for free grazing by cattle, often without supplementary feeds. Semi-intensive falls somewhere in the middle and involves a combination of above two approaches

The cattle breeds found in the area belong to crosses of European breeds, Indian breeds and local. European breeds were mainly Jersey crosses or Frisian crosses (74.94 %) and local breed were only 6.9 per cent.

3.2. Milk Producer's Income, Expenditure and Profit

There are different sources of income, obtained from both in farm and off-farm and family members often engaged themselves in more than one occupation. Total farmer expenditure includes expenses associated with farm level activities such as feed expenses, breeding and animal health expenses and cost for labour inclusive of family labour. In calculating opportunity cost for family labour, it is found to be the highest cost in all the three farming systems. Other studies also revealed results similar to results from calculating family labour. Even so, in reality farmers do not spend any money on family labour. Dairy farmers extensively use family labour for farming activities and do not use family labour in any other productive manner.

As several studies have shown, if family labour is excluded, the highest cost is incurred on the concentrate and supplement feeds [5, 6]. Current study also found that the highest variable cost in the dairy farm is incurred on purchase of concentrates and supplement feeds, if family labour is not taken into the account. Naturally, it is the highest expenditure incurred in intensive farming system and the lowest in extensive farming system. The results show that intensive system farmers spent five times more on concentrate feeds than on extensive system. As stated in Ranawana [7] farmers are compelled to use expensive concentrates and this could make dairy farming uneconomical. According to him this is deemed as the main drawback regarding increasing milk production in the country [7].

Profit margin per litre of milk is the lowest for intensive farmers without considering family labour, while extensive farmers are getting the highest profit per litre (table 1). Besides, excluding family labour, intensive farmers are still the lowest and extensive farmer's profit margin is 1.19 higher than that of the intensive farmers (P=0.045). Including family labour, the profit margin per litre on extensive farming it is 4.6 times higher than that of intensive farming.

Whilst, family labour is excluded; all three farming systems became viable and having reasonable profit margins. Therefore region has potential to increase milk productivity and efficiency to be profitable industry for the farmers if address inefficiencies properly.

Table 1. Monthly profit of dairy farmers in selected area of Uva province.

	Description	Intensive System	Semi Intensive System	Extensive System
Profit per month (Rs.) (Milk income –milk expenditure)	Without labour cost	17,993.00	16,938.00	21,522.00
	With labour cost	2,722.00	3,759.00	10,756.35
Profit per litre of milk (Rs.)	Without labour cost	44.76	47.63	53.30
	With labour cost	8.16	22.77	37.12
Dairy income–total expenditure (Rs./month)	Without labour cost	-1,231.00	346.00	7,035.00
	With labour cost	-16,502.00	-12,833.00	-3,731.00
Total income– total expenditure (Rs./month)	Without labour cost	7643.00	8,982.00	12,343
	With labour cost	-7628.00	-4.197.00	1,577.00

3.3. Technical Efficiency

Technical efficiency model was estimated only for the intensive and semi-intensive management systems farmers and didn't estimate for the extensive system farmers due to less number of farmers in the sample. Based on the estimated results, it is revealed that that technical inefficiencies do exists in intensive and semi-intensive management systems in Uva province (table 2 and table 3). The mean technical efficiency of intensive farmers is estimated as 62.9 per cent and this implies indirectly that 37.1 per cent of milk production is being lost due to inefficiency. Similarly 20 per cent of productions are being lost in semi-intensive systems due to inefficiency of production techniques.

Estimation of TE suggests that milking cows are significance for intensive and semi-intensive management systems at 1 per cent significant level with the values of 0.621and 0.668 respectively. Concentrate feed cost and supplement cost are also significant for intensive and semi intensive farmers at 1 per cent significance and having positive relationship with milk production. Although concentrate feed and supplement feeds are important, their contributions towards to increase of milk production remain considerably at a low level.

However, these findings were found to be in line with the results of a study Masuku *et al.* [8] in their study on economic efficiency of smallholder dairy farmers in Swaziland and Wubeneh and Ehui [9] in their study on technical efficiency of smallholder dairy farmers in the central Ethiopian Highlands. These studies have revealed that concentrate feed, chemicals, herd size and labour cost significantly affect the economic efficiency of dairy farmers.

Table 2. Model estimate for intensive farmers

Variable		Coefficient	Standard error	t-ratio
Input Variable	Constant	4.9279	0.6745	7.3056***
	Milking cows	0.6211	0.0298	17.5108***
	Feed cost	0.0378	0.0086	4.3915***
	Supplement cost	0.0455	0.0036	12.5518***
	Other costs	-0.0098	0.0145	-0.6816
	Labour hours	0.2197	0.1293	1.6985*
Inefficiency Variables	Constant	1.5749	0.8742	1.8014
	District	-0.0597	0.3089	-0.1931
	Gender	-0.1746	0.1591	-1.0972
	Age	-0.0104	0.0078	-1.3299
	Education	-0.1149	0.1339	-0.8582
	Membership milk society	-0.0273	0.1025	-2.6609***
	Main income dairy	-0.1628	0.1842	-1.8358*
	Main income other	0.3504	0.1757	1.9943**
	Availability of European cows	-0.9823	0.4115	-2.3878**

$\gamma = \sigma_u^2/\sigma^2$: 0.8228; MT Efficiency : 0.8054; Sum of regression coefficients: 0.9425
(***), (**), (*) denote levels of significance at one, five and ten per cent respectively

Table 3. Model estimate for semi-intensive farmers

Variable		Coefficient	Standard error	t-ratio
Input Variable	Constant	4.7120	0.5282	8.9215***
	Milking cows	0.6684	0.0566	8.2731***
	Feed cost	0.02979	0.0107	2.7716***
	Supplement cost	0.0208	0.0107	1.9469**
	Other costs	0.0044	0.0097	0.4508
	Labour hours	0.2191	0.0722	3.0350***
Inefficiency Variables	Constant	-1.4343	1.9698	-0.7282
	District	-0.0403	0.1763	-0.2286
	Gender	-0.2928	0.3509	-0.8345
	Age	0.0092	0.0087	1.0579
	Education	-0.1565	0.1154	-1.3561
	Membership milk society	-0.0951	0.1997	-0.4763
	Main income dairy	-0.3511	0.3077	-1.1408
	Main income other	0.3617	0.4429	0.8166
	Availability of European cows	-0.4311	0.3496	-1.6632**

$\gamma = \sigma_u^2/\sigma^2$: 0.6267 MT Efficiency : 0.8054 Sum of regression coefficients: 0.9425
(***), (**), (*) denote levels of significance at one, five and ten per cent respectively

While considering inefficiency variables membership of milk society, availability of European cows and dairy farming as main source of income are significant determinants of efficiency in milk production for the intensive farmers and the only significant variable for the semi- intensive farmers is availability of European cows at five per cent significant level.

The main income being dairy, it is positively significant and increases efficiency by 0.163 for intensive farmers while the main income other than dairy will reduce the efficiency of milk production by 0.350 for intensive farmers. Availability of European cows increase efficiency for all management systems and for intensive farmers it is 0.982 and it is 0.431 for semi-extensive farmers respectively (table 3).

4. Conclusion

The paper elaborated present situation of the dairy farming sector in the Uva Province of Sri Lanka, measurement of productive efficiency at farm level, and profitability of milk production in order to address opportunities to increase milk production.

It was revealed that higher proportions of lactating cows, expenses on concentrates and supplements and spending more labour hours in farm activities increase the efficiencies while availability of European crossbred animals, higher education levels, male farmers, full-time dairy farming and being a member of farming society decrease occurrence of the inefficiencies. Although milk yield per cow and the income per cow are relatively high for the intensive farmers, profit margins are lower for the intensive farmers, particularly due to high cost for concentrate feeds. Profit was the highest for the extensive farmers due to low cost for concentrate feeds, low labour, feeding of different forage types and handling a comparatively large number of animals. Although extensive farming is profitable there exist limitations for expansion of the extensive farming due to land scarcity, current applicable laws prohibiting straying of animals and usage of public places for free grazing, high probability of spread diseases etc.

Moreover, the region has the potential to increase milk productivity and efficiency to be profitable to farmer in the industry if farmer address inefficiencies properly and if the responsible authorities in livestock sector properly address findings by proper identification the UVA province would be an ideal model to be followed in other provinces of the country to achieve the goal of self-sufficiency in milk production.

References

[1] Haider M Z, Ahmed S and Mallick A 2011 Technical efficiency of agricultural farms in khulna, bangladesh: stochastic frontier approach *Int. J. of Econ. and Finance* **3** [on line] Available at: http://dx.doi.org/10.5539/ijef.v3n3p 248 Accessed on 12.03.2016

[2] Aigner D J, Lovell C A K and Schmidt P J 1977 Foundation and estimation of stochastic frontier production function models *J. of Econometrics* **6** 21-37

[3] Vitanage U Y N 2014 *Technical Efficiency and Socio-economic Status in Animal-crop Mixed Farming Systems in Dry Lowland Sri Lanka* M Phil Thesis (Sri Lanka: Postgraduate Institute of Agriculture, University of Peradeniya, Sri Lanka)

[4] Vernooij A, Houwers W and Zijlstra J 2015 *Old friends - New trends; Emerging business opportunities in the dairy sector of Sri Lanka Livestock Research Report 831* (Lelystad, Wageningen UR (University & Research centre) Livestock Research)

[5] Rathnayaka D 1992 *Empirical identification of eco-specific dairy farming systems in Sri Lanka and estimation of their economic performance* M Phil Thesis (Sri Lanka: Postgraduate Institute of Agriculture, University of Peradeniya)

[6] Mahipala M B P and Gunarathne L H P 2003 Comparison of Cattle Farming Systems in Upcountry Wet Zone: Cattle management, Social structure and Involvement in Cattle Farming *Sri Lankan J. of Animal Prod.* **2** (1) 41-55

[7] Ranawana S 2008 Dairy industry in Sri Lanka: Problems and prospects *Economic Review* **34** 2-5

[8] Masuku B B 2014 Economic efficiency of smallholder dairy farmers in Swaziland: An application of the profit fuction *J. of Agr. Studies* **2** (2)

IOP Conf. Series: Earth and Environmental Science **157** (2018) 012063 doi:10.1088/1755-1315/157/1/012063

[9] Wubeneh N and Ehui S 2006 Technical efficiency of smallholder dairy farmers in the central Ethiopian highlands *Int. Assoc. of Agr. Econ. Conf.* (Gold Coast, Australia)

Contribution and efficiency of labor allocation analysis of income in household industry using raw material of agricultural commodity in South Sulawesi.

A N Tenriawaru, Mahyuddin, M H Jamil, L Fudjaja and S Nurbaya

Department of Agriculture Socio Economic, Faculty of Agriculture, Hasanuddin University, Jalan Perintis Kemerdekaan KM 10, Makassar 90245, Indonesia.

E-mail: *nixia_gany@yahoo.com*

Abstract. In South Sulawesi, various home industry businesses have grown. This industry is actually the basis of community livelihoods that need to be developed and nurtured by the government so family income get increased and the absorption of workers will improve the regional economy in general. The purpose of this study is to analyse the contribution of income, and efficiency of labour allocation in household industries made from raw agricultural commodities. The method of determining the respondents is done by direct appointment (purposive) on the industry players made from raw agricultural commodities. The type of research is quantitative descriptive and data are analysed using income analysis, cost analysis, income contribution analysis, Working Day (HOK) analysis and efficiency analysis of labour allocation. The results showed that the average income earned per year ranged from IDR. 16,866,867.- up to IDR. 125,271,500.-. There are 2 industries that have high contribution to family income such as banana chips industry and rice milling industry with value of 96.3% and 68.7% respectively. In the meantime, there are 5 industries with high average labour allocation efficiency of IDR. 218,135.- / HOK per day and above the efficiency standard of labour allocation based on UMR in South Sulawesi Province.

1. Introduction

The micro, small and medium industries have proved to be the most resistant industry groups in the face of the economic crisis, both during the crisis of the late 1990s and the worldwide crisis in late 2008 to early 2009. In the period 1997 to 2000 contributions Small and Medium Micro Enterprises (MSMEs) are more than 46% in Indonesia's GDP. Furthermore, in 2007 Indonesia's GDP grew by 6.28%, where at the time, small and medium enterprises grew by 18.76% and specialized in the small and medium scale manufacturing sector had a growth of 16.29%, well above average GDP growth in Indonesia [1].

Besides contributing to GDP growth, small and medium micro industries have a role to play in employment and become a source of income for the wider community. For example, according to Soeratno [1], the absorption of manpower in the sector of Small Industry, Handicraft and Household (IKKR) is based on 23 Industry Classification of Indonesian Industrial Industries (KBLI) for more than 8 million workers which, when compared to manufacturing industry group it only absorbs about 4.5 million

IOP Conf. Series: Earth and Environmental Science **157** (2018) 012064 doi:10.1088/1755-1315/157/1/012064

workers. This condition further reinforces the role of small and medium industries in moving the Indonesian economy.

The same thing is shown in South Sulawesi Province, where household industries made from raw agricultural commodities have developed various businesses such as; business of making brown sugar, food processing industries and others scattered in several cities / districts. In general, these efforts become a source of livelihood of the community that continues to be developed along with government support by providing various policy facilities to encourage the role of society in the economic field. Bantaeng Regency is one of the districts in South Sulawesi which is one of the areas of small and home industry development (IKRT) which can be seen in table 1.

Table 1. Number of Household Industries (Micro), Small and Medium Based on Division District Area in Bantaeng Regency, South Sulawesi, 2012 [2]

No.	Sub-District	Number of Industries		
		Household (Micro)	Micro	Medium
1	Bantaeng	59	27	-
2	Bissappu	213	44	1
3	Eremerasa	32	2	1
4	Sinoa	7	-	-
5	Ulu Ere	6	1	-
6	Gantarangkeke	29	3	-
7	Tompobulu	17	6	-
8	Pajukukang	20	16	1
	Amount	383	99	3

Bissappu Sub-District (table 1) has the largest number of home industries and most of these industries are made from agricultural commodities. This is a consideration for researchers to choose the area purposively as a research area. Household industries made from raw agricultural commodities are generally developed by farmers, farmers' wives and other family members on the grounds that in addition to adding value to the farm products they have been producing, it also has the potential to become an additional source of family income including expanding employment opportunities. According to [2], the type of household industry made from raw agricultural commodities developed in Bantaeng Regency to date is; compost making industry, rice milling, corn flour milling, red sugar industry, banana chips and seaweed food.

Base on the various classical problems faced by the perpetrators of industries made from raw agricultural commodities such as limited business capital, marketing and product quality is still low, apparently until now the industry still survive and still be a mainstay for business actors in supporting the family economy. Based on the phenomenon, the researcher is interested to study how much income earned by the industrial business actors and the large contribution of the income to the family income and efficiency of family labor allocation in the industry made from raw agricultural commodities.

2. Method

The research was conducted in Bissappu Sub-District Bantaeng Regency of South Sulawesi Province, purposively chosen with the consideration that this sub-district has a varied household industry and generally made from agricultural commodities as well as being a population in this study (table 2).

Table 2. Population and Sample of Household Industries made from Raw Agricultural Commodity in Bissappu District, Bantaeng Regency, 2013

No.	Household Industries using Raw Material of Agricultural Commodities	Population	Sample
1	Compost	1	1
2	Rice Milling	16	2
3	Corn Flour Milling	18	2
4	Brown Sugar	64	6
5	Seaweed Food	1	1
6	Banana Crackers	2	2
	Amount	102	14

The types of data are secondary data (obtained from library studies and related institutions, such as BPS-Statistics of Bantaeng Regency, Trade and Energy Industry Office of Bantaeng Regency, and Bissappu District Office of Bantaeng Regency) and primary data (collected by field observation followed by interviews on respondents using questionnaires systematically). The data collected, then analyzed using quantitative descriptive analysis to confirm the analysis based on research objectives.

The income analysis, is done by formulation;

$$\pi = TR\text{-}TC = (\Sigma Y.Hy) - (FC + VC) \tag{1}$$

where; π = Revenue (Rp),
TR = Total Reception / Revenue (Rp),
TC = Total Cost (Rp),
ΣY = Total Production (Kg),
Hy = Production Price (Rp),
FC = Total Fixed Cost (Rp) and
VC = Total Variable Cost (Rp).

The results of income calculation are then categorized as referring to the Regional Minimum Wage (UMR) standard of South Sulawesi Province of Rp.1.400.000, - per month. If > from UMR is categorized as large, and if < from UMR is categorized as small.

Analysis of the contribution of household income [3], from agricultural raw materials to family income, using the formula;

$$K = (P1 / PtRt) \times 100\% \tag{2}$$

where; K = Contribution of household industry income (%),
P1 = Net income of home industry ladder (Rp) and
PtRt = Total household income (Rp).

Contribution rate category is low contribution = 0% - 33.3%, medium = 33.4% - 66.6% and high = 66.7% - 100% [4].

Analysis of labor allocation of family household industries [5], made from raw agricultural commodities, using the formulation;

$$Jktotal = JO \times HK \times JK \text{ and } HOK = JKtotal / JKS \tag{3}$$

IOP Conf. Series: Earth and Environmental Science **157** (2018) 012064 doi:10.1088/1755-1315/157/1/012064

where; HOK = Day of worker (Weekdays),
JO = Number of persons Person),
HK = Day of work (Day),
JK = Working hours (Hours) and
JKS = Standard working hours (8 Hours).

Furthermore, the efficiency level of family labor allocation is used formulation: Efficiency Labor Allocation's formula;

$$\text{IRT Acceptance / IRT Labor Allocation} \quad (4)$$

The efficiency level indicator of labor allocation of household industries made from raw agricultural commodities is determined by comparing the efficiency standard of labor allocation based on the Regional Minimum Wage (UMR) of South Sulawesi Province of 53.846 26 Working Days. If > from 53.846 26 Working Days = High Efficiency and if < 53.846 26 Working Days = Low Efficiency.

3. Results and Discussion

3.1. *Industrial Household Income using Raw Material of Agricultural Commodities*

Household income is income earned by the household concerned either from the basic work income or income received from the side job. Source of income received by respondents in each type of home industry made from raw agricultural commodities varied considerably. Sources of income are obtained from within the household industry made from raw agricultural commodities and some are from outside the home industry (non-industrial). The more sources of family income the greater the income earned by households. Rural and non-industrial household income as shown in table 3.

Table 3. Average Revenue of Household Industry Using Raw Material of Agricultural Commodities Sourced of Industrial and Non-Industrial Income in Bissappu District, Bantaeng Regency, 2013

No	Industries	Household Income of Industrial (IDR)	Household Income of Non-Industrial (IDR)	Total Income (IDR)
1	Corn Flour Milling	110,702,375	29,250,000	139,952,375
2	Brown Sugar	16,566,867	9,766,666	26,633,533
3	Seaweed Food	4,675,000	16,800,000	21,475,000
4	Banana Crackers	125,271,500	4,750,000	130.021.500
5	Compost	47,458,500	73,000,000	120,458,500
6	Rice Milling	52,662,750	22,850,000	75,512,750

Corn flour milling industry, the brown sugar industry, banana chips industry, compost industry and rice milling industry provide average net income of Rp 16,866,867 up to Rp 125,271,500 per year, while the seaweed food industry provides a small net income with a value of Rp 4,675,000 per year. In general, it appears that the income of the home industry is greater than that of non-domestic industries. The same thing is shown in the results of the study [6] which states that the greatest incomes of farmers generally come from agriculture (on-farm) in both raw and processed forms.

3.2. Contribution of Industrial Income using Raw Material of Agricultural Commodities to Family Income

Revenue contribution is a donation received in return for members of the working household [7]. The contribution of household income made from agricultural commodities in Bissappu District is calculated based on the comparison between household income from household income from agricultural commodity with total household income multiplied by 100%. The total income of the household is determined by the income from the household income sector made from raw agricultural commodities and income outside the sector of household industry income. The contribution of households' incomes using raw materials from agricultural commodities is shown in table 4.

There are two industries that have high income contribution level, namely banana chips industry and rice milling industry with contribution value of 96.3% and 68.7% respectively. Three industries with medium income contribution, corn flour milling industry, red sugar industry and compost industry with 66.53%, 65.6% and 39.4% contribution respectively and one industry with contribution level low income, namely seaweed food industry with a contribution value of 21.76%. The results [8] show similarly that the management of domestic corn chip industry can contribute to household income of 71.43%, while female labor can contribute to household income of 8.43%.

Table 4. Average Revenue Contribution of Household Industry Using Raw Material of Agricultural Commodities Sourced from Industry and Non-Industry Revenues To Household Income at Bissappu District, Bantaeng Regency, 2013.

No	Industries	Household Income of Industrial (IDR)	Contribution To Total Household's Income (%)	Household Income of Non-Industrial (IDR)	Contribution To Total Household's Income (%)	Total Household's Income (IDR)
1	Corn Flour Milling	110,702,375	66.53	29,250,000	32.97	139,952,375
2	Brown Sugar	16,866,867	65.60	9,766,666	34.40	26,633,533
3	Seaweed Food	4,675,000	21.76	16,800,000	78.24	21,475,000
4	Banana Crackers	125,271,500	96.30	4,750,000	3.70	130,021,500
5	Compost	47,458,500	39.40	73,000,000	60.60	120,458,500
6	Rice Milling	52,662,750	68.70	22,850,000	31.30	75,512,750

3.3. Analysis of Labor Allocation and Efficiency of Labor Allocation in Home Industry using Raw Material Agricultural Commodities.

An analysis of the efficiency of the allocation of domestic labor, preceded by knowing the amount of labor allocation and net income of the industry. Clearly the efficiency of labor allocation of agricultural commodity-based industries in Bissappu District, Bantaeng Regency can be seen in table 5.

Table 5. Average Labor Allocation and Efficiency of Household Industry Domestic Allocation Using Raw Materials of Agricultural Commodities in Bissappu District, Bantaeng Regency, 2013

No	Industries	Household Income of Industrial (IDR)	Labor Allocation (HOK/Year)	Efficiency of Labor Allocation on Household Industrial (IDR/HOK)
1	Compost	110,702,375	544.125	203,450
2	Rice Milling	16,866,867	321.30	52,495
3	Corn Flour Milling	4,675,000	78.00	59,935
4	Brown Sugar	125,271,500	561.00	231,748
5	Seaweed Food	47,458,500	151.00	314,294
6	Banana Crackers	52,662,750	215.63	281,270
	Average		65.60	190,532

There are five industries that have average efficiency of high allocation of labor, namely corn grinding industry, banana chips industry, compost industry, rice milling industry and seaweed food industry, with efficiency value greater than standard efficiency of labor allocation based on UMR of South Sulawesi Province, then only 1 industry has average efficiency level of low labor allocation, with efficiency value less than standard efficiency of labor allocation based on UMR of South Sulawesi. The results of the study [9] show that the technical efficiency of labor in technological innovation of certified rice seed programs is positive, which means that the addition of 1% of labor can still increase production by 1.014% in the pre-program period and 0.467% after the program.

4. Conclusion

It can be concluded as follows: Corn flour milling industry, red sugar industry, banana chips industry, compost fertilizer industry and rice milling industry provide average net income with bigger value than Regional Minimum Wage (UMR), while the seaweed food industry provides a small net income with less income value than UMR South Sulawesi. There are two industries that have high income contribution rate, namely banana chips industry and rice milling industry, three industries that have medium income contribution rate, corn flour milling industry, red sugar industry and compost fertilizer industry and one industry with low income contribution seaweed food industry.

Furthermore, there are five industries that have average efficiency of high allocation of labor that is corn flour milling industry, banana chips industry, compost industry, rice milling industry and seaweed food industry with efficiency value greater than standard efficiency allocation of labor based on UMR of South Sulawesi Province, and only 1 industry having average efficiency level of low labor allocation with value less than standard efficiency of labor allocation based on UMR of South Sulawesi Province.

References

[1] Soeratno 1996 *Source of Household Income* (Yogyakarta: UGM)
[2] Department of Trade and Energy Industry of Bantaeng Regency 2012 *Annual Report on Small and Medium Industry* (Bantaeng: Department of Trade and Energy Industry of Bantaeng Regency)
[3] Sajogyo 1990 *Source of Household Income* Summary of Thesis and Dissertation of IPB Graduate (Jakarta: Program Puspaswara)
[4] Kumala P 2010 *Contribution of Fishermen Income to Family Income in Tokolan Batang Tumu Village Mandah District Indragiri Hilir* (Indragiri: Economic Education Studies Program Faculty of Teacher Training and Education)
[5] Suratiyah K 2006 *Small Industry and Household (Definition, Definition, and Example)* (Yogyakarta: UGM)

[6] Burhansyah R and Azri 2010 Feasibility analysis of integrated farming and contribution of household income in dry land of farmers *J. of Agro Econ.* **17** No 2 (Yogyakarta Gadjah Mada University)
[7] Santoso 2012 *Peningkatan Daya Saing pada Industri Kecil dan Menengah* www.Bikasolusi. co.id
[8] Sulaksana J. Dinar and Ispanji RK 2014 Labor and its contribution to household income *J. of Agr. and Livestock Sci.* **2** (2)
[9] Rachmina D and Maryono 2008 Technical efficiency and income analysis for certified rice seed program: Stochastic production frontier approach *J. of Agribus. and Agr. Econ.* **2** (2)

IOP Conf. Series: Earth and Environmental Science **157** (2018) 012065 doi:10.1088/1755-1315/157/1/012065

Transdisciplinary research on local community based sago forest development model for food security and marginal land utilization in the coastal area

D A Rampisela [1, 2], R Sjahril[3], S A Lias[1], and R Mulyadi.[4]

[1]Dept. of Soil Science, Faculty of Agriculture, Hasanuddin University, Jalan Perintis Kemerdekaan KM 10, Makassar, 90245, Indonesia.
[2]Research institute for Humanity and Nature, 457-4 Kamigamo Motoyama, Kita-ku, Kyōto-shi, Kyōto-fu, 603-8047, Kyoto, Japan.
[3]Dept. of Agronomy, Faculty of Agriculture, Hasanuddin University, Jalan Perintis Kemerdekaan KM 10, Makassar, 90245, Indonesia.
[4]Dept. of Architecture, Faculty of Engineering, Hasanuddin University, Jalan Perintis Kemerdekaan KM 10, Makassar, 90245, Indonesia.

E-mail: *agnes.unhas@gmail.com*

Abstract. The most important issue in food security in Indonesia is limited land area which suitable for food crops. While production already reached its highest potency. Eastern part of Indonesia seems to have hidden answer for the food security issues. Sago plant is origin from Indonesia. Sulawesi is known by its unique semi-cultivated system in food production. This paper will describe the transdisciplinary approach in developing a model of sago plant development based on local community. The model is derived based on the research findings related to characteristic of semi-cultivation of sago plant as follows: (a) Sago in Tana Luwu only 3463,36 ha, most of the land already converted to fishpond, and plantation (b) Semi cultivated sago forest produce 24 tons dry sago/year, 4 times than paddy in dryland, (c) Transdisciplinary research method enables the triple helix collaboration of University, Industry and Government, and (d) Collaborative action is a key to bind the stakeholders, in this case the Plan and Design of Sago Science techno park worked effectively. (e) Several government regulations and local community activities have been born at different level.

1. Introduction

Food security is the most important issue in the world. Indonesia as a country with population of 258,704,900 in 2016 [1] is now struggling to increase food production. Efforts in increasing rice production are come to the limit and government of Indonesia exploring alternative foods. Land for food crop always compete with other land use and directly affecting the size of forest land. Growing sago in semi-cultivated method means producing food in parallel with increasing forest area.

Sago palm (*Metroxylon sagu* Rottb.) is mainly distributed in swamp area with a high temperature and abundant sunshine located at latitude between 10•‹ North and South, and it can grow up to an elevation of 700m [2]. According to Stanton [3] one of the first to advocate sago palm research, the advantages of the crop are that it is: (1) economically acceptable; (2) relatively sustainable; (3) environmentally friendly; (4) uniquely versatile; (5) vigorous, and (6) promotes socially stable agroforestry systems.

Sago starch traditionally is staple food for most of people in eastern part of Indonesia. If we use assumption that Indonesian people consume food in form of rice 600kg/year for a family with 6 family members, then the necessary food is equal to 3 or 4 sago trunks. This is one of the reasons why the sago plants in Papua is underutilized. If productive sago forest in Papua is about 400000 hectare, it can feed 240 million people the same number with population of Indonesia. Abundant food source forgotten by the policy makers in the country [4].

Tana Luwu is a region which is originally a royal territory under a King called Datu Luwu. The area was much larger including several more regencies in South Sulawesi and South East Sulawesi. Now Tana Luwu consists of one city (Palopo City) and three (3) regencies (Kabupaten Luwu, Kabupaten Luwu Utara and Kabupaten Luwu Timur) in South Sulawesi Province. For people of Tana Luwu Sago is not just as an important staple food but also customs, culture and tradition are very much related. The existence of the King (Datu Luwu) is a strong bond for the community, however sago function as staple food is fading away along with the popularity of rice as the symbol of prosperity.

Even though in this official data is mentioned that sago in Sumatera and Riau is classified as semi cultivated but it is believed that recently most of them are plantation type mainly for industry. Sago in Sulawesi is a true semi cultivation, still has forest ecology which make it has its own unique environment. Different with sago plantation for industry, planting sago in semi-cultivation way has many positive impacts both for (alternative) food production and forest function. Starch productivity of sago palms cultivated on Semi cultivated forest area, however, is poorly documented.

Transdisciplinary research is defined as research efforts conducted by investigators from different disciplines working jointly to create new conceptual, theoretical, methodological, and translational innovations that integrate and move beyond discipline-specific approaches to address a common problem. This approach also recognized intangible knowledge as data which has the same quality with tangible knowledge. Recently transdisciplinary research is introduced as a new research method which is simultaneously encourage local community to become a main actor in finding solution for the issues. Issues here are how to maintain the sago forest area and increasing the sago plants.

This study aims to develop a model of sustainable sago forest for food security. Local community's initiative is a must when we talk about sustainability, because they live there and the effects or impact will affect them, positively or negatively. Enabling the local community to take initiative will enrich both community and researchers. In a specific environment such as sago forest, abundant knowledge was accumulative and digging the intangible knowledge and make it work for the future prosperity is a task of transdisciplinary research team.

Objective of this study are:

- Re-discover the benefit and merit of sago as staple food and
- Support sago farmers to become main actors in sago development
- Description of the traditional intangible knowledge on sago cultivation and sago forest
- Description of roles and function of stakeholders

2. Methods

The research was conducted in Tana Luwu, administratively means Kota Palopo, Kabupaten Luwu, Kabupaten Luwu Utara and Kabupaten Luwu Timur of South Sulawesi.

The intensive survey to make profile of semi cultivated sago forest was conducted in Village (Desa) Waelawi in subdistrict (Kecamatan) Malangke Barat of District (Kabupaten) Luwu Utara in South Sulawesi Province. This village is chosen because the sago forest in this village is still exist and based on the preliminary survey, the largest sago forest area is in subdistrict (Kecamatan) Malangke Barat. Several plots size of 50 m x 50 m. is used to define the profile of semi-cultivated sago forest. Number of clumps is counted and distribution of clumps is plotted by coordinates using GPS.

The Transdisciplinary team including agronomist, soil scientist, hydrologist, public health scientist, social economist cooperate with sago farmers, local community and local government to define a concept of Development of Sago Forest based on local community.

IOP Conf. Series: Earth and Environmental Science **157** (2018) 012065 doi:10.1088/1755-1315/157/1/012065

3. Result and Discussion

3.1. Profile of Semi-cultivated Sago Forest

Semi-cultivated sago forest has its own unique profile, very much different with plantation style where the interval between trees or clumps is orderly prepared, the plant spacing in semi cultivated sago forest defined naturally by the suckers which will form new clumps. There is an argument that the suckers will extended until it found a favorable point to emerge. Sunshine maybe the dominant factor in this case. It is means that the empty space will be naturally enable new suckers to emerge. According to Pasolon *et al.* [5], vegetative grow and dry starch production were lower in dry soil condition, except with molar sago in Laosu. Sago palms growing on dry land had higher root density maximum root depth of 50-70 cm.

3.1.1. Clumps distribution. The clumps position is plotted by coordinates. Averages of 136 clumps is counted in a hectare. The distribution in plot is shown in fig 1. *This distribution is equal to the average clumps spacing of* $D = \sqrt{10000/136} = 8.57\ m$.

50 m

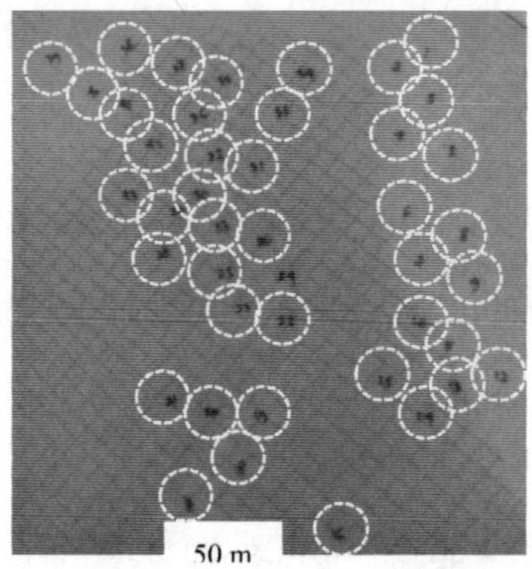

Figure 1. Distribution of sago clumps in research plot.

3.1.2. Sago clump. Sago clump consist of several suckers, with at least one sucker is starting trunk formation. It is not easy to differentiate clumps in sago forest. The existence of trunk formation stage sucker is a use to indicate an independent clump.

Table 1 shows that in average we can find 24 suckers, showed that no pruning or maintenance has been done by farmers. Traditional way of leaf-sheath in fact is giving pruning impact for plants. Yiu-Liong, *et.al.* [6] reported that the overall mean number of clumps were found to be 239. This gave an average of about 8 young suckers per clump. Number of large suckers at about trunk formation stage decreased nearly ten-fold to only 203 per hectare. Immature palms with trunk and mature, harvestable palms averaged 77 and 42 per hectare respectively.

Table 1. Age and harvesting year estimation of plants in a clump

	Average age of suckers/plants	Number of suckers	Year of harvesting
1.	6 years	1	2021
2.	5 years	1	2022
3.	1 year	5	2027
4.	2 months	6	2028
5.	1 month	8	2028
6.	1 week	3	2028
	Total number of suckers in a clump	24	

3.2. Productivity of Semi-Cultivated Sago Forest.

The criteria by which sago palms are selected for cutting are the most poorly documented aspect of sago processing. The starch reserves are apparently at their maximum just before flowering and fruiting deplete these reserves, but scientifically little more is known of the timing of starch build-up. Most sago-using peoples cut palms at varying stages of maturity [7]. Among 136 clumps in one hectare of sago forest research plots, in average there are 280 sago plants which ages more than 6 years, means these trunks have potency to be harvested in 5 years ahead. In average will be 56 trunks per/year. Based on the direct observation in the field, a trunk can produce 20 tumangs (basket made from sago leaves) with content of 20 kg wet sago starch, means total of 400kg wet sago from one trunk. Production of sago starch could be estimate as =56 trunks x 400 kg wet sago = 22400 kg wet sago or 22.4 ton ha^{-1} $year^{-1}$ of wet sago. Dry sago starch calculation = 60% of wet sago equal to 13.44 ton $ha^{-1}year^{-1}$ of dry sago

As comparison we can refer the condition in Sarawak as Yiu-Liong, *et al.* [6] mentioned the range for every category of palms varied widely from plot to plot. Harvestable palms varied from 6 to 144 per hectare while the number of immature palms with trunks ranged from 6 to 250 per hectare. The numbers of young suckers, palms at trunk formation stage and dead palms were in the range of 288-4,925, 13-475 and 0-131 per hectare respectively. According to Novarianto *et al.* [8] the productivity of sago Duri in average is 226.34 kg per trunk. Sago Duri originate from Kabupaten Kepulauan Meranti, Provinsi Riau in Indonesia is officially recognized as National Superior sago variety under the name of Selatpanjang Meranti. Natural sago forest has lower productivity. This is caused by uncontrolled growth of palnts, competition between suckers and with other plants, irregular plant spacing, and difficulties during harvesting. Plants management is necessary to increase the potency of sago in natural sago forest. The potency of sago production in Indonesia is 12.8 ton ha^{-1} $year^{-1}$, but after better management and maintenance the potency will increase up to 64.3 ton ha^{-1} $year^{-1}$ [9].

3.3. Local Community-Based Sago Development Model.

3.3.1. First step: Formation of research team. The research team is formed based on the rules of transdisciplinary which is require interdisciplinary team. The team consist of scientist focus on hydrology, soil science, agronomy, public health, medicine, forestry, food technology, law, architecture, urban planning, sociology etc including also local government staff, NGO and sago farmers

3.3.2. Second step: Collaboration between scientist team and local community. Local community-based sago development is a concept where sago farmers of local community is the main actor in initiating the sago development. Basically, they have all the indigenous knowledge in plant propagation and sago starch processing. First step of supporting them to be a confident initiator is to acknowledge their knowledge and skill. The best ways are:

- collaborate in inventory of sago forest condition,
- analyse recent issues, and
- discuss the solution to put it in the real context.
- collaborate in scientific measurement
- feedback of scientific findings to the community

This process will convert intangible knowledge into tangible form, in more communicative format. In parallel, local community aware about their local resources and its potency. This new understanding related to their pride and confident.

3.3.3. Third step: Inclusion of Local Government. The result and findings of collaboration work between local community and research team is reported and consulted with local government. In this step government become more aware and ensured about the urgency of sago development make response to follow up the result. In this study, research team and local government agree to prepare the outline of sago science techno-park. Basic design is prepared by research team. In this step intensive discussion and consultation between research team and government are expected.

3.3.4. Fourth step: Follow up action of local government. Local government of Kota Palopo make progress of follow up the concept of science technopark by initiating detail design and procurement of land, Kabupaten Luwu Utara make preparation on Sago conservation regulation and village government also prepare regulation at village level. Kabupaten Luwu also initiating the regulation.

3.3.5. Fifth step: Embracing the private company and industry. Concept of science techno-park then introduced to the private sector to get suggestion and make it more realistic and implementable plan. One of the suggestion to the private sector is to challenge the sago plant utilization for post-mining land rehabilitation. Kabupaten Luwu Timur has an urgency to the rehabilitation of this kind of land, and the related company already agreed to implement this plan, and started to prepare the plan

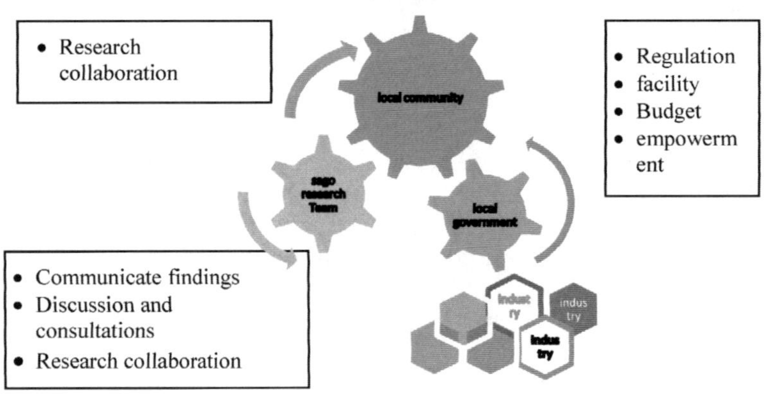

Figure 2. Local Community based Sago Forest Development Model

4. Conclusion

- Sago in Tana Luwu only 3463,36 ha, most of the land already converted to fishpond, orange, and recently to cacao and palm oil.
- Semi cultivated sago forest produce 14 ton dry sago/year. This means 4 times than paddy in dryland or marginal land. And still twice than irrigated paddy field.

IOP Conf. Series: Earth and Environmental Science **157** (2018) 012065 doi:10.1088/1755-1315/157/1/012065

- Sago forest conservation depends on active participation of local community and regulation made by government
- Transdisciplinary research method enables the triple helix collaboration of University, Industry and Government, based on solid collaboration between scientist researchers and local community as research member. Identification, description and documentation of local wisdom is a suitable topic to build up transdisciplinary research because it is triggered the active participation, trust and confidence of local community
- Collaboration action is a key to bind the stakeholders, in this case the Plan and Design of Sago Science techno park worked effectively.
- Several government regulations and local community activities have been born at different level.

References

[1] Badan Pusat Statistik 2016 *Statistical Yearbook of Indonesia* (Jakarta: BPS) ISSN: 0126-2912

[2] Flach M 1977 *Yield potential of the sago palm and its realization* (*in Sago -'76: Papers of the 1st international sago symposium* ed. M A Tankoolin (Kuala Lumpur) p 157-177

[3] Stanton W R 1993 Perspective on, and future prospects for, the sago palm *Sago Palm* **1** 2-7

[4] Bintoro M, Purwanto Y, and Amarillis S 2010 Sagu Di Lahan Gambut (Bogor: PT Penerbit IPB Press) p 61-62

[5] Pasolon Y B, Subair, Udin, Rembon F S, Mamundu M and Yamamoto Y 2002 *Growth and Starch Production of Sago Palms in Southeast Sulawesi: the Effect of Soil/Water Regimes* in K Kainuma, M Okazaki, Y Toyoda, and J E Cecil *New Frontiers of Sago Palm Studies* (Tokyo Japan: Universal Academy Press INC) p 51-59

[6] Yiu-Liong T and Eng-Tian K L 1991 *The Current Status and Future Prospects of Harvestable Sago Palm in Serawak* in Ng Thai-Tsiung, T Yiu-Liong, and K Hong-Siong *Towards Greater Advancement of The Sago Industry In The' 90s. Proceedings of the Fourth International Sago Symposium* August 6-9 (Kuching Sarawak Malaysia: Lee Ming Press) p 11-16

[7] Ruddle K and Rondinelli A D 1983 *Transforming Natural Resources for Human Development: A Resource Systems Framework for Development Policy, Resource Systems Theory and Methodology* Series no 1 (Tokyo Japan: United Nations University Press)

[8] Novarianto H, Tulalo M A, Kumaunang J, Manaroinsong E, and Sulistyowaty E 2016 *Selection and Release of Sago Palm Selatpanjang Meranti Variety for Developing Sago Palm* in Prastowo (*Indonesia Journal of Metroxylon* Vol I No. I ISSN No. 2527-8134) (Bogor Indonesia) p 5-9

[9] Bintoro M, Ahmad F, Nurulhaq M I, Fathnoer V, Alamako R P, Mulyanto M R, and Pratama A J 2016 *Pengembangan sagu di Indonesia* (Bogor: PT Penerbit IPB Press) p 29-48

IOP Conf. Series: Earth and Environmental Science **157** (2018) 012066 doi:10.1088/1755-1315/157/1/012066

Economy and political ecology perspective of Indonesian food security at South Sulawesi

I M Fahmid[1], H Harun[2], M M Fahmid[3, 4], Saadah[1] and N Busthanul[1]

[1] Department of Socio-Agriculture, Faculty of Agriculture, Hasanuddin University, Jalan Perintis Kemerdekaan KM 10, Makassar, 90245, Indonesia.
[2] University of Canberra, University Drive, Bruce, ACT 2617, Australia.
[3] INDEF (Institute for Development Economics and Finance) Jl. Batu Merah No.45, Pejaten Timur, Jakarta 12510, Indonesia.
[4] ISPEI (Institute for Social and Political Economic Issues) Jl. Ance Dg. Ngoyo No.88, Makassar, 90231, Indonesia.

E-mail: *imfahmid@mail.com*

Abstract. The purposes of this study are: firstly, to demonstrate the relations of agro-ecological function, agricultural innovation system, social-ecological system and political ecology to encourage production for Indonesian Food Security Program (PKP) in South Sulawesi. Secondly, to identify the most influential and interested stakeholders in the success of PKP program. The study conducted by applying an interdisciplinary analysis of triangulation method. The result showed, the success of PKP in South Sulawesi with the achievement of 2 million rice overstock mainly impacted by the application of agro-ecological concept, agricultural innovation system, and political ecology while disregarding the concept of social agroecology.

1. Introduction

Food security defined as the situation when people have physical and economical access to food that is sufficient, safe, and nutritious to support an active and healthy lifestyle [1, 2, 3]. Many factors in food security taken into consideration namely: climate, geography, socio-economic system, and political structure. Other than those factors, the food crisis in developing countries such as Indonesia impacted by other variables such as agroecology, agricultural innovation, social ecology system, and political ecology.

In Indonesia, rice is the staple food commodity and South Sulawesi province as the primary food buffer has been able to produce an overstock rice between 2-2.5 million tons per year within the last five years [4]. Even though in macro (provincial) level South Sulawesi has rice overstock, at the mezzo (regency/city) level may have food insecurity, or even food scarcity at the micro (district, sub-district or village) level. At this condition, the roles and influence of the stakeholders who are involved in the food security program are very crucial.

This study aims to identify the most influential stakeholders in supporting the success of Indonesia food security program at South Sulawesi province. The roles of stakeholders will then relate to the agro-ecological function, agricultural innovation system, social-ecology system, and political ecology to determine the functions of stakeholders in strengthening the program in the province of South Sulawesi.

2. Literature Review

A study by Ranis *et al.* [5] proof the relationship between human development and economic growth; stated that success in human development has a significant contribution to economic growth which cannot sustainable without improvement in human development. Furthermore, few important aspects of stakeholders in food security program that should be considered are agriculture innovation system [6, 7], agro-ecological [8], socio-ecological system [9], agriculture innovation [10], social ecology system [9], and political ecology [11, 12].

However, the success or failure of food security program in South Sulawesi not solely influenced by the role and positive relation between those aspects, but power and contestation of interests among actors and stakeholders involved in the program determined its successful. Thus, economic activities always involved means and focus on power that inherent to interpersonal networking between actors, and influenced by interest as the fundamental force that society defined.

3. Methodology

This study was conducted using interdisciplinary method of triangulation. The data were analysed qualitatively by describing the results of in-depth interviews of focus group discussion, as well as the secondary data from Statistical Bureau, RPJMD (Regional Medium-Term Development Plan) and other sources such as Poor Households Data from Department of Social Services 2013, Statistical Data of Agriculture of Ministry of Agriculture 2013, and Agricultural Census 2013.

All 24 regencies and cities in the province of South Sulawesi chosen as the location of this study. The scale used to measure the influence and interests of stakeholders in this study is a score of 1 to 5, by asking five questions to determine the level of interest and five questions to measure the degree of influence.

4. Results and Discussion

The principal components analysed to find out the role of the regencies and the city in achieving food security in South Sulawesi are rice production, people's access to food and rice distribution as well as their relation to agro-economy, agricultural innovation system, social ecology system, and political ecology referring to table 1.

Table 1 The role of the district and the city in achieving food security in South Sulawesi, 2004-2013

Regency/City	Avg. Production (Ton)	Avg. Increase of Rice Production (%)	Avg. Rice Need (Ton)	Avg. Surplus/Deficit (Ton)	Percentage of Surplus/Deficit (%)	Subsidised Rice Allocation (%)
Selayar	5,787.53	42.8	11,061	(5,273)	(48)	13.9
Parepare	3,846.43	55.1	11,218	(7,371)	(66)	7.0
Sidrap	248,364.21	14.0	23,912	224,452	939	7.7
Pinrang	275,120.31	3.0	32,356	242,764	750	9.9
South Sulawesi	2,483,394.13	5.3	738,100.44	1,745,293.69	236	11.1

There are couple of regencies and a city with more than 20% of average production rate during the selected period namely Selayar and Parepare. However, even with high production rate, both regencies could not contribute to the food security program in South Sulawesi due to their limited resources and land area. Meanwhile, such regencies as Sidrap and Pinrang, despite their production far below 20%, but they contributed more than 300% to South Sulawesi food surplus. Even more, they had even provided nearly 1000%.

4.1. The Relation between Production, Agroecology, and Innovation.

Considering the relation between the rice production factors and agroecology, it would be reliable to say that farmers from such as Sidrap, Pinrang, and a few other selected regencies had acquired science

IOP Conf. Series: Earth and Environmental Science **157** (2018) 012066 doi:10.1088/1755-1315/157/1/012066

and social practice on rice production. The result of this research showed that high production rate of some regencies (table 1) had been the outcome of applying agricultural system based on the agro-ecological concept using mechanical, commercial, and high output agricultural system combined with organic substance inputs to keep the ecosystem in check. At the same time, farmers also had used institutional reformation to increase the ratio of both internal and external inputs – they have a good awareness and upholding the local wisdom, biodiversity, innovation capacity on water and soil management, agricultural design, as well as other related functions especially those belonging to the traditional agricultural ecosystem.

Most farmers and other agriculture stakeholders in South Sulawesi had previously very limited amount of innovations. Agricultural innovation systems emergence in South Sulawesi had been more to respond the lack of technology framework transfer to the farmers due to knowledge or technology perceived as the exclusive right of scientists, then handed over to intermediaries or end users. Whereas farmers and other stakeholders who involved in agriculture totally aware that innovation can increase individual and organisational capacities to put science into social or economic agriculture practices. Moreover, creative social network of regencies had participated in encouraging faster social and economic innovations, either through farmer organisation or other valuable channels. Thus, a framework of innovation system directed the agricultural practices to capture and utilise different types of knowledge to achieve common goals.

4.2. The Relation between Production, Social, and Political Ecology

According to the Indonesian Statistical Bureau [13], the average rice production in four selected regencies increased from 1994-2003 which are Sidrap (15.1%), Soppeng (4.9%), Wajo (21.1%), and Pinrang (3.7%). Compared to the data from 2004-2013, the average rice production from those selected regencies are 14%, 4.7%, 20.4%, and 3% respectively. This data shows the decreased in the average of rice production for the past 20 years.

Even though Sidrap, Soppeng, Pinrang, and Wajo regencies have high production rates, they have not optimally overcome the problem – or changed and maintained the role of the control and function – of the structure. The capacity to regulate itself internally is also low, so that, the food production in those regencies are expected to be stagnant or decline in the future because they do not strengthen the capacity to learn and adapt. Here the problem of ecological imbalance found, also as the evidence that social ecology is tightly linked with the production factors.

The methods of social ecology are the systems based on the procedures of management of ecology and ecosystem, they also included the discussion of the theory of food security and vulnerability. The system of social ecology has a significant role in the process of environmental changes, particularly in the medium and long-term consequences of human activities on the future. Differ from this view, the concept of political ecology reviews the relationship between food security and power, politics, policy, and justice. There are, at least, two aspects observed in this study related to the political ecology, namely human ecology and critical discourse analysis.

Stakeholders running the food security program, especially the governments (of regencies or province) as actors or entities that have the power to determine policy, tend to be influenced by the interest to maintain or expand their power. The form of the expansion of power and maintain of control are awarding "gift" in the form of food security program to certain communities of a particular government, ruling family, or fellow ethnic and regional origin. Usually the process of awarding done in the form of social assistance or grants, such as the government supply some agricultural inputs like seeds, fertilizers, farm tools or machinery, and infrastructure for the agriculture sector. Out of such a pattern, the food security program in South Sulawesi was dominated by policies based on political ecology system.

4.3. Stakeholders/Actors Interest Analysis

Stakeholders involved in the food security program in the province of South Sulawesi have different level of interests and influences. The difference level of each stakeholder influenced by the form of participation, benefits, commitment, network relationships, and their role in the food security program. Table 2 indicates that the stakeholders with the highest level of interest are the farmers, as the target of the food security program. Farmers are not just acting as producers, but at the same time, also as consumers working to increase the household income and to improve the food security program itself. The rice farmers, traders, consumers, and office of food security are the stakeholders with the high score of support to the food security program; as well as the ones that benefited the most.

Table 2 Stakeholder Level of Interest

Stakeholder	Score Questions					Total
	I	II	III	IV	V	
Governor	2	3	4	3	4	16
Regent/Mayor	3	3	4	4	4	18
Provincial Office of Planning *(Bappeda)*	3	1	4	1	1	11
Regency/City Office of Planning *(Bappeda)*	3	2	4	3	1	13
Provincial Office of Food Security Affairs	4	3	5	3	4	19
Regency/City Office of Food Security Affairs	4	3	5	3	4	19
Rice Traders (Represent the market)	3	5	5	3	4	20
Farmer	3	4	5	4	4	21
Consumer	3	5	3	3	5	19

The level of influence of stakeholders is measured using and identification scoring based on total value of each stakeholder. Table 3 shows the governor and the regent score the highest level of influence with 22 and 23 pints, respectively. It means that both of them have the power and huge influence to determine the direction of food security program policy; started from preparation, execution, growth, and development, up to evaluation of the program. Paradoxically, the stakeholders with the highest interests such as farmers, traders, and consumers have no influence and authority at all to determine the policy of the program, even though they are beneficiaries from the program, and continually participate in every stage of the program.

Table 3 Stakeholder Level of Influence

Stakeholder	Score Questions					Total
	I	II	III	IV	V	
Governor	5	5	4	5	4	23
Regent/Mayor	4	5	4	4	4	22
Provincial Office of Planning *(Bappeda)*	3	1	1	3	3	11
Regency/City Office of Planning *(Bappeda)*	3	2	2	3	4	14
Provincial Office of Food Security Affairs	3	4	2	3	4	16
Regency/City Office of Food Security Affairs	2	5	2	4	4	17
Rice Traders (Represent the market)	3	1	2	2	1	9
Farmer & Informal Leaders	2	1	1	1	2	7
Consumer	3	1	1	1	1	7

This research clearly shows that the food security program in South Sulawesi province can be considered as very successful, especially when considering the capacity of rice production because the province has a very high stock of rice, *i.e.* an average of 1,735,293 tons during the year of 2004 to 2012. This success, however, is not followed by the market capacity (private sector) which responsible for *opening access to the public, to obtain food.* The same thing happen to weak distribution capacity that supposed to be controlled by the government. This indicates that boosting free market competition [14] is another potential route to stabilize food security. This is a reason why the government should prioritize the revitalization of agricultural and rural development [15] for food security system, or rural food barn [16]. In addition, training and development of leadership potency to increase capability in recognizing and solving problems of food diversification and understanding on nutrition and food is needed [17].

5. Conclusion

The success and failure of food security program primarily determined by three main things: production, distribution, and accessibility of food. The food security program in South Sulawesi shown a very high result in the aspect of production, but experienced the failure in the distribution and the access of food. It is because the stakeholders of the program do not carry out the concepts of agro-ecosystem, agricultural innovation, social and political ecology during the development of South Sulawesi food security program. The most influential stakeholders to determine the direction of food security program policies in South Sulawesi are the governor and regent. While stakeholders with the highest level of interest are the farmers, traders, and consumers. However, having a high rate of interest does not mean have any level to influence to determine the policies of food security program. Therefore, the most influential stakeholders must encourage the involvement of all stakeholders to have more influence and roles in this program through an agreement, discussion, and general consent. An actively contribution of all stakeholders to the fair market system will surely solve the distribution and accessibility problems.

References

[1] World Food Summit 1996 *Rome Declaration and Plan of Action on World Food Security* (Rome: World Food Summit)
[2] Singh S 2009 Global food crisis: magnitude, causes and policy measures *Int. J. of Soc. Economics* ***36*** (1-2) 23-36
[3] Fahmid IM 2004 *Gagalnya Politik Pangan di Regim Orde Baru (Kajian Ekonomi Politik Pangan di Indonesia)*, (Jakarta, Yayasan Studi Perkotaan-ISPEI)
[4] Badan Pusat Statistik 2017 *Buletin Statistik Perdagangan Luar Negeri Ekspor Menurut Kelompok Komoditi dan Negara, November 2017* (Jakarta: BPS).
[5] Ranis G, Stewart F and Ramirez A 2000 Economic growth and human development *World Development.* ***28*** (2) 197-219
[6] De Schutter O 2010 *Agroecology and the right to food* (Geneva: United Nation Human Rights Council)
[7] Scoones I, Thompson J and Chambers R 2008 *Farmer first revisited: Innovation for agricultural research and development* (Sussex: Institute of Develoment Studies)
[8] Pretty J, Noble A D, Bossio D, Dixon J, Hine R E and Penning de Vries F W 2006 Resource-conserving agriculture increases yields in developing countries. *Env. Sci. & Tech.* **40** (4) 1114-1119
[9] Ingram J, Ericksen P and Liverman D (Eds.) 2010 *Food security and global environmental change* (London: Earthscan)
[10] Amankwah K, Klerkx L, Oosting S J, Sakyi-Dawson O, Van Der Zijpp A J and Millar D 2012 Diagnosing constraints to market participation of small ruminant producers in northern Ghana: An innovation systems analysis *NJAS - Wageningen J. of Life Sci.* **60-63** 37-47
[11] Akram-Lodhi A H 2013 *Hungry for change: Farmers, food justice, and the agrarian question* (Halifax, Canada: Fernwood Publishing)
[12] Blaikie P and Brookfield H 1987 *Land degradation and society* (London: Mathuen & Co, Ltd.)
[13] Direktorat Pangan dan Pertanian Kementerian Negara Perencanaan Pembangunan Nasional/Badan Perencanaan Pembangunan Nasional Republik Indonesia 2006 *Profil Pangan dan Pertanian* (Jakarta: Bapenas)
[14] Thalyta E Y, Mustajab M and Arsyad M 2010 Abstracts of doctoral theses on the Indonesian economy *Bull. of Indonesian Eco. Studies* **46** (2) 251-254 DOI: 10.1080/00074918.2010.486113
[15] Ekasari Z K, Saleh S A M, Jusoff K, Salman D, Akhsan, Kasirang A, Arsyad M, Amrawaty A A and Fudjaja L 2013 communication pattern and conflict in agricultural extension *Asian Soc. Sci.* **9** (5) 27-33
[16] Bahua M I, Arsyad M and Jamil M H 2016 Farmers community empowerment through

institutional strengthening of rural food barn *Int. J. Agr. Syst.* **4** (2) 161-167

[17] Rahmadanih, Bulkis S, Amrullah A, Ruka R M and Arsyad M 2015. Strengthening institutional model of women-farmers group in developing household food diversification *Int. J. Agr. Syst.* **3** (1) 29-40

Competitiveness, production, and productivity of cocoa in Indonesia

I M Fahmid[1], H Harun[2], M M Fahmid[3, 4], Saadah[1] and N Busthanul[1]

[1] Department of Socio-Agriculture, Faculty of Agriculture, Hasanuddin University, Jalan Perintis Kemerdekaan KM 10, Makassar, 90245, Indonesia.
[2] University of Canberra, University Drive, Bruce, ACT 2617, Australia.
[3] INDEF (Institute for Development Economics and Finance) Jl. Batu Merah No.45, Pejaten Timur, Jakarta 12510, Indonesia.
[4] ISPEI (Institute for Social and Political Economic Issues) Jl. Ance Dg. Ngoyo No.88, Makassar, 90231, Indonesia.

E-mail*: imfahmid@mail.com*

Abstract. Cocoa is one of Indonesia's five foreign exchange earner, thus cocoa must stay competitive for the export market. Aims of this study are: analyze the cost structure, production and productivity of cocoa farming, the level of competitiveness, and map the types of government policies that affect the competitiveness of cocoa plants. The method used is descriptive qualitative and quantitative. Data analysis is done by using PAM (Policy Analysis Matrix). The results showed, structures are at the cost of production of cocoa farming in Indonesia almost 50 percent for wages, and 31.6 percent for land rental. The big percentage of workers wages indicates that cocoa farming is labor intensive production. In Indonesia total productive cocoa farms only 27.6%, with a productivity level of 655,515 kg per hectare. Cocoa farming in Indonesia is carried out with protective policies, the value of EPC 4.29, indicating the government's policy towards the inputs and outputs of cocoa has been effective. While the PCR value of 0.51, indicating cocoa farming has a competitive advantage, but it does not have a comparative advantage. In conclusion, productivity, out-put prices, and exchange rates should be raised, and input prices should be lowered, so that cocoa farming can provide higher net transfer values for farmers. To improve the competitiveness of cocoa farming, the islands of Sulawesi and Sumatra are two islands that require special policies, especially on out-put price policy, input prices, and productivity, as well as improvement of other cocoa commodity farming systems, as these two islands contributed more the 80 percent of Indonesia cocoa bean production.

1. Introduction

Cocoa is one of the main export commodities of Indonesia where over the past five years, Indonesia has provided about 1,951,270 hectares' area of cocoa plants. Unfortunately, its productivity level in the same period results only around 655 kg per hectare. In comparison, the productivity of cocoa in Malaysia reaches about 1800 kg per hectare, around 800 kg per hectare in Ivory coast, while in Ghana it reaches about 360 kg per hectare [1].

The top five cocoa production countries only grow about 2.01% in average in the last ten years. Meanwhile, the industrial demand for this product grows 3% per year and those big five cocoa production countries produce more than 95% of the world's cocoa needs [2]. Here, the trend of the world market needs for cocoa clearly show that the demand is much higher than the production capacity of cocoa. Unexpectedly, cocoa producers like Indonesia and Nigeria tend to have a negative trend of cocoa production instead of experiencing a positive growth. In Indonesia itself, there are various reasons support the low growth of cocoa' production and productivity, for instance, the limited knowledge of the farmers in cocoa cultivation and crop management; the limited access of cocoa's farmers to the financial institutions; the problem of marketing (price stability is not guaranteed); the decline of soil fertility; the pests and disease; the old cocoa plants which already passed the peak of production; the government policies toward the input and output prices; and the different exchange rates for different islands of each province [3].

Since the cocoa is important for the Indonesia's export, there are two basic strategies to be applied in terms to be able to compete in the global market such as: cost minimisation strategy, and product development strategy. Indonesia applies the first strategy. One of the ways to minimise the production cost is through increasing the cocoa crops' productivity since its cost is lower compared to the cost of crops which result in scant productivity level [4]

Cocoa plantations in Indonesia are dominated by the small-scale farmers and are carried out by relatively poor farmers as well [5]. This farmers' plantation is required the government supports so that the plantation product can compete in the export market. The problem is that it is not clear yet whether the government' policies are supporting the competitiveness or competing capabilities of the cocoa products of Indonesia [6]. This research is interesting because it manages to show the cause of the low competitiveness of Indonesia's farmer-based cocoa production, such as the structure of production cost of cocoa; the government policies; and the levels of productivity viewed especially from provincial, island, and national levels.

2. Literature review

2.1. The Concept of Competitiveness and Comparative Advantage

The competitiveness will determine the position of a commodity in the competition market, and one of the measurement is through the market share [7]. Increases in the market share of a commodity will lead to the augment of commodity competitiveness. Therefore, competitiveness can be determined using the approaches of market share and market growth. The approach to be used in measuring the competitiveness commodity are the degree of profits generated and the business efficiency of producing the commodity. The level of profits portrayed from the private and social advantages, while the business efficiency of producing the goods illustrated from their level of comparative and competitive advantages.

The term of comparative advantage itself used for the first time by Ricardo [8] when discussing trade between two countries. Theoretically, Ricardo proved that when two nations perform a mutual trade, and both countries concentrate on exporting their goods of comparative advantage, then both sides will benefit together. In regional economy, the term of comparative advantages in this sense applies in the context of a comparison, not for the real added value. In other words, the comparative advantage is an economic activity that is in comparison more advantageous for the development of the region.

2.2. Competitive Advantage and Policy Analysis

A competitive advantage is a tool used to measure the competitiveness of an activity based on the actual condition of the economy. Operationally, competitive advantage can be defined as the ability to supply the goods and services on the time, place, and form desired by the consumers; either in domestic or international market, at a price that is equal to or better than that offered by the competitors to earn profit.

2.3. Policy Analysis Matrix

Policy Analysis Matrix (PAM) is a model used to analyse the comparative advantages (economic analysis) and competitive advantage (financial analysis) of a commodity that was introduced by Monke and Pearson [9]. The result of the PAM analysis can be used to determine whether a country has a high or low competitiveness in a system of commodity production as seen from a particular area and technology, as well as how policy can improve the competitiveness through the creation of business efficiency and income growth. In addition, it can also show the extent of the impact of the input price policy, the impact of the output price policy, or the combination of both price policies established by the government for the manufacturers.

3. Methodology

3.1. Methods of Analysis and Data Processing

The method used in this study is qualitative and quantitative descriptive. Data analysis is carried out by using PAM model.

3.2. Types and Sources of Data

The data used in this research is the secondary data of Agricultural Census 2013 obtained from the Central Bureau of Statistics (BPS), the Directorate General of Plantations of the Ministry of Agriculture, and Food Agriculture Organization (FAO). Based on the goals that will be analysed, the data used, among others: (1) farmer-based input-output, (2) the distribution cost, (3) the input price, (4) the output price, (5) the government policy.

4. Results and Discussion

4.1. Cost Structure of Cocoa Farm in Indonesia

In 2014, the largest part of production cost was in labour fee and agricultural services, as shown in figure 1, that reached 49.7% or equivalent to IDR. 5.94 million. The rent of land also contributes a high portion which amounting to 31.6%.

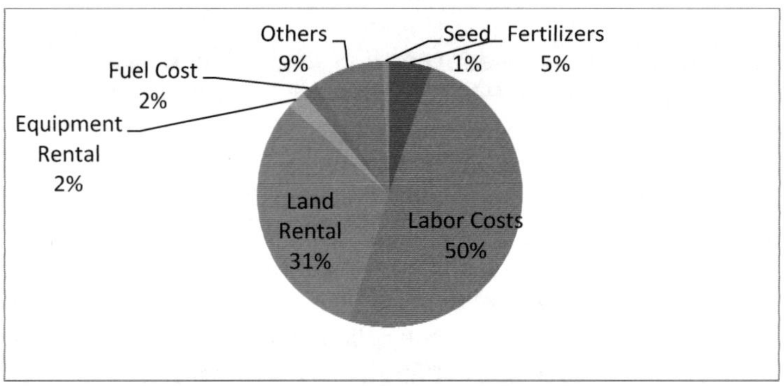

Source: Indonesia Statistical Bureau (2015)

Figure 1. Structure cost of cocoa farming per season/ha in 2014

The harvesting expenses around 40% and drying around 9.9% of the total cost. The data indicates that cocoa farming is a labour-intensive production and its cost structure, also, determine the season and place of cultivation whether it on lowland or highland. Moreover, the innovation, individual experience, and farmers' ethnics ascertain to the structure costs as well.

4.2. Productivity and Production Capacity of Cocoa

Indonesia cocoa production in 2014 was 728,418 ton [2], with the average production within the least five years was 745,845. Total cocoa farm in Indonesia approximately 1,727,437 hectares, with the productive farm merely 477,488 hectares. The highest cocoa farm productivity is on Sulawesi island with an average of 823.67 kg/ha then followed by Sumatera island of 664.7 kg/ha, next is Java island of 595.6 kg/ha. As overall, the national average of productive land is 27.67% from total 1,828,437 ha cocoa farm.

Based on the data above, two possible solutions can be taken by the government and the related stakeholders to increase the cocoa production in Indonesia. First, increase the productive land. Second, set target productivity. The main point of these two activities is the government involvement in improving infrastructure, availability, and affordability of production facilities, strengthening farmers' institution, and maintain price stability by rationalising market chain.

4.3. Cocoa Competitiveness Analysis

Nominal Protection Coefficient on Output (NPCO) is an indicator showing the level of the government protection against domestic output. Policy considers being protective of output if NPCO > 1, and vice versa. Based on Table 1, all cocoa policies in Indonesia indicates the level of the government protection against domestic output. The island of Sumatera has the highest level of protection to domestic output, with 2.16, and the lowest is on the islands of Bali-Nusa Tenggara at 1.25.

Table 1. Policy Analysis Matrix of Cocoa Farm in Indonesia, 2014

Island	NPCO	NPCI	Net Transfer	EPC	PCR	DRC
Bali-Nusa Tenggara	1.25	0.27	5,155,550	1.37	0.51	0.62
Java	1.99	0.58	113,736,862	2.34	0.50	1.23
Kalimantan	1.35	0.33	66,348,480	2.39	0.37	0.88
Maluku-Papua	1.55	0.20	61,472,673	1.87	0.54	0.98
Sulawesi	1.65	0.49	105,310,756	2.73	0.62	1.87
Sumatera	2.16	0.58	139,859,010	15.04	0.51	3.23
Indonesia	1.66	0.41	81,980,555	4.29	0.51	1.47

Source: Agricultural Census (2013), data processed

Nominal protection coefficient on input (NPCI) is an indicator that shows the level of the government protection against domestic agricultural input prices. The policy is protective of input if NPCI < 1, meaning there is a policy of subsidy on the input of tradeable, and vice versa. Table 1 shows that the cocoa farming in Indonesia is protected of inputs, since the NPCI is 0.41. Papua-Maluku islands are the highest level of protection against input prices with the NPCI value of 0.20. In contrast, the islands of Java and Sumatera, are the lowest.

Net Transfer (NT) represents the difference between the net gain earned by the manufacturer and its net social gain. Value NT > 0, indicates an additional surplus of producers caused by the government policies applied to inputs and outputs, and vice versa. NT cocoa farming business nationally amounted to 81,980,555.

Illustration of the government' policy on cocoa farming, whether it is simultaneously protective of outputs and inputs, or not, measured by the Effective Protection Coefficient (EPC) approach. If the policy is protective then the value of EPC > 1. The greater the value of EPC means the higher the level of government protection against domestic cocoa beans commodity. Table 1 shows the national EPC value of cocoa farming is 4.29, with the highest EPC on the island of Sumatera 14.04, and the lowest EPC of 1.37 in the islands of Bali-Nusa Tenggara.

Private Cost Ratio (PCR) is a private profitability indicator that shows the ability of the commodity system to pay the cost of domestic resources and remain competitive. If PCR < 1, means cocoa commodity system has a competitive advantage and otherwise. Nationally, the value of cocoa commodity PCR in Indonesia is 0.51, means Indonesia has the competitive advantage in the cocoa farming unit.

IOP Conf. Series: Earth and Environmental Science **157** (2018) 012067 doi:10.1088/1755-1315/157/1/012067

Domestic Resource Cost (DRC) is a formula for the indicator of comparative advantage that shows some domestic resources that can be saved to produce one unit of foreign exchange. Cocoa commodity system in Indonesia has a comparative advantage if DRC < 1, and vice versa. Based on the data in Table 1 shows that cocoa farming in Indonesia has a competitive advantage but does not have a comparative advantage.

Profitability Coefficient (PC) and Net Transfer (NT) values are used to measure the impact of the government policy on farmer output, input and income as a whole. National value of EPC is 4.29, meaning that government policy on cocoa input and output has been effective. A positive net transfer value indicates that government policy on inputs and outputs will increase cocoa farmers' surplus by Rp 81,980,555 per hectare.

4.4. Sensitivity Analysis of Cocoa Competitiveness in Indonesia

Switching value analysis aims to find out the extent to which changes in output prices, input prices, productivity, and exchange rates affect the competitiveness of cocoa farming in Indonesia. Sulawesi and Sumatera are the two main islands where has the concentration of production and the most area of cocoa farmland in Indonesia. The cocoa farmland on the island of Sulawesi, accounted for 62.5% of the total cocoa farming land in Indonesia with the total production of 61.5% of the total production of cocoa beans Indonesia. While the island of Sumatera has total land area of 416,832 hectares, with cocoa beans production of 155,842.

Table 2. Switching Value Analysis of Cocoa Commodity in Indonesia (%), in 2013

Island	Scenario					
	Output Price	Input Price	Productivity	Exchange Value	Average Production in 5 years (ton)	Average Land Size in 5 years (ha)
Sumatera	(0.74)	(1.73)	(0.74)	28.69	155,842	416,832
Java	(19.67)	943.27	(19.67)	(1.67)	33,614	88,618
Bali-NT	(51.43)	226.22	(44.77)	(32.56)	17,059	75,392
Kalimantan	(53.76)	1349.22	(53.76)	(17.76)	9,118	36,674
Sulawesi	4.99	(216.45)	3.49	53.19	495,306	1,220,857
Indonesia	(23.32)	378.36	(23.71)	4.92	745,845	1,951,270

Source: Agricultural Census (2013), data processed

To increase competitiveness Indonesia cocoa farming, the price of output, productivity, and exchange rate should be increased, while input prices should be lower, especially those on Sulawesi and Sumatera, which covers 83.8% of the total land area cocoa farming in Indonesia. Just like [10] found that the agricultural economic activity variable that has the strongest positive direct effect on poverty reduction is cocoa production, including agricultural extension sustainability [11].

5. Conclusions

The cost structure of cocoa farming production in Indonesia is almost 50% for wages which indicated a labour-intensive farming, and 31.6% for land rental. Stagnant production over the past five years, and declined in productivity, due to the traditional management of cocoa farm by the majority of poor farmers. Under these conditions, the cocoa commodity farming system in Indonesia, generally has no financial or economic competitiveness or lacks comparative advantage, but has a competitive advantage. Therefore, productivity, output prices, and exchange rates should be increased, and input prices should be lower, resulting in higher net transfer value for the farmers.

To improve the competitiveness of cocoa farming in Indonesia, the islands with 80% contribution of cocoa beans in Indonesia namely Sulawesi and Sumatra are two islands that require special policies, especially on output and input price policy, productivity, and improvement of other cocoa commodity farming systems.

IOP Conf. Series: Earth and Environmental Science **157** (2018) 012067 doi:10.1088/1755-1315/157/1/012067

References

[1] Dormon E N A, Van Huis A, Leeuwis C, Obeng-Ofori D and Sakyi-Dawson O 2004 Causes of low productivity of cocoa in Ghana: Farmers' perspectives and insights from research and the socio-political establishment *NJAS - Wageningen J. of Life Sci.* **52** (3–4) 237-259
[2] FAO 2015 *Statistical Pocket Book 2015 World Food and Agriculture* (Rome: Food and Agriculture Organization of United Nations) p 236
[3] BPS 2013 *Buku D Perkebunan Sensus Pertanian* (Jakarta: Publikasi Badan Pusat Statistik)
[4] Fahmid I M 2013 Cocoa Farmers Performance at Highland Area in South Sulawesi, Indonesia *Asian J. of Agr. and Rur. Dev.* **3** 360-370
[5] Fahmid I M 2013 Social Innovation Among Ethnics in Cocoa Farming at Sulawesi, Indonesia *J. of Biol. Agr. and Healthcare* **3** (15) 122-131
[6] Fahmid I M 2004 *Gagalnya Politik Pangan di Bawah Rezim Orde Baru: Kajian Ekonomi Politik Pangan di Indonesia.* (Jakarta: Yayasan Studi Perkotaan (Sandi-Kota))
[7] Westgren M L and Van Duren E 1991 Agribusiness competitiveness across national boundaries *American J. of Agr. Economics* **73** (5) 1456-1464
[8] Ricardo D 1917 *On the Principles of Political Economy and Taxation* (London: John Murray)
[9] Monke E A and Pearson S 1995 *The Policy Analysis Matrix for Agricultural Development* (London: Cornell University Press)
[10] Thalyta E Y, Mustajab M, Arsyad M 2010 Abstracts of doctoral theses on the Indonesian economy *Bull. of Indonesian Eco. Studies* **46** (2) 251-254 DOI: 10.1080/00074918.2010.486113
[11] Ekasari Z K, Saleh S A M, Jusoff K, Salman D, Akhsan, Kasirang A, Arsyad M, Amrawaty A A and Fudjaja L 2013 communication pattern and conflict in agricultural extension *Asian Soc. Sci.* **9** (5) 27-33

The strategy of women in facing agrarian land conflict: Case of female farmers of Makassar Ethnics

D A T Pulubuhu[1, 2], A N Eryani [2], M E Fachry[1] and M Arsyad[3]

[1] Study Program of Gender and Development, Graduate School, Hasanuddin University, Jl. Perintis Kemerdekaan KM 10 Makassar, 90245, Indonesia.
[2] Department of Sociology, Faculty of Social and Political Science, Hasanuddin University, Jl. Perintis Kemerdekaan KM 10 Makassar, 90245, Indonesia.
[3] Department of Agricultural Socio-economics, Faculty of Agriculture, Hasanuddin University. Jl. Perintis Kemerdekaan KM 10 Makassar, 90245, Indonesia.

E-mail: *dwiatn@unhas.ac.id; dwiatn@yahoo.com*

Abstract. One of the crucial issues in developing country is land conflict. It occured since 1978 and still unsolved. This research aimed to investigate and analyze (1) the social and cultural values of Makassarese Women (2) the impact of the land conflict on Makassarese Women, (3) the strategy of female farmers in facing land conflict. The research was conducted in the villages of Timbuseng and Kampung Beru, North Polongbangkeng Sub-District, Takalar Regency, Indonesia. The research employed qualitative method combined with ethnography case study with in-depth interviews, observation, document study, and focus group discussion (FGD). The results indicate that the land conflict happened in North Polongbangkeng Sub-District had an impact on the socio-economic and cultural life of the community. The women had to play a role and the same position as the man in productive, social and domestic matters. For 25 years, the community had to avoid and wait. Since 2007, the community had thought that the contact with PTPN XIV land had ended and the community could take over the land they owned. The strategy done by women was to join the man in their fight to get back their land. They started with confronting actions to forming an organization called Polongbangkeng Farmers Unity. In 2014 the partnership program was established as an effort to settle the land conflict, though until now this program had not answered the primary needs of the community in the region.

1. Introduction

Intensity of conflict in natural resource management has continued to increase within the last few years. In Indonesia one of the causes of agrarian conflict and natural resources is the inequality of ownership of natural resources. Vast lands controlled by the state or private sector while the peasant, majority owned only small farm [1]

The disparity and scarcity of land for the farmers is a rational reason behind farmers struggle that lead to the many agrarian conflicts spread throughout Indonesia. In 2016 released by KPA noted that there have been at least 450 agrarian conflicts throughout 2016, with an area of 1,265,027 hectares and involving 86,745 households spread across provinces in Indonesia. In 2015 there were 252 agrarian conflicts, thus it was a significant increase, almost double the figure. If on average, every day there is an agrarian conflict and 7,756 hectares of land are involved in the conflict; then the community lost about nineteen times the size of the province of DKI Jakarta [2].

There are two conditions that contributed to the widespread poverty in rural areas in Indonesia [3]. The first is the increasing percentage of small farmers, whose proportion is already majority since 1963 to 2003. The second condition is the growth in the number of farmers' households without land. Two of the many in-depth studies conducted by SDP-SAE (Survey Agro Economy) in the 80s concluded that rural poverty mostly occurs in landless farmer households [4].

The change from nature to "natural resources" has resulted very bitterly for the peasants who had to be ousted from their own land and partly forced to transform into wage labor. This is a forced process of creating people who no longer tied to the land and nature. These people rely only on their strength, then become free workers [5].

Farmers and the most rural community have difficulty to meet the educational and health needs of their families. They are constantly experiencing economic downturns and quality of life due to the insufficient wages, the high cost of production and the low price of their products. This as a result of the monopoly and control of the lands, agricultural production (work tools and agricultural input) even the market as well as the price of agricultural products by imperialism and their accomplices in the country [6].

Women are most affected by the agrarian conflict. They have to experience the doubled oppression and the burden of poverty. According to Rozaki [7] women experience two forms of violence at once, namely structural violence resulting from agrarian conflict and cultural violence in the form of family and community habits that do not empower women. These situtations encouraged women to take a role in fighting for their rights.

The role of women in fighting for the source of livelihood directly involved in the various of mass action, namely women of Sugapa hugging trees that will be torn down by PT. Inti Indorayon main in Porsea, North Sumatra. Women in Bojong, West Java along with men blockade the machinaries of the companies to enter their reclaiming lands that was taken by the company. Similar action done by Mrs. Werima who fought against PT.Inco in South Sulawesi [8].

The role of women is also seen in the land conflicts in Sub-District of North Polongbangkeng, Takalar. Economic pressure and the character of Makassarese women who are brave are the factors of women farmer involvement in the peasant struggle movement to reclaim their land in North Polongbangkeng. This study aims to analyze the strategy of women farmers land conflict in Sub-District of North Polongbangkeng, Takalar.

2. Materials and Method

2.1 Types of Research

The type of research used in this study was qualitative descriptive research by combining case studies and ethnography. Case study is a research that is used and done intensive, detailed, and deeply to a problem that became the object of research, whereas narrow ethnography can be interpreted as a description of a particular ethnic in space and in a certain period.

2.2 Time and Research Site

The research was conducted from September to December 2017. The location of this research is in two villages, namely Timbuseng and Kampung Beru, Sub-District of North Polongbangkeng, Takalar Regency, Indonesia.

2.3 Informant

Informants in this study were selected in accordance with the interests and objectives of the study. The informants in this study are; 1). Women farmers in Timbuseng Village and Kampung Beru Village are experienced direct impact of the agrarian conflict. They are small farmers and farm labore who used to own land claimed by Sugar Cane Plantation. 2). The head

IOP Conf. Series: Earth and Environmental Science **157** (2018) 012068 doi:10.1088/1755-1315/157/1/012068

of the Polongbangkeng Farmers' Union organization represents both men and women. Polongbangkeng Farmers Unionis an organization built by Farmers of Polongbangkeng as a means of peasant struggle since 2009. Informants are considered have deep understanding about the conflict, and are also considered able to provide all the required information.

2.4 Data Collection and Analysis
The data in this study were obtained through in-depth interviews, field observation, document review and focus group discussion. Data analysis in this research use interactive model of Miles and Huberman, a model consists of three processes, which take place interactively. First, data reduction where researchers will sort, focus, simplify and abstract the data those researchers get from interviews, observation, document review and FGD. Secondly, compiling and presenting existing data is needed for easy understanding. Researchers will use tables, and images to present existing data. Thirdly, drawing the initial conclusions from the initial data that researchers have and compose final conclusion at the end of the study.

3. Results
Land conflicts in North Polongbangkeng Sub-District have been around since 1978 until now. Various problems have arised and never resolved. The people of North Polongbangkeng have been dependent on agriculture for a long time. They have farmed the land and lived separately on their land since 1942 during the Dutch colonial period. This is evidenced by the discovery of old graves, the former wells and settlements inside the sugar cane plantations.

3.1 The Beginning of Land Conflict
The presence of sugarcane plantation in Takalar Regency started in 1974. Initially as a form of gratitude Sultan Hamengku Buwono IX for Takalar people's for their assistance in defending Jogjakarta from the Dutch attack during the war. In 1978 PT. Madu has just started the land acquisition, managed by the committee of 9 people and the decree of KDH Takalar Regent number 24/Kpts/1978, dated March 1, 1978 on stipulation of the amount of compensation. The decree stipulates the amount of compensation: (1) for the first grade paddy field of IDR. 350,000 per hectare, (2) paddy field belonging to class II and III amounting to IDR. 275,000/ hectare (3) land / moor of IDR. 200,000 / ha, (4) arable land or P2 land of IDR. 107,000 / ha.

Upon the publication of the Decree of the Regent of Takalar, the community spontaneously refused and since then the community began to make efforts to defend continues farming activities on their land. On the other hand, PT. Madu Baru was assisted by the government along with some members of army, forced the community to accept the settlement by in various ways such as intimidation, arrest, detention and firing the firearms against people who still do farming activities on land they owned. Communities were also perceived to be inhibiting development and accused of being member of communist if they do not want to give up their land.

To facilitate land acquisition, the committee promised that their land would only be rented for 25 years and would be returned after the lease term ended. Land acquisition conducted by the committee of 9 has many fraudulant transactions, thus PT. Madu Baru releases its assets in Takalar to PTP XXIV and XXV. In 1996, the management of sugarcane plantations was handed over to PTPN XIV. Until now, the area of sugarcane plantation managed by PTPN-XIV Takalar sugar factory, based on 10 temporary land used certificates, is an area of 6,546.22 ha. In addition, land tenure is based on temporary building used certificate, which is 181.93 ha. Therefore, the total land area controlled by PTPN-XIV PG.Takalar is an area of 6.728.15 ha, located in Sub-District of North Polongbangkeng and South Polongbankeng, Takalar Regency.

3.2 The Impact of Land Conflicts
Conflict in Sub-District of North Polongbangkeng has an impact on socio-economic conditions of the community. Losing land makes it difficult for the community to meet the economic

needs of the family. Fulfillment of nutritional intake is no longer important, as long as they have food to consume. People often consume corn, yams and bananas instead of rice because they can not afford rice. Many children drop out of school due to lack of funds. The culture of mutual cooperation in managing the land has been abandoned.

Another perceived impact is violence and threat since the inception of sugar cane plantations. On land acquisition the community will be accused of being a communist for impeding the construction if they refused to surrender their land. From WALHI data in South Sulawesi, there were at least 14 victims of violence since the sugar cane plantation until 2009. 13 farmers were detained in 2009 and 23 people received a summons from the police.

3.3 The Social and Cultural Value of Makassarese Ethnic Women Farmers

In Sub-District of North Polongbangkeng, Takalar, men and women have the same roles and positions. In the division of domestic roles, productive roles, and social roles, even in family decision making, men and women have equal positions. In fact, often found the portion of work taken by women in the land much more than men. Men also did not hesitate to do domestic work such as cooking or doing housechores for the family.

When land conflicts between PTPN XIV and communities at peak, women also took many roles. Some of the Polongbangkeng Farmers Union (*STP*) members are women, and frequently in a family women who join and become members of Polongbangkeng Farmers Union (*STP*). Their husbands do not object to his wife's activeness to the organization.

Buginese-Makassarese cultural values known as the "*sulapaq appaq*" philosophy are "*caraddeq*" (smart), "*lambusuq*" (honest), "*kalumangnyang*" (rich), "*barani*" (brave) do not distinguished between men and women. In this study the researchers found that the value of the philosophy is also described on women farmers in Sub-District of North Polongbangkeng. Out of four philosophies, the value of "*caraddeq*" (smart) and "*barani*" (brave) most prominent than others. While the value "*kalumangnyang*" (rich) is not found because the people of North Polongbangkeng has lost their livelihood since the involvement of PTPN XIV.

3.4 Farmers' Strategies Face Land Conflicts or Farmer's Strategy in Dealing with Land Conflicts

In the face of conflict with PTPN XIV Takalar Sugar Factory, Polongbangkeng Farmers Union (*STP*), run the general strategy of the struggle they call "Master - Keep - Maintain". The mobilization of Polongbangkeng Farmers Union (*STP*) members to work together in the field also as a show of strength of the peasants in conveying the message to PTPN XIV, that Polongbangkeng farmers have strong unity and fighting power to regain their land. With the collective work of land tenure not only facilitates and accelerates the completion of work, it also strengthens the confidence and spirit of the peasants in taking over their land.

In carrying out the strategy of maintaining the land of the farmers of Polongbangkeng incorporated in the Takalar Polongbangkeng Farmers Union (*STP*), divide the work into three parts of the job field, namely organizational work, politics, and alliance work. In the organization section, *STP* Takalar carries out educational work both at the level of group members and at the leadership level. Political work is done with mass campaigns and propaganda through mass action. While alliances work is done by strengthening forces in external Polongbangkeng Farmers Union (*STP*), mainly student organizations, Non Governmental Organizations (NGOs), and Legal Aid Institutions (LBH) located in the city of Makassar.

After the mass action carried out in early 2014 the concept of partnership back in the idea by PTPN XIV. In a letter sent to *STP*, PTPN XIV is willing to grant 2 Ha / KK area to *STP* Takalar members. Takalar *STP* members generally agree with the proposed partnership pattern with the record that all *STP* Takalar members get the land to be managed. However, until now the partnership program is still unclear. PTPN XIV in the end only gives the land area of 125 Ha to be managed as people's sugar cane. The lack of land provided by PTPN XIV to *STP*

Takalar to be managed as a sugar cane makes only a small proportion of *STP* members who can grow sugarcane. Average *STP* members who manage partnerships between 40 and 60 are.

4. Discussion

This study finds that the strategy undertaken by women in facing agrarian conflict is by joining men to fight for the right to land. The loss of land has made people lose their livelihood. The absence of skills other than farming has impoverished society and deprived them of the agrarian culture they have been practicing for decades. Women farmers as part of the community do not abide with the conditions that occur in the socio-economic conditions of the family. Women to deal with the impact of the conflict implement various strategies. The presumption that women only live in the shore, wells and rooms does not apply.

The study also found that the cultural values contained in the "sulapaq appaq" philosophy of "*caraddeq*" (smart), "*lambusuq*" (honest), "*kalumangnyang*" (rich) and "*barani*" (brave) can be found in this study. The value of "*caraddeq*" (smart) and "*barani*" (brave) is very prominent compared to other values. The value of "*kalumangnyang*" (rich) in the sense of physically rich can not be found until 2007. Improvement of economic conditions in *STP* members can be found after the tenure of land for 5 years, but again declined since 2014. The value of "*caraddeq*" (smart) was done. They put their cattle into a sugar cane plantation that they claim to own. This strategy is done subtly and very effectively in the process of community land acquisition. Incorporating cattle into sugarcane plantations keeps the community protected from the threat of land acquisition reports. The strategy of inclusion of more massive livestock is done after the establishment of *STP* because the organization considers that the strategy is very effective and can avoid intimidation and violence from PTPN XIV and security apparatus. While the value of "*barani*" (brave) can be seen since the beginning of manifest conflict until now, women are always at the forefront when dealing with security apparatus or with PTPN XIV.

In the conflict that took place in Takalar, in addition to sharing productive roles by doing various jobs to make money for families, the women of North Polongbangkeng also took over social roles by engaging directly with the Polongbangkeng Peasant Union either as a member or a leader. Even before the arrival of the Polongbangkeng Peasant Union women had begun the resistance in taking over their land. The role of women in North Polongbangkeng is in line [9] the Folklore text describing the activities of women and men in carrying out their natural (reproductive role) roles and cultural roles (domestic role, productive role, and social role) shows that there are no restrictions on the rights and powers for men and women in various areas of life.

In latent conflicts, communities were forcibly accepted land contracts for 25 years. Women choose avoidance strategies until their land contract expires. To survive the community one by one left his hometown. Men become construction workers, pedicab drivers, and freelance workers in other areas to become migrant workers abroad. While women living in villages are also looking for strategies to not fully rely on their husbands' incomes.

In this period there was no female resistance to defend the land. Women choose to do a strategy to survive with "*mappulung-pulung*" (looking for remnants of rice) and "*makkai*" (rice harvest workers) in Wajo Regency, Pinrang Regency, and Sidrap Regency. A sugarcane farm worker faces the risk of violence and threats from sugarcane plantation workers, *mattesang* fields and livestock, and looking for work outside Takalar.

At the beginning of the conflict, the community considers that their land contract has ended. In addition to the basic ownership claims that led to farmers' resistance to land tenure by PTPN XIV, another factor that contributed to the resistance of farmers was the socio-economic factor where since land tenure by PTPN XIV had many negative impacts on socio-economic community [10]. In this period women are more involved in fieldwork in mastering and maintaining land. Women still lack confidence to speak in meetings and forums.

Women's role and involvement became more visible after the formation of Polongbangkeng Farmers Union. The consolidation work garnered the strength of the alliance for the struggle of Polongbangkeng Farmers Union, many played by women by visiting NGO offices and visits to campuses in Makassar. After delivering testimony regarding the land conflicts they face, visits to offices and campuses will end with a request to join the alliance and give support to the Polongbangkeng peasant struggle.

The mode used by companies to silence people protesting land grabs is by using law enforcement such as police and army, they also hired thugs [11]. This condition also occurs in Sub-District of North Polongbangkeng, from the land acquisition process until reclaimed the land with expired contract by the community. In order to avoid violence or physical clashes in the field during the action to defend the land, prioritizing the strength of women farmers to survive in the good location as a negotiator while arguing with the PTPN XIV and police officers, as well as a pioneer in the formation of defensive action against police officers, employees and PTPN XIV's tractors that will go to cultivate farmers' land.

Women also began to emerge as a leader figure after the establishment of *STP*. In the First Major Consultative Assembly, there were no women who became heads. At the Second Deliberation II was elected 1 woman from 9 collective leaders of *STP*, and III Deliberation III was elected 3 women from 13 *STP* collective leaders.

The efforts to resolve this conflict have not been clear. Partnership program initiated by PTPN XIV and local government still proceed to this day, but not yet able to solve the main problem that happened in Sub-District of North Polongbangkeng. The concept of partnership can actually be a middle ground for land conflicts that occur in Sub-District North Polongbangkeng. Where by submitting the land to be managed as sugarcane of the people to *STP* Takalar then both parties in conflict will be equally benefited. Farmers will return with their livelihoods and PTPN XIV will get raw materials for its sugar industry. The idea of conflict resolution is to hand over the responsibility of producing raw materials to farmers [12] to enhance agricultural and rural development [13] and improving income through agriculture [14]. Production of raw materials is basically not difficult to be implemented by farmers; the problem is the unavailability of land for farmers. So the way out is to hand over land utilization to farmers to produce industrial raw materials. Furthermore, the company focuses on running the industry or processing raw materials produced by farmers.

5. Conclusion

Land conflicts occurring in Sub-District of North Polongbangkeng between farmers and PTPN XIV has had an impact on the economic, social and cultural life of the community. Women farmers in the face of the conflict took the strategy of joining men in fighting for their land rights. This strategy has advanced women and made women stronger and more confident. The value of *caraddeq* (intelligence) and *barani* (brave) in the philosophy of *sulapaq appaq* affect women farmers in determining strategy. This research hopes to solve conflicts that are just for the community. The government must find a middle ground to solve the problem of land conflict in Sub-District North Polongbangkeng without harming the community.

References

[1] Fahmid I M 2013 Typology of Rural Agrarian Conflicts, *Journal of Economic and Sustainable Development* **4** (3) 171-178

[2] Konsorsium Pembaruan Agraria (KPA) 2016 Catatan akhir tahun. Available from: http//www.kpa.or.id accessed on October 2016.

[3] Bachriadi D and Gunawan W 2011 *Enam Dekade Ketimpangan Masalah Penguasaan Tanah di Indonesia* (Bandung: ARC, Bina Desa & KPA).

[4] Mintoro A 1984 *Distribusi Pendapatan Dalam Prospek Pembangunan Ekonomi Pedesaan Indonesia*. (Jakarta: Yayasan Obor Indonesia).

[5] Fauzi N 2013 Rantai Penjelas Konflik-Konflik Agraria yang Kronis, Sistemis dan Meluas *Jurnal Bhumi* **33** 1-14

[6] Aliansi Gerakan Reforma Agraria (AGRA) 2016 *Catatan Akhir Tahun*. Available from: http//www.agraindonesia.org. accessed on July 2017.

[7] Rozaki A 2016 Konflik Agraria, Perempuan, dan Kemiskinan Di Desa *Jurnal Musawa* **15** 1

[8] Fatinaware I 2010 *Perempuan dalam Pengelolaan Sumber Daya Alam dan Lingkungan* (Jakarta: WALHI)

[9] Iswary E 2010. *Perempuan Makassar: Relasi gender dalam folklor*. (Yogyakarta: Penerbit Ombak).

[10] Budianto 2015 *Perlawanan petani dalam konflik agraria (Studi kasus konflik agraria masyarakat takalar dengan PTPN XIV di Kabupaten Takalar)* (Thesis) (Makassar: Universitas Hasanuddin)

[11] Widiyanto 2013 Potret konflik agraria di Indonesia *Jurnal Bhumi* **33** 15-27

[12] Hamid I 2015 *Konflik Agraria dan Jalan Keluarnya (Studi Kasus Konflik antara PTPN XIV dengan petani Polongbangkeng Utara Takalar dan Keera Wajo)* (Thesis). (Makassar: Universitas Hasanuddin)

[13] Ekasari Z K, Saleh S A M, Jusoff K, Salman D, Akhsan, Kasirang A, Arsyad M, *Amrawaty* A A and Fudjaja L 2013 Communication Pattern and Conflict in Agricultural Extension *Asian Social Science* **9** (5) 27-33

[14] Thalyta E Y, Mustajab M and Arsyad M 2010 *Abstracts of Doctoral Theses on The Indonesian Economy* Bulletin of Indonesian Economic Studies **46** (2) 251-254 DOI: 10.1080/00074918.2010.486113

IOP Conf. Series: Earth and Environmental Science **157** (2018) 012069 doi:10.1088/1755-1315/157/1/012069

Stakeholder analysis in the management of irrigation in Kampili area

Jumiati[1], M S S Ali[2], I M Fahmid[2], Mahyuddin[2]

[1] Graduate Student, Hasanuddin University, Jalan Perintis Kemerdekaan KM 10, Makassar, 90245, Indonesia.

[2] Department of Agriculture Socio-Economics, Faculty of Agriculture, Hasanuddin University, Jalan Perintis Kemerdekaan KM 10, Makassar, 90245, Indonesia.

E-mail: *jumiati.amin@unismuh.ac.id*

Abstract. Irrigation has appreciable contribution in building food security, particularly rice crops. This study aims to analyze the role of stakeholders involved in distributing of irrigation water. The study was conducted in the Kampili Irrigation Area in South Sulawesi Province Indonesia, the data were obtained through observation and interviews with stakeholders involved, and analysed by stakeholder analysis, based on the interests and power held by the actors. This analysis is intended to provide an optimal picture of the expected role of each stakeholder in the management of irrigation resources. The results show that there were many stakeholders involved in irrigation management. In the arrangement of irrigation distribution there was overlapping authority of the stakeholders to its management, every stakeholder had different interests and power between each other. The existence have given positive and negative values in distributing irrigation water management, then in the stakeholder collaboration there was contestation between them. This contestation took place between the agriculture department, PSDA province, the Jeneberang River Region Hall, the Farmers Group and the P3A.

1. Introduction

Food sovereignty can be achieved if supported by good irrigation. Irrigation is a requirement for paddy fields therefore in order to sustain water supply for irrigation, it needs improvement from upstream to downstream, from the presence of rivers and dams as the vital infrastructures [1]. One of the existing irrigation sources In South Sulawesi is coming from Jeneberang River. Jeneberang River is a legendary river in Gowa regency. Jeneberang is one of 15 major rivers in Gowa. Started from Mount Bawakaraeng, flowing through the area of Gowa Regency and disembogue into Barombong and Tanjung Bayang. Jeneberang River which has a length of 75 km with the extent (coverage area) of 727 km^2 flowing from Mount Bawakaraeng, this river often overflows during the rainy season that occurred in December to January. The most severe condition occurred in 1976 almost 2/3 city of Ujung Pandang (Makassar) was inundated. This water comes from the overflowing of the Jeneberang River in the downstream area of the Sungguminasa Bridge, and drainage channels like Sinrijala, Jongaya and Panampu are inadequate in catching water, on the other hand during the dry season unable to meet the needs of irrigation and drinking water [2, 3, 4, 5].

The dams that handle irrigation are Bili - Bili, Bissua and Kampili irrigation. Service area capacity, Bili - Bili is 2,350 ha, Bissua is 10,758 ha and Kampili area is 10,545 ha. In the technical and operational implementation in the management of Kampili Irrigation Area, there are many stakeholders involved, from the central to the local level [4].

The objectives to be achieved in this research were: 1) to identify stakeholders in the management of Kampili irrigation; 2) to get an explanation of interest and influence of each stakeholder in the management of Kampili irrigation

2. Research Method
This research paradigm used constructivism paradigm, this research type was a descriptive by using case study approach. The research was conducted at Kampili Irrigation Area in South Sulawesi Province of Indonesia

3. Data collection
Data collection was done through observation and interview to a number of informants [6, 7]. Interview activities intended to gain an explanation of the interests and influence of each stakeholder, as well as the role of stakeholders in accommodating the interests of farmers in obtaining and fulfilling their needs in obtaining good irrigation.

4. Data analysis
Analysis of data used stakeholder based on interest and power [8, 9, 10] (figure 1).

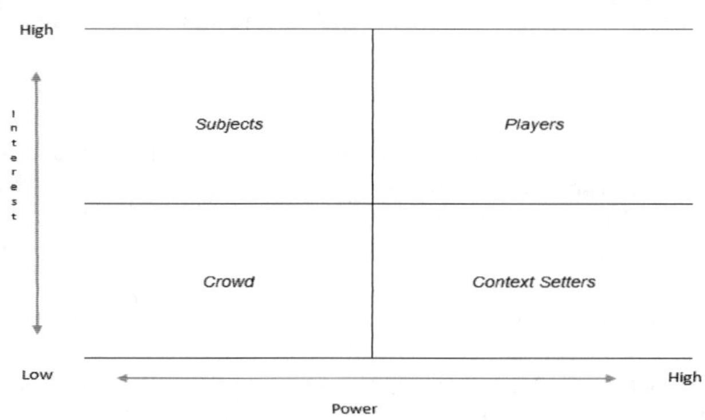

Figure 1. Matrix of Stakeholder *Gird*

5. Results and Discussion
Mitchell and Wood [11] define stakeholders as groups or individuals who can influence and or be influenced by a particular goal achievement. While Fletcher [12] briefly define stakeholders is a person with an interest or attention to the problem.

These stakeholders have an interest and power in each owned positions of importance and influence of each stakeholder [12, 11, 10]. So that stakeholders can be grouped into three namely primary, secondary and key stakeholders in mapping of Stakeholder of Kampili Irrigation Area. Stakeholders generally have high interest and power over the management of the Kampili Irrigation Area.

This analysis intended to provide an optimal picture of the expected role of each stakeholder in the management of Kampili Irrigation Area. The meaning of interest here is the interest / concern of stakeholders in the management of Kampili Irrigation Area while power is the strength/ability/authority/influence of stakeholders to implement the management of Kampili Irrigation Area. All parties involved in the management of the Irrigation Area are grouped in four quadrants (two-by-two matrix) as in figure 2.

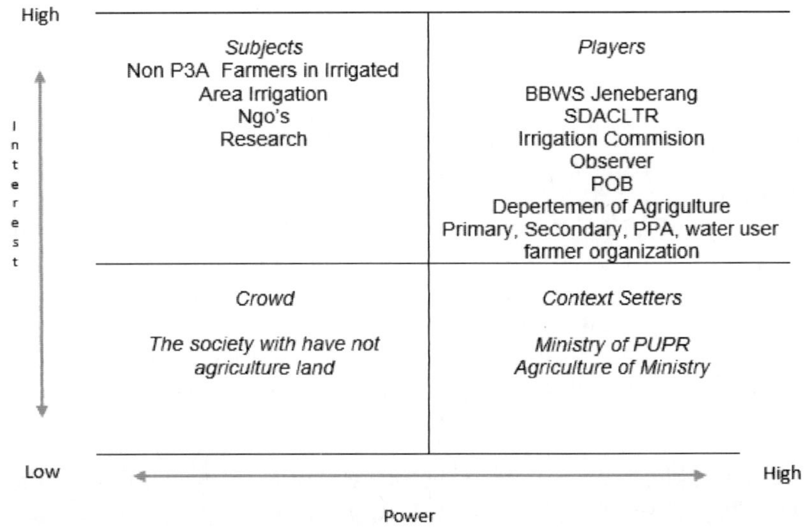

Figure 2. Mapping of stakeholders based importance and inteligance

6. Subjects

In Kampili irrigation area management, the matrices of interest and power of stakeholders for irrigation distribution management are grouped into high interest - low power. Subjects are the stakeholders who are consists of non P3A farmers, NGOs and researchers. These stakeholders have a high interest but have no influence or authority in regulating the management of water according to their need. Non P3A. farmers or farmers around the channel or irrigation buildings at upstream and downstream but outside the in the irrigation map, have no right to obtain water distribution from irrigation. However, they are in dire need of water for the farm, thus these farmers sabotage in the main irrigation channel to meet their needs. While for NGOs and researchers have an interest in improving the distribution of irrigation but they have no right and authority in the regulation or management of Kampili irrigation area as one of the instrument for agricultural and rural development [13] to make a direct contribution to reducing poverty [14].

7. Players

Stakeholders who have high interest and power in the management of Kampili irrigation area.

7.1. BBWS Jeneberang, Irrigation Management Unit

Has the most authority and interest in the management of the river, dam, weir, primary and secondary irrigation, having a considerable source of funding BBWS Jeneberang has not been able to produce improved conditions as farmers hope in the operation and maintenance of the Kampili irrigation area. Where in the management of Irrigation Area there is still a lot of damage to irrigation buildings, garbage disposal, illegal buildings operating in the upstream area and no sanctions given so that many farmers' needs are not met well in accordance with the schedule and the amount of water. Likewise, with the operation of the irrigation, where in the distribution of water resources prioritize the PDAM and PLTA so that the supply for irrigation is still lacking.

7.2. Irrigation Commission

Is an agency that handles when the dam door is operated for water distribution / order of opening of the weir gate, the determination of planting schedule, but this institution is always slow in the implementation. Often times after planting irrigation water not flowing, so the farmer pumped the water or submitted request to IP3A to coordinate with Primary Interpreter and POB.

7.3. POB, Staf and PPA

Have a high interest and importan role in the management of Irrigation Area, because they have authority in giving permission in opening of water doors.

IOP Conf. Series: Earth and Environmental Science **157** (2018) 012069 doi:10.1088/1755-1315/157/1/012069

7.4. Water User Farmer Organization (P3A, GP3A and IP3A),
Have the authority to coordinate from irrigation to tertiary in terms of distribution of farmer's needs. These stakeholders are working in accordance with their respective responsibility. Some water gates that are sometimes not maintained in accordance with the schedule especially the secondary doors, so that farmers through the foremen open the gates which should not be done.

8. Context Setters

The management of the Kampili Irrigation Area which is located under the central shelter because it has a distribution area above 3000 hectares, it is included in context setters is the stakeholders who have a big role in determining the direction and policy of Irrigation management. However, Irrigation management in the Jeneberang River Region is not the only area but many areas in it are in addition to Kampili Irrigation Area Management. Likewise, with the ministry of agriculture that has the authority in providing assistance facilities and infrastructure in irrigation management, especially tertiary areas.

9. Crowds

There are many stakeholders do not care about the condition of Kampili irrigation area, let alone to keep and repair, they even contributed in eroded the building, by littering into irrigation channel, making distribution less substandard and damaging the condition of the building. Communities living around the channel consider that irrigation channels are the safest place to dispose of waste because the water flows so that in the downstream area the has tons of garbage which obstructing the distribution of water. This is due to the lack of sacntions enforcement are given to the people who destroy the building despite the fact that the rules and sanctions are written around the irrigation areas

10. Conclusion

There are many stakeholders involved in irrigation management. In the arrangement of irrigation distribution there is overlapping of authority of stakeholder, every stakcholder has different interests, and power different from each other. The existence of stakeholders has given positive and negative value, in the management of irrigation water distribution, and then in the stakeholder collaboration there is contestation among the stakeholders.

Reference

[1] Fahmid I M 2004 *Gagalnya politik pangan di bawah rezim orde baru: kajian ekonomi politik pangan di Indonesia* (Jakarta: Yayasan Sandi Kota-ISPEI)

[2] Anonim 2017 Kabupaten Gowa Dalam Angka. https://gowakab.bps.go.id/ *diakses 5 Januari 2018*

[3] Rampisela D A 2016 *Buku Pegangan Pengaturan Air Irigasi Wilayah Irigasi Bili - Bili DAS Jeneberang - Sulawesi Selatan* (Makassar: Universitas Hasanuddin)

[4] BBWS 2018 Profil Balai Besar Sungai Pompengan Jeneberang

[5] Rampisela D A, Marunding R and Bulaeng 1999 *Croping Pattern in Working Area of FWUA Sirannuang Bili - Bili Irrigation System* (Makassar)

[6] Creswell J W 2013 *Research Design : Qualitative, and mixed Methods Approaches* (California: Sage Publications)

[7] Creswell J W 2007 *Chapter 3: Designing a Qualitative Study* (California: Sage Publications)

[8] Crosby B 1991 Stakeholder Analysis : A Vital Tool for Strategic Managers *U.S Agency Int. Dev.* 1–6

[9] Bryson J M 2004 What to do when stakeholders matter: a guide to stakeholder identification and analysis technique *Public Manag. Rev.* 9–11

[10] Wilayah S and Citarum S 2010 For Integrated Water Resources Management (IWEM) In The 6 Ci â€™ S River Basin Territory - Package B 'Citarum Stakeholders Analysis'

[11] Mitchell R K and Wood D J 2017 Toward a Theory of Stakeholder Identification and Salience : Defining the Principle of Who and What Really **22** 853–86

[12] Fletcher A, Guthrie J, Steane P, Roos G and Pike S 2003 Mapping stakeholder perceptions for a third sector organization *J. Intellect. Cap.* **4** 505–27

[13] Ekasari Z K, Saleh S A M, Jusoff K, Salman D, Akhsan, Kasirang A, Arsyad M, Amrawaty A A, Fudjaja L 2013 Communication pattern and conflict in agricultural extension *Asian Soc. Sci.* **9**

(5) 27-33
[14] Thalyta E Y, Mustajab M, Arsyad M 2010 Abstracts of doctoral theses on the Indonesian economy, Bulletin of Indonesian Economic Studies, 46:2, 251-254, DOI: 10.1080/00074918.2010.486113

IOP Conf. Series: Earth and Environmental Science **157** (2018) 012070 doi:10.1088/1755-1315/157/1/012070

The emerging roles of agricultural insurance and farmers cooperatives on sustainable rice productions in Indonesia

C Lopulisa[1], Rismaneswati[1], A Ramlan[1] and I Suryani[2]
[1]Deparment of Soil Science, Hasanuddin University, Jalan Perintis Kemerdekaan KM 10, Makassar, 90245, Indonesia.
[2]Faculty of Agriculture, Cokroaminoto University, Jalan Perintis Kemerdekaan KM 10, Makassar, 90245, Indonesia.

E-mail: *riesma76@yahoo.com*

Abstract. Rice is the main staple food of most Asian countries including Indonesia. Most of the rice producers are constituted by small individual farmers characterized with mostly landless, have a less farming capitals and less access to pool resources and of course are confronted with various risk. Agriculture is faced with a lot of uncertainly most of which are not within the control of farmers. Global climatic change, climatic disasters, fluctuation of global economic and competitiveness of multinational company make difficulties of farmers to pursue his sustainable farming activity. The challenge and the role of government is to reduce uncertainly and to improve resiliency of the small farmer. Agriculture insurance shall focus on risk factors that are difficult to manage or cannot be managed by small farmers and it is should be viewed as just one aspect of the "**holistic**" risk management strategy. Technology, market, consumer, behaviour, development will always move forward, and no individual farmers can adapt this change alone, so small farmers need to corporate with each other that can optimized the resources they have. Cooperative could create possibilities, value added, shortening the supplied chain, made a product more effective and efficient, and finally can complete in domestic and global markets. Therefore, agriculture insurance as well a farmer cooperative may play an important role on sustainability of rice production in Indonesia. Nowadays and in the future agriculture sustainability is a not merely of technology problems but also a matter of economic-social-culture and politic issues within local, national, and international context.

1. Introduction

1.1. Fact About Rice Farmers and Farming in Indonesia

The contribution of agriculture to product domestic bruto of the country reach 10.28% at the year of 2017 which around 3.18% of it given by seasonal crops including rice. Total rice production amounted to 75.36 million ton of unhulled rice (9kg). Total number of the rice farmer were 17.73 million farmers households or 67.83% of the total farmer households of the country. Most of the rice farmers were small farmers; land holdings of less than 0.5 ha per farmer households, very poor access to institutional credit: poor capital formation, still taking loan from traders/money lenders. Most of the rice farming activities were done by individual farmers: rice cultivation, purchase of farming inputs (seeds, fertilizer, pesticide), produce procession and marketing, low adaption of technology, poor forward and backward linkage. Despite this rice's productivity average are still low as compared to potential yield that can be obtained by using recommended fertilizer and seeds [1].

IOP Conf. Series: Earth and Environmental Science **157** (2018) 012070 doi:10.1088/1755-1315/157/1/012070

1.2. Farmers Groups and Farmer's Group Associations

Farmers groups in Indonesia were developed since in the early 1970s with the objective as media for extension services to meet the farmers for dissemination of technology and delivery farming input the "Bimas Program" for increasing rice production and receiving facilities from government of the country [1]. A farmer group consist of 25 to 50 neighbourhood farmers and 15 to 25 ha rice fields at subvillage level. After around 30 years, there is no attempted has been made to develop a farmer group, since 2004-2009 the government (ministry of agriculture) did amalgamation of a farmer groups within village as called "Gapoktan" which united all farmer groups in a village. Gapoktan may consist of 10-20 farmer groups which involved 250 to 500 ha rice field and 250 to 500 farmers members. Till now, no significant effort has been made to established a Gapoktan to become a farmer cooperates.

2. Agriculture Insurance and Smallholder Farming Dimension

2.1 Introduction

Small holder farming was characterized by a family operated firm often living in poverty, not diversified production and mean of living, more vulnerable in front of climatic risk, and economical fluctuations, difficult to reach geographically, living with lack of infrastructure, low level of education and awareness and they were not a target-group for big financial institution because of micro financing scale. Such circumstances made small farmer were confronted with various risk. Risk is a probable loss due to unfavourable factors such as: (1) weather like a periodic fluctuations in rainfall and strong winds, (2) natural disasters as floods, droughts, earth quakes, volcanic eruptions, (3) biological and environmental like pest, disease, farm contamination, degradation of resources, (4) market such change in supply and demand, changing standard, supply chain failures, (5) logistical and infrastructure cost and reliability, infrastructure destruction, conflict, labour disputes, (6) management and operational like poor management decisions, asset allocation, resistance to change, (7) Policy and institutional, monetary, tax labour and (8) political : domestic and international security concerns, trade disputes, nationalisation, and confiscation.

The challenge and the role of government is to reduce the uncertainly and to improve the resilience of the small farmers. Agriculture insurance should be view as just one aspect of the "holistic" risk management strategy. Agricultural Insurance plays an important role in reducing vulnerability of global food system to acute food shocks there by contributing to resilience and sustainability. The objectives of agricultural insurance are: (1) Protection for livelihoods and smoothing income during mayor events, (2) protection against default and (3) safety net, hence the allied objectives of agricultural insurance as a medium of risk transfer are: reducing poverty, reducing the severity of events and ecologically sustainable production of good. Insurance is not a magic wand-insurance should be (1) Economically effective, sustainable overtime, not risk distorting, socially responsible, systemic effect (BookP, 2016). Some critical points, of agricultural insurance are: (1) technically complex and one of the most expensive lines more satisfied when the standard of agricultural are of insurance, (2) farmer are more satisfied when the standard of insurance the same across the whole country, (3) insurance cannot cover 100% of farmers risk; the risk need to be layered and retained by different stake holders, (4) insurance is only one of multiple possible risk management tools, (5) insurance product and infrastructure development requires financial investment, which is beyond insures possibility, (6) the functions of different stake holder with PPP (Public Private Partnership) vary from country to country, but only their combination assures the whole system functionality and sustainability.

The role of the government on the establishment of agricultural insurance in many countries were highly significant. Type of government support to agricultural insurance program were: (1) Premium subsidy, state reinsurance protection, price support mechanism, tax rebates, (2) crop and weather – data, data processing, actuarial calculation, product design and insurance performance analysis, (3) informational and educational activities, legislative regulation capital requirements, (4) infrastructure; weather stations, irrigation system [2].

IOP Conf. Series: Earth and Environmental Science **157** (2018) 012070 doi:10.1088/1755-1315/157/1/012070

There are world well known agri-insurance PPP managing agents, federal crop insurance coorporation USA, Agro Segoro, Spain, Alberta Finance service corporation, Monitoba, agricultural service corporation, Agricurp - Canada, Tarsim - Turkey : African Risk Capacity, Mongolion Agricultural Reinsurance joint stock company, PCIC, Philippines crop insurance corporation, PCIC): IBLIP – Mongalion – Pool of insurance under the umbrella of Mongolian Agriculture Reinsurance Company Providing Mortality index based livestock insurance. Agriculture is faced with a lot of uncertainty, most of which are not within the control of farmers risk can be controllable, uncontrollable, catastrophically (systematic) risk management strategies may include; saving, help insurance risk retention, risk avoidance, diversification, production, fragmentation, future contracts, crop modelling, maintaining farm health of plants and animals, insurance, linking insurance to credits, innovations, optimization, job outside of farm, government support (emergency relief) international financial support (donors funds).

Some risks can be insurable or not depends on the circumstances (a) Mutuality (risk community, big number of participants, (b) Accessibility (loss should be assessable, relation between the risk and the loss) (c) Randomness (the event occurrence time must not be predictable, the event is independent from the insured's will) (d) Economic viability (the loss-related financial needs can be covered on a planned basis) (e) Similarity of threat (all insured are exposed to the same threat). Which is not insurable goes into cover exclusions?

Functions of insurance were: (a) Provides protection against probable loss (b) Removes the uncertainties (sense of security) (c) Stabilized the farm income (d) Enable capital borrowing from bank and investment in Equipment / technology (e) Helps farm efficiency, production growth and development (innovation) (f) Attracts service of skilled insurance specialists to the insured, sometimes helps to prevent loss. (g) Disciplines (insurance terms and requirements).

2.2 USA Agro-insurance Schema and Regulations

Successful models and schemes and best practices in the management of agricultural insurance schemes/programs.

- The agro-production represents 1% of GDP, the world largest arable land, (b) Largest crop insurance market with a premium above 10 billion dollars, (c) The Program is available from 1915, first as a hail insurance, since 1980. MPCL was established after passing the crop insurance Act, (d) The federal crop insurance corporation (FCIC) administer the whole system, (e) Main crops covered are corn, cereal, soya and cotton, (f) The largest premium is written in Illinois, Dakota, Iowa, Minnesota, Nebraska, Kansas, and Texas.
- PCIP, Philippines: Philippines Coop Insurance Corporation 2015:
- No of farmers 1:194.932, Amount of cover 41.078.791, (Peso Million) Area (Ha) 855.504,5 Claim Paid, No of Farmer 154.224, Indemnity (em)1,378.57, Damage 3.36, Loss ratio (%) 0,52.
- Fondos Mexico: 370 small cooperative-like fondos accept the farmer risk and obtain proportional reinsurance from Agroasemex, Insurance
- Tarsim, Turkey: Pool of 23 insurance companies that cede the premium to a common tarsim insurance portfolio, which reinsured private, Insurance can retention, Insurance premium is close to 300 mn Euro [3].

3. A Farmer Cooperative

3.1. Definition of Cooperative

Development will always move forward and change the technologies market and consumer behaviour and will no individual farmers can adapt this change alone. Therefore, small farmers need to cooperate with each other that can be optimized the resources they have, cooperative could create possibilities, value added, shortening the supplied chain and made a product more effective and efficient and finally can compete in domestic as well as global market.

IOP Conf. Series: Earth and Environmental Science **157** (2018) 012070 doi:10.1088/1755-1315/157/1/012070

Co-operative is an enterprise which is jointly owned and democratically controlled by members; an autonomous association of persons united voluntarily, to meet their common economic, social and cultural needs and aspiration [4, 5]. According to USDA, 1987 co-operative are enterprises owned and controlled by the users which have:

1. User-owner principle: Those who own and finance the co-operative are those who use the co-operative
2. User-control principle: Those who control the co-operative are those who use the co-operative
3. User-benefit principle: The co-operative sole purpose is to provide and distribute benefits to its users based on their use

Social aspects of co-operatives [4, 5, 6]:

- To provide members with services or economic institutions which is beneficial to them and socially desirable
- Is a hybrid organization which is balanced between management of enterprise and social interest?

Why agricultural co-operative?

1. Market failure, (a) to build countervailing power against spatial monopsony (oligopsony) power in downstream or monopoly (oligopoly) power in upstream agro-food market. (b) To avoid expose market power or opportunistic behavior and (c) To avoid cheating based on information asymmetry
2. Market imperfections: (a) To supply missing services avoided by private company due to low profitability, (b) to gain from better coordination in horizontal or/and vertical dimension

Therefore, establishing and sustaining an agricultural co-operative is an effective means for farmer to save various type of transaction costs or to create additional value through better coordination.

Aggregation of farmer may improve bargaining power such as access to markets, inputs, technology and investment and enables small and medium farmers to exploit emerging opportunities domestic and global by integrating with SMES, retail chain and other organisations. Aggregation facilitates effective targeting of subsidies to farmers and aggregated platforms for efficient delivery of agri-inputs, insurance, etc for climate change coping mechanisms and risk mitigation [6, 7].

Typology of agricultural co-operative

- Co-ops that directly engaged in farming activity (production co-op)
- Co-ops that provide all kinds of goods and services to the farmer (supply co-op); farm inputs, farm machinery services, credit, insurance etc
- Co-ops that have taken over the sales activities of the farmer (marketing/processing co-op)

3.2. Successful Models of Agricultural Co-Operatives

3.2.1. India. The cooperative movement in India has witnessed substantial growth in diverse areas of the economy [7]. The cooperative movement in India has emerged as one of the largest in the world [9]. At present total number of cooperatives is amounted to 610020 with total number of members around 249.3 million. Total number of primary agricultural cooperative society (PACS) which are the foundations of Indian cooperative movement and agriculture is 147001 with number of members 181.15 million with rural network (village covered) 100% [10].Some examples of India's most successful cooperatives are:

- Sugar cooperatives;
    - 280 cooperative sugar mills out of 340 mills, 30 million sugarcane grower-members, 50% of world sugar, Extensive diversification alcohol, paper, manufacture, power generation irrigation, large number of social institutions, strong welfare schemes

- Dairying cooperative;
    - 120.000 primary dairy cooperative, 12 million individual dairy-farmers, largest milk productions, Large producers like "AMUL", 45% share of co-op ice cream, animal-care services, cattle-feed plants, extensive marketing network, 80% dairy farmers are women
- Cooperatives Banking;
    - Large, medium, small credit through 63.000 functional PACS, 32 state cooperative Banks, 850 district/subdistrict banks with 1860 branches, World's largest agri-credit system, Central financing through NABARD (National Board Agriculture Rural Development), 8% lending rate (2% subsidy by government), PACS major users for agri production investment in assets, land improvement

Movement and development of cooperative in India started since several decades ago [11]. For instance, AMUL Gujarat Co-operative milk marketing federation Ltd (GCMMF) was funded in 1946, under voluntary action taken by Bhuwan Das Patel and Sardar Vallabhbhai Patel against sale of milk to Maharashtrian Got, with two villages only. Dr.VergheseKurien, arriving in Anand in 1949 as a government employee to manage a dairy, he went from helping farmers repair their machinery to revolutions sing the Indian dairy industry by scripting operation flood corporative movement that turned India from a net importer of milk into one the world's two largest producer to day.

3.2.2. Philippines. Cruzian multipurpose cooperative is a local cooperative organized and existing under the laws of the Republic of Philippines. It is located of Brgy.Sta.Cruz.Magelang.Pampanga.

The cooperative started its operation in 1974 as a SamahangNayon under the trade name Sta.CruzSamahangNayon during the general assembly held on June 6, 1940, the SamahangNayon graduated as a multipurpose cooperative and then it was registered as a full-fledged cooperative wish amended name CRUZIAN multipurpose cooperative, INC on June 14, 1990 is one of the most successful farmer's cooperative in central Luzon. It has gained recognition not only among its peers but to the national government as well. It was a title holder as the most outstanding cooperative in region II for three consecutive years, 1999 to 2001 from then on, it continuously shown competence in its operation [12].

3.2.3. Bangladesh. Integrated farm management component (IFMC) is a 43 million USD project funded by Danida (75%) and government of Bangladesh (25%). It is implemented by Department of Agriculture Extension (DAE) and runs from July 2013 up to June 2018. It covers 373 upsila (sub-district). It has target of implementing 20.000 farmer field schools (FFS) for increased yield and diversification aim to work with more than 1000 farmer organization on marketing [13].

Jathian farmer organization date back to 2008. It was formed after completing a farmer field school on integrated co-op management. Today, it has 54 members (26M-28F). The executive committee has 13 members (9M-4F). Secretary and cashier are females. Jathian FO has collected weekly saving since 2008. They run a loan scheme for members and collective marketing on garlic and lentils.

In 2015, Jathian FO was selected to participate in the marketing and FO capacity building activities of IFMC. Four business Focal Resources (BFPS), two men and two women, attended a 10-day residential training course on farming, as a business, business planning and marketing. After ward's,the BFPS conducted 8 training session for 35 members of their organization and got them mobilizedfor the idea of collective marketing. Later three leaders (chairman, secretary and cashier) attended a 3-day training course on vision development, good governance and book keeping [13].

4. Agriculture Insurance and cooperatives in Indonesian: Perspectives, challenges and opportunities

A farmer groups in Indonesia had been established since 1970's which were main objectives as a farmer group consist of 50 to 100 farmers groups association at village level (5 to 10 farmer groups) was established. At present called Gapoktan administration in national level called were amount to around 70.000 Gapoktan. Due to global climatic change and economic global fluctuating to increase rice production and reduction of poverty the government with the parliament passed the law on year 2013 number 19 on the protection and empowerment of the farmer. The law deals with among other were agricultural insurance (article 37) and on the development of BUMP (farmer owned, and control cooperate (FOCC).Nationally some earlier attempt has been made for agricultural insurance and farmers corporate. On the rice crops year 2015-2016 ministry of agriculture provide ministry of agricultural provide fund of Rp. 150 Billion for agriculture insurance particularly rice on 1.0 million ha of field rice through JASINDO (Insurance Company Owned and Controlled Government) [14]. It is believed that agriculture insurance will encourage production. The orientation of agricultural production ('agricultural sector' in a broad sense) is strongly and directly associated with rural poverty reduction [15], including communication pattern and agricultural extension [16]. In South Sulawesi Province some initial action made to establish a corporate farmer by Hasanuddin University, supervision focus on: (1) the establishment of the farmers' cooperative at village level, (2) services and assistance to development small, medium enterprises at village level that can supply farm inputs on time at minimal cost. This effort indeed shows promising results unfortunately due to limited time and financial capacity the project should ended only in one year.

5. Conclusion

Established farmer group to a farmer corporate in necessary in order to make a small farmer more productive, profitable and sustainable. Services to increasing the productivity and sustainability of rice farming. Intensify guidance and supervising were requires from professionals/researches/extension.A long-term funding agency or sponsorships was not easily to find so International/ regional technical / expertized and finance were highly needed.

References

[1] Fahmid I M 2004 *Gagalnya Politik Pangan Di Bawah Rezim Orde Baru: Kajian Ekonomi Politik Pangan Di Indonesia* (Jakarta: Yayasan Sandi Kota-ISPEI)

[2] Mahul O, and Stutley C J 2010. *Government Support to Agricultural Insurance: Challenges and Options for Developing Countries.* World Bank. https://openknowledge.worldbank.org/handle/10986/2432 License: CC BY 3.0 IGO Accessed on February 16, 2017

[3] Bora B 2016 *Subsidised Agricultural Insurance in Turkey (TARSIM). General Manager of TARSIM.* Turkey. http://www.mapfre.com/ccm/content/documentos/mapfrere/fichero/en/subsidised-agricultural-insurance-in-Turkey.pdf. Accessed on February 16, 2017

[4] Jongik J 2017 *Basic of Cooperatives, Principles and fundamentals; Characteristics and Types of Cooperatives* The Workshop on Emerging Roles of Producer's Associations and Farmers Cooperatives Dhaka Bangladesh

[5] Jongik J 2017 Management *of Cooperatives, Organizational and Financial Perspective*

[6] Sharma M 2017 *Successful Models; Agriculture Extension-Rural SME.S in Cooperatives Increasing Productivity and Profitability/India's Case*

[7] Sharma M 2017 *Changing Roles of Cooperatives to Prepare Future Challenges In Agriculture, Cares in Indian Rural Community*

[8] Gupta J and Jain S 2012 A Study on Cooperative Banks In India with Special Reference to Lending Practices *Int. J. of Sci. and Res.* **2** (10)

[9] Misra B S 2010 Credit Cooperative in India, Past, Present And Future *Routledge Taylor and Francis Group, London ISBN 13:978-0-203-85493*

[10] Yashoda 2017 Role of Primary Agricultural Co-operative Society (PACS) In Agricultural Development In India. *Glob. J. of Manag. and Bus. Res.: C Finance* ***17*** *(3) Version 1.0*

[11] Sandhu H K 2012 Agricultural Development in India and China: A Comparative Study *Int. J. of Soc. Sci. & Interdiscip. Res.* ***1*** *(6) ISSN 2277 3630*

IOP Conf. Series: Earth and Environmental Science **157** (2018) 012070 doi:10.1088/1755-1315/157/1/012070

[12] Philippines Statistics Authority 2004 *A Review of the Agriculture Sector in Central Luzon*
[13] IFMC (Integrated Farm Management Component) Department Agriculture Extension 2016 *IFMC Annual Report 2015-2016*. Ministry of Agriculture
[14] Muthmainah D A 2017 *Membaca Masa Depan Asuransi Pertanian di Tangan Jasindo* https://www.cnnindonesia.com/ekonomi/20171120093212-84-256780/membaca-masa-depan-asuransi-pertanian-di-tangan-jasindo. Accessed on February 16, 2017.
[15] Thalyta E Y, Mustajab M and Arsyad M 2010 *Abstracts of Doctoral Theses on The Indonesian Economy* Bulletin of Indonesian Economic Studies **46** (2): 251-254 DOI 10.1080/0074918.2010.486113
[16] Ekasari Z K, Saleh S A M, Jusoff K, Salman D, Akhsan, Kasirang A, Arsyad M, Amrawaty A A, and Fudjaja L 2013 Communication Pattern and Conflict in Agricultural Extension *Asian Soc. Sci.* **9** (5) 27-33

Application of *in ovo* injection of L-Glutamine for improving productivity of Indonesian native chicken: hatchability and hatching time

D P Rahardja, A R Hakim and V Sri Lestari

Dept. Animal Production, Fac. Animal Agriculture, Hasanuddin University Jalan Perintis Kemerdekaan KM 10, Makassar, 90245, South Sulawesi, Indonesia.

E-mail: *djonipra@gmail.com*

Abstract. A study was aimed to improve productivity of the Indonesian Native Chicken (INC) by the technique of *in ovo* injection (IOI) of L-Glutamine (Gln), which hatchability and hatching time were measured. A 300 fertile egg (45.26±3.65 g) of INC were used and arranged as Randomized Block Design within 3 different incubators as blocks, and 5 treatment groups of 20 eggs: negative control (P0), positive control (P1-0.5 ml sterilized saline), 2.50 (P2), 5.0 (P3) and 7.5 (P4) mg Gln in 0.5 ml sterilized saline. The injections were conducted into the albumen at d 7 of incubation. *IOI* of Gln resulted in a significantly heavier weight of the newly hatched chickens (NHC), Hatchability of the P1 (60.83 ± 7.52%) and P2 (62.50 ± 2.89 %) was significantly higher ($P<0.01$) compared to those of the other groups; On the other hand, the embryo mortality in P1 and P2 were significantly lower compared with the other three groups, which mostly occurred before d 17 of incubation. It is concluded that *IOI* of 2.5 mg Gln of INC increased hatchability, decreased the duration of incubation time and the heavier weight of NHC would apparently have a better performance compared to that of the untreated eggs.

1. Introduction

In Indonesian, chicken is the most widely distributed livestock. In addition to exotic breed farmed commercially, at least 31 unidentified breeds of local chicken were reported [1]. These local chickens usually called "*Ayam Kampung* or *Ayam Buras*" or Indonesian native chicken having different morphologic characteristics. In spite of having the potency to be developed, native chickens have a major constraint in production and reproduction. In meat production, they are growing very slow and having bad feed conversion. In reproduction, they are having low egg productions. In addition to their genetic background, low performance of the native chicken may be attributed with low management input, free scavenge for a feed around the farmer's house during the day, as mostly raised extensive traditional by the farmers for their additional income.

In ovo injection of exogenous materials may apparently be an alternative approach in order to improve the performance of native chicken. Previous investigations mostly on exotic breeds indicated that *in ovo* injection of nutrient into the hatching eggs is to supply adequate nutrient which resulted in improving the hatchability of the egg and subsequent post-hatch performance, body weight at marketing age and carcass quality of the chicken [2, 3, 4, 5, 6, 7, 8]. Moreover, post-hatch growth of the chick is determined by the nutrients present in the yolk remaining in the peritoneal cavity. Egg

IOP Conf. Series: Earth and Environmental Science **157** (2018) 012071 doi:10.1088/1755-1315/157/1/012071

contains an excess of fat and moisture but not protein [9, 10]. Previous research in this laboratory [10] indicated that IOI of Arginine (2.5, 5.0 and 7.5 mg in 0.5 ml sterilize saline) into the *Kampung* chicken resulted in a better performance of pre- up to 42 d post-hatching, including embryo development and body weight gain.

L-Glutamine (Gln) has usually been considered as a non-essential amino acid since most animal cells can synthesize it, and primary function is to store nitrogen in the muscle and to transport it between organ; however during period of rapid growth, the cellular demand for Gln outstrip its supply, and Gln become essential, hence its named as a 'conditionally' essential amino acid. As an illustration, during the first two weeks of chicken embryo growth and development, proliferating cells shows an intense appetite for Gln, reflecting its flexibility as a nutrient and mediator of other processes. Although it contributes only 4% of the amino acid in muscle protein, Gln accounts for more than 20% of the free amino acid pool in plasma and more than 40% in muscle [12, 20]. Attributed to the low input management for Indonesian native chickens, application of in ovo injection of Gln in this research aimed to elucidate the effects on hatchability and hatching time.

2. Material and Method

A total of 400 native chicken eggs were collected from traditional farmers around Bantaeng Regency – South Sulawesi. Eggs were numbered and weighed individually. Eggs were set in a semi-automatic incubator at 37.5 °C (99.5 °F) and 55% relative humidity. The eggs were turned every 3 h. Eggs were examined by candling at d 7 of incubation; the infertile eggs and the eggs containing dead embryos were removed.

After examination, 300 eggs were selected and arranged into a Randomized Block Design consisted of 3 different incubators as the block and 5 treatment groups of 20 eggs having equal weight; In accordance with the treatment group, the eggs were injected with Gln (2.5, 5.0 or 7.5 mg dissolved in 0.5 sterilized saline) as P2, P3, and P4, respectively. Whereas P0 was a control group which without injection (negative control) and P1 was a positive control which treated only with sterilized saline injection. Before injection, each egg was candled to identify the point of the inserted needle, which then cleaned the area with ethyl alcohol (70%) and a piece plastic type of 1 x 1 cm covered the point; a pointed tip of a small scissor was used to make a small hole in the middle of the type. A 0.5 ml solution was injected into the albumen of each egg using automatic syringe equipped with 22 gauge needle to a depth of 10 -13mm. At the end, the hole was sealed with nail paint and placed back into the incubator. The amino acid was provided by Merck, Darmstadt, Germany.

Hatchability percentage was calculated based on the number of hatched chicks as a percentage of the incubated eggs per replication of every treatment. Hatching time was recorded every 12 h. Chicks were observed at 12-h intervals from d 19 to d 21 (456 to 528 h) of incubation, and the weight of newly hatched chicken was recorded using a digital balance to the nearest 0.01 g. Resulted data were analysed by using Statistical Packages of the 13.1. version of Systat.

3. Results and Discussion

The results are summarized in table 1 and table 2. The initial weights of all incubated eggs were not significantly different. However, IOI of Gln increased ($P<0.05$) the body weights (g and % of egg) of newly hatched chick significantly; while the hatchability (%) of positive control (P1) and 0.25 mg Gln (P2) groups were significantly higher compared to those of negative control (P0), 5.0 mg Gln (P3) and 7.5 mg Gln (P4). The Embryo mortality indicated a reverse result and the most mortality occurred before d 17 of incubation. The length of the incubation period of the eggs treated Gln and the positive control group was significantly faster compared with that of the negative control eggs, and hatching times of all egg groups were less than 21 days.

Previous research of IOI nutrients resulted in decreasing [6, 13, 14], increasing [15], or un-affecting [5, 13, 14, 16] the hatchability. These all previous research of IOI were conducted on the eggs of the exotic chicken breeds which have superior genetic traits of the eggs. These contradictive results about hatchability might be attributed to the different breed, nutrition and injection techniques, depth and position of injection, nutrient and osmolality of injected solution, the day of incubation, etc.

Table 1. Effect of in ovo injection of L-Gln on hatchability and newly hatched body weight of Indonesian Native Chicken

Parameter	*In ovo* group				
	P0 = Control (-)	P1 = Control (+)	P2 = 2.5 mg Gln	P3 = 5.0 mg Gln	P4 = 7.5 mg Gln
Egg weight (g)	45.57 ± 5.14 ^{a}	46.21 ± 3.65 ^{a}	45.14 ± 3.28 ^{a}	45.51 ± 3.30 ^{a}	46.51 ± 2.66 ^{a}
NHCW (g)	31.86 ± 0.41 ^{a}	33.32 ± 0.99 ^{b}	33.78 ± 1.31 ^{b}	33.57 ± 1.49 ^{b}	33.73 ± 1.21 ^{b}
NHCW (% egg)	69.92 ± 1.42 ^{a}	72.11 ± 0.73 ^{b}	74.83 ± 2.08 ^{b}	73.76 ± 1.76 ^{b}	72.54 ± 1.14 ^{b}
Hatchability (%)	34.80 ± 4.58 ^{a}	60.83 ± 7.52 ^{b}	62.50 ± 2.89 ^{b}	32.22 ± 7.81 ^{a}	26.11 ± 12.73 ^{a}
Mortality < d17 (%)	46.00 ± 4.73 ^{a}	20.00 ± 8.66 ^{b}	19.92 ± 5.54 ^{b}	41.67 ± 7.55 ^{a}	48.33 ± 5.77 ^{a}
Mortality >d 17 (%)	19.20 ± 4.57 ^{a}	19.17 ± 5.23 ^{a}	17.58 ± 3.58 ^{a}	26.11 ± 5.00 ^{b}	25.56 ± 2.88 ^{b}

^{a,b}Means in the same row with different superscript differ significantly (P<0.05).
NHCW : New Hatched Chick Weight

Tabel 2. Length of Incubation period and hatching distribution over the incubation period of the Indonesian native Chicken treated in ovo injection of L-Gln

Parameter		In ovo group				
		P0 = Control (-)	P1 = Control (+)	P2 = 2.5 mg Gln	P3 = 5.0 mg Gln	P4 = 7.5 mg Gln
Incubation period (hours)		496.14 ± 4.39 ^{a}	482.80 ± 0.65 ^{b}	484.88 ± 3.18 ^{b}	485.33 ± 3.65 ^{b}	484.28 ± 6.53 ^{b}
Hatching time Distrib. (%)	465-480	6.25 ± 2.87 ^{a}	26.47 ± 4.11 ^{b}	18.52 ± 4.76 ^{c}	18.18 ± 4.77 ^{c}	7.69 ± 0.86 ^{a}
	481-504	81.25 ± 4.65 ^{b}	73.53 ± 3.76 ^{c}	81.48 ± 6.83 ^{b}	72.73 ± 7.88 ^{c}	92.31 ± 6.73 ^{a}
	505 -528	12.50 ± 2.46	0	0	9.09 ± 3.74	0

^{a,b}Means in the same row with different superscript differ significantly (P<0.05)

The present study was conducted on eggs of unknown breeds of local chicken; there were some interesting results to be discussed. Firstly, the hatchability of all egg groups in this experiment was lower than those resulted from eggs of the exotic breeds. In addition to genetic background, these differences might be attributed with different managements, particularly for daily feed (quality, quantity, and continuity) intake in the condition of the traditional farming system. Candling data (unpublished) at d 7 of incubation soon before IOI indicated that the fertility of the eggs was about 75%.

Secondly, different levels of IOI of Gln resulted in different hatchability responses of eggs; hatchability of the positive control and IOI of 2.5 mg Gln was significantly higher compared with that of the negative control. When the level of IOI of Gln increased to 5.0 and 7.5 mg, hatchability was to decrease significantly to the levels as that of the negative control. In these levels, IOI of Gln, as "conditionally essential amino acid", may result in amino acid imbalance as the rate of it metabolism is lower than its supply. However, a search of the literature provided no information to support this explanation. Thirdly, the egg size of the local chickens was about 10-20 g smaller than that of the exotic breeds, it appears that IOI of 2.5 mg Gln in 0.5 ml sterilized saline is an optimal dose for the local chicken eggs. Fourthly, the weights of the newly hatched chickens (g or % of egg) of the treated egg groups (including positive control) were significantly higher than that of the negative control. In general, IOI of amino acids is primarily aimed to provide amino acid for gluconeogenesis and then to reduce albumin catabolism and sparing for muscle and tissue protein synthesis, and consequently improving body weight at hatch [17, 18]. IOI of a mixed solution of some amino acids at early incubation resulted in increasing the weights of newly hatched chickens by 2-3.4% [15]. Moreover, in ovo feeding to the late term embryos also increased hatching weight by 5–6% over controls [18]; IOI of all 20 AA increased newly chick weight 2.1% [19]. In the present study, IOI of 2.5 to 7.5 mg Gln in 0.5 ml increased the weight of the newly hatched chicken by 2.62 – 4.91% over the negative control.

Overall, the results of the present study suggested that IOI of 2.5 mg Gln in 0.5 ml sterilized saline to the eggs of Indonesian native chicken increased hatchability, the weights of the newly hatched chicken and shorten the length of incubation period.

Acknowledgment
This work was supported by a grant from Hasanuddin University through the research scheme of The World Class University – Acceleration of International Publication or WCU-PPI – 2017.

References

[1] Nataamijaya A G 2000 *Bul. Plasma Nutfah* **6** (1) 1 – 6
[2] Tako E, Ferket P R and Uni Z 2004 *Poult. Sci* **83** 2023 – 28
[3] Shafey T M, Alodan M A, Al-Ruqaie I M and Abouheif M A 2012 *S. Afr. J. Anim. Sci* **42** (2) 210 – 20
[4] Shafey T M, Sami A S and Abouheif M A 2013 *J.Anim. Vet.Adv* **12** (1) 135 – 39
[5] Shafey T M, Mahmoud A H, Alsobayel A A and Abouheif M A 2014 *South African J. of Anim. Sci* **44** (2) 123 – 30
[6] Salmanzadeh M, Ebrahimnezhad Y, Shahryar H A, and Ghaleh-Kandi J G 2016 *Arch. Anim. Breed* **59** 235–42
[7] Gao T, Zhao M M, Li Y J, Zhang L, Li J L, Yu L L, Gao F and Zhou G H 2017 *J Anim Physiol Anim Nutr* **10** 1–10
[8] Selim Sh A, Gaafar K M and El-ballal S S 2012 *Emir. J. Food. Agric.* **24** 264–71 .
[9] Al-Murrani W K 1982 *Br. Poult. Sci* **23** 171–74
[10] Mueller C A, Burggren W W., and Tazawa H 2015 The Physiology of the Avian Embryo *Sturkie's Avian Physiology* ed C.G. Scanes 6th ed., (Amsterdam : Academic Press Elsevier,) chapter 32 pp. 739 – 66
[11] Azhar M, Rahardja D P and Pakiding W 2016 *Media Peternakan* **39** (3)168–72
[12] Curthoys N P and Watford M 1995 *Ann. Rev. Nutr* **15** 133–59
[13] Zhai W, Gerard P D, Pulikanti R and Peebles E D 1999 *Poult Sci* **90** 2134–43
[14] Mcgruder B M, Zhai W, Keralapurath M M, Bennett L W, Gerard P D and Peebles E D 2011 *Poult. Sci* **90** 1058-66
[15] Bakyaraj S, Bhanja S K, Majumdar S and Dash B 2012 *J. Sci. Food Agric.* **92**313–20
[16] dos Santos T T, Corzo A, Kidd M T, McDaniel C D, Torres F R A and Araújo L F 2010 *J. Appl. Poult. Res* **19** 1–12
[17] Stevens L 2004 *Avian Biochemistry and Molecular Biology* (Caambridge, UK: Cambridge University Press) p 195 - 211
[18] Uni Z, Ferket P R, Tako E and Kadar O 2005 *Poult.Sci* **84** 764–70
[19] Bhanja S K and Mandal A B 2005 *J. Anim. Sci* **18** 524 – 31
[20] Newsholme P, Procopio J, Lima M M R, Pithon-Curi T C and Curi R 2003 *Cell Biochem. Funct* **21** 1–9

Mapping of land tenure institution rotating patterns in the highlands

N Mappa[1], **D Salman**[2], **AR Siregar**[2] **and M Arsyad**[2]

[1]Muhammadiyah University, Makassar, Jalan Sultan Alauddin, No. 259 Makassar, 90222, Indonesia.
[2]Graduate School of Hasanuddin University, Jalan Perintis Kemerdekaan, Km. 10, 90245, Makassar, Indonesia.

E-mail: *nurdin.mappa@unismuh.ac.id*

Abstract. Tenure of farmers in each community has institutional. In general, land tenure institutions regulate the utilization permanently, but there are also institutions that regulate land tenure in rotation. This study aims to map the pattern of land tenure institution in rotation. The study used case study method, with case unit of a District in Gowa District, South Sulawesi Province, Indonesia. The result of the research shows that the institution of land cultivation in rotation is not single pattern but consists of various patterns. There are four patterns of land tenure in rotation found in Tombolopao sub-district, namely inheritance system, cultivation ownership system, management rights purchasing system and cooperation system purchasing of land. Each institutional land tenure rotates, has its own pattern characteristics.

1. Introduction

1.1 Background

The land tenure institutions found in the middle of the farming community are some forms of ownership, rent, capable (profit sharing), and pawning, where by property rights can be individually controlled, and in groups, while rent, pledge and profits result in the transfer of rights cultivation from landowners to others. This institution is a dynamic rural society [1]

The dynamic institutional character has created institutional change, especially in Indonesia after the green revolution, particularly on the field rice [2] institution. Institutional changes then result in changes in land tenure for farmers, due to the incentives for non-agricultural land use, such as industry, settlements, toll roads, which lead to land conversion that implication changes in agricultural production and unemployment due to decreased assets owned by farm households [3, 4]. Even inheritance has also contributed to the fragmentation of land, so the land is narrower, let alone land acquisition through inheritance, is the most common land acquisition, although this condition is still seen from the local level [5]. On the other hand, the availability of suitable land for paddy fields is also increasingly difficult to find, even if there is a need for very expensive investments [6].

In some countries, land tenure has evolved from communal to private parties in private, this is due to the migration of the population in the villages, resulting in private ownership tends to be higher [7], [8]. Specifically Indonesia changes in the use of wetland or wetland conversion have reached an alarming stage. The rate of conversion of agricultural land to non-agricultural land according to the Ministry of Agriculture of the Republic of Indonesia with reference to BPS data for 1998-2002 of about 110,000 ha year^{-1}. Wetland conversions are exacerbated by the adoption of RTRW by district / city governments that are less favorable to agriculture [9].

In general, the land tenure of agricultural land in Indonesia, especially Java within the village community has its own system in the division of land, one of the ownership system belongs to an individual hereditary derived from the land of inheritance, the system of land discovery and communal ownership, from the results of renting and the ability of others (gemeen bezit). While in Button, Southeast Sulawesi,land tenure system based on a local tradition known as *parabola* that is related to local leaders and rules of play in general agricultural management and land tenure in particular [10].

One of the land holdings that contribute to land fragmentation is the ownership of an individual derived from inherited land, which has resulted in land fragmentation, resulting in narrower land ownership for farmers, an inheritance system for, where inherited land is broken down and distributed to the heirs. This pattern causes land ownership from generation to generation narrower (marginalization of land ownership). Splitting of land ownership at a certain level of land is considered inefficient to be cultivated in agriculture so that farmers tend to sell it or convert it into housing [5]. Meanwhile, the income of farmers from various farming efforts that tends to decrease [11]. One reason is that the land managed by farmers is narrower, resulting in smaller farmer income, this is due to the development of ownership and control of land in the countryside, especially in the agro-ecosystem of dynamic movement and tends towards increasingly narrow land ownership, especially in villages dominantly paddy rice, this implies to the income of farmers who follow the pattern of land ownership [12].

Land tenure in the highlands of Tombolopao sub-district, Gowa regency has its own characteristic compared to other land tenure, which is land ownership of rotating pattern obtained through inheritance system, management rights purchasing system, cultivation system and capital joint land buying system. These four patterns of land tenure do not emphasize private ownership that tends to make the land fragmentation, so that the land in Tombolopao sub-district is not subject to size change (Stagnan), this implies on the economic scale of sustainable farming, furthermore it is expected that the land ownership rotate contribute to the sustainability of agriculture, especially on the economic dimension.

This research will focus on mapping of rotational pattern of land inheritance system, purchasing management rights system, ownership cultivation system, and land purchase relationship as well as institutional arrangement of rotational land tenure with economically sustainable agriculture.

1.2 Research Methods

The study used case study method, with case unit of a District in Gowa District, South Sulawesi Province, Indonesia. Data collection is done through in-depth interviews with landlords, farmers, community leaders, local government. The data obtained are then analyzed and then grouped into matrix in accordance with the required data, which includes data on the rotating pattern of land tenure system, and the production, income and income data that will be used to analyze the relationship of the rotational land holding pattern institution with the economical sustainability of agriculture.

2. Results and Discussion

2.1. Condition of Land Tenure in Tombolopao District

Tombolopao sub-district is located in Gowa District, with a stretch of 251 km^2 area. This area is partly used as agricultural land that is 4,996 ha which is divided into rice field 4,296 ha the remaining 700 ha is used for cultivation [13].

There are two forms of land tenure institutional in Tombolopao sub-district, which is permanent land tenure institution and land tenure institution in rotation. The institution of permanent tenure regulates private ownership derived from purchases or grants. Institutional land tenure rotating pattern of land ownership in rotation that is divided into four systems, namely land rotation pattern system, inheritance system, land tenure rotating pattern of ownership system obtained through cooperation between landowners and tenants, land tenure patterns rotate the purchasing system of rights acquired through the purchase of management rights to the owner of the right of management and land tenure rotating pattern of land acquisition system, acquired by cooperating in buying land.

IOP Conf. Series: Earth and Environmental Science **157** (2018) 012072 doi:10.1088/1755-1315/157/1/012072

2.2. Institutional Land ownership rotation of inheritance system

Rotating land ownership of the inheritance system is the acquisition of land acquired by a person from the inheritance of parents to their offspring, where the land inherited does not suffer from fragmentation, since the land is not divided but the share is the right of land tenure in rotation each year. The pattern of control depends on how many heirs, if the number of siblings is only 2 people then there is a 1: 1 rotation pattern is the first year is managed by the first child, the second year is run by the second child, then back to the first child. This cycle continues, until they pass on the right of management, to the next heirs.

This case can be found in the case of Doda Anwaris which manages a land area of 20 acres. Derived from his grandmother named Cendang, then handed on the rotating management rights, to his three children, namely Damin, Racing and Naiman, Damin then passed down the rotating management rights to 10 of his children, Racing passed down his 3 children, and Naiman lowered to his two children. Case of Doda Anwaris being in Damin descent. The rotation pattern of land management is *bali tallu taun* (a three-year cycle), meaning that in the first year it is managed by Damin descendants, the second year is managed by Racing descendants, and the third year is managed by the descendants of Naiman. Among the descendants of the first offspring, land management rights were also excluded, so in the case of Doda Anwaris got management rights, in the 13th year in a cycle of rotating management. The structure of land tenure can be seen in figure 1 below:

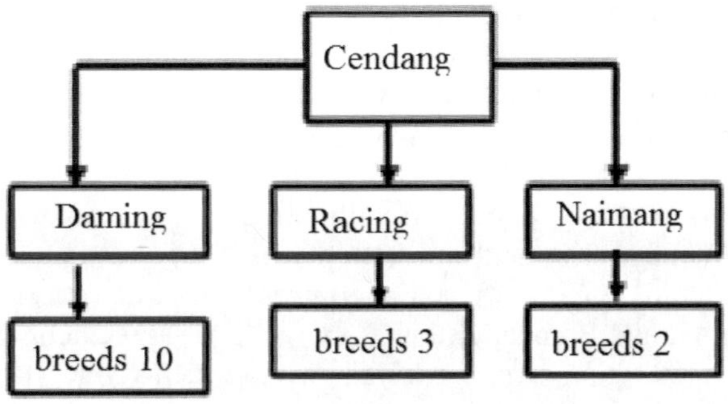

Figure 1. Rotation Structure of Land Mastery of Rotating Patterns of Inheritance System Case of Doda Anwaris

In the rotation structure of land tenure rotating pattern indicates that the land comes from grandmother named Cendang, then passes the land to the three descendants named Daming, Racing and Naiman to be managed in rotation, even the right of land ownership in rotation of the system of inheritance has been passed down again to the children. so that those who manage the land are now in second generation.

2.3. Institutional Land Rights Mastery Purchase Rights Management System

The rotating land tenure system of the management rights purchasing system is the rotating land tenure pattern that a person acquires by purchasing the management rights of a person who has rotating management rights, which may be purchased management rights, derived from the owner of the management rights of the inheritance system, the ownership system or the derived from the purchase of land system. The case of controlling land tenure of the management rights, management system found by the researcher is the case of Puang Baso, who bought the management rights from Sapia who controlled the rotating land with his brother named Sangkala, so that the continuation of land holding in rotation is Puang Baso with Sangkala. The rotation pattern is *bali taun* which is one year managed by Puang Baso, one year managed by Sangkala. This pattern can be seen in figure 2 below:

IOP Conf. Series: Earth and Environmental Science **157** (2018) 012072 doi:10.1088/1755-1315/157/1/012072

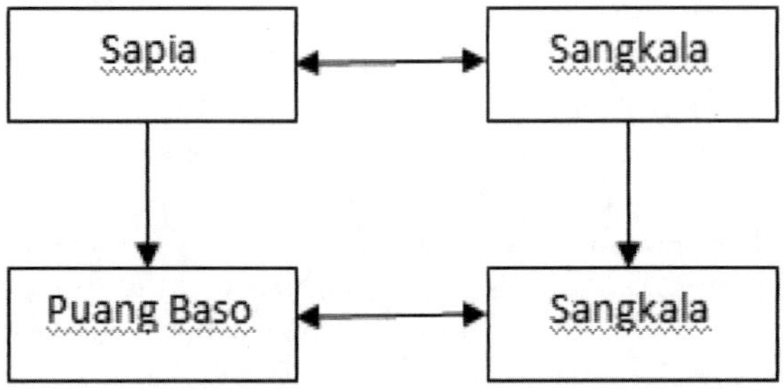

Figure 2. Revolving Structure of Purchase Rights Management System In Puang Baso Case

2.4. Institutional Land Mastery Pattern of Cultivation and Ownership System

Institutional land tenure rotating pattern of cultivation and ownership system is the institutional governing land tenure rotating pattern between landowners and Cultivators. Institutional land tenure rotating pattern of ownership system, born as a result of cooperation between landowners and Cultivators. The landowner hands over the land to be managed to become a wetland to the tenants by agreement, after the land is managed to become a paddy field, landowners and tenants will manage the land in rotation. The rotation pattern of land ownership is determined by deliberation between landowners and land tenants, taking into account the difficulty level faced by the farmers when the land is managed, if the land is managed easily, such as land is not too tilted, the land is not too much vegetation to be cleaned then the pattern the rotating landlord is usually 2: 1 two years managed by one-year owners managed by Cultivators, otherwise if land is difficult to manage then the pattern of land tenure becomes 1: 1 one year managed by the owner of one year managed by cultivators. The provisions of this pattern will be maintained until they transfer the right of tenure to their respective heirs.

The case of land tenure rotating pattern of cultivation system and ownership can be found in the case of M. Saleh, who manages the land in rotation which originated from the cooperation between the landowner and the cultivator, rotating started from the cultivation of land owned by Muma 'Ma'lalang by his cousin named Bombang, then they made an agreement to manage the land that has been worked with the system of *bali taun,* one year worked by Muma 'Ma'lalang one year worked by Bombang (rotation cycle 1: 1). The rights of land management were later handed down to their descendants, Muma Ma'lalang lowered the management rights to his son named M. Pious, as well as Bombang lowered the management rights to Sunara, so that the current rotation between Sunar and M. Saleh. The pattern of land tenure structure M. Saleh can be seen in figure 3 below:

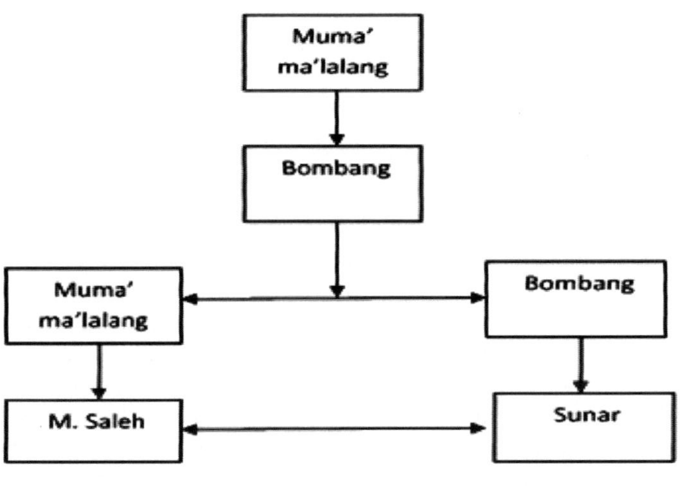

Figure 3. Pattern ofland tenure structure M. Saleh

The owner of the land originally was Muma 'Ma'lalang then worked on by Bombang, after they had agreed to divide the two management ie one year managed by Muma' Ma'lalang one year managed by Bombang, rotation provisions remain valid after being handed down to their respective children M. Saleh and Sunar.

2.5. Land tenure institution, rotating pattern of land purchasing cooperation system

Land tenure institution, rotating pattern of land purchasing cooperation system, is a rotating land tenure system governing land tenure in rotation, on land purchased by means of sharing between two or more persons. Purchase of land in a joint, done in general because the land to be bought cannot be redeemed itself, so they invite other families, or close friends to share the freed of the land. The approved rotation pattern based on the contribution given, if the funding is 50%: 50% then the agreed rotation pattern, is *bali taun* (one year cycle) means the first party manages one year, the second one manages one year, if the contribution of the given fund compared to 75%: 25%, the agreed rotation pattern is 2: 1 means the contributing 75% will manage two consecutive years, then the third year is managed by the second party contributing 25%.

Land ownership case of land purchase cooperation system, can be found in the case of Budi, who share the purchase of land with his brother named Bonggole. They both contribute 50%: 50% of the total funds used are IDR. 2,000,000. Land cleared for 10 acres is owned by Jappari. The rotation pattern on Budi's is bali taun (one year cycle), ie the first year of land is managed by Budi, and the second year is managed by Bonggole. This pattern will continue to be maintained until they die and the land is inherited to their offspring. Structural Pattern Land tenure rotating buying system pattern in the case of Budi, can be seen in figure 4 below:

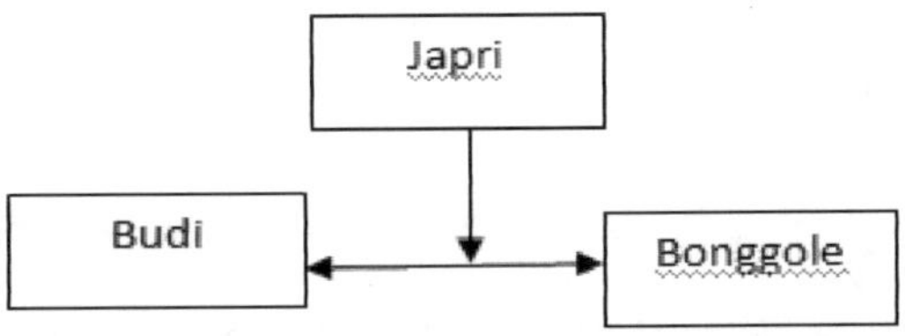

Figure 4. Pattern of land tenure structure Japri

In figure 4 above shows that Japri as the first landowner has no rotating land management pattern but after purchased by Budi share with Boggole, start the management of land is bred by the pattern of *bali taun*.

3. Discussion

There are two forms of land tenure institutions at the farm level, which are permanent, and not permanent. The institution of non-permanent land tenure is the tenure of land that is not permanently owned, but the rotation of the land, as a form of local wisdom, which has been passed down from generation to generation.

There are four rotational land holding institutions found in Tombolopao sub-district, namely the rotation of land ownership, the pattern of inheritance, the rotation of land ownership, the pattern of management rights purchasing, the rotating lid ownership institution, the ownership and cultivation pattern and the rotation of land ownership, the land purchase cooperation pattern. The four institutions of rotational land tenure, in each pattern have an equation, each of which spin the land to be managed, on communal members incorporated therein, and has been going on for generations, this is different from the institute found by [2] who found an institutional change, in rice paddy farming in Indonesia after the green revolution. The difference between these four institutions lies in the foundation base, as well as the structure of land tenure, the length of the rotation cycle (bali) depends on the agreement of communal members who have land tenure in each system. The pattern of land tenure is similar to land tenure in Java which is known as communal ownership such as what is conveyed by [14], but there is a difference side because communal ownership by [14] belongs to society in general and used the results for the public interest whereas in the control of land rotation pattern in Tombolopao District controlled only the rights of management and the owner is only communal members incorporated in the institutional, the result is enjoyed by the owner of the land ownership rights. Differences in land tenure can also be found elsewhere such as in Buton which has a local land-based holding institution called *parabela*. Where parabela is a leader of local leaders who are assigned to maintain the land but the management for the community as a result of farming is also for the people who manage [10].

4. Conclusion
The rotation of land ownership in Tombolopao sub-district can be mapped into four patterns, namely the institutional of rotational land tenure, the rotating land ownership pattern of the management rights, the rotating tenure ownership and the land ownership management. Each institution has its own pattern in spinning the management of land and has tangible rules of the game that prevailed from generation to generation and cannot be changed by anyone.

References

[1] Pakpahan A, Saliem H P, Suhartini S H and Syafa'at N *1993 Penelitian tentang Ketahanan Pangan Masyarakat Berpendapatan Rendah* Monograph Series No. 14 (Bogor: Pusat Penelitian Sosial Ekonomi Pertanian)
[2] Yunus A, Salman D, Demmallino E B and Viantika N M 2016. Sociotechnical change and institutional adjustment in paddy rice farming during post green revolution in indonesia *Int. J. of Agr. Syst.* **4** (2)
[3] Nuhung I A 2015 Faktor-faktor yang memotivasi petani menjual lahan dan dampaknyadi daerah suburban *J. Agro Ekonomi* **33** 117–33
[4] Fahmid I M 2013 Typology of rural agrarian conflicts *J. of Eco. and Sust. Dev.* **4** 3
[5] Susilowati S H, Purwantini T B, Supriyati, Sejati W K, Hidayat D, Maulana M, Ar-Rozi A M, and Yofa R D 2012 *Panel Petani Nasional (Patanas)*: *Dinamika Indikator pembangunan Pertanian dan Perdesaan di Wilayah Agroekosistem Lahan Kering Berbasis Perkebunan.* Laporan Penelitian (Bogor: Pusat Penelitian Sosial Ekonomi dan Kebijakan Pertanian)
[6] Rachmat M and Muslim C 2013 *Peran dan tantangan implementasi uu 41/2009 dalam melindungi lahan pertanian pangan berkelanjutan. Dalam kemandirian pangan indonesia dalam perspektif kebijakan MP3I* (Jakarta: Badan Penelitain dan Pengembangan Pertanian)
[7] Mwesigye F T, Matsumoto and Otsuka K 2017 Population pressure, rural-to-rural migration and evolution of landtenure institutions: The case of Uganda *Land Use Policy* **65** June 1-14
[8] Basupi V L, Claire Q H and Dougill J A 2017 Using participatory mapping and a participatory geographic information system in pastoral land use investigation: Impacts of rangeland policy in Botswana *Land Use Policy* **64** 363-373
[9] Anonym 2013 *Indonesia Tanah Air Beta....Tanah Petani Yang Mana?,* Sub Direktorat Pengendalian Lahan Dit. PPL. *Direktorat Jenderal Prasarana dan Sarana Pertanian, Kementerian Pertanian Republik Indonesia.* http://psp.pertanian.go.id/index.php/page/newsdetail/18 Accessed June 16 2017
[10] La Nalefo M K, Yusoff, Ali M S S, Salman D, Demmallino E B, Meisanti, Muhidin, Ismail I Y, Kamaluddin M and Nirwati 2013 Towards an institutional sustainable agriculture in Parabela *World App. Sci. J.* ***26***
[11] Dahar, Darmiati and Fatmawati 2016 Analisis sosial ekonomi masyarakat petani kecamatan Randangan kabupaten Pohuwato *J. Ilmiah Ilmu Ekonomi* **5** 9
[12] Winarso and Bambang 2012 Dinamika pola penguasaan lahan sawah di wilayah pedesaan di Indonesia. *J. Penelitian Pertanian Terapan* **12** 3
[13] Anonym 2016 *Statistik Penggunaan Lahan Kabupaten Gowa.* (Indonesia: BPS Kabupaten Gowa)
[14] Tjondronegoro S M P and Wiradi G 2008 *Dua abad penguasaan tanah, pola penguasaan tanah pertanian di Jawa dari masa ke masa* (Indonesia: Yayasan Obor Indonesia)

IOP Conf. Series: Earth and Environmental Science **157** (2018) 012073 doi:10.1088/1755-1315/157/1/012073

Food expenditure share analysis of household: Case study of food reserved garden area program in Bone Bolango regency of Gorontalo province

N Buri and Z Mantau

Gorontalo Assessment Institute for Agricultural Technology (AIAT) Jl. Muh Van Gobel (Kopi) No.270, Kec. Tilongkabila, Kab. Bone Bolango, Gorontalo, 96183, Indonesia.

E-mail: *nanangburi@gmail.com*

Abstract. The share of food expenditure is one of food security indicator in communities. It also can be used as an indicator of the success of rural development. The aim of this research was to find the share of food expenditure of farm households before and after the program of Food Reserved Garden Area (KRPL/ FRGA) in Suwawa and Tilongkabila districtat Bone Bolango Regency of Gorontalo Province. Analysis method used share of food expenditure method. The method measure the ratio of food expenditure and total expenditure of household for a month. Statistical test used a non-parametric method, especially The Wilcoxon Test (two paired samples test). The results found that KRPL program in Ulanta Village of Suwawa district did not significantly affect the share of food expenditure of farm household. While in the South Tunggulo village of Tilongkabila district, FRGA program significantly affected the share of food expenditure.

1. Introduction

The Republic of Indonesia Government Regulation No. 68 at 2002 year about Food Security [1], explained that food security was a fulfillment condition of food for household which reflected by availability of food, quantity and quality, secure, prevalent and affordable. Furthermore, the availability of food was food availability from domestic production and other sources. This is macro indicator, because probably food can be available but it can not be accessed by people. Meanwhile, food diversification is the effort to increase consumption of variety food with balanced nutrition principles. These things that could be basic rules of the formation of Food Reserved Garden Area (FRGA).

Good quality of nutrition is very important for food consumption. Therefore, there's a policy for food consumption diversification. It's called Desirable Dietary Pattern (DDP), which applied toFRGA. DDP is group of food which based on energy contribution regarding to total energy of main group of food. DDP is real form of food security, because commonly, DDP was known as a pattern consumption of variety food, balanced nutrition and secure [2].

In 2011 year, Agriculture Ministry of Republic of Indonesia released a program to maintain the food security. It's called The Pattern of Food Reserved Garden Area. In 2012 year, the program is going to massification to the whole of Indonesia. The principles of FRGAare: 1). Utilization of home yard for food security and self-sufficiency. 2). Food diversification based on local resources, 3). Conservation of food genetic resources (plant, cattle and fish) and 4). Sustainability of village's seed garden to get high income and society prosperity [3].

FRGA development and improvement have purposes for food diversification to get the food security and self-sufficiency which impact to increase the prosperity of society. In additional, food security reflected by several indicators, such as : 1) the damage level of plant, cattle and fish; 2) the decreasing of food production; 3) food availability in the household; 4) the share of food expenditure; 5) food price fluctuation which consumed by household; 6) a changes of social life, such as migration, sell assets; 7) eating habits, quantity and quality of foods; and 8) nutrition status [4].

The share of food expenditure can be use too for indicator of succeeded of villages' development. High share of food expenditure shows that household income is still for basic needs (subsistence). Vice versa, high share of non-food expenditure indicates that there has been a shift of household's needs from basic needs (subsistence) to the commercial needs. It means that the primary needs have been fulfilled or there has been a shift of food expenditure pattern. [5] argues that till 2006 year, the share of paddy-rice expenditure was very high. This condition was change after 2008 (economy crisis in Indonesia), where the share of ready meals expenditure (include instant noodles) was higher than paddy-rice (main carbohydrate source). It' indicates that there has been a shift of dietary habit in the society. The people were more choice spends their money and time to buy ready food in the restaurant, cafe or other food stores. This habits can be indicates to the decreasing of rice consumption in the society.

Susenas Data in September 2014 shows that the share of food expenditure per capita in the rural area of Gorontalo province was 50.12%. It is higher than non-food expenditure per capita which just 49.88% was. In the urban area, the share of food expenditure was just 39.90%, but the non-food expenditure per capita was reach about 60.10% [6]. Those values indicate that the life behaviour priority of people in the rural area was fulfilment the raw food needs. In additional, the lacks of infrastructure facilities were the other factors that caused of low of prosperity in the rural area. It is very different with the condition of people in the urban area. They are more spending their income to buy the non-food goods. Even though, they bought a food for eat, so they will be spend their money for but a ready food in the restaurant or cafe than buy some raw materials of food and cook it. They but the food was not just for satisfied, but more than it, such as nutrition, healthy factors. The important is quality of food than quantity. So that, the FRGA or KRPL program did in Gorontalo Province, especially in Bone Bolango Regency. The program purposes to increase a Desirable Dietary Pattern (DDP), self-sufficiency of food and an alteration in the share of food expenditure, especially in the rural area.

The aim of this research is want to know about the share of food expenditure of farm household, before and after FRGA program in Suwawa and Tilongkabila district of Bone Bolango Regency.

2. Materials and methods

2.1. Place and time

The research conducted at Ulanta Village of Suwawa District and South Tunggulo Village of Tilongkabila District in Bone Bolango Regency. The both of villages are the location of FRGA program in Bone Bolango Regency during 2013 – 2014.

2.2. Source of data

This research took a data from beginning of the program at 2013 till the end at 2014 (before and after). The research was more focused on the survey activity. The data took with a questionnaire and direct interview. Kinds of data are:

1. Farm household characteristics
2. Kinds and total of household expenditure for food consumption (before and after FRGA)
3. Kinds and total of household expenditure for non-food (before and after FRGA).

2.3. Analysis method

The share of food expenditure (SFE) is an expenditure ratio for food and non-food (total) for a month. The share of food expenditure was obtained by the data at the farm household level and then divided by the number of farm household's members. The formula as follow as, where:

$$SFE\ (\%) = \frac{EF}{ET}\ x\ 100 \quad (1)$$

SFE= The share of food expenditure (%)
EF=an expenditure for food (Rp/month)
ET = Expenditure total of farm household (Rp/month)

The hypotesis :
1. H0 = there's no difference to the share of food expenditure between before and after FRGA program.
2. H1 = there's significant difference between before and after FRGA program to the share of food expenditure.

Non parametric method of statistic used to test that hypothesis. Non-parametric method using The Wilcoxon test (2 related sample test). This method was used because the data were not normally distributed based on Kolmogorov-Smirnov normality test. The test showed that the significance value is less than 0.05 (0.000) which means that the sample was not normally distributed. Although the data type was ratio data [7].

All of the respondents were the whole of member of group that followed the FRGA program between 2013 – 2014 in the Ulanta and South Tunggulo villages. All statistical analysis were performed using IBM SPSS Statistics ver.20.0.

There are two criteria in the Wilcoxon test for decision making:
1. Compare the value of z calculate and z table:
   - If z calculate $<z$ table, H0 accepted
   - If z calculate $>z$ table, H0 rejected or H1 accepted
2. Probability values :
   - If probability value >0.05, H0 accepted
   - If probability value < 0.05, H0 rejected atau H1 accepted.

To calculate the z value using a following formula, [7]:

$$z = \frac{T-[\frac{1}{4N}(N+1)]}{\sqrt{1/24(N)(N+1)(2N+1)}} \quad (2)$$

where:
T = a smallest difference
N = the total sample

3. Result and discussion

Descriptively, the percentage of share of food expenditure each respondents shows at table 1. Table 1 shows that after FRGA program, SFE increased in average 1% after FRGA program at Ulanta Village and an average 4% in South Tunggulo Village. This result argues that commonly, more of household income spent for food (groceries) or primary needs than secondary or tertiary needs (luxury goods). Wilcoxon test performed to test the relationship of SFE between before and after FRGA program (table 2).

Negative ranks in the Table 2 shows that there are 6 (six) negatives data difference in Ulanta Village. Meanwhile, there are 4 (four) negatives data difference in South Tunggulo Village. *Negative ranks* are data differences between "after" and "before" which had negative value or value of "after" was smaller than "before" value. This result explain that frequency of food purchases or SFE value before FRGA program was greater than after FRGA program.

Positive ranks are values of "after" will be greater than "before" values. Table 2 shows that there are 9 (nine) datas in Ulanta village which showed that SFE value in after FRGA program was greater than before FRGA program. Meanwhile, there are 11 (eleven) datas in South Tunggulo village which showed

IOP Conf. Series: Earth and Environmental Science **157** (2018) 012073 doi:10.1088/1755-1315/157/1/012073

that SFE value after FRGA was greater than before FRGA program. In addition, "Ties" is there's no same value or there's same value of SFE in the whole of respondent data.

Tabel 1. Recapitulation data of the share of food expenditure in Ulanta and South Tunggulo Villages, before and after FRGA Program.

Respondents	Village	District	Share of Food Expenditure (SFE) (%)	
			Before KRPL	After KRPL
1	Ulanta	Suwawa	22%	11%
2	Ulanta	Suwawa	12%	15%
3	Ulanta	Suwawa	28%	22%
4	Ulanta	Suwawa	8%	15%
5	Ulanta	Suwawa	9%	11%
6	Ulanta	Suwawa	12%	13%
7	Ulanta	Suwawa	16%	5%
8	Ulanta	Suwawa	8%	5%
9	Ulanta	Suwawa	16%	9%
10	Ulanta	Suwawa	21%	13%
11	Ulanta	Suwawa	5%	12%
12	Ulanta	Suwawa	4%	11%
13	Ulanta	Suwawa	2%	17%
14	Ulanta	Suwawa	10%	14%
15	Ulanta	Suwawa	8%	21%
16	South Tunggulo	Tilongkabila	99%	90%
17	South Tunggulo	Tilongkabila	5%	16%
18	South Tunggulo	Tilongkabila	9%	16%
19	South Tunggulo	Tilongkabila	6%	16%
20	South Tunggulo	Tilongkabila	9%	14%
21	South Tunggulo	Tilongkabila	8%	11%
22	South Tunggulo	Tilongkabila	12%	11%
23	South Tunggulo	Tilongkabila	12%	11%
24	South Tunggulo	Tilongkabila	8%	16%
25	South Tunggulo	Tilongkabila	10%	9%
26	South Tunggulo	Tilongkabila	11%	17%
27	South Tunggulo	Tilongkabila	9%	11%
28	South Tunggulo	Tilongkabila	8%	12%
29	South Tunggulo	Tilongkabila	9%	14%
30	South Tunggulo	Tilongkabila	8%	15%

Source: primary data (2013 and 2014)

Tabel 2. The result of Wilcoxon test of SFE, before and after FRGA program in Ulanta Village and South Tunggulo Village (2014).

Ranks				
		N	Mean Rank	Sum of Ranks
Desa Ulanta (PPP after – PPPbefore)	Negative Ranks	6[a]	9.00	54.00
	Positive Ranks	9[b]	7.33	66.00
	Ties	0[c]		
	Total	15		
Desa Tunggulo selatan (PPP after – PPPbefore)	Negative Ranks	4[a]	4.75	19.00
	Positive Ranks	11[b]	9.18	101.00
	Ties	0[c]		
	Total	15		

a. P3after < P3before
b. P3after > P3before
c. P3after = P3before

Source : output IBM SPSS ver.20.0 (2015)

Tabel 3. The result of statistic test of Wilcoxon, before and after FRGA program in Ulanta and South Tunggulo Villages.

Test Statistics[a]		
	Ulanta Villages (PPPafter – PPPbefore)	South Tunggulo Villages (PPPafter – PPPbefore)
Z	(-)0.341[b]	(-)2.329[b]
Asymp. Sig. (2-tailed)	0.733	0.020

a. Wilcoxon Signed Ranks Test
b. Based on negative ranks.
Z table : 0.6368 ($\alpha = 0.05$) *Z* table : 0.9906 ($\alpha = 0.05$)

Source : output IBM SPSS ver.20.0 (2015)

Table 3 shows that in Ulanta village, *Z* calculate value (*sign ignored*) was smaller than *Z* table value.So that, the final result is accept H0. There are same value of SFE between before and after FRGA program in Ulanta village. The meaning is there's no influence of FRGA program to the share of food expenditure in Ulanta Village, Suwawa District. There's same result if we use probability criteria.

Meanwhile, according to table 3, *Z* calculate value was greater than *Z* table value in South Tunggulo village. So that, the final result is reject H0 or H1 accepted. There's same result if we use probability criteria. The meaning is there's significant difference of SFE between before and after FRGA program in South Tunggulo Village. Therefore, It's important to continue the FRGA program in that village, especially for improve the Desirable Dietary Pattern (DDP) through food diversification.

As a food security indicator of the region, [4] said that if share of food expenditure (SFE) is higher, so the food security is more reduce. As a prosperity indicator, if SFE value is smaller, so the prosperity level is higher. This is because the income of each household has just small part for buy food (groceries) and largely to spend for secondary and tertiary needs (luxury goods).

According to that statement, it can be argued that the prosperity and food security level of household in South Tunggulo is better than the household in Ulanta village.

Ilham and Sinaga [4] found that there's a relationship between share of food expenditure and food security. This can be indicated by higher income, share of food expenditure decrease, but energy and protein consumption increase. The increasing of consumption of energy and protein indicated that Desirable Dietary Pattern was improve. People was more choice the quality of food for healthy reason than quantity of food for satisfied reason.

The differences of income and education were the one of decided factor for food expenditure pattern. Kirpatrik and Tarasuk [8] argued that the household with low education level spend considerably less income to buy food in restaurants or food stores. They tend to spend their money to buy a cheaper material food in the traditional market and then cook it by themselves. Pusposa [9] found that income variable was significantly influence to the consumption pattern. Kirpatrik and Tarasuk [8] also found that food expenditure pattern depend on education level. Household with high level of education have a high income either. So that, it is very influence to the purchasing power and food expenditure pattern of household. The household which have a high income usually able to access the restaurant or food stores which sell the premium quality food, easily.

Moreover, Rachman and Ariani [10] said that food consumption pattern in Indonesia (rural and urban area) dominated by share of carbohydrates source. In the other hand, Indonesian people consume more carbohydrates (satisfied factor) than the other nutrition source. Whereas, the urban income is higher than the rural income. So that, purchasing power factor of each household is depending factor of food consumption pattern of society. Commonly, people who live in the urban area are more concerned to the nutrition ingredients of food (quality factor) than the quantity of food (satisfied factor) [10].

4. Conclucion

- The implementation of FRGA program in Ulanta Village does not significantly influence to the share of food expenditure of each respondent household.
- FRGA program in South Tunggulo Village is significantly influence to the share of food expenditure of household.
- Several researches showed that education and income level are factors which significantly influence to the food expenditure pattern and food consumption pattern of each household.
- Preferably, Food Reserved Garden Area (FRGA) program should be continued. Because, It can be influence the share of food expenditure (SFE) and Diserable Dietary Pattern (DDP) , directly or inderictly.

References

[1] Presiden Republik Indonesia 2002 *Peraturan Pemerintah Republik Indonesia Nomor 68 Tahun 2002 Tentang Ketahanan Pangan* (Jakarta: Sekretariat Negara Republik Indonesia)

[2] Badan Ketahanan Pangan Provinsi Bengkulu 2011 *Analisis Konsumsi Pangan Provinsi Bengkulu 2011* (Bengkulu: BKP)

[3] Kementerian Pertanian Republik Indonesia dan SIKIB 2012 *Panduan Kawasan Rumah Pangan Lestari Kementan* (Jakarta) p 27

[4] Ilham N and Sinaga B 2007 Penggunaan Pangsa Pengeluaran Pangan sebagai Indikator Komposit Ketahanan Pangan *J.SOCA*.**73** 122

[5] Ariani M 2010 Analisis Konsumsi Pangan Tingkat Masyarakat Mendukung Pencapaian Diversifikasi Pangan *J. Gizi Indonesia* **3** 31 20 28.

[6] Badan Pusat Statistik (BPS) Nasional 2014 *Pengeluaran untuk Konsumsi Penduduk Indonesia per Provinsi. Survei Sosial Ekonomi Nasional (Susenas)* Buku 3 (Jakarta: BPS Republik Indonesia) p 129.

[7] Santoso S 2014 *Statistik Non Parametrik. Konsep dan Aplikasi dengan SPSS. Edisi Revisi* (Jakarta : PT. Elex Media Komputindo) p 279

IOP Conf. Series: Earth and Environmental Science **157** (2018) 012073 doi:10.1088/1755-1315/157/1/012073

[8] Kirpatrik S and Tarasuk V 2003 The Relationship between Low Income and Household Food Expenditure Patterns in Canada *Pub. Health Nutrition J.* **66** 589-597

[9] Pusposa F 2012 *Analisis Pola Konsumsi Pangan Masyarakat di Provinsi Maluku* Thesis of Magister Sains Planning and Public Policy (Jakarta: Fakultas Ekonomi. Universitas Indonesia) p 101

[10] Rachman H P S and Ariani M 2008 Penganekaragaman Konsumsi Pangan di Indonesia : permasalahan dan implikasi untuk kebijakan dan program *J. Analisis Kebijakan Pertanian* **6** (2) 140-154